T0138304

The Mind of
the Chimpanzee

The Mind of the Chimpanzee

ECOLOGICAL AND
EXPERIMENTAL PERSPECTIVES

Elizabeth V. Lonsdorf,
Stephen R. Ross,
& Tetsuro Matsuzawa

The University of Chicago Press
Chicago and London

Elizabeth V. Lonsdorf is the director of the Lester E. Fisher Center for the
Study and Conservation of Apes at the Lincoln Park Zoo in Chicago and a
faculty member of the Committee on Evolutionary Biology at the University
of Chicago. Stephen R. Ross supervises behavior and cognitive research at the
Fisher Center and chairs the Chimpanzee Species Survival Plan of the Association of Zoos and Aquariums. Tetsuro Matsuzawa directs the Primate Research
Institute at Kyoto University in Japan.

The University of Chicago Press, Chicago 60637
The University of Chicago Press, Ltd., London
© 2010 by The University of Chicago
All rights reserved. Published 2010
Printed in the United States of America

18 17 16 15 14 13 12 11 10   1 2 3 4 5

ISBN-13: 978-0-226-49278-0 (cloth)
ISBN-13: 978-0-226-49279-7 (paper)
ISBN-10: 0-226-49278-8 (cloth)
ISBN-10: 0-226-49279-6 (paper)

Library of Congress Cataloging-in-Publication Data
The mind of the chimpanzee : ecological and experimental perspectives / [orga-
nized by] Elizabeth V. Lonsdorf, Stephen R. Ross, and Tetsuro Matsuzawa.
        p.   cm.
    Based on the papers presented at a conference held in Chicago in 2007.
    Includes bibliographical references and index.
    ISBN-13: 978-0-226-49278-0 (cloth : alk. paper)
    ISBN-13: 978-0-226-49279-7 (pbk. : alk. paper)
    ISBN-10: 0-226-49278-8 (cloth : alk. paper)
    ISBN-10: 0-226-49279-6 (pbk. : alk. paper)   1. Chimpanzees—
Behavior—Congresses.   2. Chimpanzees—Psychology—Congresses.
3. Chimpanzees—Ecology—Congresses.   4. Chimpanzees—
Conservation—Congresses.   5. Cognition in animals—Congresses.
6. Social behavior in animals—Congresses.   I. Lonsdorf,
Elizabeth.   II. Ross, Stephen R.   III. Matsuzawa, Tetsuro, 1950–
IV. Goodall, Jane, 1934–
    QL737.P96M57   2010
    599.885'15—dc22
                                                        2009042882

♾ The paper used in this publication meets the minimum requirements of the
American National Standard for Information Sciences—Permanence of Paper
for Printed Library Materials, ANSI Z39.48-1992.

Contents

Foreword

Jane Goodall

This volume is based on the papers presented at the third conference in the spirit of "Understanding Chimpanzees." What a lot has happened since Paul Heltne and I began discussing ideas for the first of these conferences! Held in 1986, it was to mark the publication of *The Chimpanzees of Gombe: Patterns of Behavior,* in which I brought together some of the results from the first 20 years of observations at Gombe National Park, Tanzania, set against an overview of published data from other chimpanzee studies in the wild and in captivity. We planned to bring together, for the first time, all (or as many as possible) of the field biologists from the different chimpanzee study sites across Africa, along with some of the scientists conducting noninvasive studies on captive groups. Back then it was a novel idea to have a conference on the behavior of one species. In fact, at the time we included those working with bonobos, *Pan paniscus,* which were still known as pygmy chimpanzees; the others were called common chimpanzees.

First Conference: Understanding Chimpanzees

That first conference was held in Chicago and organized and supported by the Chicago Academy of Sciences. Invited speakers presented papers about research results from the field sites of Gombe and Mahale Mountains in Tanzania, Bossou in Guinea, and Kibale in Uganda. The topics included population dynamics, feeding ecology, vocal communication, and interactions between resident and immigrant females. These papers and the resulting discussions provided a great deal of information about the behavior of our closest relatives in the wild. An important—and exciting—aspect of the conference was the opportunity for dialogue between Western researchers from the various African study sites and our colleagues from Japan. Only one African scientist was present at that conference: Gilbert Isabirye-Basuta, from Uganda.

Some of the field biologists present, the "old timers" or "silverbacks" as we referred to ourselves, had been consistently in the field since the sixties. In addition to myself there were Adriaan Kortlandt, Junichiro Itani, Yukimaru Sugiyama, and Toshisada Nishida studying the common chimpanzee. Bonobo field studies had begun in the 1970s and Takayoshi Kano, Suehisa Kuroda, and Nancy Thompson-Handler joined us at that first conference. There were a number of silverbacks in the field of captive studies also: Emil Menzel, Allen and Beatrix (Trixie) Gardner, Duane Rumbaugh and Sue Savage-Rumbaugh. Throughout the conference there were many animated conversations about differences between chimpanzees and bonobos—and I remember making a passionate plea for dropping the name "pygmy chimpanzee" since, clearly, they were so very different. This was substantiated not only by the work of Frans de Waal with captive bonobos, but also by the research at field sites in the Democratic Republic of Congo (known at that time as Zaire).

A fascinating session that focused on the mind of the

chimpanzee opened with a discussion of that great pioneer, Wolfgang Köhler. Emil Menzel discussed his work on chimpanzee intelligence in a variety of captive experimental situations. Language acquisition research, involving the teaching of American Sign Language to chimpanzees, was discussed by Allen and Trixie Gardner (who pioneered this work) and by Roger and Deborah Fouts. Duane Rumbaugh and Sue Savage-Rumbaugh reported on their research teaching chimpanzees and bonobos to use lexigrams, which comprised a language they called Yerkish. And it was at this conference that the accomplishments of chimpanzee Ai in Japan were presented by Tetsuro Matsuzawa. He called her "his partner in exploring the upper limits of chimpanzee intellect" and her abilities at the keyboard amazed most of those present.

There were two sessions that, for me, were extremely significant. One of these was on the plight of chimpanzees in Africa, and the second was on the treatment of chimpanzees in a variety of captive situations. These two sessions changed the course of my life, and I shall return to them at the end of this foreword. Here let me say that Understanding Chimpanzees was a landmark conference for many of us; it revealed the rich diversity of chimpanzee behavior in different areas, and reinforced my long-held conviction that chimpanzees have their own primitive cultures. By the end of the four days there was one thing we all agreed on: the conference was a huge success, greatly increasing our understanding of chimpanzees. In fact, so many new studies were underway that we unanimously decided it would be of great value for us to get together again after five years to share and review the new data that would then be available. The book *Understanding Chimpanzees,* an outcome of the conference, was published in 1989.

Second Conference: Chimpanzee Cultures

This second conference, organized by Paul Heltne and Linda Marquardt, also took place in Chicago in 1991. It focused on variation in chimpanzee behavior from one study site to another. We had felt that video taken in the field would provide a significant way of evaluating such differences, and all participants had been asked to bring film with them. Among the most remarkable aspects of this conference were the proliferation of behavioral and ecological studies, the number of new study sites across Africa, and of course the many new young researchers who were working in the field of chimpanzee ecology and behavior.

Richard Wrangham, Bill McGrew, and Suehisa Kuroda discussed the various new sites across Africa where chimpanzees and bonobos were being observed, and summed up the status of the research there. Much new research was presented, including the work of Colin Chapman and John Mitani in Kibale National Park, of Christophe and Hedwidge Boesch in Taï Forest, and of Caroline Tutin in Gabon. There was a section on the ecology of chimpanzees and bonobos that included speculation on the role of the environment in determining party size, types of tools and purposes for which they were used, different hunting strategies and so on that had been observed in different populations. Many of the papers discussed differences in social organization and social interactions between chimpanzees and bonobos—and for the first time gorilla research was also included.

During this second conference there were more papers resulting from studies of chimpanzees in zoo populations, including comparisons between behavior seen in wild and in captive situations (stemming in part from the Jane Goodall Institute's ChimpanZoo program which I had briefly discussed in 1986). There had been major developments in captive research in cognitive and cultural behaviors, and Matsuzawa and his students were not only working on cognitive issues in his lab in Inuyama, but had begun observing tool-using behavior in the Bossou population in Guinea. The explosion of research into many aspects of behavior since the first conference meant that it would be possible to investigate cultural variation in much greater detail than before. And as we had predicted, the many video sessions on different aspects of behavior—ranging from tool use to social interactions—made it possible to compare behaviors that truly differed from one study site to another.

Increasing collaboration between wild and captive studies was valuable in many ways. For one thing, the ever-increasing understanding of behavior in the wild meant that the adaptive value of certain behaviors long observed in captivity became apparent. Behaviors seen in the wild as rare events could, in some cases, be further investigated through noninvasive testing in captive situations. And, of course, captive chimpanzees can be encouraged to use their minds in new situations. This ever-increasing understanding of the cognitive abilities of chimpanzees and bonobos helped to give credence to anecdotal examples of intelligent behavior in the wild. The book *Chimpanzee Cultures,* an outcome of the conference, was published in 1994.

Third Conference: The Chimpanzee Mind

This most recent conference, held in 2007 in Chicago and organized by Elizabeth Lonsdorf and Steve Ross of the Lincoln Park Zoo, focused on the chimpanzee mind. Since 1985, when I wrote the chapter on this subject for *The Chimpanzees of Gombe,* a positive explosion of fascinating studies on animal minds has taken place. At this third conference there were, in addition to scientists who had participated in one or both of the first two (some of whom during the intervening years had changed from "blackback" to "silverback"—that is, from students to professors), there were many new, young researchers representing the different study sites in Africa and captive research around the world. Early observations of tool use in the Goualougo Triangle chimpanzees had been greatly expanded through the use of camera traps by Crickette Sanz and Dave Morgan. While Ai, whose scientific debut in the United States had so amazed the participants in 1986, continued to perform brilliantly a variety of extraordinarily complex tasks, she was all but overshadowed by her six-year-old son, Ayumu, with his astounding photographic memory. The advance of technology was enabling new studies of the kind scarcely envisioned when I had begun in 1986. For example, DNA profiling from fecal samples enabled researchers to determine the paternity of particular chimpanzees for the first time, and geographic information systems and satellite imagery were helping us to better understand ranging patterns. Moreover, there was much better collaboration between researchers from different sites and different universities, and more willingness to share information. Andrew Whiten, who had been zealously polling the various study sites as to the presence or absence of a whole variety of proposed cultural behaviors, was able during the conference to corner recalcitrant respondents and get the final data that would enable him to complete his survey of chimpanzee diversity in these respects. This volume presents all these findings and more.

Conservation, Ethics, and Well-Being

In this third conference, major concerns for many participants were conservation issues in the wild, and ethical issues regarding captive situations in laboratories, non-accredited zoos, and entertainment as well as in the African sanctuaries for orphaned great apes. The final session of the conference was dedicated to these concerns.

As I mentioned earlier in this introduction, there had been a session on conservation in the first Understanding Chimpanzees conference, organized by Geza Teleki and Richard Wrangham. The picture painted by the various researchers was grim: from across Africa there were reports of human population growth, habitat destruction, forest fragmentation, chimpanzees caught in wire snares, and the hunting of chimpanzees for the wild animal trade and the "bushmeat trade"—the commercial hunting of wild animals for food. All this had clearly had a major impact on chimpanzee numbers across their range. No one really knew how many there were, but it was clear that the more than one million who had ranged across 25 African nations at the turn of the century had been reduced to fewer than 300,000 by 1986. In four countries they were already extinct. Photographs by Karl Ammann and Michael "Nick" Nichols showed agonizing images of chimpanzees who had lost a hand or a foot after being caught in a wire snare, chimpanzees and gorillas butchered for food, and infants whose mothers had been shot being offered for sale in the marketplace or by the roadside.

Another significant session at the 1986 conference had been on the treatment of chimpanzees in medical research laboratories, zoos, and the entertainment industry. Again, images obtained by Nick and others showing the stark conditions in research labs (followed up by secretly filmed footage at the SEMA lab in Rockford, Maryland) made a deep impression on many of those present.

It was those two sessions that had such an impact on my own life. In 1986 I was planning to continue collecting data at Gombe, and I had already begun work on a sequel to *The Chimpanzees of Gombe*—a volume that was to include analysis of my own particular interest, the mother-infant study (infant development, the changing relationship between mother and her child, the development of relations between siblings), and changes in behavior over the life cycle. Instead I became an activist, determined to do my share to raise awareness about the plight of the chimpanzees and their forests.

Today, as I travel some 300 days a year, I seldom get the chance to spend more than three or four days twice a year at Gombe. But the research continues. It is clear that in many countries there are increasing numbers of young people determined to work with chimpanzees, not only as scientists pursuing different aspects of behavior but also as conservationists fighting for the protection of the rain forests and the delineation of new national parks and reserves, as caretakers in the sanctuaries, and as advocates for improved conditions in captivity around the world.

In another few years there will, I hope, be a fourth conference in the Understanding Chimpanzees series. And as our knowledge increases, as the line between humans and chimpanzees—biologically, intellectually and socially—becomes ever more blurred, it is to be hoped that those fighting to save the species in the wild and ensure that those in captivity are treated with appropriate respect will be increasingly supported by the scientific establishment.

Paul Heltne, for your vision all those years ago, we thank you. I know that bringing together all those involved in chimpanzee research and advocacy has advanced both our knowledge of these magnificent creatures and our ability to provide for their protection both in and out of the wild.

Jane Goodall, PhD, DBE
Founder, the Jane Goodall Institute
and UN Messenger of Peace
www.janegoodall.org

Acknowledgments

This volume represents a collection of work that was first presented at the Mind of the Chimpanzee conference, hosted by Lincoln Park Zoo's Lester E. Fisher Center for the Study and Conservation of Apes, in March 2007. The editors are extremely grateful to the staff and volunteers at Lincoln Park Zoo, without whom the conference would not have been possible. The conference was generously supported by the Wenner-Gren Foundation for Anthropological Research and the David Bohnett Foundation, with in-kind support provided by United Airlines and Lincoln Park Zoo. Thanks are also due to the Leo S. Guthman Foundation for core support of the Fisher Center. The editors are grateful to the conference steering committee, including Frans de Waal, Brian Hare, Tanya Humle, Lisa Parr, Crickette Sanz and Andrew Whiten. In addition, Jane Goodall, Paul Heltne, and Richard Wrangham provided much appreciated support and advice. Finally, we are grateful to the chimpanzees themselves and to the many people who work tirelessly to study, protect, and care for them in the field and in captive settings.

I

The Chimpanzee Mind: Bridging Fieldwork and Laboratory Work

Tetsuro Matsuzawa

Ayumu is the five-year old son of chimpanzee Ai. Since his birth, he has observed his mother performing cognitive tasks using a computer touch screen. When he reached the age of four years, Ayumu learned to touch Arabic numerals in ascending order. Within six months he had mastered the ability to touch the nine numerals in the correct order as they appeared randomly arranged on the screen. We then proceeded to present Ayumu with a greater challenge: the masking task. In this task, after the first numeral on the touch-sensitive monitor is touched, all other numerals are replaced by white squares. The subject therefore must remember which numerals appeared in which location, and then touch them in ascending order. Ayumu's performance on this task was remarkable. He could remember the location of nine numerals, in the proper order, at a glance. Even after training for the same task, university students still could not match his performance. This is a case in which chimpanzees performed better than humans in a cognitive task.

The human mind is an evolutionary product, just like the human body. It is located in a soft tissue, the brain, which cannot fossilize like teeth and bones. Therefore, the fossil record of our evolutionary neighbors, such as Neanderthals (*Homo neanderthalensis*), does not reveal anything about how their minds worked. However, a comparison between the human mind and the mind of our closest living relative, the chimpanzee, may help us better understand the similarities and differences in cognitive functions between the two species, the shared characteristics that were present in our common ancestor, and those derived through species differentiation. Cognitive science has revealed various functions and capabilities of the human mind, but many unanswered questions remain regarding its evolutionary history. Therefore, an understanding of the chimpanzee mind can provide important insights into the evolutionary heritage of the mind of our own species.

Chimpanzees are remarkably similar to humans in many ways. The taxonomic group known as hominoids comprises humans and the great apes, which in turn include three genera: chimpanzees (including bonobos), gorillas, and orangutans. Recent advances in the study of the human and chimpanzee genome (Chimpanzee Sequencing and Analysis Consortium 2005) have revealed that chimpanzees are the closest evolutionary living relatives of humans. The difference in DNA between the two species is only 1.23%—so in other words, humans are 98.77% chimpanzee. This genetic difference is comparable to that between horses and zebras, estimated to be about 1.5% (Walner et al. 2003). You may think that zebras are horses with black and white stripes. If so, then you should also think that chimpanzees are humans with long black body hair.

Chimpanzee Life in the Natural Habitat

Chimpanzees live in the tropical rain forests and surrounding savannas of Africa. They probably once spanned most of equatorial Africa—including the land area of what are now at least 25 countries—and numbered more than a million just 100 years ago. Today, however, they are known to occur in only 22 countries, and an estimate by the World Conservation Union (IUCN) in 2003 put their numbers in Africa at between 172,700 and 299,700. This sudden decrease is linked to various human activities such as deforestation, poaching, and the bushmeat trade, as well as disease transmission (see chapter 28).

The common chimpanzee is usually classified into four subspecies (Fisher et al., 2004; Gonder et al. 1997): the eastern subspecies (*Pan troglodytes schweinfurthii*), the central subspecies (*Pan troglodytes troglodytes*), the Nigerian subspecies (which has only one small population, *Pan troglodytes vellurosus*), and the western subspecies (*Pan troglodytes verus*). The eastern subspecies is relatively small, and usually appears to have a paler face. The western subspecies is relatively large, and the face looks masked because of characteristic black patches around the eyes. However, there are large differences in physical appearance among individuals, so that even experts cannot readily recognize the different subspecies. Thanks to genetic analysis from blood, hair, and feces, it is now possible to differentiate individuals at the subspecies level.

Chimpanzees also include another species, known as the bonobo (*Pan paniscus*). About 600,000 years ago there existed the common ancestor of two species of hominids, *Homo sapiens* and *Homo neanderthalensis*; the Neanderthal man subsequently survived until only 30,000 years ago. Similarly, there are now two *Pan* species living in Africa, although they do not overlap in their geographical range. Bonobos live only in the densely forested regions south of the Congo River, in the Democratic Republic of Congo (former Zaire), central Africa, where neither the common chimpanzee nor the gorilla occur (Kano 1999; Boesch and Hohmann 2002; Furuichi and Thompson 2007). Once you recognize bonobos, it is easy to discriminate them from common chimpanzees: bonobos are more slender, and their voices are more tonal and richer in frequency modulation. The two species shared a common ancestor a little less than a million years ago (Becquet et al. 2007) and are thus equally close to humans (figure 1.1).

There are currently several major chimpanzee research sites in Africa (see appendix). Long-term observations spanning three to four decades and sometimes more have taken place at Gombe (Goodall 1986) and Mahale (Nishida

1990) in Tanzania, Budongo (Reynolds, 2005) and Kibale (Wrangham et al. 1996) in Uganda, Taï (Boesch and Boesch-Achermann 2002) in Côte d'Ivoire, and Bossou (Sugiyama 2004) in Guinea. There are also some young but important research sites such as the Goualougo triangle in the Republic of Congo (see chapter 11), Ngogo in Uganda (see chapter 15), Kalinzu in Uganda (Hashimoto and Furuichi 2006), and Fongoli in Senegal (Pruetz and Bertolani 2007). All of their information on the ecology, life history, and social life of chimpanzees comes from the collective efforts of people who have dedicated their lives to understanding chimpanzees in their natural habitat.

Chimpanzees may live in primary forests, secondary forests, gallery forests, and even in savanna. Long-term records from Bossou (Sugiyama and Koman 1992) indicate that they consume about 200 plant species, a third of the species available in their habitat. These include various plant parts such as fruit, leaves, flowers, bark, stem, roots, gum, etc. Since chimpanzees are omnivorous, they also eat insects, eggs, birds, and mammals. However, their communities differ remarkably in feeding repertoire. For example, the chimpanzees at Taï, Mahale, Gombe and Ngogo often engage in hunting for meat (Boesch 2002; see also chapters 15 and 18) while Bossou chimpanzees seldom eat meat. Chimpanzees also exhibit various tool uses unique to each community (McGrew 1992; Whiten et al. 1999; see also chapters 10, 11, and 12). A behavioral assessment based on time sampling revealed that Bossou chimpanzees use tools for getting food during 16% of their feeding time (Yamakoshi 1999). This was much higher than expected, and it suggests that chimpanzees in some communities really need tools to survive.

Chimpanzees live in groups termed communities or unit groups. Each community consists of multiple males

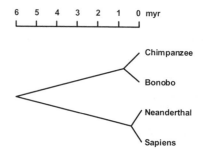

Figure 1.1 Phylogeny of chimpanzees and humans

Figure 1.2   An old female chimpanzee; her age is estimated as approximately 52 years. Photo by Gaku Ohashi.

and females, numbering together about 20 to 100 individuals. There are infants (less than 4 years old), juveniles (4 to 7 years old), adolescents (8 to 11 years old), young adults (12 to 35 years old) and old adults (36 years or older). These age classes may slightly differ between the sexes and among communities.

Thompson et al. (2007) examined mortality and fertility patterns in six free-living chimpanzee populations. They compared age-specific fertility patterns calculated from 534 chimpanzee births with equivalent demographic data from two well-studied human foraging populations, the !Kung of Botswana (Howell 1979) and the Ache of Paraguay (Hill and Hurtado 1996). The longevity of chimpanzees was about 50 years (figure 1.2). Age-specific fertility formed an inverted U shape, characterized by lower birth rates at the beginning and end of the reproductive life span. Compared with humans, chimpanzees reproduced more broadly across their life cycle. Reproductive performance began to decline at a similar age in chimpanzees and humans (25–35 years old) and approached zero at approximately the same age (50 years old).

Figures 1.3 and 1.4 show the mortality and fertility patterns of captive chimpanzees compared to wild ones (Kurashima and Matsuzawa, forthcoming). The dataset was based on the record of chimpanzees in Japan:

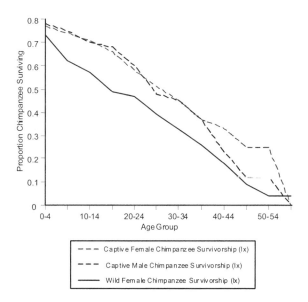

Figure 1.3   Mortality curve of chimpanzees

286 males and 388 females for the mortality curve, and 360 chimpanzee births by 118 females for the fertility curve. The mortality curves of the wild and captive chimpanzees were parallel (figure 1.3), although the mortality rate is lower in captivity. Fertility in chimpanzees declines at a pace similar to the decline in survival probability, whereas in humans it nearly ceases at a time when mortal-

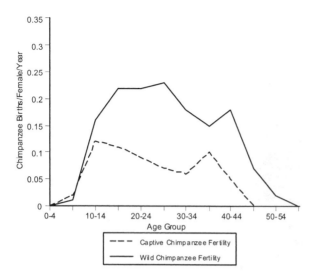

Figure 1.4 Fertility curve of chimpanzees

ity is still comparatively low. Thus, the post-reproductive period of the lifespan was almost absent in the wild chimpanzees, and less marked in the captive chimpanzees (Videan, et al. 2006), while marked in humans and known as "menopause." Therefore, during the course of human evolution, the role of grandmothers in the care of their grandchildren may have been an important selective pressure (Pavelka and Fedigan, 1991).

According to the fertility pattern, the inter-birth interval of chimpanzees is about five years (although this varies among field sites and among individuals). This means that a mother typically gives birth to a single infant on average once every five years. Chimpanzees have twins rarely in comparison to humans, who have twins in one per every 100 births. Weaning in chimpanzees occurs at about four years of age, which means a long period of dependence. After weaning, a mother resumes the menstrual cycle (about 35 days) until the next pregnancy occurs. Upon pregnancy the gestation period is about 230 days in chimpanzees, in contrast to about 270 days in humans. At birth a chimpanzee baby weighs a little less than 2 kg, while the human baby weighs about 3 kg (see the details in table 1.1 described later). During the first five years of life until a younger brother or sister is born, the chimpanzee infant is fully taken care of by his/her mother, with little to no paternal care.

Around the age of five, sex differences in behavior, including tool use, may appear in chimpanzees. Female juveniles have a tendency to spend more time with their

mothers, taking care of younger siblings, and partaking in tool-use activities such as termite fishing (Lonsdorf et al., 2004). Male juveniles have a tendency to associate more with adult males; they often patrol at the periphery of the territory (see chapter 15), and also follow estrus females, who advertise their reproductive status by the presence of a large pink ano-genital swelling.

Chimpanzees typically exhibit male philopatry, whereby males remain in their native community while females tend to emigrate at puberty. Chimpanzee society is therefore more patrilineal—characterized by the presence of multiple male generations including grandfathers, fathers, sons, and so on—compared to matrilineal societies which are characterized by the presence of multiple female generations with the presence of grandmothers, mothers, daughters, and so on. Most mammalian societies are matrilineal with female philopatry. A female chimpanzee in estrus receives ample attention from males of most age classes, including juveniles, adolescents, and adult males.

The History of Understanding Chimpanzees

As detailed above, the study of chimpanzees helps us understand the evolutionary foundation of human nature. The pioneer of the study of wild chimpanzees was Jane Goodall, although at the same time there were also several attempts by other scientists including Adrian Kortlandt and Kinji Imanishi (figure 1.5; Matsuzawa and McGrew 2008). This early fieldwork brought us much information on the life of chimpanzees in their natural habitat.

The pioneer of the study of the chimpanzee mind is Wolfgang Köhler (1925), known as one of the founders of Gestalt psychology. A Russian psychologist, Ladygina-Kohts (1935; 2002), also carried out groundbreaking work in comparing the development of chimpanzee infants with that of human infants. Other early attempts were focused on the rearing of chimpanzees by humans in a home setting, and on teaching spoken language to chimpanzees (Hayes 1951; Hayes and Hayes 1952; Kellogg and Kellogg 1933).

In parallel to the rise of fieldwork in Africa in the 1960s, there was a corresponding attempt to understand the chimpanzee mind in the laboratory. These were the ape-language studies in which researchers tried to teach American Sign Language to home-reared chimpanzees. In their pioneering work, Allen and Beatrix Gardner trained

Figure 1.5   In 1958, Kinji Imanishi and Jun'ichro Itani went to Africa to study gorillas and then chimpanzees in the wild. This was two years before Jane Goodall arrived at Gombe. Imanishi and Itani continued to send young students to the field. One of those students, Toshisada Nishida, began observations of the wild chimpanzees at Mahale, Tanzania, in 1965. Another, Takayoshi Kano, succeeded in observing wild bonobos at Wamba, the Democratic Republic of Congo (formerly Zaire), in 1973.

an infant chimpanzee named Washoe and claimed that she learned at least 132 kinds of ASL signs (Gardner and Gardner 1969). David Premack (1971) tried to teach a different system of artificial language, using colorful plastic chips, to a chimpanzee named Sarah. Duane Rumbaugh and his colleagues later used a lexigram system to teach Lana the chimpanzee (Rumbaugh 1977). Sue Savage-Rumbaugh aimed to teach both spoken language and its visual equivalents to Kanzi, a bonobo (Savage-Rumbaugh 1993). ASL was taught not only to chimpanzees (Fouts and Mills 1998) but also to lowland gorillas (Patterson 1978) and orangutans (Miles 1990). This paradigm of teaching human language to nonhuman animals has also been applied to non-primate species such as dolphins (Herman et al. 1984), dogs (Kaminski et al. 2004), and parrots (Pepperberg 1999). Criticism of these early studies came from Herbert Terrace, who taught ASL to an infant chimpanzee named Nim Chimpsky (Terrace 1979). He argued that the apes were not using the signs to communicate. After detailed analysis of video recordings of the ASL behavior, Terrace concluded that the results achieved by the Gardners could be explained by associative learning rather than by comprehension of the symbols' meaning.

Almost two decades later, we can now summarize the results of the ape-language studies that took place be-

tween the 1960s and the 1980s. It is clear that chimpanzees and other great apes can master language-like skills to some extent. However, there are clear constraints in their abilities to demonstrate the semantics, syntax, phonetics, and pragmatics central to human language (Matsuzawa 2009). Regarding semantics, chimpanzees can learn the use of symbols, such as ASL, plastic signs, or letters to represent objects, colors, numbers, and so on. However, their ability is limited in several ways. The number of signs or "words" learned was several hundred at most, and never exceeded 1,000. The rate of word acquisition did not increase but reached a plateau in the apes while it increases exponentially in humans. Regarding syntax, chimpanzees showed very little evidence for learning grammatical rules. They seldom used multiple signs in communicative contexts except for the simple repetition of signs within their repertoire. The mean length of "utterances" (MLU) was less than two words or signs, and even in the case of multiple sign use there was no clear evidence that syntactical rules were being followed. Regarding phonetics, it is apparent that no research succeeded in making the apes use their vocal tracts to produce various sounds with frequency modulation, even though humans and chimpanzees share the common developmental process of laryngeal descent (Nishimura et al. 2005). Regarding pragmatics, the language-like skills mastered by the apes faced the criticisms of Terrace and others. The apes often showed imitation of the human signers and replication of their signs. The paradigm of ape-language studies thus suffered two major problems: one was social cueing and the other was the removal of chimpanzee infants from their natural mothers, which is ethically unacceptable (see ethical note below).

Communication is based on intentionality and social interaction. After having faced numerous criticisms in the 1980s, the ape-language studies were stopped and research shifted towards understanding social intelligence as shown in chimpanzees' daily interactions with conspecifics (Byrne and Whiten 1988; de Waal 1982; Hauser et al. 2002). These studies did not require discrimination learning or intensive training such as was used in the ape-language studies. The landmark paper that stimulated this paradigm shift was Premack and Woodruff's article (1978) entitled "Do chimpanzees have a theory of mind?" Theory of mind became a central focus of research even in the field of human cognitive development (Wimmer and Perner 1983). The details of these studies on social intelligence are well documented (Call and Carpenter 2001; Hare et al. 2000; Povinelli and Eddy 1996; Tomasello 1998), and are discussed in chapters 7, 19, and 21. The study of the social aspects of the chimpanzee mind became more popular, and began to be linked with studies focused on other species such as vervet monkeys (Cheney and Seyfarth 1985) and baboons (Cheney and Seyfarth 2007).

In my laboratory, we developed another research paradigm called comparative cognitive science (CCS). This paradigm arose from studies based in psychophysics, a classic psychological study of measuring human sensation and perception, and it aims at understanding the perceptual world of nonhuman animals. The studies began with Ai, who was the first chimpanzee to learn to discriminate the 26 letters of the alphabet. The letters were then used to quantitatively compare her visual acuity with that of humans (Matsuzawa 1990). Ai also mastered visual symbols for color, so that her ability for color naming and classification could directly be compared with that of humans using identical materials and testing methods (Matsuzawa 1985a). In addition, Ai was the first chimpanzee to master the use of numerals to represent numbers (Matsuzawa 1985b). She learned both cardinal and ordinal aspects of the number system, and used numerals to label the numbers of real-life items (such as "five red toothbrushes") shown in a display window. She also mastered the skill of touching Arabic numerals in ascending order (Biro and Matsuzawa 1999, 2001), And her numerical knowledge was then used for testing her short-term memory (Kawai and Matsuzawa 2000). In this discipline, the studies focused not on teaching human language to the apes but on testing through language-like media the relative cognitive skills of humans and nonhuman animals using the same apparatus and following the same procedure (Fujita and Matsuzawa 1990; Hayashi 2007; Matsuzawa 2001, 2003; Matsuzawa and Tomonaga 2002; Tomonaga et al. 1993).

Humans as a Primate Species

Humans are one of the 350 or so primate species. A comparison of the human primate with other primate species may be the best way to understand human nature and to answer such questions as "What is uniquely human?" and "Where did we come from?" To get the answers, we should first understand the physical differences between primates so that we can appreciate the unique features of humans. Table 1 shows such parameters as adult body mass, neonatal body mass, adult brain mass, neonatal brain mass, gestation period (intrauterine life), weaning age, age at reproductive maturity, and life span. While it is well known that humans have a large brain (about three times as large as that of chimpanzees), the following three points are worth noting.

First, the postnatal growth of the brain is roughly the same in humans and chimpanzees. The neonatal brains of both species triple in size before reaching adult size (expanding 3.26 times in humans and 3.20 times in chimpanzees). This suggests that chimpanzees may experience cognitive developmental changes similar to those in humans. During postnatal development humans learn through experience, as do chimpanzees. The cognitive world of the infant chimpanzee is markedly different from that of juveniles, adolescents, and adults, which is why we need to study not only cognition in chimpanzees, but also the developmental changes in their cognitive abilities (Matsuzawa et al. 2006).

Second, humans do not have a large brain in relation to body size as is often assumed. Their adult body size shows clear sexual dimorphism as well as huge individual differences. Therefore, let us focus on brain size and body size at birth. As the indices show in table 1.1, there is not much

Table 1.1 Characteristics of human growth in primate perspective. Representative species for macaques, capuchins, and lemurs, respectively. Data based on Hamada and Udono (2006); age at weaning taken from Bogin (1999).

Species Scientific name	Human Hs	Chimpanzee Pt	Gorilla Gg	Orangutan Pp	Gibbon Hl	Macaque Mf	Capuchin Ca	Lemur Lc
Body mass, male (kg)	48	42	160	69	6	12	3	2.9
Body mass, female (kg)	40	31	93	37	5	9	2	2.5
Neonatal body mass (g)	3300	1756	2110	1728	411	503	248	88
Adult brain mass (g)	1250	410	506	413	108	109	71	25
Neonatal brain mass (g)	384	128	227	170	50	55	29	9
Adult/neonatal brain mass ratio	3.26	3.2	2.23	2.43	2.16	1.98	2.45	2.78
Neonatal brain/body mass ratio	0.12	0.07	0.11	0.1	0.12	0.11	0.12	0.1
Gestation period (days)	267	228	256	260	205	170	160	135
Weaning (days)	730	1460	1583	1095	730	182	270	105
Reproductive maturity (months)	198	118	78	84	108	60	43	10
Life span (years)	60	42	39	50	32	25	40	27

Note: Hs: *Homo sapiens*. Pt: *Pan troglodytes*. Gg: *Gorilla gorilla*. Pp: *Pongo pygmaeus*. Hl: *Hylobates lar*. Mf: *Macaca fuscata*. Ca: *Cebus apella*. Lc: *Lemur catta*.

difference among the primate species, including humans, in neonatal brain/body mass ratio; they all fall within the range of 0.07 to 0.12. Of course the absolute volume of the brain may be important and critical in cognitive processing, but it is also possible to downsize its volume while keeping the same functions. For example, a recent survey of caves on the island of Flores in Indonesia reported a new hominid fossil, *Homo florensiensis* (Brown et al. 2004). This species is estimated to have survived until only 18,000 years ago. They were about 1 m tall and only had a brain capacity of 500 ml, comparable to that of the chimpanzee. However, there is also evidence that they used complex tools, as well as fire. Thus, absolute volume of the brain may not be the critical factor in determining cognitive function. This view is also supported by recent studies showing that the behavior of capuchin monkeys is comparable to that of apes in many respects (Fragaszy et al. 2004a; Perry et al. 2003). Capuchins show adeptness in the area of combinatorial object manipulation, as do chimpanzees (Fragaszy and Adams-Curtis 1991; Matsuzawa 2001; Torigoe 1985). In addition, recent studies have reported that wild capuchins adapting to arid habitats feed on hard-shelled nuts and use stone tools to crack open palm nuts just as chimpanzees do (Fragaszy et al. 2004b), and also perform well on tasks requiring collaboration between two individuals (Hattori et al. 2005).

Third, physical separation of mother and infant shortly after birth is a feature unique to humans (Matsuzawa 2007). Human infants can be stable in a supine posture while great ape infants cannot. The human infant is fatty (about 20% fat, possibly as an adaptation to the colder air temperature at ground level), while the chimpanzee infant is not (about 4% fat). Human mothers do not always carry their infants, but chimpanzee mothers continue to hold their infants 24 hours a day for at least during the first three to four months of life. Weaning is also earlier in humans than in chimpanzees and other great apes. Chimpanzee infants suckle until they are at least four to five years old and the average inter-birth interval is longer in chimpanzees than in humans. Thus a long socialization process is not uniquely human, but rather a unique characteristic of hominoids. Instead, humans are characterized by physical separation of the mother-infant dyad soon after birth.

In sum, the existing data on human morphology and life history and our increasing understanding of nonhuman primates tell us the following three important points. First, developmental changes should be explored to understand the nature of cognition in nonhuman primates. Second, nonhuman primates with smaller brain volume may develop unique cognitive skills that are adaptive to their own way of life. Third, the mother-infant relationship in humans is somewhat unique in terms of physical separation, stable supine posture, and early weaning as compared to that of other hominoids.

Chimpanzee Cognitive Development

Numerous studies have examined chimpanzee cognition in captivity with regard to such capacities as tool use, insightful problem solving, and rudimentary forms of collaboration. However, very little literature has paid attention to the developmental changes in chimpanzee cognition. To address this deficit, my colleagues and I developed a research paradigm known as "participation observation" (Matsuzawa et al. 2006). In our laboratory we had three chimpanzee infants born in the year 2000 and raised by their biological mothers. The researchers observed the behavior of the mother-infant pairs and tested them through participation in their everyday life. Thanks to our long-term relationship with the mother chimpanzees, we can test the cognition of the infants with their assistance (figure 1.6). The three mother-infant pairs are the members of a community of 14 chimpanzees consisting of three generations living in an enriched environment (figure 1.7); we believe that to understand cognitive development we should provide social contexts equivalent to those in the wild.

A series of cognitive experiments has been completed in our lab that reveals the similarity between human and chimpanzee cognitive development. Chapter 2 details these similarities in the area of neonatal imitation. Face recognition is one of the key issues of cognitive development in humans (Johnson and Morton 1991), and our studies of facial recognition, gaze, and attention are described by Tomonaga (see chapter 4). Tool use is another characteristic shared by humans and chimpanzees, and Hayashi (see chapter 3) details her investigation of object

Figure 1.6 Participation observation, a unique way to test the cognitive development of infant chimpanzees with the assistance of their mothers. A tester is opening his own mouth to elicit neonatal imitation in the infant chimpanzee. A tiny video camera records the infant's facial expression. Photo by Nancy Enslin.

Figure 1.7 Outdoor compound at the Kyoto University's Primate Research Institute (KUPRI). The 15m-high climbing frame is surrounded by vegetation. Details are available at http://www.pri.kyoto-u.ac.jp/ai/. Photo by Tetsuro Matsuzawa.

manipulation, which is the precursor of tool use, in young chimpanzees. The summary of this body of work is that chimpanzees show cognitive development similar to that of humans in some areas. Their developmental process is, however, different from that of humans in some key respects. The following section will focus on the unique features of chimpanzee cognition.

A Breakthrough: Working Memory of Numerals in Chimpanzees

The chimpanzee mind has been extensively studied, and the general assumption is that it is inferior to that of humans (Premack 2007). However, Inoue and Matsuzawa (2007) showed for the first time that young chimpanzees have an extraordinary working memory capability for numerical recollection. As described in the anecdote that opens this chapter, it was better than that of human adults tested on the same apparatus. The subjects of the study were six chimpanzees who belonged to three mother-offspring pairs. One of the mothers, Ai, was the first chimpanzee who learned to use Arabic numerals to label sets of real-life objects with their corresponding numbers (Matsuzawa 1985b). The other five chimpanzees had also participated in many previous studies (Matsuzawa 2001; Matsuzawa et al. 2006), but they were naive to tasks employing numerals.

In 2004, when the three young chimpanzees reached the age of four years, the mother-offspring pairs started to learn the sequence of Arabic numerals from 1 to 9, using a touch-screen monitor connected to a computer. In the task, each trial was unique; the nine numerals always appeared in different locations on the screen. Accurate performance with 1–2–3–4–5–6–7–8–9 spontaneously transferred to nonadjacent sequences such as 2–3–5–8–9. All naive chimpanzees successfully learned this numerical sequence task.

A memory test called the "masking task" was introduced at around the time when the young turned five. In this task, after touching the first numeral, all other numerals were replaced by white squares. The subject had to remember which numeral appeared in which location, and then touch all the numerals in correct ascending order. All five naive chimpanzees mastered the masking task just like Ai (Kawai and Matsuzawa 2000). It must be noted that the chance level of this task is extremely low: $p = 1/24$ for four numerals, $1/120$ for five numerals, and $1/362,880$ for nine numerals respectively. In general, the performance of the three young chimpanzees was better than that of the three mothers. Ayumu, Ai's son, was the best performer among the three subjects (figure 1.8). Response time in humans was slower than in the three young chimpanzees.

Next, a new test called the "limited-hold memory task" was invented. This was a novel way of comparing the working memory of chimpanzee and human subjects. In this task, after the subject touched the initial white circle on a screen, the numerals appeared only for a certain limited duration, and were then automatically replaced by white squares. We tested both chimpanzees and humans on three different hold duration conditions: 650, 430, and 210 ms. The duration of 650 ms was equivalent to the average initial latency of the five-numeral masking task described above. The shortest duration, 210 ms, was close to the frequency of occurrence of human saccadic eye movement (Bartz 1962), thus not leaving subjects enough time to explore the screen by eye movement. The limited-hold memory task provided a means of objectively comparing the two species under exactly identical conditions. We compared Ai (the best mother performer), Ayumu (the best young performer), and human subjects (n = 9, all were university students) in this task.

Figure 1.9 shows the results of the comparison between two chimpanzees and human subjects in the limited-hold

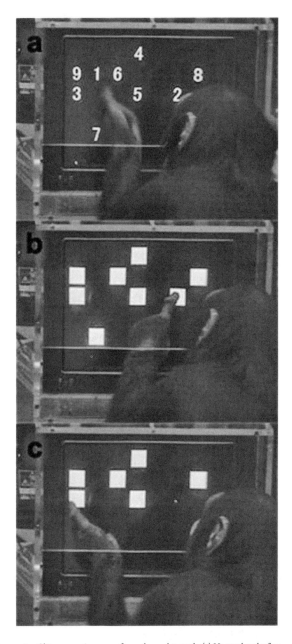

Figure 1.8 Chimpanzee Ayumu performs the masking task: (a) He touches the first numeral in the sequence, (b) the remaining numerals are immediately replaced by white squares, and (c) he remembers which numbers have appeared in which locations on the screen.

memory task. The number of numerals was limited to five. For example, the numerals 2, 3, 5, 8, and 9 might appear very briefly on the screen and then all five numerals were replaced by white squares at the same time. Subjects had to touch the squares in the correct ascending order that had been indicated by the original numerals. In human subjects, the percentage of correct trials decreased as a function of hold duration: the shorter the duration, the worse

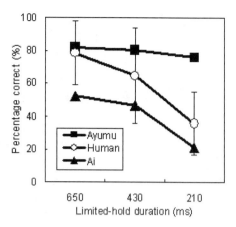

Figure 1.9 Performance in the limited-hold memory task by Ai, Ayumu, and human subjects (n = 9; the bars represent the SD). The *x* axis shows the three different limited-hold durations tested; the y axis shows the percentage of trials correctly completed under each condition. Each session consisted of 50 trials. Each chimpanzee received 10 sessions and each of nine humans received a single test session. A two-way ANOVA revealed that both main effects were significant (subjects: F2, 29 = 29.50, p < .001, hold duration length: F2, 29 = 121.45, p < 0.001), as was the interaction between them (F4, 58 = 20.10, p < 0.001). Post-hoc tests revealed that Ayumu's performance did not change as a function of hold duration (F2, 58 = 2.07, p = 0.136), whereas Ai and the human subject's performance decreased with shorter duration lengths (F2, 58 = 58.12, p < 0.001, F2, 58 = 101.45, p < 0.001, respectively). Pairwise multiple comparisons by Ryan's method showed significant differences in performance between Ayumu and human subjects at the 430ms and 210ms hold durations (p < 0.001, respectively).

the accuracy. Ai's performance was below that of the human subjects' average, and it showed the same tendency. From the very first session, however, Ayumu's performance remained at almost the same level irrespective of hold duration, and showed no decrement as recorded in the other subjects. These data show that chimpanzees can memorize, at a glance, Arabic numerals scattered on the touch-screen monitor and that Ayumu outperformed all of the human subjects in both speed and accuracy.

The results may be reminiscent of the phenomenon known as "eidetic imagery" found by Jaensch (1930). Eidetic imagery has been defined as the memory capability to retain an accurate, detailed image of a complex scene or pattern. It is known to be present in a relatively high percentage of normal children, and it declines with age. The present study demonstrated that a young chimpanzee was capable of retaining the location of many numerals at a glance and presented no decrement in performance irrespective of hold duration. Moreover, the young chimpanzees performed better than adults in the memory task. Taken together, the study showed for the first time that young chimpanzees have an extraordinary working mem-

ory capability for numerical recollection—better than that of human adults. Therefore, the results fit well with what we know about eidetic imagery in humans.

The Trade-off Theory of Memory and Language

The study described above showed that young chimpanzees were better than human adults in a memory task. Why do chimpanzees have better immediate memory than humans? One plausible explanation is that Ayumu's performance was reminiscent of that of "autistic savants" who possess extraordinary skills not exhibited by most persons. The data can be interpreted according to an evolutionary trade-off hypothesis. The common ancestor of humans and chimpanzees may have possessed an extraordinary memory capability, much like that exhibited by Ayumu. At a certain point in evolution, because of limitations on brain capacity, the human brain may have acquired new functions in parallel with losing others—such as acquiring language while losing the visuospatial temporal storage ability shown in the present task.

Previous studies have revealed that there are some other cognitive tasks in which chimpanzee performance exceeds human performance. Chimpanzees are good at identifying pictures of faces presented upside down, which is very difficult for humans (Tomonaga et al. 1993). They are also good at voice-face matching-to-sample tasks, in which a chimpanzee's voice is presented as a sample and then the subject must choose a photo of the vocalizer from among alternatives. This is also a very difficult task for human subjects (Izumi and Kojima 2004; Martinez and Matsuzawa 2009). In sum, chimpanzees are good at tasks that seem to have ecological validity for them—such as associating an individual's face with their voice, which is how chimpanzees communicate over a wide area.

The trade-off theory can be valid not only in phylogeny but also in ontogeny. Human youngsters can be better than adults at certain memory tasks—for example, at such chimpanzee-like memory tasks as remembering cards turned upside-down. In the course of their cognitive development, they may acquire linguistic skills while losing their chimpanzee-like photographic memory. This may be due to the time lag in the myelination of neuronal axons in each part of the brain. It is known that the association cortex responsible for complex representation and linguistic skills develops more slowly than the other primary areas. Further study of the mind-brain relationship will illuminate this kind of hypothesis.

Field Studies to Understand Chimpanzee Cognition

After several years of work with my chimpanzee partners in the laboratory, the accumulation of knowledge about chimpanzee cognition elicited novel questions in my mind. I had learned that chimpanzees were astonishingly intelligent in laboratory cognitive tests—but how was such intelligence actually used in their natural habitat? Upon taking a two-year sabbatical in David Premack's laboratory at the University of Pennsylvania in 1986, I decided to go to Africa to visit my senior colleague from Kyoto University's Primate Research Institute (KUPRI), Yukimaru Sugiyama, who has been exploring the ecology and behavior

of wild chimpanzees at Bossou, Guinea, since 1976. In the 23 years since my first visit, I have returned to Bossou at least once a year.

Each chimpanzee community in the wild has its own unique set of cultural traditions (Matsuzawa 2003; McGrew and Tutin 1978; Whiten et al. 1999, see also chapters 8 and 9). Chimpanzees seem to pass knowledge and skills from one generation to the next. In our study, my colleagues and I have been focusing on tool use and cultural behavior in the chimpanzees of Bossou. These chimpanzees are well known to use pairs of stones as hammers and anvils to crack open the hard shells of oil palm nuts (figure 1.10). Their repertoire of tool manufacture and use also includes a variety of unique examples such as pestle pounding, algae scooping, hyrax toying, the use of leaves as cushions and of folded leaves for drinking, and so on (Matsuzawa 1999)—all of which are limited to the Bossou community. Another behavior, ant dipping, is known

Figure 1.10 Stone tool use by wild chimpanzees at Bossou. Photo by Etsuko Nogami.

in several communities across Africa, but with features unique to each community regarding techniques, target ant species, and length and material composition of tools used (see chapter 10).

Thanks to the efforts of many colleagues, we have continued to illustrate the unique aspects of the Bossou chimpanzees' material and social culture. In addition to traditional fieldwork practices, we have established a unique way of studying chimpanzee cognition in the wild. We refer to this approach as "field experiments" in tool use (see chapter 12). In the core of the chimpanzees' ranging area, we set up an open-air "laboratory" site by laying out stones and nuts in a clearing. We then simply waited for chimpanzees to pass by and use the objects provided. Here their use of stone tools could be directly observed and video-recorded from behind a vegetation screen located about 15 m away. This unique experimental setup allows us to directly compare our findings in the field with those in the lab.

For example, as outlined by Hayashi (see chapter 3), participation observation at the indoor laboratory revealed that the chimpanzees' developmental changes in object manipulation and tool use corresponded to what we had observed in their natural habitat at the field "laboratory" (Biro et al. 2003; Humle and Matsuzawa 2002; Inoue-Nakamura and Matsuzawa 1997; Matsuzawa 1994, 1999). Tool use can be classified into several levels by focusing on how the objects used as tools are related to each other (Matsuzawa 1996). At around two years of age, wild chimpanzees show level 1 tool use—relating one object to another, as in ant fishing, algae scooping, and use of leaves for drinking water. At around four years, they start to show level 2 tool use—relating three objects to each other in a hierarchical fashion, as in using moveable stone tools to crack nuts. For example, the chimpanzees in Bossou use pairs of stones as hammers and anvils to crack open oil palm nuts. They place a nut on the anvil stone and then hit it with a hammer stone to crack the hard shell and get the edible kernel within. At around six years, they develop level 3 tool use—relating four objects in hierarchical order, as in the use of a wedge stone to stabilize the anvil stone, on which the nut is placed and then hit with a hammer stone. There is no evidence that the chimpanzees can reach level 4 tool use—relating five objects in hierarchical order. There is a limit to how many objects they can combine in a hierarchical order. In contrast, humans can develop hierarchical combinatorial manipulations at infinite levels.

Human technology is characterized by the self-embedding and recursive structure of tool use (Greenfield 1991; Hayashi 2007; Matsuzawa 1991, 1996).

Since 1997 we have also been recording intensively the use of leaves for drinking water at the same field experimental site. We drilled a hole in the trunk of a large tree and filled the hollow with fresh water. This setup created a unique opportunity to compare two different kinds of tool use at the same site at the same time (Sousa et al. 2009). As a result, it has become clear that the use of leaves to drink water is acquired at the age of about two years, much earlier than stone tool use. This may be due to the complexity of the tools involved: nut cracking requires a pair of tools (hammer and anvil) in comparison to the single tool (leaves) needed for drinking. Nevertheless, there are also characteristics that are common to the acquisition of both types of tool-use. Just like the chimpanzees in our laboratory simulation, infants like to carefully observe the actions of their mothers and other older members of the community, especially just before they begin using tools or just after they fail in using them. Infants also have a strong tendency to use "leftover" tools: sets of stones or drinking tools that have been used and discarded by elder community members.

Education by Master-Apprenticeship

The parallel efforts in the laboratory and in the wild have revealed a unique kind of social learning in chimpanzees, called "education by master-apprenticeship" (Matsuzawa et al. 2001). "Education by master-apprenticeship" is characterized by (1) prolonged exposure based on the mother-infant bond, (2) no teaching (no formal instruction and no positive/negative feedback from the mother), (3) intrinsic motivation of the young to copy the mother's behavior, and (4) high tolerance by the mother towards her infant.

Let us consider stone tool use as an example of master-apprenticeship social learning. First, the learning is based on the mother-infant bond because the infant chimpanzees are always carried by their mothers. Thus it is the mothers who provide the learning context. It is important to note that learning does not depend on the infant observing the mother's actions in any single case. A prolonged pre-exposure to the nut-cracking situation may be important for the later emergence of imitative processes in the chimpanzee infant.

Second, there is no overt teaching behavior in chimpanzees. In contrast, a human mother might mold the hand of her child to teach him or her how to hold the hammer stone or how to use it. She might provide a good nut to be cracked, or a good stone to use as a hammer. She might stabilize the anvil stone and put a nut on it to facilitate and stimulate nut cracking in her offspring. However, this is not the case in chimpanzees (although Boesch and Boesch-Achermann reported two such anecdotal cases in their long-term observation over 10 years). Thus, field observations show that chimpanzee mothers in the wild do not teach as human mothers do.

Let us look at this from a different perspective—that is, from the viewpoint of the infant chimpanzee. Consider a human child facing a nut-cracking task. She may observe her mother before starting. The mother may look at the infant to see what is going on. If the child succeeds in opening the nut, then she may look back at the mother to confirm her success. She may smile at the mother and the mother may in turn smile back to encourage her. This, however, is not the case in chimpanzees. In humans there is a triadic relationship between the mother, the infant, and the object(s) that does not exist in chimpanzees. Infant chimpanzees carefully observe the behavior of others, which may help them improve their own attempts (Hirata and Morimura 1999; Ueno and Matsuzawa 2005), but their problem-solving is based only on a direct and dyadic subject-behavior relationship, not on social referencing and feedback.

Third, infant chimpanzees have a very strong intrinsic motivation to copy their mothers' behavior. Chimpanzee infants typically start manipulating nuts or stones at less than one year of age: touching, mouthing, pushing, etc. At around two years of age they start to combine several objects in various ways: stacking, pushing one onto the other, etc (Inoue-Nakamura and Matsuzawa 1997). These attempts do not result in the infant obtaining the edible kernel, because the hard shell is not cracked. Moreover, an infant chimpanzee is allowed to steal the kernel from the mother after the mother cracks open the nut. Simple learning theory may predict an increase in this stealing behavior as it is reinforced by obtaining the edible reward. However, the stealing actually decreases over time as the infant continues her effort of combining the nuts and stones in a proper order.

Finally, the chimpanzee mothers are highly tolerant of their infants; they never scold their infants as humans

Figure 1.11   Chimpanzee mother extends a hand to her 2.5-year-old daughter to help her climb a tree. Photo by Tetsuro Matsuzawa.

sometimes do. They are also highly protective of and altruistic towards their infants. Food sharing is common between mother and infant in chimpanzees (Ueno and Matsuzawa 2004). The infant takes the food from the mother, and not the reverse.

Education by master-apprenticeship may be the underlying basis of human education, which in many societies has now evolved to be based on active teaching. A chimpanzee, on the other hand, may facilitate an infant's behavior in some cases (see figure 1.11) but such facilitation is unidirectional from mother to infant. Similarly, in chimpanzees food sharing is unidirectional while in humans it is bidirectional. For example, human infants often try to put edible items in the mouths of their mothers or other persons. This phenomenon has never been observed in chimpanzees. Therefore, reciprocity may be the key difference between humans and chimpanzees. Finally it must also be noted that social praise is absent in chimpanzees, while frequent in humans. Thus human education is characterized by the triadic and bidirectional nature of the behavioral interaction between master and apprentice when facing a problem.

Evolutionary Scenario for the Uniqueness of Human Cognition

This chapter summarizes recent progress in the study of chimpanzee cognition, focusing on both laboratory work and fieldwork, and also stresses the importance of developmental change. Cognitive development in chimpanzees can illuminate unique features of human cognition and its evolution during development. This last section will discuss an evolutionarily plausible scenario for the emergence of unique aspects of human cognition. It highlights the importance of stable supine posture, rather than bipedal locomotion, as a unique human feature (Matsuzawa 2007; Takeshita et al. 2009).

Let us look at the evolutionary stages of mother-infant relationships. Among mammals, primates developed a unique mother-infant relationship, clinging and embracing (Ross 2002; Matsuzawa 2007). Primates have potentially four limbs to grasp objects. Then, based on the continuous ventro-ventral contact between mother and infant, mutual gazing and smiling developed in the common ancestors of humans and chimpanzees.

In contrast, the physical separation of mother and infant soon after birth is unique to humans. The mother chimpanzee has to raise offspring on her own and in succession, one at a time. Humans mothers differ from this system of rearing which is common to hominoids; they have evolved to have shorter inter-birth intervals, and therefore give birth to successive offspring who still need to be taken care of after weaning—thus requiring help from the spouse, grandmother, and others. Because of their need to care for multiple youngsters, human mothers had to put their infants aside on the ground. To get the help from their spouses, human females occluded their sexual cycle from males, whereas chimpanzee females display large, external sexual swelling.

Humans are the terrestrial hominoids who have adapted to the savanna, where day-night temperature differences are relatively large. In the forest, the temperature on the ground is only one to two degrees lower than that 10 m above the ground (Takemoto 2004). Thick fat in the human infant might be an adaptation both to the cold night environment and to the mother-infant separation. Human infants can lie on their backs, adopting a stable supine posture, but chimpanzee infants cannot; when placed in that position they slowly move their contralateral limbs and try to cling. Whereas chimpanzee infants almost exclusively clung to their mothers, in humans mother-infant separation and the stable supine posture facilitated vocal exchange and provided freedom of limb movement for object manipulation and gesture. Through these changes, humans developed a unique way of communicating, incorporating multiple signals such as facial expression, manual gesture, and vocalization.

The need for this separation of mother and infant may have been due to changes in their social life. There might have been several interrelated factors including collaboration and division of labor by males and females, the existence of helpers or grandmothers because of the long post-reproductive lifespan, and the strong pair bonding which resulted in family-type group formations. The early life of the human infant is characterized by a dense social network that includes parents, siblings, and other members of the community. Although allomothering—maternal behavior shown by the members of a group towards infants—occurs in other primate societies, the allomothering tendency and mother-infant separation were mutually reinforced in humans. This led to early weaning in humans in comparison to the other hominoids, and was enforced by the development of infant provisioning through food processing and food donation.

Accumulation of knowledge and technology is the key to modern human life. Humans have the family network in addition to the mother-infant bond, so that there are multiple models available to the infant. The supine stable posture not only provided the basis for learning in a face-to-face situation right after birth, but also enhanced facial, gestural, and vocal exchange. These conditions may have favored human infants' ability to acquire skills through true imitation, or generalized imitation, in ways not seen in chimpanzees (Byrne and Russon 1998; Myowa and Matsuzawa 1999, 2000; see also chapter 2). This imitative capacity allows the apprentice to learn skills even after a single demonstration by the master. Humans and chimpanzees share the common intrinsic motivation to copy other individuals, but in the case of humans this motivation is channeled through mechanisms of social praise, teaching, and imitation, yielding more faithful cross-generational transmission of values, knowledge, and technology. Thus, the scenario proposed above highlights how the similarities and differences between the human and chimpanzee minds may have arisen.

Ethical Note: Chimpanzee Infants Should Be Raised by Chimpanzee Mothers

For about 100 years since the beginning of the twentieth century, humans have explored the mind of the chimpanzee. Comparative psychologists have examined the similarities and differences between the two species. Thanks to laboratory work, we have learned a great deal about

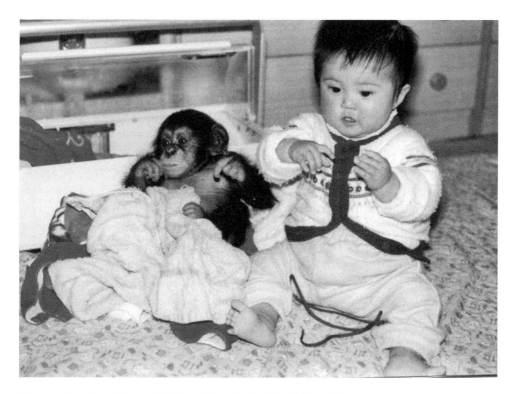

Figure 1.12 Comparison of human and chimpanzee by "cross-fostering." Photo by Tetsuro Matsuzawa.

various aspects of the chimpanzee mind, such as mirror self-recognition (Gallup 1970), cognitive mapping (Menzel 1973), symbol use (Savage-Rumbaugh 1986), numerical competence (Boysen and Berntson 1989), physical-causal understanding (Povinelli 2000), and so on. However, at present in the twenty-first century, it is time to think of the most ethical way to study chimpanzees in captivity.

The ape-language studies relied on "cross-fostering" (Gardner and Gardner 1969). The essential point of this method was to keep the two species in the same environment—i.e., rearing chimpanzees in a human home environment. At a glance this seems to have been a good method for comparison, because the environment was exactly the same for both species. In the logic of cross-fostering, the observed behavioral differences—such as the emergence of speech in human infants—should be due to the species differences.

I myself had the opportunity to raise an infant chimpanzee at home (figure 1.12). The biological mother had rejected her infant by screaming and running away as soon as she gave birth. I decided to take the female infant home, and at the time I happened to have a daughter of the same

age. Through this direct experience with cross-fostering I immediately recognized that this was not a fair comparison. My daughter had her parents, but the infant chimpanzee had no parents at all.

Infant chimpanzees separated from their mother often show such symptoms of depression as clasping their knees, rocking, and losing the shine in their eyes. The infant is separated from his or her biological mother, and also from the other conspecific members of his or her community. Home-raised chimpanzees in the cross-fostering studies were forced to adapt to the environment of a different species. A real cross-fostering study would require a human infant to be raised by chimpanzees as the corresponding control condition. But if this were done, could the human infant grow up as a human? It is time to stop this kind of comparison: We should not raise and keep infant chimpanzees in human home environments. Chimpanzee infants should be raised by their biological mothers. Cross-fostering is not a fair or scientific way to compare the two species, or to understand their cognitive development. In tests of their cognitive functions, chimpanzees should be reared by their mothers and kept in a group with conspecifics, living in an enriched environment that provides trees and other climbing structures to encourage their natural developmental and locomotor abilities.

There is one final point relating to the appropriate care of chimpanzees in captive environments. In many countries around the world, chimpanzees can still be privately owned by individuals as pets and for commercial enterprises such as television, movies, and advertising. Invariably, chimpanzees in these situations do not receive appropriate care in enriched environments such as those described above. Likewise, when chimpanzees are portrayed inaccurately and inappropriately in the media, it creates false impressions of their endangered status in the wild (Ross et al. 2008) and in some cases may convince people to seek out chimpanzees as pets. I raise these issues to provide context for some of the images throughout this volume that depict chimpanzees and humans in close contact as part of various cognitive tests. The chimpanzees portrayed in the following chapters are housed with other chimpanzees in facilities regulated by the appropriate federal and local agencies. For instance, the chimpanzees housed at KUPRI live in large social groups, and although they have opportunities to interact directly with humans, these are always in finite, controlled experimental settings with trained and experienced personnel. Chimpanzees do not make appropriate pets in any circumstance, and there are many reasons for this, not the least of which is the inability of private citizens to provide adequate mental stimulation for these cognitively advanced animals. Readers will have ample opportunity to learn more about the range of these impressive cognitive abilities throughout this volume.

The Mind of the Chimpanzee: Ecological and Experimental Perspectives

The following chapters collectively represent the multidisciplinary, multigenerational, captive-field synthesis approach that my colleagues and I have employed over the past several decades. Because the chimpanzee mind and the human mind are evolutionary byproducts that leave behind no physical trace, we are best served by using any and all available methods to determine their origins. Though elusive, the insights gained by studying our closest living relatives may provide some of the most valuable information to be gathered in the study of humankind. We begin with an investigation of some key mechanisms at play in chimpanzee cognition. Concepts such as imitation, ges-

tural and facial communication, and social cognition are examined from a neurological, developmental, and experimental perspective. These examinations are important to lay the groundwork for discussion of complex behaviors and concepts in later chapters.

Much of the following section is devoted to the most exciting revelations from the study of chimpanzee culture and, specifically, tool-use behavior. Unlike the state of the field just a decade ago, the study and appreciation of chimpanzee culture is now widely accepted. Data both from the field and from captive facilities provide a rich tapestry of results that continues to inspire future investigations and seed new questions about the ways in which chimpanzee sociality and traditions are expressed across multiple generations.

The next two sections highlight the complimentary nature of studies on chimpanzees living in the wild and those living in zoos and research centers. The social nature of chimpanzees is examined both through observations in the field and through controlled experiments in captive settings, and only by synthesizing results from both of these approaches are we able to approach a holistic view of the capabilities and tendencies of the chimpanzee mind.

Finally, it is crucial to consider some of the broader issues in our examination of and partnership with this complex and threatened species. Chimpanzees in captive settings require special consideration to address their psychological needs, and their endangered status in the wild deserves our attention as well. An examination of how these welfare and conservation issues overlap with our interest in cognitive ability provides a final perspective into the mind of the chimpanzee.

Whether your interests lie in the evolutionary origins of human cognition or in the intrinsic value of the chimpanzee mind itself, this volume presents a wide breadth of information to consider. Like my studies of the chimpanzees living at KUPRI and my observations of the Bossou chimpanzees, this volume examines related questions from multiple perspectives. Sometimes the answers seem contradictory, and at other times they align perfectly; often it is simply a matter of asking the right questions. Knowledge and appreciation of chimpanzees, both as close relatives of humankind and as a species unique and special in their own right, is requisite as we continue to find new and exciting ways to investigate the mind of this fascinating species.

Acknowledgments

The present study was financially supported by grants from MEXT (#07102010, 12002009, 16002001), as well as by the following grants: JSPS Global COE program for biodiversity (A06), JSPS core-to-core program HOPE. I would like to thank my colleagues, students, and the administration staff at KUPRI. Special thanks are due to Masaki Tomonaga, Masayuki Tanaka, and Misato Hayashi. I would also like to thank the veterinary staff and caretakers of the chimpanzees: Kiyoaki Matsubayashi, Juri Suzuki, Shunji Gotoh, Takako Miyabe, Akino Kato, Akihisa Kaneko, Kiyonori Kumazaki, Norihiko Maeda, Shohei Watanabe, Shino Yamauchi, and others. I am also very grateful to Michiko Sakai, Sumiharu Nagumo, Sana Inoue, Tomoko Takashima, and Suzuka Hori for their help in the laboratory study. Without their efforts we could not continue our study of chimpanzees at KUPRI. I also want to acknowledge the help by our field assistants in Bossou, Seringbara, and Yeale in Africa. Thanks are also due to the government of the Republic of Guinea—especially the Direction Nationale de la Recherche Scientifique et Technologique (DNRST) and the Institut de Recherche Environmentale de Bossou (IREB)—and the government of Côte d'Ivoire. Finally, I also would like to thank Tatyana Humle for English corrections.

Literature Cited

Bard, K., M. Myowa-Yamakoshi, M. Tomonaga, M. Tanaka, A. Costall, and T. Matsuzawa. 2005. Group differences in the mutual gaze of chimpanzees (*Pan troglodytes*). *Developmental Psychology* 41:616–24.

Bartz, A. E. 1962. Eye-movement latency, duration, and response time as a function of angular displacement. *Journal of Experimental Psychology* 64:318–24.

Becquet, C., N. Patterson, A.C. Stone, M. Przeworski, and D. Reich. 2007. Genetic structure of chimpanzee populations. *PLoS Genetics* 3:e66.

Biro, D., N. Inoue-Nakamura, R. Tonooka, G. Yamakoshi, C. Sousa, and T. Matsuzawa. 2003. Cultural innovation and transmission of tool use in wild chimpanzees: Evidence from field experiments. *Animal Cognition* 6:213–23.

Biro, D., and T. Matsuzawa. 1999. Numerical ordering in a chimpanzee (*Pan troglodytes*): Planning, executing, monitoring. *Journal of Comparative Psychology* 113:178–85.

———. 2001. Use of numerical symbols by the chimpanzee (*Pan troglodytes*): Cardinals, ordinals, and the introduction of zero. *Animal Cognition* 4:193–99.

Boesch, C. 2002. Cooperative hunting roles among Taï chimpanzees. *Human Nature* 13:27–46.

Boesch, C., and H. Boesch-Achermann. 2000. *The Chimpanzees of the Taï Forest*. Oxford: Oxford University Press.

Boesch, C., and G. Hohmann. 2002. *Behavioural Diversity in Chimpanzees and Bonobos*. Cambridge: Cambridge University Press.

Boysen, S. T., and G.G. Berntson. 1989. Numerical competence in a chimpanzee. *Journal of Comparative Psychology* 103:23–31.

Brown, P., T. Sutikna, M. J. Morwood, R. P. Soejono, W. Jatmiko, E. Saptomo, and Rokus Awe Due. 2004. A new small-bodied hominin from the Late Pleistocene of Flores, Indonesia. *Nature* 431:1055–61.

Byrne, R., and A. Whiten, eds. 1988. *Machiavellian Intelligence: Social Expertise and the Evolution of Intellect in Monkeys, Apes, and Humans*. New York: Oxford University Press.

Byrne, R., and A. Russon. 1998. Learning by imitation: A hierarchical approach. *Behavioral and Brain Sciences* 21:667–721.

Call, J., and M. Carpenter. 2001. Do Apes and Children Know What They Have Seen? *Animal Cognition* 4:207–20.

Cheney, D. L., and R. M. Seyfarth. 1985. *How Monkeys See the World*. Chicago: University of Chicago Press.

———. 2007. *Baboon Metaphysics: The Evolution of a Social Mind*. Chicago: University of Chicago Press.

De Waal, F. 1982. *Chimpanzee Politics: Power and Sex among Apes*. London: Jonathan Cape.

Fischer, A., V. Wiebe, S. Paabo, and M. Przeworski. 2004. Evidence for a complex demographic history of chimpanzees. *Molecular Biology and Evolution* 21:799–808.

Fouts, R., and S. T. Mills. 1998. *Next of Kin: My Conversations with Chimpanzees*. New York: Harper Paperbacks.

Fragaszy, D., and L. E. Adams-Curtis. 1991. Generative aspects of manipulation in tufted capuchin monkeys (*Cebus apella*). *Journal of Comparative Psychology* 105:387–97.

Fragaszy, D., E. Visalberghi, and L. Fedigan. 2004a. *The Complete Capuchin: The Biology of the Genus Cebus*. Cambridge University Press.

Fragaszy, D., P. Izar, E. Visalberghi, E. B. Ottoni, and M. G. de Oliveira. 2004b. Wild capuchin monkeys (*Cebus libidinosus*) use anvils and stone pounding tools. *American Journal of Primatology* 64:359–66.

Fujita, K., and T. Matsuzawa. 1990. Delayed figure reconstruction by a chimpanzee (*Pan troglodytes*) and humans (*Homo sapiens*). *Journal of Comparative Psychology* 104:345–51.

Furuichi, T., and J. Thompson, eds. 2008. *The Bonobos: Behavior, Ecology, and Conservation*. New York: Springer.

Gardner, R. A., and B. T.Gardner. 1969. Teaching sign language to a chimpanzee. *Science* 165:664–72.

Gallup, G. G. 1970. Chimpanzees: Self-Recognition. *Science* 167:341–43.

Gonder, M. K., J. F. Oates, T. R. Disotell, M. R. Forstner., J. C. Morales,and D. J. Melnick. 1997. A new west African chimpanzee subspecies? *Nature* 388:337.

Goodall, J. 1986. *The Chimpanzees of Gombe: Patterns of Behavior*. Cambridge, MA: Harvard University Press.

Green, R. E., J. Krause, S. Ptak, A. Briggs, M. Ronan, J. Simons, Lei Du, M. Egholm, J. Rothberg, M. Paunovic, and S. Paabo. 2007. Analysis of one million base pairs of Neanderthal DNA. *Nature* 444:330–36.

Greenfield, P. 1991. "Language, tools, and brain: The ontogeny and phylogeny of hierarchically organized sequential behavior." *Behavioral and Brain Sciences* 14:531–95.

Hare, B., J. Call, B. Agnetta, and M. Tomasello. 2000. "Chimpanzees know what conspecifics do and do not see." *Animal Behavior* 59:771–85.

Hashimoto, C., and T. Furuichi. 2006. Frequent copulations by females and high promiscuity in chimpanzees in the Kalinzu Forest, Uganda. In N. E. Newton-Fisher, H. Notman, J. D. Paterson, V. and Reynolds, eds., *Primates in Western Uganda*. New York: Springer, 247–57.

Hattori, Y., H. Kuroshima, and K. Fujita. 2005. Cooperative problem solving by tufted capuchin monkeys (*Cebus apella*): Spontaneous division of labor, communication, and reciprocal altruism. *Journal of Comparative Psychology* 119(3): 335–42.

Hauser, M. D., N. Chomsky, and W. T. Fitch. 2002. The faculty of language: What is it, who has it, and how did it evolve? *Science* 298:1569–79.

Hayashi, M. 2007. A new notation system of object manipulation in the nesting-cup task for chimpanzees and humans. *Cortex* 43:308–18.

Hayashi, M., and T. Matsuzawa. 2003. Cognitive development in object manipulation by infant chimpanzees. *Animal Cognition* 6:225–33.

Hayashi, M. 2007. A new notation system of object manipulation in the nesting-cup task for chimpanzees and humans. *Cortex* 43:308–18.

Hayes, C. 1951. *The Ape in Our House*. New York: Harper.

Hayes, K. J., and C. Hayes. 1952. Imitation in a home-raised chimpanzee. *Journal of Comparative Psychology* 45:450–59.

Herman, L. M., D. G. Richards, and J. P. Wolz. 1984. Comprehension of sentences by bottlenosed dolphins. *Cognition* 16:129–219.

Hill, K., and A. M. Hurtado. 1996. *Ache Life History: The Ecology and Demography of a Foraging People* (New York: Aldine de Gruyter).

Hirata, S., and M. Celli. 2003. Role of mothers in the acquisition of tool-use behaviours by captive infant chimpanzees. *Animal Cognition* 6:235–44.

Hirata, S., and N. Morimura. 2000. Naive chimpanzees' (*Pan troglodytes*) observation of experienced conspecifics in a tool-using task. *Journal of Comparative Psychology* 114(3): 291–96.

Howell, N. 1979. *Demography of the Dobe !Kung*. New York: Aldine de Gruyter.

Humle, T., and T. Matsuzawa. 2002. Ant-dipping among the chimpanzees of Bossou, Guinea, and some comparisons with other sites. *American Journal of Primatology* 58(3): 133–48.

Inoue, S., and T. Matsuzawa. 2007. Working memory of numerals in chimpanzees. *Current Biology* 17:R1004–5.

Inoue-Nakamura, N., and T. Matsuzawa. 1997. Development of stone tool-use by wild chimpanzees (*Pan troglodytes*). *Journal of Comparative Psychology* 111(2): 159–73.

Izumi, A., and S. Kojima. 2004. Matching vocalizations to vocalizing faces in a chimpanzee (*Pan troglodytes*). *Animal Cognition* 7:179–84.

Jaensch, E. R. 1930. *Eidetic Imagery and Typological Methods of Investigation*. Second edition, trans. by Oscar Oeser. New York: Harcourt, Brace, and Company.

Johnson, M. H., and J. Morton. 1991. *Biology and Cognitive Development: The Case of Face Recognition*. Oxford: Blackwell.

Kano, T. 1999. *The Last Ape: Pygmy Chimpanzee Behavior and Ecology*. Stanford, CA: Stanford University Press.

Kawai, N., and T. Matsuzawa. 2000. Numerical memory span in a chimpanzee. *Nature* 403:39–40.

Kellogg, W. N., and L.A. Kellogg. 1933. *The Ape and the Child: A Comparative Study of the Environmental Influence upon Early Behavior*. New York, Hafner.

Koehler, W. 1925. *The Mentality of Apes*. New York, Harcourt, Brace.

Ladygina-Kohts, N. N. 2002. *Infant Chimpanzee and Human Child: A Classic 1935 Comparative Study of Ape Emotions and Intelligence*. New York: Oxford University Press.

Lonsdorf, E. V., L. E. Eberly, and A. E. Pusey. 2004. Sex difference in learning in chimpanzees. *Nature* 428:715–16.

Martinez, L., and T. Matsuzawa. 2009. Effect of species-specificity in auditory-visual intermodal matching in a chimpanzee (*Pan troglodytes*) and humans. *Behavioral Processes* 82:160–63.

Matsuzawa, T. 1985a. Color naming and classification in a chimpanzee (*Pan troglodytes*). *Journal of Human Evolution* 14:283–91.

———. 1985b. Use of numbers by a chimpanzee. *Nature* 315:57–59.

———. 1990. Form perception and visual acuity in a chimpanzee. *Folia Primatologica* 55:24–32.

———. 1991. Nesting cups and meta-tool in chimpanzees. *Behavioral and Brain Sciences* 14(4): 570–71.

———. 1994. Field experiments on use of stone tools by chimpanzees in the wild. In R. Wrangham, et al., eds., *Chimpanzee Cultures*. Cambridge, MA: Harvard University Press, 351–70.

———. 1996. Chimpanzee intelligence in nature and in captivity: isomorphism of symbol use and tool use. In W. McGrew, et al., eds., *Great Ape Societies*. Cambridge: Cambridge University Press, 196–209.

———. 1999. Communication and tool use in chimpanzee: Cultural and social contexts. In M. Hauser and M. Konishi, eds., *The Design of Animal Communication*. Cambridge: Cambridge University Press, 645–71.

———, ed. 2001. *Primate Origins of Human Cognition and Behavior*. Tokyo: Springer-Verlag, 587.

———. 2003a. The Ai project: Historical and ecological contexts. *Animal Cognition* 6:199–211.

———. 2003b. Koshima monkeys and Bossou chimpanzees: Culture in nonhuman primates based on long-term research. In F. de Waal and P. Tyack, eds., *Animal Social Complexity: Intelligence, Culture, and Individualized Societies*. Cambridge, MA: Harvard University Press.

——— 2007. Comparative cognitive development. *Developmental Science* 10:97–103.

——— 2009. Symbolic representation of number in chimpanzees. *Current Opinion in Neurobiology* 19:92–98.

Matsuzawa, T., D. Biro, T. Humle, N. Inoue-Nakamura, R. Tonooka, and G. Yamakoshi. 2001. Emergence of culture in wild chimpanzees: Education by master-apprenticeship. In T. Matsuzawa, ed., *Primate Origins of Human Cognition and Behavior*, Tokyo: Springer, 557–74.

Matsuzawa, T., M. Tomonaga, and M. Tanaka. 2006. *Cognitive Development in Chimpanzees*. Tokyo: Springer.

Matsuzawa, T., and W. C. McGrew. 2008. Kinji Imanishi and 60 years of Japanese primatology. *Current Biology* 19:R587–R591.

McGrew, W. C., and C. E. G. Tutin. 1978. Evidence for a social custom in wild chimpanzees? *Man* 12:234–51.

McGrew, W.C. 1992. *Chimpanzee Material Culture: Implications for Human Evolution*. Cambridge: Cambridge University Press.

Meltzoff, A. N., and M. K. Moore. 1977. Imitation of facial and manual gestures by human neonates. *Science* 198:75–78.

Menzel, E. W. 1973. Chimpanzee spatial memory organization. *Science* 182:943–45.

Miles, L. 1990. The cognitive foundations for reference in a signing orangutan. In *"Language" and Intelligence in Monkeys and Apes*, edited by S. T. Parker and K. R. Gibson. Cambridge: Cambridge University Press. 511–39.

Mizuno, Y., H. Takeshita, and T. Matsuzawa. 2006. Behavior of infant chimpanzees during the night in the first four months of life: Smiling and suckling in relation to behavioral state. *Infancy* 9:215–34.

Myowa-Yamakoshi, M., and T. Matsuzawa. 1999. Factors influencing imitation of manipulatory actions in chimpanzees (*Pan troglodytes*). *Journal of Comparative Psychology* 113(2): 128–36.

———. 2000. Imitation of intentional manipulatory actions in chimpanzees (*Pan troglodytes*). *Journal of Comparative Psychology* 114:381–91.

Myowa-Yamakoshi, M., M. Tomonaga, M. Tanaka, and T. Matsuzawa. 2003. Preference for human direct gaze in infant chimpanzees (*Pan troglodytes*). *Cognition* 89:113–24.

————. 2004. Imitation in neonatal chimpanzees (Pan troglodytes). *Developmental Science* 7(4): 437–42.

Myowa-Yamakoshi, M., M. Yamaguchi, M. Tomonaga, M. Tanaka, and T. Matsuzawa. 2005. Development of face recognition in infant chimpanzees (*Pan troglodytes*). *Cognitive Development* 20:49–63.

Nishida, T. 1990. *The Chimpanzees of the Mahale Mountains.* Tokyo: University of Tokyo Press.

Okamoto, S., M. Tomonaga, K. Ishii, N. Kawai, M. Tanaka, and T. Matsuzawa. 2002. An infant chimpanzee (*Pan troglodytes*) follows human gaze. *Animal Cognition* 5:107–14.

Patterson, F. G.. 1978. The gestures of a gorilla: Language acquisition in another pongid. *Brain and Language* 5:72–97.

Pavelka, M. S. M., and L. M. Fedigan. 1991. Menopause: A comparative life history perspective. *Yearbook of Physical Anthropology* 34:13–38.

Pepperberg, I. M. 1999. *The Alex Studies: Cognitive and Communicative Abilities of Grey Parrots.* Cambridge, MA: Harvard University Press.

Perry, S., M. Baker, L. Fedigan, J. Gros-Louis, K. Jack, K. C. MacKinnon, J. H. Manson, M. Panger, K. Pyle, and L. Rose. 2003. Social conventions in wild white-faced capuchin monkeys. *Current Anthropology* 44:241–58.

Povinelli, D. J., and T. J. Eddy. 1996. What young chimpanzees know about seeing. *Monographs of the Society for Research on Child Development* 61(247): 1–189.

Povinelli, D.J. (2000). *Folk Physics for Apes.* Oxford, Oxford University Press.

Premack, D. 1971. Language in chimpanzees? *Science* 172:808–22.

————. 2007. Human and animal cognition: Continuity and discontinuity. *PNAS* 104:13861–67.

Premack, D., and G. Woodruff. 1978. Does the chimpanzee have a theory of mind? *Behavioral and Brian Sciences* 1:515–26.

Pruetz, J., and P. Bertolani. 2007. Savanna chimpanzees, *Pan troglodytes verus*, hunt with tools. *Current Biology* 17:412–17.

Reynolds, V. 2005. *The Chimpanzees of the Budongo Forest: Ecology, Behavior, and Conservation.* Oxford: Oxford University Press.

Ross, C. 2002. Park or ride? Evolution of infant carrying in primates. *International Journal of Primatology* 22:749–71.

Ross, S. R., K. E. Lukas, E. V. Lonsdorf, T. S. Stoinski, B. Hare, R. Shumaker, and J. Goodall. Inappropriate use and portrayal of chimpanzees. *Science* 319:1487.

Rumbaugh, D., T. V. Gill, et al. 1973. Reading and sentence completion by a chimpanzee. *Science* 182:731–33.

Savage-Rumbaugh, E. S. 1986. *Ape Language: From Conditioned Response to Symbol.* New York: Columbia University Press.

Savage-Rumbaugh E. S., J. Murphy, R. A. Sevcik, K. E. Brakke, S. L. Williams, and D. Rumbaugh. 1993. Language comprehension in ape and child. *Monographs of the Society for Research in Child Development*, series 233. 58(3–4): 1–254.

Sousa, C., D. Biro, and T. Matsuzawa. 2009. Leaf tool-use for drinking water by wild chimpanzees (*Pan troglodytes*): Acquisition patterns and handedness. *Animal Cognition* DOI 10.1007/s10071-009-0278-0.

Sugiyama, Y. 2004. Demographic parameters and life history of chimpanzees at Bossou, Guinea. *American Journal of Physical Anthropology* 124:154–65.

Sugiyama, Y., and J. Koman. 1992. The flora of Bossou: Its utilization by chimpanzees and humans. *African Study Monographs* 13:127–69.

Takeshita, H., M. Myowa-Yamakoshi, and S. Hirata. 2009. The supine position of postnatal infants: Implications for the development of cognitive intelligence. *Interaction Studies* 10:252–69.

Terrace, H. 1979. *Nim.* New York: Alfred A. Knopf.

The Chimpanzee Sequencing and Analysis Consortium. 2005. Initial sequence of the chimpanzee genome and comparison with human genome. *Nature* 437:69–87.

Thompson, M. E., J. H. Jones, A. E. Pusey, S. Brewer-Marsden, J. Goodall, D. Marsden, T. Matsuzawa, T. Nishida, V. Reynolds, Y. Sugiyama, and R. W. Wrangham. 2007. Aging and fertility patterns in wild chimpanzees provide insights into the evolution of menopause. *Current Biology* 17:1–7.

Tomasello, M. 1998. Emulation learning and cultural learning. *Behavioral and Brain Sciences* 21:703–4.

Tomonaga, M., S. Itakura, and T. Matsuzawa. 1993. Superiority of conspecific faces and reduced inversion effect in face perception by a chimpanzee. *Folia Primatologica* 61:110–14.

Tomonaga, M., and T. Matsuzawa. 2002. Enumeration of briefly presented items by the chimpanzee (*Pan troglodytes*) and humans (*Homo sapiens*). *Animal Learning and Behavior* 30:143–57.

Tomonaga, M., M. Tanaka, T. Matsuzawa, M. Myowa-Yamakoshi, D. Kosugi, Y. Mizuno, S. Okamoto, M. Yamaguchi, and K. Bard. 2004. Development of social cognition in infant chimpanzees (*Pan troglodytes*): Face recognition, smiling, gaze, and the lack of triadic interactions. *Japanese Psychological Research* 46(3): 227–35.

Torigoe, T. 1985. Comparison of object manipulation among 74 species of non-human primates. *Primates* 26:182–94.

Ueno, A., and T. Matsuzawa. 2004. Food transfer between chimpanzee mothers and their infants. *Primates* 45:231–39.

————. 2005. Response to novel food in infant chimpanzees: Do infants refer to mothers before ingesting food on their own? *Behavioral Processes* 68:85–90.

Videan, E. N., J. Fritz, C. B. Heward, and J. Murphy. 2006. The effects of aging on hormone and reproductive cycles in female chimpanzees (*Pan troglodytes*). *Comparative Medicine* 56:291–99.

Walner, B., G. Brem, M. Muellwer, and R. Achman. 2003. Fixed nucleotide differences on the Y chromosome indicate clear divergence between *Equus prizewlskii* and *Equus caballus*. *Animal Genetics* 34:453–56.

Whiten, A., J. Goodall, W. McGrew, T. Nishida, V. Reynolds, Y. Sugiyama, C. Tutin, R. Wrangham, and C. Boesch. 1999. Cultures in chimpanzees. *Nature* 399:682–85.

Wimmer, H., and J. Perner. 1983. Beliefs about beliefs: Representation and constraining function of wrong beliefs in young children's understanding of deception. *Cognition* 13:103–28.

Wrangham, R., C. Chapman, A. Clark-Arcadi, and I. Isabirye-Basuta. 1996. Social ecology of Kanyawara chimpanzees: Implications for understanding the costs of great ape groups. In W. McGrew, L. Marchant, and T. Nishida, T. eds., *Great Ape Societies*. Cambridge: Cambridge University Press, 45–57.

Yamakoshi, G.. 1998. Dietary responses to fruit scarcity of wild chimpanzees at Bossou, Guinea: Possible implications for ecological importance of tool use. *American Journal of Physical Anthropology* 106:283–95.

PART I

Cognitive Mechanisms

2

Early Social Cognition in Chimpanzees

Masako Myowa-Yamakoshi

Shino, a female chimpanzee neonate, sleeps in an incubator, clasping several towels in her arms. She has just been rejected by her biological mother. She strongly resembles a human neonate; her hair is sparse, and while her head is relatively large, her trunk and limbs have a delicate build. It is obvious that she will not survive without help from others. She is gently removed from the incubator and held by her human caretaker. Suddenly, her eyes open wide and are directed toward the human's face in a vigorous and unflinching gaze. The caretaker notices the gaze and smiles at Shino, who also smiles. Unconsciously, the caretaker once again returns the smile to Shino. After a short time, he slowly protrudes his tongue at her. Shino continues staring at his face and then begins to open her mouth slightly. Slowly, the tip of her small tongue appears through the opening of her lips.

Early Development of Social Cognition in Chimpanzees

Over the last 20 years, social cognition in nonhuman primates has attracted much attention. In the mid-1980s, a group of primatologists hypothesized that the intelligence of primates evolved to deal with the special complexities of primate social life, such as frequent competition and cooperation with conspecifics in the social group (Byrne and Whiten 1988; Whiten and Byrne 1997). This hypothesis

is called the Social Intelligence Hypothesis or Machiavellian Intelligence Hypothesis, and throughout the 1990s it gained not only observational but also empirical support. The findings have revealed various aspects of social cognition in nonhuman primates (especially the great apes) such as imitation, tactical deception, observational learning of cultural behaviors including tool use, and inference of another conspecific's mental state (e.g., Tomasello and Call 1997; Whiten and Byrne 1997; see also chapters 17, 19, and 20). Imitation in particular is regarded as a typical social-cognitive skill that supports high-level intelligence in humans and apes. It contributes to the acquisition of adaptive nongenetic skills in a complex environment and plays a key role in the transmission of knowledge and skills—in an accurate and stable form—from one generation to the next to produce "cultural" traditions in great apes (see chapters 8, 9, and 13).

Comparative studies on social cognition across ape taxa contribute to our understanding of the evolutionary basis of such skills. Another important perspective is gained by taking a comparative developmental approach to illuminate which cognitive features might be considered unique to humans. Not surprisingly, chimpanzees have much in common with humans, particularly during the early stages of ontogeny. Several studies have indicated that there are similarities between the early competence of humans and chimpanzees when they are measured with the same cognitive tests (Mathieu and Bergeron 1981; Hallock et al. 1989; Bard et al. 1992). Traditionally, such results have been derived from chimpanzees raised by humans, and we might expect that a human-like social environment would influence the development of the social-cognitive domains in chimpanzees (Tomasello et al. 1993; Russell et al. 1997). In order to truly understand primitive predisposition and the development of social cognition in chimpanzees, we must investigate their natural emergence during the course of development.

In 2000, the Primate Research Institute of Kyoto University (KUPRI) initiated a project involving a longitudinal study of chimpanzee development (Matsuzawa 2003, 2007; Tomonaga et al. 2004). That year, three chimpanzee infants were born at KUPRI. Fortunately they were raised by their biological mothers, which helped eliminate undue human influence. Given the long-term relationship between the chimpanzee mothers and human researchers, we were able to closely test the cognition of infant chimpanzees with the assistance of their mothers. The research project provided several important findings, ranging from the physiological to cognitive domains (also see Matsuzawa et al. 2006). In this chapter I focus on the development of early social cognition based on mother-infant bonds. I will first discuss early social responses in chimpanzee neonates, and explore similarities with and differences from human neonates. Next I turn to results suggesting that chimpanzees undergo a "two-month revolution" in cognitive development, as is commonly found in humans. The question of whether chimpanzees show a parallel "nine-month revolution" is then discussed. Subsequently, I summarize recent research that suggests developmental differences in imitative abilities between chimpanzees and humans. Finally, I present new findings on prenatal cognitive development in chimpanzees.

Neonatal Imitation and Smiling: Social-Cognitive Competence in Newborns

When do we witness the origin of human social cognition? Such questions have intrigued many developmental psychologists, and the findings of Meltzoff and Moore (1977) have attracted exceptional attention. Their study revealed that human newborns possess the ability to imitate certain facial expressions they cannot even see themselves, such as those involving a protruding tongue or an open mouth.

Since then, many follow-up studies have been conducted in many cultures and there is no dearth of researchers who challenge the premise of Meltzoff et al. that neonatal imitation is the origin of later imitation, which appears at approximately 8 to 12 months of age. These researchers' criticisms are based on the assertion that the neonatal imitative response disappears or is lessened at approximately two months of age and then reappears at one year (Abravanel and Sigafoos 1984; Fontaine 1984). Moreover, since only one type of imitative response toward a facial expression (i.e., tongue protrusion) was observed, some researchers assert that neonatal imitation is not imitation at all, but merely a reflection, an innate releasing mechanism, or a form of exploratory behavior in response to interesting or arousing stimuli (Jacobson 1979; Jones 1996).

Is neonatal imitation really the origin of the more complex imitation that becomes active from around the infant's first year of life? Or is it a sensory system with origins that differ from the form of later imitation? There has been no clear answer to this puzzle. To investigate these questions, I explored the mechanism and functions of neonatal imitation from a comparative perspective.

Neonatal Imitation in Chimpanzees

There is a scarcity of information pertaining to the existence and development of neonatal imitation in nonhuman primates. In 1996 I demonstrated that an infant chimpanzee could first imitate tongue protrusion and mouth opening at the age of five to ten weeks (Myowa 1996). However, this experiment used a five-week-old subject who had been raised by humans shortly after her birth. Thus, it is possible that the subject's capacity for facial imitation was the result of postnatal socializing with humans (see Bard 2006, who provided similar findings related to human-reared infant chimpanzees aged 7 to 15 days). It was only in 2000 that we were able to systematically investigate the imitation of facial expressions in two infant chimpanzees who had been reared by their biological mothers from the time of birth (Myowa-Yamakoshi et al. 2004). The test was conducted once a week through the first 16 weeks of life for both chimpanzees. In each case a human tester and the infant chimpanzee, held by its mother, sat face-to-face. Each infant was then shown the following three gestures: tongue protrusion, mouth opening, and lip protrusion. At age one to eight weeks, both infants successfully pro-

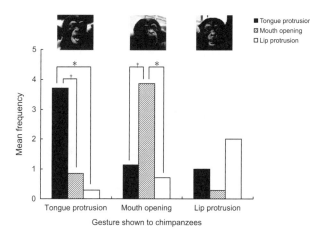
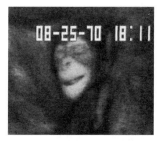
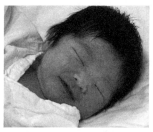

Figure 2.1 Frequencies of the three gestures (tongue protrusion, mouth opening, and lip protrusion) between one and eight weeks of age (data obtained from Pal, one of the two subjects). The x axis represents the facial gestures shown to the chimpanzee; $^*p < .05$; $^{\dagger}p < .10$.

Figure 2.2 Neonatal smiling in chimpanzee Pal, 16 days after birth (photo by Yuu Mizuno) and in a human four days after birth. Photo © Chukyo TV.

duced a greater number of tongue-protruding and mouth-opening gestures when these gestures were demonstrated to them (figure 2.1). However, their imitative responses of tongue protrusion and mouth opening disappeared once they reached the age of nine weeks.

We also conducted experiments on neonatal imitation in the squirrel monkey (*Saimiri sciureus*), the Japanese monkey (*Macaca fuscata*), and a lesser ape, the gibbon (*Hylobates agilis*), from the time of birth. However, since no clear evidence for imitation was found among them (Myowa-Yamakoshi 2006), it is plausible that neonatal imitation is limited to humans and chimpanzees (no data has been provided regarding the other apes). Recently, Ferrari et al. (2006) presented evidence of neonatal imitation in rhesus macaques. Thus the evolutionary origin of neonatal imitation is a matter of persistent debate.

Neonatal Smiling in Chimpanzees

A similar developmental change was also observed in the smiling behavior of chimpanzees. Mizuno et al. (2006) discovered the phenomenon of spontaneous smiling in chimpanzee neonates, which is similar to that in human neonates (figure 2.2). Although neonatal smiling is morphologically similar to smiling, it occurs during REM sleep without external stimulation and is characterized by closed eyes.

Social smiling differs from neonatal smiling in that it is characterized by open eyes. Interestingly, neonatal smiling in chimpanzees disappeared within the first two months, just as it does in humans. This developmental shift was similar to the disappearance of neonatal imitation (Myowa-Yamakoshi et al. 2004). In contrast, social smiling—elicited by the presentation of interesting objects and occurring during face-to-face interactions with other individuals—showed the opposite tendency and increased after the age of eight weeks. Around that time, for example, we frequently observed open-mouthed responses that could be considered social smiling (e.g., play faces) directed at the changeable facial expressions of the experimenter.

As mentioned earlier, it is still unclear whether neonatal imitation in humans and chimpanzees is reflexive or can be regarded as the foundation of imitation. However, such primitive competence is likely to play a significant role in survival during the early stage of life. Regardless of whether or not neonatal smiling and imitation are reflexive, they are beneficial to infants in attracting the attention of mothers for as long as possible (Myowa-Yamakoshi 2006).

The Two-Month Revolution in Chimpanzees

As evidenced by the early behavioral changes described above, key developmental transitions in the human mind occur at an early stage of ontogeny. When they reach two months of age, infants become actively engaged in probing the environment. Rochat (2001) viewed these transitions as the first important juncture of human development and termed the change the "two-month revolution." Evidence of such a revolution also appears in the social-cognitive domain of chimpanzees; a significant qualitative change occurs in communicative activities during the second month.

Face-to-face interactions are one of the typical styles of social communication among chimpanzees. Infant chimpanzees, both in the wild and in captivity, engage in face-to-face interactions with their mothers, especially during the first three months of life (Bard et al. 2005; Plooij 1984). At KUPRI, at around two months of age the interactions of the three infant chimpanzees gradually developed a social character, as observed in the emergence of social smiling. Their responses to social stimulation changed remarkably from being reflexive (nonsocial) to being social; they began gazing at each other in the eye, recognizing different faces and, as discussed earlier, exchanging social smiles.

Mutual Detection of Eyes and Gazing

From the evolutionary perspective, detecting direct eye gaze in another individual's face must be essential for survival. The recognition of the self as the focus of another organism's attention enables an animal to predict the possible risk of the other animal aiming to attack, or to notice its being interested for some other reason. In fact, the ability to detect direct eye gaze has been widespread among various species in evolution, including jewelfish (Coss 1979), iguanas (Burger et al. 1992), and various bird species (Scaife 1976; for details see Emery 2000). Developmental studies have revealed that human infants are extremely sensitive in their perception of others' eyes. As evidence, Batki et al. (2000) demonstrated that human neonates who were less than two days old looked longer at a photograph of a face with open eyes than they did at a photograph of the same face with closed eyes. Baron-Cohen (1994, 1995) has insisted that humans have developed a neural mechanism devoted to gaze processing–"an eye-direction detector (EDD)" or gaze module.

We examined the gaze sensitivity of infant chimpanzees at KUPRI (Myowa-Yamakoshi et al. 2003). The infants were shown two photographs of a human face wherein (a) the eyes were open or closed and (b) the eye gaze was direct or averted. We found that all the chimpanzee infants preferred looking at faces whose eyes were open or whose gaze was directed at them, rather than at faces whose eyes were closed or whose gaze was averted from them. At the age of 10 weeks each chimpanzee had developed the ability to perceive another's gaze. We do not yet know, however, whether the gaze perception ability is present at birth (for more on gaze and joint attention, see chapter 4).

In close connection with the ability to perceive direct eye gaze, "mutual" gaze between mother and infant chim-

Figure 2.3 Mutual gaze between mother Ai and one-month-old infant Ayumu. Photo by Nancy Enslin, produced by Yomiuri Shinbun.

panzees also increased when the latter reached the age of two months (figure 2.3). This kind of mutual gaze is observed during social interactions when mother and infant simultaneously look at each other's faces. Bard et al. (2005) used 24-hour video recording to conduct detailed observations on the development of mutual gaze; they found that at the age of nine weeks, the rate of occurrence in mother-infant pairs increased dramatically—on average, from 11 to 28 times per hour. These findings revealed that from the age of two months, the infant chimpanzees increased their daily face-to-face interactions with their mothers or with others. This increase in mutual gaze corresponded to a decrease in the mothers' cradling behavior. Thus, the frequency of face-to-face interactions with mutual gaze appears to be negatively correlated with that of physical contact between mothers and infants.

Recognition of the Mother's Face

With extensive visual experience of face-to-face interaction, young chimpanzees begin to recognize their mothers' faces. We investigated this ability from the first week after birth (Myowa-Yamakoshi et al. 2005). We prepared photographs of the mother of each infant and of an "average" chimpanzee face using computer graphics technology. A human tester sat face-to-face with an infant who was held by his/her mother. The infant was shown one photograph. As soon as the infant fixated on it, the photograph was slowly moved to the left and to the right five times each. The appropriate eye- or head-turning movements were defined as "tracking responses," and the numbers of such responses for each photograph were compared.

The results illustrated that before the age of one month, the chimpanzees indicated few tracking responses and no

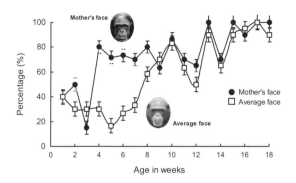

Figure 2.4   Developmental change expressed as a mean percentage of the tracking scores for each of the two faces for each week of age, plus standard error

difference in response to the three photos. From four weeks of age they began to discriminate their mothers' faces from among other individual faces, and to demonstrate a preference for them. From eight weeks onward they revealed no preference for their mothers' faces, but exhibited frequent tracking responses for all the photographs (figure 2.4).

A Nine-Month Revolution in Chimpanzees?

The two-month revolution in the social-cognitive domain underlines the "dyadic" social interaction (or "self-other" relationship) that is prototypical at this age, both in humans and chimpanzees. Later, from around the age of nine months, humans begin to develop interactions based on the "triadic" relationship of "self-object-other" (Trevarthen and Hubley 1978). This kind of interaction involves both objects and people, resulting in the formation of a referential triangle comprising the infant, adult, and object or event. For example, infants begin to look where adults are looking, to act on objects in the way adults are acting on them, and to point communicatively to request something or when they see interesting event or object. Further, these interactions help infants understand the intentions of others (Tomasello 1999). Some researchers refer to this second remarkable change that occurs at around nine months of age as the "nine-month revolution" (Rochat 2001; Tomasello 1999).

Does this revolution also occur in the social-cognitive domain of chimpanzees? In line with this question, I focused on their development in an attempt to ascertain whether the imitative ability that disappeared at the age of two months would reappear at a later age. I found that at around nine months of age chimpanzees once again began to produce imitative responses to several facial ges-

tures, much as humans do. The chimpanzees in our study differentially imitated at least three demonstrated actions: tongue protrusion, mouth opening, and lip protrusion. In addition to these facial gestures, they also produced imitative responses in the form of simple body movements such as hitting objects (Myowa-Yamakoshi 2006). However, the imitative responses differed somewhat from human imitation in that the chimpanzees' reproduction of observed actions was always accompanied by physical contact with the experimenter, which was not necessarily the case with human infants (figure 2.5).

Other comparative studies have provided conflicting evidence as to whether chimpanzees can imitate detailed

Figure 2.5   Imitative responses of the three demonstrated facial gestures by nine-month-old Ayumu, one of the two subjects—(a) tongue protrusion, (b) mouth opening, and (c) lip protrusion—as cited in Myowa-Yamakoshi (2006)

body movements (Myowa-Yamakoshi and Matsuzawa 1999, 2000; Nagell et al. 1993, Custance et al. 1995; Whiten et al. 1996). In an early experiment, Tomasello et al. (1993) observed that "enculturated" captive chimpanzees and bonobos were capable of imitating novel actions on objects as efficiently as two-year-old human children and much better than their mother-reared conspecifics. They also indicated that a human-like sociocultural environment may enhance the development of chimpanzee imitative abilities.

Myowa-Yamakoshi and Matsuzawa (1999) later conducted a study to investigate the degree of imitation in adult chimpanzees as compared to imitation in humans. This provided three important findings. First, the imitation of demonstrated actions was extremely rare in the chimpanzees (less than 6% of the overall number of actions that were presented to them). Second, the chimpanzees found it more difficult to perform actions that involved manipulating "one object" than to perform those involving the manipulation of "one object toward another object" or "one object toward oneself." Third, they found it more difficult to perform actions involving novel motor patterns than to perform those involving familiar motor patterns. It is noteworthy that they seldom imitated even those demonstrated actions that involved the motor patterns already part of a chimpanzee's normal repertoire. These data suggest that chimpanzees' style of recognizing others' actions differs from that of humans. When chimpanzees imitate the actions of others, they seem to pay less attention to body movements and more attention to the object being manipulated.

In a related investigation, Tomonaga et al. (2004) observed the social interactions of mother-reared chimpanzees at KUPRI during the first two years after birth; they concluded that it was extremely rare for the chimpanzees to interact triadically in this timeframe. For example, the infant chimpanzees never jointly interacted with both an individual manipulating an object and the object being manipulated. Instead, they interacted with either the individual or the object. While doing so, they never displayed "object showing" or "object giving," behaviors indicative of referential communication in a triadic relationship in human infants. These types of triadic interactions seem to develop in human infants at approximately nine months of age, and differences between chimpanzees and humans in imitative tendencies may be closely related to

the species-specific development of their social interactions. To confirm this view, further studies must focus on identifying the essential factors that constitute the unique human sociocultural environment.

New Perspectives on Cognitive Development in Chimpanzees: Observations from the Prenatal Period

Finally, we would like to introduce our ongoing research on cognitive development in chimpanzees in the prenatal period. As described above, both chimpanzee and human neonates employ perceptual-motor mechanisms that from birth enable them to be oriented toward social stimulation and to respond to it. Traditionally, such neonatal predispositions have been explained as "innate." However, it is possible that their primitive form might develop under the influence of experiences in utero. Human fetuses distinguish their own bodies from other entities in the womb, and demonstrate well-coordinated movements that appear to be intentional actions (Myowa-Yamakoshi and Takeshita 2006; Zoia et al. 2007). Furthermore, their mouth movements show differentiation between those made in response to the mother's voice and those made in response to the voices of strangers (Myowa-Yamakoshi and Takeshita, submitted).

In order to explore the ontogenetic roots of social cognition in chimpanzees, we have begun conducting experiments based on a new paradigm wherein we observe chimpanzee fetuses using a four-dimensional (4D) ultrasonographic technique. This new technique enables the continuous monitoring of dynamic fetal and motor activities (Kurjak et al. 2003). It is difficult to secure the participation of pregnant chimpanzee subjects in the study without administering anesthesia, but we have succeeded in doing so. Before the first test session, it is necessary to familiarize chimpanzee mothers with the experimental settings that require contact with gel on the probe and demand that they be kept stationary. This is why only a few previous studies have been conducted on chimpanzee fetal behavior. After long habituation training, however, we were able to observe the behaviors of a chimpanzee fetus (figure 2.6; for details see Takeshita et al. 2006; also see Kawai et al. 2004 for a different method). Tsubaki, a nine-year-old female chimpanzee who lives at the Hayashibara Great Ape Research Institute (GARI), participated in the

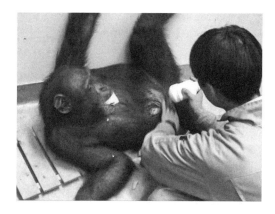

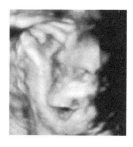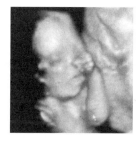

Figure 2.6 Experimental situation with 4D ultrasonography for a chimpanzee mother, Tsubaki, as cited in Takeshita et al. (2006). Photo © Great Ape Research Institute.

Figure 2.7 A chimpanzee fetus at 23 weeks of gestational age (left) and a human fetus at 21 weeks of gestational age (right)

study. We observed her fetus between 22 and 32 weeks of gestation.

Our preliminary observations showed that the fetus often touched its own body parts and other entities. A study by Rochat and Hespos (1997) indicated that in humans, such repeated explorations may lead the fetus to be able to distinguish between self-produced tactile stimulation (self-stimulation or "double-touch") and non-self or external tactile stimulation (allostimulation). Similarly, chimpanzee fetuses may also begin to acquire perceptual knowledge of their own bodies by distinguishing them from other entities in the womb (the "ecological" self; Neisser 1991, 1995). On the other hand, two distinct differences emerged between the patterns of fetal forelimb movements in chimpanzees and in humans. The first difference pertained to fewer observed limb movements in a chimpanzee fetus subject than in a human. Second, the chimpanzee fetus demonstrated frequent forelimb contact with the head, whereas human fetuses are known to show relatively more frequent forelimb contact with other parts of the face, including the eyes, nose, and mouth

(figure 2.7; Kurjak et al. 2004). We do not yet know how these differences might reflect on the variations in both species' postnatal cognitive development.

Another noteworthy issue relates to the fetal facial movements that correspond to the neonatal smiling and neonatal imitation observed in chimpanzees. Such data are important for exploring the development of these abilities from the prenatal to postnatal period as well as their functions during those respective periods. However, we were unable to obtain high-quality video images of face and mouth movements, perhaps due to the low volume of amniotic fluid in chimpanzees. Therefore, it remains unclear whether facial movements corresponding to smiling and imitation in chimpanzee neonates are found in chimpanzee fetuses as well. In humans, several studies revealed that facial expressions, which are a part of these activities, emerge during fetal life and continue through the postnatal period (Kurjak et al. 2004). It is possible that the facial expressions corresponding to these activities may be found in chimpanzee fetuses as well.

Implications and Further Directions

How have human and nonhuman primate minds developed, and what has caused the developmental differences that exist between species? To address these questions, I have taken an evolutionary approach to the development of social cognition and focused on a comparison of chimpanzee and human abilities in this realm. Chimpanzees seem to share similar social-cognitive skills with humans in the early stages of life, but diverge in their development in differential social environments very shortly after birth or even before.

Few studies have examined the relative effects of genetic predisposition and environmental influence on the development of social-cognitive skills. I suggest two key approaches to addressing this issue. The first is to explore the ontogeny of social cognition from the very early stages of life including the prenatal period. Study of these early stages of development may help elucidate the primitive roots of species-specific predispositions with minimal influence by experience and environment. The second approach involves the identification of key experiences that may play a critical role in the development of social cognition in both chimpanzees and humans. As discussed earlier, one of the essential factors leading human infants into the second cognitive transition (the "nine-month revolution") is the human-unique environment that includes active teaching and molding. Furthermore, human infants are often led or compelled by adults to participate in triadic communicative activities in ways that chimpanzee infants are not. In this light, it is plausible that human environments may have the ability to modify the emergence of key social cognitive skills in nonhuman primates, as has been demonstrated with chimpanzees and Japanese monkeys (Tomasello et al. 1993; Kumasiro et al. 2003). Further studies are warranted to address the differential effects of rearing environments and life history on subsequent development of cognition in the social domain.

In addition to traditional observational and experimental techniques to address these questions, I advocate the application of new methodological frameworks for understanding social cognition mechanisms in primate species. For example, new noninvasive techniques have allowed the monitoring of brain activities of infants using near-infrared spectroscopy and event-related potential methods. Ueno et al. (2008) have succeeded in measuring the event-related potential of a fully awake chimpanzee during auditory stimulation. Using the results from these and other diverse research fields such as neuroscience, genetics, computer science, and robotics offers promise. A convergent, multidisciplinary approach seems the most promising way to unravel the intricacies surrounding the social-cognitive development of the chimpanzee mind.

Acknowledgments

The research reported here was financially supported by Grants-in-Aid for Scientific Research from the Japan Society for the Promotion of Science (JSPS); the Ministry of Education Culture, Sports, Science and Technology (MEXT) (nos. 12002009 and 16002001 to T. Matsuzawa, 13610086 to M. Tomonaga, 09207105 to G. Hatano, 10CE2005 to O. Takenaka, 16203034 to H. Takeshita, and 16683003 and 19680013 to M. Myowa-Yamakoshi); the MEXT Grant-in-Aid for the 21st Century COE Programs (A2 and D2 to Kyoto University); the research fellowship to M. Myowa-Yamakoshi from JSPS for Young Scientists (No. 3642); the JSPS Core-to-Core Program, HOPE; the JSPS Grant-in-Aid for Creative Scientific Research for the "Synthetic Study of Imitation in Humans and Robots" to T. Sato; and the Cooperative Research Program of the Primate Research Institute, Kyoto University. The authors are grateful to all the coauthors, students, and staff, and also to the mother and infant chimpanzees and humans who participated in the studies at the PRI, the GARI, and the "Umikaze" Infant Laboratory of the University of Shiga Prefecture.

Literature Cited

Abravanel, E., and A. D. Sigafoos. 1984. Exploring the presence of imitation during early infancy. *Child Development* 55:381–92.

Batki, A., S. Baron-Cohen, S. Wheelwright, J. Connellan, and J. Ahluwalia. 2000. Is there an innate module? Evidence from human neonates. *Infant Behavior and Development* 23:223–29.

Bard, K.A. 2006. Neonatal imitation in chimpanzees (*Pan troglodytes*) tested with two paradigms. *Animal Cognition* 10:233–42.

Bard, K., M. Myowa-Yamakoshi, M. Tomonaga, M. Tanaka, A. Costall, and T. Matsuzawa. 2005. Group differences in the mutual gaze of chimpanzees (*Pan troglodytes*). *Developmental Psychology* 41:616–24.

Bard, K. A., K. A. Platzman, B. M. Lester, and S. J. Suomi. 1992. Orientation to social and nonsocial stimuli in neonatal chimpanzees and humans. *Infant Behavior and Development* 15:43–56.

Baron-Cohen, S. 1994. How to build a baby that can read minds: Cognitive mechanisms in mindreading. *Current Psychology and Cognition* 13:513–52.

———. 1995. *Mindblindness: An essay on autism and theory of mind.* Cambridge, MA: MIT Press.

Burger, J., M. Gochfeld, and B.G. Murray. 1992. Risk discrimination of eye contact and directness of approach in black iguanas (*Ctenosaura similes*). *Journal of Comparative Psychology* 106:97–101.

Byrne, R. W., and A. Whiten, eds. 1988. *Machiavellian Intelligence: Social Expertise and the Evolution of Intellect in Monkeys, Apes, and Humans.* New York: Oxford University Press.

Coss, R. G.. 1979. Delayed plasticity of an instinct: Recognition and avoidance of 2 facing eyes by the jewel fish. *Developmental Psychology* 12:335–45.

Custance, D. M., A. Whiten, and K. A. Bard. 1995. Can young chimpanzees (*Pan troglodytes*) imitate arbitrary actions? Hayes and Hayes (1952) revisited. *Behaviour* 132:839–58.

Emery, N. J. 2000. The eyes have it: The neuroethology, function and evolution of social gaze. *Neuroscience & Biobehavioral Reviews* 24:581–604.

Ferrari, P. F., E. Visalberghi, A. Paukner, L. Fogassi, A. Ruggiero, and S. J. Suomi. 2006. Neonatal imitation in rhesus macaques. *PLoS Biology* 4:1501–8.

Fontaine, R. 1984. Imitative skills between birth and six months. *Infant Behavior and Development* 7:323–33.

Hallock, M.B., J. Worobey, and P.S. Self. 1989. Behavioral development in chimpanzee and human newborns across the first month of life. *International Journal of Behavioral Development* 12:527–40.

Jacobson, S. W. 1979. Matching behavior in the young infant. *Child Development* 50:425–30.

Jones, S. S. 1996. Imitation or exploration? Young infants' matching of adults' oral gestures. *Child Development* 67:1952–69.

Kawai, N., S. Morokuma, M. Tomonaga, N. Horimoto, and M. Tanaka. 2004. Associative learning and memory in a chimpanzee fetus: Learning and long lasting memory before birth. *Developmental Psychobiology* 44:116–22.

Kumashiro, M., H. Ishibashi, Y. Uchiyama, S. Itakura, A. Murata, and A. Iriki. 2003. Natural imitation induced by joint attention in Japanese monkeys. *International Journal of Psychophysiology* 50:81–99.

Kurjak, A., G. Azumendi, N. Veček, S. Kupešic, M. Solak, D. Varga, and F. Chervenak. 2003. Fetal hand movements and facial expression in normal pregnancy studied by four-dimensional sonography. *Journal of Perinatal Medicine* 31:496–508.

Kurjak, A., M. Stranojevic, W. Andonotopo, A. Salihagic-Kadic, J. M. Carrera, and G. Azumendi. 2004. Behavioral pattern continuity form prenatal to postnatal life: A study by four-dimensional (4D) ultrasonography. *Journal of Perinatal Medicine* 32:346–53.

Mathieu, M. and G. Bergeron. 1981. Piagetian assessment on cognitive development in chimpanzee. In A.B. Chiarelli and R.S. Corruccini, eds., *Primate Behavior and Sociobiology*, pp.142–47. Berlin: Springer-Verlag.

Matsuzawa, T. 2003. The Ai project: Historical and ecological contexts. *Animal Cognition* 6:199–211.

———. 2007. Comparative cognitive development. *Developmental Science* 10:97–103.

Matsuzawa, T., M. Tomonaga, and M. Tanaka, eds. 2006. *Cognitive Development in Chimpanzees*. Tokyo: Springer-Verlag.

Meltzoff, A. N., and M. K. Moore. 1977. Imitation of facial and manual gestures by newborn infants. *Science* 198:75–78.

———. 1992. Early imitation within a functional framework: The importance of person identity, movement, and development. *Infant Behavior and Development* 15:479–505.

Mizuno, Y., H. Takeshita, and T. Matsuzawa. 2006. Behavior of infant chimpanzees during the night in the first four months of life: Smiling and suckling in relation to behavioural state. *Infancy* 9:215–34.

Myowa, M. 1996. Imitation of facial gestures by an infant chimpanzee. *Primates* 37:207–13.

Myowa-Yamakoshi, M. 2006. How and when do chimpanzees acquire the ability to imitate? In T. Matsuzawa, M. Tomonaga, and M. Tanaka, eds., *Cognitive Development in Chimpanzees*, pp. 214–32. Tokyo: Springer-Verlag.

Myowa-Yamakoshi, M., and T. Matsuzawa. 1999. Factors influencing imitation of manipulatory actions in chimpanzees (*Pan troglodytes*). *Journal of Comparative Psychology* 113:128–36.

———. 2000. Imitation of intentional manipulatory actions in chimpanzees (*Pan troglodytes*). *Journal of Comparative Psychology* 114:381–91.

Myowa-Yamakoshi, M., and H. Takeshita. 2006. Do human fetuses anticipate self-directed actions? A study by four-dimensional (4D) ultrasonography. *Infancy* 10:289–301.

Myowa-Yamakoshi, M., M. Tomonaga, M. Tanaka, and T. Matsuzawa. 2003. Preference for human direct gaze in infant chimpanzees (*Pan troglodytes*). *Cognition* 89:B53–B64.

———. 2004. Neonatal imitation in chimpanzees (*Pan troglodytes*). *Developmental Science* 7:437–42.

Myowa-Yamakoshi, M., M. K. Yamaguchi, M. Tomonaga, M. Tanaka, and

T. Matsuzawa. 2005. Development of face recognition in infant chimpanzees (*Pan troglodytes*). *Cognitive Development* 20:49–63.

Nagell, K., R. Olguin, and M. Tomasello. 1993. Processes of social learning in the imitative learning of chimpanzees and human children. *Journal of Comparative Psychology* 107:174–86.

Neisser, U. 1991. Two perceptually given aspects of the self and their development. *Developmental Review* 11:197–209.

———. 1995. Criteria for an ecological self. In P. Rochat, ed., *The Self in Infancy: Theory and Research. Advances in Psychology*. Amsterdam: Elsevier Science.

Plooij, F. X. 1984. *The Behavioral Development of Free-living Chimpanzee Babies and Infants: Monographs on Infancy, Vol. 3*. Norwood, NJ: Ablex.

Rochat, P. 2001. *The Infant's World*. Cambridge, MA: Harvard University Press.

Rochat, P., and S. J. Hespos. 1997. Differential rooting response by neonates: Evidence for an early sense of self. *Early Developmental and Parenting* 6:105–12.

Russell, C. L., K. A. Bard, and L. B. Adamson. 1997. Social referencing by young chimpanzees (*Pan troglodytes*). *Journal of Comparative Psychology* 111:185–93.

Scaife, M. 1976. The response to eye-like shapes by birds II: The importance of staring, pairedness and shape. *Animal Behaviour* 24:200–206.

Takeshita, H., M. Myowa-Yamakoshi, and S. Hirata. 2006. A new comparative perspective on prenatal motor behaviors: Preliminary research with four-dimensional ultrasonography. In T. Matsuzawa, M. Tomonaga, and M. Tanaka, eds., *Cognitive Development in Chimpanzees*, 37–47. Tokyo: Springer-Verlag.

Tomasello, M. 1999. *The Cultural Origins of Human Cognition*. Cambridge, MA: Harvard University Press.

Tomasello, M., and J. Call. 1997. *Primate Cognition*. New York: Oxford University Press.

Tomasello, M., S. Savage-Rumbaugh, and A. C. Kruger. 1993. Imitative learning of actions on objects by children, chimpanzees, and enculturated chimpanzees. *Child Development* 64:1688–1705.

Tomonaga, M., M. Tanaka, T. Matsuzawa, M. Myowa-Yamakoshi, D. Kosugi, Y. Mizuno, S. Okamoto, M.K. Yamaguchi, and K. A. Bard. 2004. Development of social cognition in chimpanzees (*Pan troglodytes*): Face recognition, smiling, mutual gaze, gaze following and the lack of triadic interactions. *Japanese Psychological Research* 46:227–35.

Trevarthen, C., and P. Hubley. 1978. Secondary intersubjectivity: Confidence, confiding and acts of meaning in the first year. In A. Lock, ed., *Action, Gesture and Symbol*, 183–229. London: Academic Press.

Ueno, A., S. Hirata, K. Fuwa, K. Sugama, K. Kusunoki, G. Matsuda, H. Fukushima, K. Hiraki, M. Tomonaga, and T. Hasegawa. 2008. Auditory ERPs to stimulus deviance in an awake chimpanzee (*Pan troglodytes*): Towards hominid cognitive neurosciences. *PLoS ONE* 3:e1442. doi:10.1371/journal.pone.0001442.

Whiten, A., and R. W. Byrne, eds. 1997. *Machiavellian Intelligence II: Extensions and Evaluations*. New York: Cambridge University Press.

Whiten, A., D. M. Custance, J. C. Gomez, P. Teixidor, and K. A. Bard. 1996. Imitative learning of artificial fruit processing in children (*Homo sapiens*) and chimpanzees (*Pan troglodytes*). *Journal of Comparative Psychology* 110:3–14.

Zoia, S., L. Blason, G.. D'Ottavio, M. Bulgheroni, E. Pezzetta, A. Scabar, and U. Castiello. 2007. Evidence of early development of action planning in the human foetus: A kinematic study. *Experimental Brain Research* 176:217–26.

3

Using an Object Manipulation Task as a Scale for Comparing Cognitive Development in Chimpanzees and Humans

Misato Hayashi

An infant chimpanzee named Pal is playing with blocks beside her mother, who is participating in a cognitive task with a human experimenter. Pal drags a set of blocks around the floor while holding an extra one in her mouth. She takes the block from her mouth and places it on top of one of the blocks on the floor, adjusting it with her right hand and aligning it perfectly with the block below. She then picks up another block and stacks it on top of the first two. She looks around for another block to stack, then moves around her three-block tower, picks up a new block, and carefully and gently stacks it on top of the other three. Suddenly she starts jumping repeatedly over the four-block tower with a play face, as if enjoying the result of her stacking effort. For the next several minutes she continues stacking blocks, finally constructing a tower seven blocks high.

You may have had the experience of playing with blocks during your childhood, since it is one of the most common activities among human children, who regularly stack blocks to make high towers or assemble various constructions. Chimpanzees also possess the ability to stack blocks, although this behavior is not typically within the natural repertoire of nonhuman animals. This chapter presents ways to test the cognitive development of both chimpanzees and humans using blocks. The task of stacking blocks requires both manual skill and high levels of cognition; it allows for controlled variation in the level of difficulty,

and also for testing different domains of physical and social cognition with differently designed tasks.

In recent years a major focus of research on chimpanzees has been their social intelligence. However, their material intelligence—that is, their cognitive ability to manipulate materials to achieve specific goals—has received much less attention, especially in experimental studies conducted in captivity. This is despite the fact that both humans and nonhuman primates manipulate various kinds of objects in everyday life and use their intelligence in problem-solving situations. In addition, advanced physical cognition is an essential condition for the emergence and performance of complex tool-use skills, characterizing the impressive material culture of chimpanzees in their natural habitat (see chapters 8, 10, 11, and 14).

Significance of Object Manipulation Tasks to Tool-Use Studies

Tool use has been a topic of intensive research as an indicator of material intelligence (Van Schaik et al. 1999; Yamakoshi 2004) and tool-use abilities have been investigated across a range of animal species (Beck 1980) including primates and fossil hominids (Van Schaik and Pradhan 2003; Matsuzawa 2001). Among nonhuman primates, chimpanzees are the most frequent tool users in the wild. There exist a variety of patterns in their tool use that can be explained by ecological factors (see chapter 10; Yamakoshi 1998), as well as cultural factors (Whiten et al. 1999, 2001). There are also different levels of complexity in their tool use (see chapters 1 and 11; Matsuzawa 1996), defined by the number of objects and actions that are combined and coordinated to achieve a goal. Developmental studies on the process of tool-use acquisition in chimpanzees have been conducted in their natural habitat (Biro et al. 2006; Inoue-Nakamura and Matsuzawa 1997; Lonsdorf et al. 2004; Lonsdorf 2005). Tool-use behavior has also been studied in captive chimpanzees, focusing on repertoire (Takeshita and Van Hooff 1996), complexity (Hayashi et al. 2005), development (Hirata and Celli 2003), and diffusion (see chapter 9).

The fundamental structure of tool use is that an object—the tool—is combined with a target to achieve a goal. In this sense, tool use is based on combinatory manipulation, which has been investigated as an indicator of cognitive development in primates, including humans (Fragaszy and Adams-Curtis, 1991; Tanaka and Tanaka, 1982; Potì and Spinozzi 1994; Takeshita 2001; Westergaard 1993). Combinatory manipulation can be defined as relating an object to another object; stacking blocks is one example. It is rarely observed in primates (Torigoe 1985). In human development, its onset occurs at around 10 months of age (Tanaka and Tanaka 1982).

Comparative Investigations of Combinatory Manipulation: Stacking Blocks

One common feature of primates, including humans, is their manual dexterity. A primate can hold an object with the five digits of the hand—an adaptation to an arboreal life that requires them to hold onto branches. This morphological characteristic also allows primates to control their hands finely to grasp or hold objects. Great apes have opposable thumbs, which allow them even greater control over their object manipulation abilities (Crast et al. 2009). Since humans and chimpanzees share this skill, one can compare their material intelligence by simply analyzing their patterns of manipulation, without requiring any verbal response.

Human children love to produce various structures (Forman 1982) by combining multiple blocks. While stacking blocks is just a simple game for children, it is a difficult task for most animal species. Up to now, only the four species of great apes—the chimpanzee, bonobo, gorilla, and orangutan—are known to stack blocks as humans do (Parker and McKinney 1999; Hayashi et al. 2006). There are no reports of stacking blocks by other primates—although some, such as capuchin monkeys (Westergaard and Suomi 1994) and baboons (Westergaard 1993), are known to exhibit combinatory manipulation or even to use tools in their natural habitat (Fragaszy et al. 2004). These findings suggest that stacking blocks requires not only dexterous manipulatory skill but also high levels of physical cognition. The question that naturally follows is: Why is stacking behavior so difficult for primates, and only observed in the evolutionary lineage of humans and great apes? I will attempt to shed some light on this question by focusing on the onset of this behavior in chimpanzees and humans.

Approach

Tasks using blocks as testing material can be applied to comparative studies of humans and chimpanzees focusing on species difference as well as cognitive development. Having provided chimpanzee subjects with blocks under testing conditions identical to those in tests with humans, I present findings from four different lines of inquiry. The first condition was purely observational, to assess spontaneous block-stacking ability. In the second condition, chimpanzees were actively trained to stack successively taller towers of blocks. In the third portion of the study, I investigated chimpanzee understanding of the physical rules determined by gravity and the properties of objects. Finally I used stacking behavior to test the chimpanzees' capacity for social referencing by investigating their ability to copy a model built by another individual.

The sections below present the results of a series of comparative studies in both chimpanzees and humans using blocks. There is a group of 14 chimpanzees, including three generations, living in an enriched semi-natural environment at the Primate Research Institute at Kyoto University. The group includes three immatures (named Ayumu, Cleo, and Pal) who were born in 2000 and were all reared by their biological mothers. This setting has yielded many studies on the cognitive abilities of chimpanzees from many perspectives under both individual or group conditions (chapter 2; Matsuzawa et al. 2006). Thanks to the chimpanzees' close long-term relationship with humans, experimenters can enter the same room with them to conduct face-to-face tasks to assess their cognitive development, just like the testing conducted with human children. Human data was collected from 29 participants at the Umikaze Infant Laboratory of the University of Shiga Prefecture. The children were tested once a month longitudinally, in the presence of their parents, and the materials and testing procedure were identical to those used with chimpanzees. Unless otherwise specified, all the results concerning humans presented below are based on this ongoing project, focusing on human children for a comparison with chimpanzees.

Spontaneous Development of Block Stacking

A developmental study of mother-reared chimpanzees reported that preliminary combinatory manipulation (object-object combinations) started at 8 to 11 months of age, almost the same age as in humans (Hayashi and Matsuzawa 2003). However, the onset pattern of different types of combinatory manipulations differed from that of humans. Based on the human data gathered by Ikuzawa (2000), the age of onset of inserting a rod into a hole and of stacking blocks was almost identical: 13 months. Chimpanzees start to insert a rod into a hole at an age comparable to that of humans, but the onset of block stacking is delayed in chimpanzees as compared to humans (Hayashi and Matsuzawa 2003). In this study, the chimpanzee mothers served as demonstrators to their offspring, thus simulating a naturalistic learning context. Human experimenters provided no instructive guidance, no food reward, and no social praise.

The infants were provided ample opportunities to observe the stacking behavior of their mothers and to manipulate the blocks on their own. During the three years of free observation, only one infant—a female named Pal—began stacking blocks at two years and seven months of age. The first phase of stacking consisted of relating one block to another. Then, a subject had to release the block to complete the stacking action. This was initially difficult, however: instead of releasing the block, the subjects persisted in holding it and making repeated contact with other blocks, or even started hitting other blocks while holding the block firmly in one hand. The difficulty of performing the releasing action might be one reason for the inability

Figure 3.1   A chimpanzee (Pal) stacks seven blocks spontaneously.

of primates other than great apes to demonstrate stacking behavior.

The chimpanzee mothers never encouraged or guided their infants in stacking, although human mothers may do so under naturalistic conditions. Moreover, the chimpanzee mothers showed no signs of social praise or reinforcement following successful stacking by their infants. During the first 16 days of testing, Pal stacked only two to three blocks. Then, during the 17th session, which occurred 24 days after the onset of her stacking behavior, a sudden change took place: she spontaneously tried to construct higher towers, and finally succeeded in making one seven blocks high (figure 3.1).

In order to stack many blocks, each block must be placed accurately to maintain the tower's stability. This act requires fine manual control. The human children started stacking blocks at around one year of age, although their motor control of their hands was not yet fully mature. They were able to stack seven blocks successfully at one and-a-half years of age. Pal, in contrast, needed only one month to stack seven blocks after her first success at stacking two. This difference in time lag between humans and chimpanzees is likely due to the fact that when Pal first started stacking blocks, her motor control was already fully mature. Having noticed the interesting outcome of her stacking behavior, she was ready to stack more blocks into a higher tower, relying on her fine motor control and adjustment. Thus, in chimpanzees the motor control of their hands developed before their cognitive ability to perceive the goal of their stacking behavior. Without any reinforcement, the frequency of their block stacking gradually decreased after several months. Moreover, the two remaining infants showed no sign of stacking behavior during the first three years of free observation. Thus, the next step of the study began only when the infants were three years old.

Training Subjects to Stack Blocks

Previous studies have shown that both adult and juvenile chimpanzees possess the ability to stack blocks (Matsuzawa 1987; Potì 2005). These studies used "enculturated" chimpanzees, who were actively guided by their human caretakers. Thus, human experimenters began actively training the infants in order to explore their cognitive ability in stacking behavior. Before the training started, only the mother chimpanzees had been required to stack blocks, and had received food rewards from human experimenters in a face-to-face situation. When each infant was three years and one month of age, he or she was invited to sit in front of the experimenter, who then placed one block on the floor and gave another to the infant chimpanzee. The experimenter prompted the infant to orient one block towards the other on the floor by pointing at it or tapping it. When the infant successfully stacked one block onto another, the human experimenter handed him or her a food reward, such as a slice of apple or orange, accompanied by verbal praise.

Pal, the first infant who stacked blocks spontaneously, resumed doing so when she realized that she could receive a food reward. Ayumu, the second infant, also had no difficulty in learning the stacking behavior with guidance from the human experimenter. Cleo, the third infant, first tried to return the block to the experimenter, who then held one hand close to the top of a tower of blocks so that Cleo would have to orient the block towards the tower when trying to return it to the experimenter's hand. If Cleo oriented or actually touched a block to the tower, she was rewarded with food. With the use of this positive-reinforcement training (see chapter 25), Cleo gradually learned to stack blocks. Once she learned that the behavior would be rewarded, she continued to stack them in the same way as the other two infants. Whenever the infants successfully stacked one block onto another on the floor, human experimenters gave them another block to stack. The infants thus began constructing increasingly high towers. The maximum height was 13 blocks, stacked by Pal. The course of spontaneous development and training is reported in Hayashi (2007).

Food rewards were given after the collapse of a tower around five or more blocks high. This meant that if a subject was motivated to get a food reward as quickly as possible, the best way to do so would be to stack five blocks and then knock the tower down. Just like human children, however, the chimpanzees did not like it when their towers collapsed. When a tower got taller, they took great care in placing the next block. When a tower wobbled or was unstable, the infant sometimes scratched their head or body, thus indicating stress or anxiety (Itakura 1993; Kutsukake 2003; Leavens et al. 2001). In some cases they tried to support a tower with their free hand. They also showed adjusting behavior by removing the top block and carefully replacing it to adjust the tower's balance.

Physical Understanding: Stacking Differently Shaped Blocks

Based on the block-stacking ability of infant chimpanzees, a new set of tasks was introduced to focus on their physical understanding. Povinelli (2000) proposed a paradigm to test their understanding of the physical world. Povinelli and Dunphy-Lelii (2001) also reported the behavior of chimpanzees and humans in a task requiring the placement of an oblong block onto a platform, and analyzed the behavior they engaged in after their failure in performing it, such as investigating the block's slanted end. This task can be categorized as a test of understanding the physical causality of objects in relation to gravity and their supporting surface.

The task was invented to test physical understanding in the context of object-object combination. Blocks of various shapes were first introduced to the chimpanzees, who had already demonstrated the ability to stack cubic blocks. By using blocks of differing shapes, the task becomes more complex and requires higher levels of cognitive ability for problem-solving. The subject must consider differences in orientation and choose the appropriate orientation for each block in order to stack them efficiently. Since each block functions as a support for the next one in the stack, the subject's physical understanding in the context of object-object combination can readily be assessed.

It must be noted that this task requires the subject to actively solve the problem. In the previous task by Povinelli and Dunphy-Lelii (2001), the subjects' reaction to failure had been the main variable analyzed to infer their causal understanding. The present study, however, allowed them to manipulate blocks freely to find a solution to the task. If a block was not oriented appropriately for stacking, the subject had to actively change that orientation. This active manipulative solution may thus directly reveal the subject's physical intelligence in a problem-solving situation. It might also be informative in helping us understand how chimpanzees relate an object's properties to its function—a capacity they demonstrate in the wild when manufacturing and modifying tools for various purposes (see chapter 11). The behavioral strategy adopted to solve a problem by actively changing the environment could be assessed in the present task by using differently shaped blocks.

Stacking Cylindrical Blocks

For the cubic blocks that were exclusively used in the previous phases, there was no substantial difference in their orientation and all surfaces were suitable for stacking. Cylindrical blocks, however, can be positioned in two different orientations relative to a support surface such as the floor. One orientation, in which the flat surface touches the floor, is named "upright." In this position the block is suitable for stacking, and will support another block. Another orientation, in which the rounded surface touches the floor, is named "sideways"; it neither results in stable stacking nor provides support for another block.

The actual setting of the task was simple: a subject was required to stack four blocks, including two cubes and two cylinders. One of the two cylinders was placed on the floor in an upright orientation, and the other was placed in a sideways orientation. If the subject understood the physical properties of the cylindrical blocks, he/she would choose the upright cylinder for stacking and might actively change the orientation of the sideways cylinder to an appropriate one.

Among the three infant chimpanzees, who were all around 3.5 years old at the beginning of the test, Pal was the only one to stack the cylinders efficiently from the first session. She rarely contacted a sideways cylinder with another block. This might indicate that she recognized its inappropriate orientation and changed it from the beginning. The other two infants, Ayumu and Cleo, tried to stack sideways cylinders or even to stack another block on top of a sideways cylinder (figure 3.2).

Two mother chimpanzees, Ai and Pan, stacked cylin-

Figure 3.2   A chimpanzee (Ayumu) tries to stack a block on top of a sideways cylinder.

drical blocks efficiently from the first session. Both had ample experience in face-to-face cognitive tasks, including cubic block stacking, since their youth. The other adult, named Akira, who had had limited exposure to the stacking of cubic blocks, tried frequently in the first session to stack sideways cylinders. His behavior was slightly different from that of the infants, however, as he did not try to stack a block on top of a sideways cylinder, but instead removed the sideways cylinder from the stack soon after its placement, to try other possible ways of stacking it. This may suggest that he understood that the sideways cylinder was not suitable as a support surface, although he did not try changing its orientation before stacking it.

The determinant of success or failure was not the subject's age, but more likely the richness of his or her previous experience. Individuals with plenty of experience at stacking cubic blocks were more likely to succeed in stacking the new cylindrical blocks, thus transposing their knowledge to the requirements of the new task.

After the first introduction of cylindrical blocks, the stacking task was continued with the same three infants to monitor their improvement. They started to show less contact with sideways cylinders after around 15 sessions (75 trials). They also started changing the cylinders' orientation. These changes indicated that chimpanzees can learn how to stack cylindrical blocks efficiently through the experience of manipulating them (see more details in Hayashi 2007). Finally, all chimpanzee subjects stacked cylindrical blocks either from the beginning or after an individual learning process involving successive exposures to the task. This result suggests that they may possess a rudimental form of physical understanding. Human infants were also tested in an identical face-to-face situation with blocks of the same size and shape. Children who already knew how to stack cubic blocks had no difficulty in stacking cylindrical ones with the appropriate orientation.

Stacking Other New Shapes

The follow-up studies were conducted by using differently shaped blocks in the same stacking context. After cylindrical blocks, the chimpanzees were given blocks of two new shapes in succession: first triangular, and then slanted. During the first session, none of the chimpanzee subjects succeeded in stacking triangular blocks efficiently. However, those who later learned how to stack them efficiently then also succeeded immediately in stacking the new slanted blocks. Their experience with triangular blocks had given them knowledge that was then generalized to the slanted blocks. They might have learned rules such as "a flat horizontal surface should be at the top" or "a block of inappropriate orientation should be adjusted before stacking." This series of stacking task experiments revealed the high cognitive ability of chimpanzees in understanding the basic physical rules in object manipulation.

Human children also showed difficulties in stacking the two new shapes, although they had readily stacked the cubic and cylindrical blocks. They showed patterns similar to those of the chimpanzee subjects, in that they sometimes tried to stack one block on top of the slanted surface of another. Some children first succeeded stacking triangular and slanted shapes at around two years of age. Others, however, continued to fail at that task even after age three. Thus even for human infants, stacking blocks with flat surfaces was a difficult task to solve.

The previous sections have focused on the cognitive abilities of chimpanzees and young humans in the domain of material intelligence governed by physical rules. The results suggest that the ability of chimpanzees to manipulate objects with different properties is similar to that of human children. We now move to a related issue: the ability of chimpanzees to understand social rules governing physical manipulation.

Copying a Model: Understanding Social Rules through Use of the Block-Stacking Task

Using blocks in a different task setting enabled us to assess social aspects involved in object manipulation. A commonly used task in human developmental studies requires the subject to copy an arrangement of shapes constructed by the experimenter. For example, the experimenter generates the shape of a "truck" by aligning three blocks horizontally and placing another block on top of one end of the line. The subject is then given four more blocks and required to copy the structure. These kinds of tasks can be categorized as "copying the shape" of a model. Another well-known task is the Kohs Block Design Test (Kohs 1923; Wechsler 1939), which requires a subject to copy a modeled color pattern arranging a set of colored blocks to match it. Ikuzawa (2000) used this task to assess the cognitive development of human children. Humans start to copy the shape of a truck at around 2.5 years, and to copy a simple pattern of two color patches with four blocks at around 4.5 years.

Although three adult chimpanzees were tested at these tasks in the past (Hayashi and Matsuzawa 2003), they did not succeed in copying the model of a truck or the simplest color pattern (one color for all four blocks). During these trials the researchers recognized that for the chimpanzees, aligning blocks in a horizontal plane was more difficult than stacking them vertically. Humans start to stack blocks vertically, then proceed to align them horizontally, and finally combine the two actions to produce three-dimensional structures. For chimpanzees, however, it might be difficult to understand the goal of aligning blocks in a straight line on the floor. In stacking, each block is automatically fixed on top of the last. In aligning them on the floor, however, each block can be positioned anywhere, and this seems particularly difficult for chimpanzees.

The negative results in these copying tasks did not allow us to explore the social aspects of object manipulation. The cause of the chimpanzees' failure was not clear— whether it was the difficulty of copying a model generated by another individual, or just that of aligning the blocks horizontally. Thus I developed a new version of the test using blocks to assess the social aspect involved in object manipulation.

The new task required the subject to refer to a model and to copy its order of colored blocks. For example, the experimenter produced a model by stacking two blocks in a specific order: blue on top of yellow. Then, the experimenter gave the subject the same set of blocks. If the subject succeeded in copying the model by stacking the two blocks in the correct color order, he or she received a food reward.

Two adult chimpanzees (Ai and Pan) participated in the copying task. We started by using the same model with two colors in consecutive trials; then in the next phase the models appeared randomly. New colors were then introduced, and finally the actual construction procedure of the model stack was hidden from the subjects' view by a paper screen. The two chimpanzees eventually succeeded in copying the model stack of two blocks. However, it took more than 150 sessions and multiple successive approximations for the chimpanzees to learn the requirements of this task.

The next big step was to increase the number of blocks to be stacked from two to three. The beginning of a trial was identical to that in the previous phase. When the experimenter removed the screen, the subject was faced with a model stack of three blocks and was provided with the same set of three blocks. The two chimpanzees who had already learned to copy the model stack of two blocks after a long training process were used again in this experiment.

Both subjects failed to copy the model of three blocks. One of the subjects, Ai, did not show any improvement over the course of testing with three blocks; she performed at chance level throughout the sessions. Another subject, Pan, performed similarly in the beginning, but then her percentage of correct performance gradually rose up to around 50% (figure 3.3 shows one of the correct trials). This does not mean, however, that she completely mastered the task. There were huge fluctuations in her performance, and her average level never rose above 50% even after 150 sessions (Hayashi et al. 2009).

Human children also had difficulty performing this task. It can take at least one or more years from the onset of simple block stacking of blocks to reach first success in copying a model of two colored blocks. Human children first tried to stack the blocks on top of the model stack, or to disassemble it. Then they started stacking only the two blocks provided, but failed to copy the model's color

Figure 3.3   A chimpanzee (Pan) stacks three blocks in the correct order, following a model.

order. Finally they were successful in copying the model stack. Thereafter, they sometimes failed to copy it after the number of blocks was increased to three or four.

In conclusion, the task of copying a model stack was difficult both for adult chimpanzees and for human children under two years of age.

Implications for Cognitive Development in Chimpanzees

Chimpanzees are known for their social and material intelligence. It should be noted, however, that their pattern of cognitive development is different from that of humans, as evidenced by the results of the above studies. The tasks using differently shaped blocks revealed that they possess a rudimentary form of physical causal understanding comparable to that of human children. Although they initially had difficulty in stacking blocks of some shapes, they then adjusted their stacking behavior by choosing the appropriate orientation to avoid failure. This result suggests that chimpanzees perceive the outcome of their own manipulations by observing the result of each previous manipulation. They can adjust their methods of manipulation through interaction with objects, especially after failure. The only obstacles they need to surmount are physical relationships, such as gravity or object property. The set of studies presented here indicates that chimpanzees have little difficulty in learning "physical rules."

In contrast, chimpanzees had more difficulty in copying a model by following "social rules." They live in social groups and thus have social rules that govern their interactions, such as how to interact with individuals of different dominance rank within the group. When it comes to learning social rules in the context of object manipulation, however, the difficulty for chimpanzees seems to increase dramatically. This might be relevant to previous studies on imitation in chimpanzees (Myowa-Yamakoshi and Matsuzawa 1999, 2000; Tomasello et al. 1993; Whiten et al. 1996), some of which reported their fundamental difficulties in copying the action of others (but see Lonsdorf 2005).

Chimpanzees rarely show evidence for triadic relationships involving the subject, another individual, and an external object (see chapter 2; Tomonaga 2006; Tomonaga et al. 2004). Triadic interaction is common in humans from an early age. Human mothers often try to enhance the object manipulation abilities of their infants by handing objects to them or by demonstrating how to manipulate them. Infants, in turn, present their mothers with interesting or novel objects, thus seeking social referencing. This kind of interaction between individuals mediated by an object is rarely seen in chimpanzees. Chimpanzee mothers sometimes interact with their infant using objects (personal observation), but the frequency of this is much lower than in humans. Moreover, chimpanzee mothers do not actively guide the infants' manipulation or give them social praise (Hirata and Celli 2003; Lonsdorf 2006; Matsuzawa 2001). If a mother chimpanzee and her infant find an interesting object, both may try to monopolize it and prevent the other from touching it.

Imitation in the context of object manipulation may require the ability to divide one's attention between two targets at the same time: the demonstrator and the object. While manipulating an object, chimpanzees are likely to concentrate only on the object. This tendency might be the reason why copying of the model was such a difficult task for chimpanzees to learn. They were required to consider a model arbitrarily made by others while simultaneously manipulating their own blocks. Moreover, chimpanzees may focus only on the action of stacking, and may easily neglect the color dimension that is not essential in the context of stacking.

Future Directions

The present studies indicate that the block-stacking task is useful for testing cognitive development in both humans and chimpanzees. More precise analysis of error patterns during physical understanding tasks, and the subsequent correction strategies, may reveal further important characteristics of the minds of chimpanzees and humans. Additionally, chimpanzee subjects in this set of studies showed greater difficulty in learning "social rules" that required them to copy a model constructed by other individual. Future work should fill this discrepancy between "physical rules" and "social rules" in the context of object manipulation. The former are determined by physical causality, which is universal and constant under most circumstances, and thus easy to understand for chimpanzees. The social rule, in contrast, can be arbitrarily determined by another individual. New tasks should be designed to answer the

Figure 3.4   A mother chimpanzee in the Bossou community uses a hammer stone to hit a nut placed on an anvil stone. Photo by E. Nogami.

question of what kinds of rules are used by chimpanzees while manipulating objects. One possibility is to test the "internal rules" that may be spontaneously generated by individuals. Human children start to make their own rules during the course of their development, such as spontaneously categorizing objects into groups or generating straight lines by lining up various objects.

Comparable analysis of object-manipulation skill in tool use in the wild might help us understand more comprehensively the development of material intelligence in chimpanzees. In this sense, nut cracking is an ideal candidate, as it is the only behavior in the wild that requires stacking in its flow of execution: placing a nut stably upon an anvil stone (see figure 3.4). However, detailed observations and analyses—especially of errors and successive corrections—for tool-use skills such as termite fishing (Lonsdorf 2005 and chapter 11) would be equally informative.

Precise analysis of object manipulation can be a useful measure for comparing the intelligence of different species of great apes, both in captivity and in the wild. Block stacking might be a unique feature of the physical cognition of great apes and humans. Currently, only chimpanzees and orangutans are known to use tools customarily in the wild. Byrne (1995) showed, however, that in wild gorillas the manipulative sequence in their food-processing is comparable to the hierarchical complexity observed in some of their tool use behaviors. Hayashi et al. (2006) reported block-stacking behavior in captive bonobos and gorillas. Future studies in object manipulation should further reveal the different levels of physical cognition among primates as assessed through experiments and observations, and how these cognitive abilities translate into the emergence of tool use.

To examine the foundations of material intelligence, another area of future interest might be the interaction between mother and offspring during object manipulation. As reported in this chapter, there were species differences: chimpanzees did not show the kind of joint attention or scaffolding with objects that is frequently observed in humans. The social factors promoting the development of object manipulation in both chimpanzees and humans should be systematically analyzed in the future. Its precise analysis in various settings will shed further light on the evolution of the chimpanzee and human mind.

Acknowledgments

The study was supported by grants from the Ministry of Education, Science, and Culture in Japan (#16002001 and #20002001 to Matsuzawa and #19700245 to the author) and from the Benesse Corporation. The human data was collected in collaboration with Hideko Takeshita and Masako Myowa-Yamakoshi with the support of mothers and children participating in the Umikaze Infant Laboratory at the University of Shiga Prefecture. Special thanks are due to Tetsuro Matsuzawa, Masaki Tomonaga, Masayuki Tanaka, Wataru Sato, Sana Inoue, Tomoko Takashima, Etsuko Nogami, Kiyonori Kumazaki, Norihiko Maeda, Shohei Watanabe, Juri Suzuki, Akino Watanabe, Akihisa Kaneko, and Takako Miyabe for their great advice and support in conducting the daily work with chimpanzees at the Primate Research Institute at Kyoto University.

Literature Cited

Beck, B. B. 1980. *Animal Tool Behavior: The Use and Manufacture of Tools by Animals*. New York: Garland STPM Press.

Biro, D., C. Sousa, and T. Matsuzawa. 2006. Ontogeny and cultural propagation of tool use by wild chimpanzees at Bossou, Guinea: Case studies in nut cracking and leaf folding. In T. Matsuzawa, M. Tomonaga, and M. Tanaka, eds., *Cognitive Development in Chimpanzees*. Tokyo: Springer, 476–508.

Byrne, R. 1995. *The Thinking Ape*. Oxford: Oxford University Press.

Crast, J., D. Fragaszy, M. Hayashi, and T. Matsuzawa. 2009. Dynamic in-hand movements in adult and young juvenile chimpanzees (*Pan troglodytes*). *American Journal of Physical Anthropology* 138:274–85.

Forman, G. E. 1982. *Action and Thought: From Sensorimotor Schemes to Symbolic Operations*. New York: Academic Press.

Fragaszy, D. M., and L. E. Adams-Curtis. 1991. Generative aspects of manipulation in tufted capuchin monkeys (*Cebus apella*). *Journal of Comparative Psychology* 105:387–97.

Fragaszy, D., P. Izar, E. Visalberghi, E. B. Ottoni, and M. G.. Oliveira. 2004. Wild capuchin monkeys (*Cebus libidinosus*) use anvils and stone pounding tools. *American Journal of Primatology* 64:359–66.

Hayashi, M. 2007. Stacking of blocks by chimpanzees: Developmental processes and physical understanding. *Animal Cognition* 10:89–103.

Hayashi, M., and T. Matsuzawa. 2003. Cognitive development in object manipulation by infant chimpanzees. *Animal Cognition* 6:225–33.

Hayashi, M., Y. Mizuno, and T. Matsuzawa. 2005. How does stone-tool use emerge? Introduction of stones and nuts to naïve chimpanzees in captivity. *Primates* 46:91–102.

Hayashi, M., S. Sekine, M. Tanaka, and H. Takeshita. 2009. Copying a model stack of colored blocks by chimpanzees and humans. *Interaction Studies* 10:130–49.

Hayashi, M., H. Takeshita, and T. Matsuzawa (2006). Cognitive development in apes and humans assessed by object manipulation. In T. Matsuzawa, M. Tomonaga, and M. Tanaka, eds., *Cognitive Development in Chimpanzees*. Tokyo: Springer, 395–410.

Hirata, S., and M. L. Celli. 2003. Role of mothers in the acquisition of tool use behaviors by captive infant chimpanzees. *Animal Cognition* 6:235–44.

Ikuzawa, M. 2000. *Developmental Diagnostic Tests for Children* (in Japanese). Kyoto: Nakanishiya.

Inoue-Nakamura, N., and T. Matsuzawa. 1997. Development of stone tool use by wild chimpanzees (*Pan troglodytes*). *Journal of Comparative Psychology* 111:159–73.

Itakura, S. 1993. Emotional behavior during the learning of a contingency task in a chimpanzee. *Perceptual and Motor Skills* 76:563–66.

Kohs, S. C. 1923. *Intelligence Measurement: A Psychological and Statistical Study Based upon the Block-design Tests*. Macmillan, New York.

Kutsukake, N. 2003. Assessing relationship quality and social anxiety among wild chimpanzees using self-directed behaviour. *Behaviour* 140:1153–71.

Leavens, D. A., F. Aureli, W. D. Hopkins, and C. W. Hyatt. 2001. Effects of cognitive challenge on self-directed behaviors by chimpanzees (*Pan troglodytes*). *American Journal of Primatology* 55:1–14.

Lockman, J. J. 2000. A perception-action perspectives on tool use development. *Child Development* 71:137–44.

Lonsdorf, E. V. 2005. Sex differences in the development of termite-fishing skills in the wild chimpanzees, *Pan troglodytes schweinfurthii*, of Gombe National Park, Tanzania. *Animal Behaviour* 70:673–83.

———. 2006. What is the role of mothers in the acquisition of termite-fishing behaviors in wild chimpanzees (*Pan troglodytes schweinfurthii*)? *Animal Cognition* 9:36–46.

Lonsdorf, E.V., L. E. Eberly, and A. E. Pusey. 2004. Sex differences in learning in chimpanzees. *Nature* 428: 715–16.

Matsuzawa, T. 1987. Stacking blocks in chimpanzees (in Japanese with English summary). *Primate Research* 3:91–102.

———. 1996. Chimpanzee intelligence in nature and captivity: isomorphism of symbol use and tool use. In W. C. McGrew, L. F. Marchant, and T. Nishida, eds., *Great Ape Societies*. Cambridge: Cambridge University Press, 196–209.

———, ed. 2001. *Primate Origins of Human Cognition and Behavior*. Tokyo: Springer.

Matsuzawa, T., M. Tomonaga, and M. Tanaka, eds. 2006. *Cognitive Development in Chimpanzees*. Tokyo: Springer.

Myowa-Yamakoshi, M., and T. Matsuzawa. 1999. Factors influencing imitation of manipulatory actions in chimpanzees (*Pan troglodytes*). *Journal of Comparative Psychology* 113:128–36.

Myowa-Yamakoshi, M., and T. Matsuzawa. 2000. Imitation of intentional manipulatory actions in chimpanzees (*Pan troglodytes*). *Journal of Comparative Psychology* 114:381–91.

Parker, S. T., and M. L. McKinney. 1999. *Origins of Intelligence*. Baltimore: Johns Hopkins University Press.

Potì, P. 2005. Chimpanzees' constructional praxis (*Pan paniscus, P. troglodytes*). *Primates* 46:103–13.

Potì, P., and G. Spinozzi. 1994. Early sensorimotor development in chimpanzees (*Pan troglodytes*). *Journal of Comparative Psychology* 108:93–103.

Povinelli, D. J. 2000. *Folk Physics for Apes: The Chimpanzee's Theory of How the World Works*. Oxford University Press, Oxford.

Takeshita, H. 2001. Development of combinatory manipulation in chimpanzee infants (*Pan troglodytes*). *Animal Cognition* 4:335–45.

Takeshita, H., and J. A. R. A. M. van Hooff. 1996. Tool use by chimpanzees (*Pan troglodytes*) of the Arnhem Zoo community. *Japanese Psychological Research* 38:163–73.

Tanaka, M., and S. Tanaka. 1982. *Developmental diagnosis of human infants: From 6 to 18 months* (in Japanese). Tokyo: Ootsuki.

Tomasello, M., S. Savage-Rumbaugh, and A. C. Kruger. 1993. Imitative learning of actions on objects by children, chimpanzees, and enculturated chimpanzees. *Child Development* 64:1688–1705.

Tomonaga, M. 2006. Development of chimpanzee social cognition in the first 2 years of life. In T. Matsuzawa, M. Tomonaga, and M. Tanaka, *Cognitive Development in Chimpanzees*. Tokyo: Springer, 182–97.

Tomonaga, M., M. Tanaka, T. Matsuzawa, M. Myowa-Yamakoshi, D. Kosugi, Y. Mizuno, S. Okamoto, M. Yamaguchi, and K. Bard. 2004. Development of social cognition in infant chimpanzees (*Pan troglodytes*): Face recognition, smiling, gaze, and the lack of triadic interactions. *Japanese Psychological Research* 46:227–35.

Torigoe, T. 1985. Comparison of object manipulation among 74 species of non-human primates. *Primates* 26:182–94.

Van Schaik, C. P., R. O. Deaner, and M. Y. Merrill. 1999. The conditions for tool use in primates: Implications for the evolution of material culture. *Journal of Human Evolution* 36:719–41.

Van Schaik, C. P., and G. R. Pradhan. 2003. A model for tool-use traditions in primates: Implications for the coevolution of culture and cognition. *Journal of Human Evolution* 44: 645–64.

Wechsler, D. 1939. *The Measurement of Adult Intelligence*. Williams and Wilkins, Baltimore.

Westergaard, G. C. 1993. Development of combinatorial manipulation in infant baboons (*Papio cynocephalus anubis*). *Journal of Comparative Psychology* 107:34–38.

Westergaard, G. C., S. J. Suomi. 1994. Hierarchical complexity of combinatorial manipulation in capuchin monkeys (*Cebus apella*). *American Journal of Primatology* 32:171–76.

Whiten, A., D. M. Custance, J. C. Gómez, P. Teixidor, and K. A. Bard. 1996. Imitative learning of artificial fruit processing in children (*Homo sapiens*) and chimpanzees (*Pan troglodytes*). *Journal of Comparative Psychology* 110:3–14.

Whiten, A., J. Goodall, W. C. McGrew, T. Nishida, V. Reynolds, Y. Sugiyama, C. E. G. Tutin, R. W. Wrangham, and C. Boesch. 1999. Cultures in chimpanzees. *Nature* 399:682–85.

———. 2001. Charting cultural variation in chimpanzees. *Behaviour* 138: 1481–1516.

Yamakoshi, G.. 1998. Dietary responses to fruit scarcity of wild chimpanzees at Bossou, Guinea: Possible implications for ecological importance of tool use. *American Journal of Physical Anthropology* 106:283–95.

———. 2004. Evolution of complex feeding techniques in primates: Is this the origin of great ape intelligence? In A. E. Russon and D. R. Begun, eds., *The Evolution of Thought: Evolutionary Origins of Great Ape Intelligence*, 140–71.

4

Do the Chimpanzee Eyes Have It?

Masaki Tomonaga

Ai, the well-known chimpanzee from the Primate Research Center at Kyoto University, moves slowly away from the touch-screen apparatus, having just completed a series of cognitive tasks. Looking through the glass, she notices that a research intern has dropped her silver pen near a small opening to the enclosure. Ai's gaze draws longingly to the shiny object, and her head turns—but before she can move, her young son Ayumu scurries over to the opening, reaches through it with an outstretched arm, and grabs the object of her affection. What cued Ayumu to notice the pen? He was too far away to see it on his own, and wasn't present when it fell from the researcher's pocket. Barely two years old, Ayumu appears well able to infer the location of objects simply from the direction in which his mother looks.

In daily life, our attention is captured by various kinds of social stimuli, such as another individual's face. Captured attention is then readily shifted by other types of social stimuli, such as when we follow another individual's gaze to a specific location. The capture and shift of social attention are important bases for social interactions in humans—for example, in joint attention and triadic (or triangular) interactions, in which two individuals interact with each other and with objects or events in the surrounding environment, sharing their attention through exchanges of gaze. Furthermore, since the ability to understand another's mental state may contribute to social interaction in humans, studies of social attention in nonhuman primates

are critical for understanding the evolution and development of theory of mind (Tomasello 1999). In this chapter, social attention is defined as attention triggered by social stimuli such as faces, gaze, and gesture, and attention in social situations, including joint attention.

Many studies on joint attention or gaze following in nonhuman primates have been conducted during the last decade (for review see Emery 2000). Prosimians rarely show gaze following and joint attention (Itakura 1996); New World and Old World monkeys, while showing slightly better abilities than prosimians, have shown limited capacities for gaze following in experiments that required them to follow the researcher's gaze (Itakura 1996). In contrast, in more naturalistic or ecologically valid settings, Old World monkeys and chimpanzees reliably followed the gaze of a conspecific (Tomasello et al. 1997). Chimpanzees were able to follow the eye directions of humans whether or not the humans also turned their heads (Povinelli and Eddy 1996). When the chimpanzees were tested in a setting in which the experimenter glanced at a target location that the subjects could not see directly (since the view was blocked by an opaque panel), the subjects tried to check the obstructed location (Povinelli and Eddy 1996; Tomasello et al. 1999).

Though many researchers agree that great apes have the ability to follow another's attention, the underlying mechanisms for these behaviors remain unclear. Consequently, along with other researchers I have been studying chimpanzees' attention both in social and nonsocial domains (see Tomonaga 2001, 2006; Tomonaga et al. 2004) from the comparative–cognitive–developmental perspective. In this chapter I summarize a series of studies investigating the chimpanzees' social attention as observed in the laboratory. The most common and most important social stimuli for both humans and great apes are faces, and in this chapter I focus mainly on face and gaze as sources of social attention. It is well known that faces are processed in a special manner by humans (e.g., Yin 1969). Therefore, before exploring social attention in chimpanzees I briefly summarize studies concerning how chimpanzees process facial stimuli. The latter half of the chapter is divided into two parts explaining the roles of the face and gaze in the social attention of chimpanzees—that is, the "capture" and "shift" of attention. It describes my experiments on the capture of attention in chimpanzees by direct gaze—which, in humans, plays a critical role for capturing an observer's attention. The other aspect of gaze is its ability to shift an observer's attention to the place where the gazer is looking. This fundamental ability is the key for joint attention and more complex social interactions in humans. However, only a few experimental studies have explored the underlying mechanisms of attention shift triggered by gaze—or other social stimuli—in nonhuman primates (e.g., Deaner and Platt 2003; Fagot and Deruelle 2002; Vick et al. 2006). In a series of experiments described below, I examine mechanisms responsible for attention shift by gaze and differences between humans and chimpanzees that may be critically related to the qualitative differences in their respective social interactions (Tomonaga 2006; Tomonaga et al. 2004). In total, my intention is to use a comparative developmental approach to address the roles of attention and gaze in humans and chimpanzees and to provide insight into the importance of these mechanisms in hominid social interactions.

Face Perception in Chimpanzees

Before reviewing comparative studies on social attention in chimpanzees, I will briefly discuss the face-processing abilities of chimpanzees, as faces are one of the most important social stimuli in their daily life (see chapter 5). Face processing is quite different from the processing of other visual stimuli. Several of the most influential human studies on this topic showed impaired recognition when faces were presented in an inverted orientation (Valentine 1988; Yin

1969). This "inversion effect" is generally limited to facial stimuli (cf. Diamond and Carey 1986), and is considered specific to faces because we process the face configurally (Diamond and Carey 1986). That is, even though faces consist of several separate features such as eyes, eyebrows, nose, and mouth, we recognize them not on the basis of each feature separately, but on their configuration as a whole.

Some studies have reported the facial inversion effect in nonhuman primates including chimpanzees, although these findings are controversial (Martin-Malivel and Fagot 2001; Overman and Doty 1982; Parr et al. 1998; Phelps and Roberts 1994; Tomonaga 1994, 1999a). For example, using a delayed matching-to-sample task, I found that a chimpanzee clearly showed the inversion effect in perception of human faces but not of houses (Tomonaga 1999a). Parr and colleagues (1998) obtained similar results in chimpanzees when using the faces of chimpanzees, capuchin monkeys, and humans as stimuli. Furthermore, I used a visual search task to examine chimpanzees' ability to detect an upright face among disoriented face distractors (Tomonaga 1999b, 2007a). In this experiment, an adult female chimpanzee was trained to touch a target stimulus displayed among distractors on a monitor (figure 4.1a). The target face was distinguished from the distractors by its orientation (upright, horizontal, or inverted). The chimpanzee searched more efficiently—that is, she exhibited a faster response times—when the target face was upright than when it was horizontal or inverted (Tomonaga 1999b). Although human faces were used as stimuli in the first study, this "upright-face superiority effect" was also observed for chimpanzee faces, dog faces, and schematic faces, but not for chairs or hand shapes (Tomonaga 2007a; figure 4.1b). Further examination clarified the relationships between this effect and the configural processing of faces. Efficient detection of upright targets was observed when those targets were presented with other inverted stimuli, but not when the facial features such as eyes, nose, and mouth were presented alone. Thus, the effect observed in these experiments can be considered a variation of the inversion effect in the face-recognition task (Tomonaga 1999a).

Capture of Attention by Social Stimuli

Gaze Preferences

Some kinds of social stimuli apparently capture our visuospatial attention automatically. Von Grünau and Anston

(A)

(B)
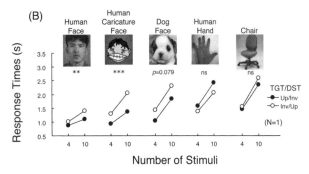

Figure 4.1 Visual search for orientation of faces (adapted from Tomonaga 2007a). (a) An adult chimpanzee, Chloe, performing the task. Photograph courtesy Primate Research Institute, Kyoto University. (b) Mean response times for each condition as a function of the number of stimuli. TGT: target stimulus. DST: distractors. Up: upright. Inv: inverted. ***: $p < .001$. **: $p < .01$. ns: not significant.

(1995) reported that a gaze directed at a human subject is detected more efficiently among multiple averted gazes than an averted gaze is detected among multiple direct gazes. This is called the "stare-in-the-crowd" effect, and it may account for the familiar feeling of being watched when one is walking in a crowd. Likewise, nonhuman primates often exhibit "gaze aversion," in which some individuals react aggressively to the gazer (Emery 2000). This can be construed as evidence that the direct gaze is salient and captures the subject's attention in nonhuman primates as well.

In contrast to the large number of such studies conducted with human infants (Farroni et al. 2002), very few studies have investigated the developmental origins of the sensitivity to direct gaze in nonhuman primates, especially in non-hominoid species (cf. Mendelson et al. 1982). In one study, gibbon and chimpanzee infants were tested under the preferential-looking paradigm using schematic or human faces as stimuli (Myowa-Yamakoshi and Tomonaga 2001; Myowa-Yamakoshi et al. 2003). Each infant was presented with a pair of faces: one looking directly at the viewer and one with an averted gaze. The subjects' duration of looking and the number of tracking responses to each face were recorded. Infants of both species looked at the direct-gaze face for a significantly longer time than at the averted-gaze face (figure 4.2). For chimpanzees, this dif-

ference was evident as early as two months of age, at which time they also increased the frequency of mutual gaze with their mothers (Bard et al., 2005, see also Okamoto-Barth et al., 2007b). The emergence of the sensitivity to direct gaze at around two months might be the basis for early dyadic social interactions in chimpanzees (Tomonaga, 2006; Tomonaga et al., 2004, see also chapter 2)

In addition to the infant studies, I also conducted visual-search experiments with an adult chimpanzee in order to demonstrate the stare-in-the-crowd effect (Tomonaga and Imura 2008b). I prepared a set of front-view photographs of human faces with both direct and averted gaze. These pictures were sequentially presented upright, inverted, or with the facial elements spatially scrambled. Figure 4.3 shows the mean response times as a function of the number of stimuli in each search display: that is, the time needed to detect one direct-gaze face among three, five, or eight averted-gaze faces. The chimpanzee detected an direct-gaze face among averted-gaze faces significantly faster than an averted-gaze face among direct-gaze faces, thus demonstrating the stare-in-the-crowd effect. Interestingly, this direct-gaze superiority was maintained when the faces were presented upside down or with the facial elements spatially scrambled. These results might suggest that the configurational processing of faces and the processing of gaze are relatively independent (Campbell et al. 1990).

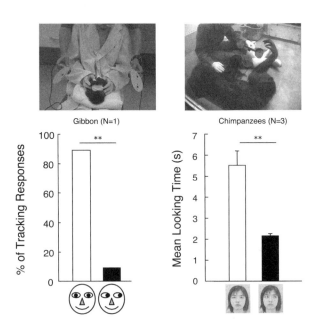

Figure 4.2 Sensitivity to human direct gaze in apes during early infancy (adapted from Myowa-Yamakoshi and Tomonaga 2001 and Myowa-Yamakoshi et al. 2003). (a) The results of a gibbon infant who was tested under the preferential tracking paradigm. (b) The mean looking time for each stimulus averaged across three infant chimpanzees tested under the forced-choice preferential-looking paradigm. **: $p < .01$.

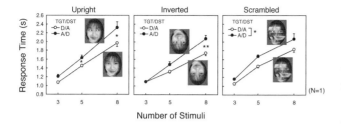

Figure 4.3 Visual search for human gaze direction in the chimpanzee (adapted from Tomonaga and Imura 2008b). Mean response times and standard errors of means for each condition are given as a function of the number of stimuli. For upright and inverted conditions, the results of analyses of variances indicated the significant target-distractor combination multiplied by the number of stimuli interactions. Asterisks indicate the results of post hoc multiple comparisons. For the scrambled condition, the main effect of the target-distractor combination was significant. **: $p < .01$. *: $p < .05$. D: directed gaze. A: averted gaze.

Efficient Detection of Faces

Some kinds of faces (or of information from faces) are efficiently detected in a crowd, but this efficiency can be considered relative. For example, in chimpanzees the detection of an upright or direct-gaze face was more efficient than that of an inverted or averted-gaze face—but both kinds of detection require serial processing, as is suggested by the linear increase in response time as a function of the number of faces seen (figure 4.3). The detection of faces is efficient but it might not be classified as automatic, preattentive, or "pop-out." However, it is quite plausible that faces *per se* may "pop out" from among non-face objects. Recently it has been reported that human observers can very quickly detect a human face displayed among non-face objects (Hershler and Hochstein 2005). Is similarly efficient detection of faces also observed in chimpanzees?

I tested three adult chimpanzees in a visual-search task to address this question (Tomonaga and Imura 2006a, 2007, and unpublished data). Each participant was initially trained to a visual-search task in which the chimpanzee was required to select a target stimulus from among an array of non-target distractors. In a preliminary homogeneous condition, all distractors were the same as each other but different from the target stimulus, which was drawn from one of four perceptual categories (chimpanzee faces, bananas, cars, and houses). Following this training phase, the task was changed so that the distractors were from different categories (see figures 4.4a and 4.4b), and the target was fixed to one of those four categories within a given session. Half of the trials thus used homogeneous distractors while the other half used heterogeneous distractors. The number of stimuli varied between 5, 10, and 18. Each participant underwent eight sessions in each category.

Figure 4.4c shows the mean response times. Like humans, chimpanzees very quickly detected the chimpanzee face among the non-face distractors. The search slope (increment in response time as a function of the number of items) for the chimpanzee face target was 14 ms/item. This was comparable to the data for humans performing a similar task (9 ms/item) as shown by Hershler and Hochstein (2005). The participants did not detect artificial object targets as quickly; the search slope in those cases averaged 45 ms/item.

Interestingly, the chimpanzees detected bananas as quickly as they did faces. There may be two explanations for this. One is that since food items have ecological significance, these items "pop out." The other possibility is that the chimpanzees paid attention only to the color (yellow) during the banana-target session. To examine possible confounding effects I conducted additional testing, manipulating the stimuli. Using chimpanzee faces, bananas, and cars as target categories, each stimulus was inverted, reduced in color to grayscale, or randomly scrambled. If the chimpanzees correctly processed the face target as a "face," inversion of the target would severely impair their test performance but color reduction would have no effect. If, however, they were detecting the banana target on the basis of color, its inversion and scrambling

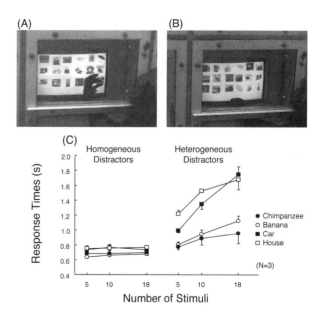

Figure 4.4 Visual search for a face among nonface objects (adapted from Tomonaga and Imura 2006a, 2007). In the top panels, the chimpanzee Chloe performs the chimpanzee-face target trial (a) and car target trial (b). Below (c), the mean response times (with SEM) are given for each distractor condition.

would not affect their performance but color reduction would affect it severely. The results showed that all chimpanzees exhibited the inversion effect only for the face target. Color reduction impaired their performance only for the banana-target trials, thus suggesting that a reliance on color matching accounted for their earlier success with those stimuli. In sum, these results clearly indicate that efficient face detection in chimpanzees is based on face-specific processing (cf. Tomonaga 2007a).

To what extent can this efficiency be generalized? In the next experiment I presented the faces of other species: human babies, human adult females, and Japanese macaques. As described earlier, chimpanzees processed dog faces similarly to the way they processed chimpanzee and human faces. Therefore it was predicted that they would detect non-chimpanzee face targets as efficiently as chimpanzee faces. However, the results only partially supported those predictions. When the ten items were presented in the search display, the chimpanzees detected the human baby (816 ms) and human female (835 ms) as quickly as they did the chimpanzee face (906 ms). But the recognition of the Japanese macaque target (1,308 ms) was as slow as for the artificial object target (1,460 ms on average) in the previous experiment. Familiarity or visual experience cannot explain these results, since the chimpanzees saw Japanese macaques living adjacent to their own outdoor compound every day and there were almost no opportunities for them to see human babies. Hershler and Hochstein (2005) also reported that animal faces did not pop out for human observers. For chimpanzees, macaques could be regarded simply as "animals" and distinct somehow from humans and chimpanzees.

Shift of Attention by Social Stimuli

Underlying Mechanisms of Attention Shift

In the previous sections I described how the attention of chimpanzees is captured by social stimuli such as face and gaze. In this section, I summarize the other properties of their social attention: disengagement and shift of attention by social and nonsocial stimuli.

Attention captured by social stimuli is readily disengaged and shifted by other types of social cues, such as gaze. Such shifts of attention are also very common in nonsocial contexts. As noted, human visuospatial attention is readily shifted toward a location where a visual

stimulus is abruptly presented (Müller and Rabbitt 1989; Posner 1980). Furthermore, it is also shifted to the location signaled by the precue presented at the center of a visual field (such as arrows). These phenomena are often tested under the cueing task, in which a cue signaling the location of a forthcoming target stimulus is followed by the target itself. In some trials—called "valid" trials—the cue signals the correct location of the target, while in the other "invalid" trials, the cue signals a location opposite to that of the target. If the observer's visuospatial attention is shifted by the cue, the response time would be faster in a valid trial than in the invalid condition. Thus, the cueing effect is usually defined as the difference in response time between valid and invalid trials.

Many researchers agree that there are two distinct mechanisms underlying the orientation of attention (Müller and Rabbitt 1989; Tomonaga 2007b). One is called reflexive (automatic, exogenous, bottom-up) and the other is called voluntary (controlled, endogenous, top-down). Reflexive orienting is triggered mainly by peripheral cues, such as abrupt onset of the cue stimulus, and it acts very quickly but is transient: the effect also disappears quickly. Voluntary orienting, on the other hand, is triggered by mainly central, symbolic cues such as arrows; it operates slowly and then is maintained at an optimum level. In addition to the temporal differences between these mechanisms, the influence of cue predictability also appears dissimilar. Reflexive orienting is relatively resistant to the manipulation of cue predictability, while voluntary orienting is readily affected by changes in predictability. These characteristics of the cues can be used as diagnostics for identifying the underlying mechanisms of attention shift.

In humans, gaze cues presented at the center of a visual field cause very quick orienting of the subject's attention to the side of the field toward which the gaze cues are directed (Friesen and Kingstone 1998; Langton and Bruce 1999; Senju et al. 2003; see also Frischen et al. 2007, for an extensive review). Even if the subjects are explicitly instructed that the gaze cue is not informative, or if the gaze cues are counterpredictive—that is, they signal that the target will appear at a different location—this cueing effect is still effective (Friesen et al. 2004; Senju et al. 2003). These results clearly suggest that the attention shift caused by gaze cues is mainly governed by the reflexive mechanism in humans. The question then arises of whether chimpanzee attention triggered by gaze is also operated by a reflexive mechanism.

Joint Attention and Shift of Attention

As mentioned earlier, primates—including chimpanzees and humans—readily follow another's gaze (for review see Emery 2000; Itakura 1996; Itakura and Tanaka 1998; Okamoto et al. 2002, 2004). Chimpanzees can follow gaze cues given both by humans (Call et al. 1998; Itakura 1996; Itakura and Tanaka 1998; Okamoto-Barth et al. 2008; Tomasello et al. 1999) and by conspecifics (Tomasello et al. 1997). These abilities emerge at about one year of age in an enriched experimental context (Okamoto et al. 2002), but the ability to shift their attention outside the visual field develops later (at around 2.5 years; Okamoto et al. 2004) than that of humans (which develops at less than 2 years).

Are there similarities or differences between the underlying mechanisms of gaze following behavior in nonhuman primates and humans? Two different mechanisms have been proposed: a low-level model, in which another's cue triggers an automatic, primitive orienting reflex, and a high-level model or evaluation system, in which subjects calculate the imaginary line of another's sight and evaluate the other's mental state (Emery 2000; Okamoto-Barth et al. 2007a; Povinelli and Eddy 1996). The fact that chimpanzees can follow another's gaze toward a hidden location or to a point outside the visual field implies that they calculate the line of sight and evaluate the other's mental state. Below I describe three experiments which address the relative effects of factors such as cue predictability, cue type, and stimulus onset asynchrony (SOA).

Effects of Cue Predictability

To examine the mechanisms of attention shift triggered by gaze cues in chimpanzees, two adult females were tested using a simple discrimination task (Tomonaga 2007b). During this task they were trained to discriminate between the letters F (target) and T (distractor), which were presented simultaneously. Before the pair of letters, various kinds of cues—such as eye gaze with a front-view face and a profile—were presented (figure 4.5a). In 50% of the trials (valid trials) the cue signaled the target's location correctly, but in the other half (invalid trials) it did not. The hypothesis was that signals that worked on a reflexive, involuntary basis would result in relatively short response times for valid cues and longer times for invalid cues. This large difference in response time, known as the cueing effect, is an important measure for assessing the mechanisms at play in these tests of attention shift. We found very little difference in response times between valid and invalid cues (i.e., a low cueing effect). These results differed from results from tests performed of humans (e.g., Langton and Bruce 1999), but were similar to the results of tests with baboons and macaques (Fagot and Deruelle 2002; Vick et al. 2006; but see Deaner and Platt 2003), and they may imply that in chimpanzees, attention shift caused by gaze is not reflexive but voluntary.

To further address this point, I manipulated the cue predictability variable. In one condition, the orientation of the head cue predicted the correct target location in 80% of trials, while in the other condition it predicted the cor-

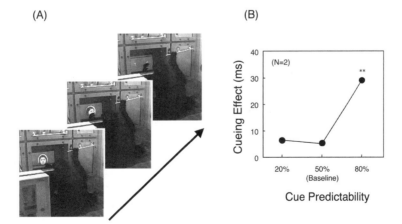

Figure 4.5   Simple letter discrimination task with head rotation cue (adapted from Tomonaga 2007b). (a) Chloe performs the task (valid trial). (b) Mean cueing effect is averaged across participants as a function of cue predictability. Cueing effect was only significant under the high predictability condition. **: $p < .01$.

rect location in only 20% of trials. If chimpanzees' attention is governed mainly by a voluntary mechanism, the cueing effect would be observed only in the condition of higher predictability. This was indeed the case: only when the cues were highly predictive of the target location (80%) were valid cues significantly more effective than invalid cues in eliciting a correct response (figure 4.5b, Tomonaga 2007b).

In an associated experiment, we focused on the ability of different types of cues to shift attention. Given that eye gaze was unable to elicit a voluntary reaction in the first task, would other forms of cue, both social and nonsocial, differ in their effectiveness? The task in question was a simple target detection, which required two young chimpanzees to touch a blue circle presented either to the left or the right of a cueing signal (figure 4.6a). Various types of cue (eye gaze, head rotation, pointing, arrow, and peripheral cue) were presented 200 ms before the onset of the target. Again, gaze did not elicit a significant cueing effect when cue predictability was random (50% of instances). Only the head-rotation cue caused a significant cueing effect (Tomonaga and Imura 2005; see figure 4.6b), suggesting that gaze defined by head direction (rather than iris position) seems to reflexively trigger the visual attention of chimpanzees. Interestingly, the young chimpanzees' performance was readily affected by a variety of cues, including nonsocial cues such as arrows, but only when those cues were highly predictive of the target's location. These results suggest that shifts in visuospatial attention are not governed exclusively by reflexive mechanisms. In fact, voluntary mechanisms may be stronger in chimpanzees than in humans. To confirm these hypotheses, I turn now to our

examination of another aspect of attention shift, namely the role of time course.

Another way to assess the mechanism at play in attention shift is to examine the importance of the task's temporal properties. Reflexive shifts in attention occur over a very short time as compared to voluntary shifts: 100 ms of delay between cue and stimulus will suffice for reflexive shifts, but more than 500 ms are required for voluntary shifts (Müller and Rabbitt 1989; Tomonaga 2007b). To investigate these temporal properties, I conducted a target-detection experiment with adult chimpanzees (Tomonaga and Imura 2003), and in this experiment I used photographs of gestural cues—such as pointing at, grasping, and looking at an object—instead of simple gaze cues (figure 4.7a). The task was again the simple target detection, in which the cue was presented for 100 ms followed by the target with either a short (100-ms) or long (500-ms) delay between the cue and the target stimulus. Cue predictability was set at 50%. Two adult chimpanzees and ten adult humans were tested in an identical experimental setting. Again, reflexive mechanisms would be indicated by a larger cueing effect in the short-delay condition than in the long-delay condition, while voluntary mechanisms would be indicated by the opposite effect. Mean response times for each species and each stimulus-onset-asychrony (SOA) condition are plotted in figure 7b. For humans, the cueing effect was significantly more effective under the short-SOA condition, while for chimpanzees it was significantly more effective under the long-SOA condition. Since both humans and chimpanzees showed significant cueing effects in the short SOA condition when nonsocial "peripheral" cues were given as control trials, the

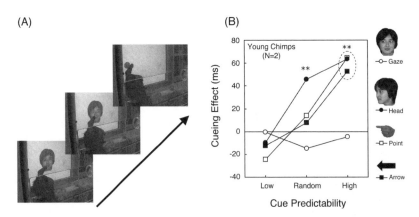

(A) (B)

Figure 4.6   Shift of visuospatial attention by the social cues in young chimpanzees (adapted from Tomonaga and Imura 2005, 2008a). (a) Chimpanzee performs the target detection task with eye-gaze cue (invalid trial). (b) Mean cueing effect for each cue type, given as a function of cue predictability. Significant cueing effects were observed for all cues except for eye gaze when cue predictability was high. When the predictability was random, however, only the head-rotation cue yielded a significant cueing effect. **: $p < .01$.

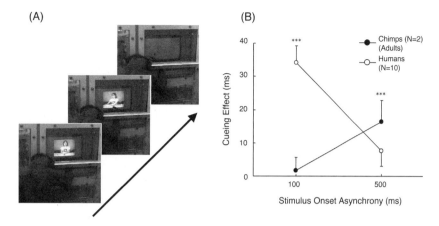

Figure 4.7 Shift of visuospatial attention by human directional gestures (adapted from Tomonaga and Imura 2003). (a) The chimpanzee Pendesa performs the task (valid trial). (b) Mean cueing effect (with SEM) for humans and chimpanzees as a function of SOA. ***: $p < .001$.

results with gestural cues cannot be explained by characterizing the cognitive processes of chimpanzees as simply being slower than in humans.

Together with the results of predictability manipulation, these results clearly imply that social attention in chimpanzees is operated by rather different mechanisms than in humans. Many studies have suggested that for humans the reflexive mechanisms are more dominant (e.g., Friensen and Kingstone 1998; Langton et al. 1999), while for chimpanzees the voluntary components play more important roles in attention shift. To date, it remains unclear whether there are fundamental relations between the current tentative conclusion and the differences in social interactions based on joint attention (Tomonaga 2006; Tomonaga et al. 2004) in these two species. Voluntary control of attention is operated by symbolic cues such as arrows, the meanings of which are generally learned through everyday experiences. Thus, there is a possibility that captive chimpanzees learn some human social cues such as pointing, head orientation, and eye gaze during their extensive interactions with humans, and that these cues can then cause the voluntary shift of visuospatial attention. Since there is still controversy concerning social attention both in humans (Frischen et al. 2007) and in nonhuman primates (Deaner and Platt 2003; Tomonaga 2007), further investigations of various aspects of social attention should be conducted from the comparative-cognitive-developmental perspective.

Implications and Future Directions

In this chapter I have briefly summarized the processing of social stimuli such as face, eyes, and gestures in chim-

panzees of various ages from the comparative-cognitive-developmental standpoint. Table 4.1 summarizes the results of studies discussed in this chapter. I note that there is a point at which humans and chimpanzees diverge developmentally in the social-cognitive domain: chimpanzees seem not to engage in subject-subject-object interactions (Tomonaga 2006; Tomonaga et al. 2004). In laboratory experiments I have shown that chimpanzees share the perceptual-

Table 4.1 Summary of laboratory studies on social attention in chimpanzees.

Topic		Task	Major finding
Face perception		Matching to sample	• Inversion effect
		Visual search	• Efficient search for upright faces → Related to inversion effect
Capture of social attention	Face detection	Visual search	• Very efficient search for faces among nonface objects • Inversion effect
	Gaze perception	Preferential looking	• Direct-gaze preference in early infancy
		Visual search	• Efficient (but not parallel) search for direct gaze face • No inversion effect • No generalization from eye to head directions
Shift of social attention		Cueing task	• No evidence for reflexive shift of attention • Dominant contribution of voluntary mechanism → Effective for Highly predictable cues → Effective under long SOA • Evident both in adult and young chimps

cognitive mechanisms required for dyadic interactions such as face processing and sensitivity to direct gaze, but that the mechanisms underlying social attention differ from those in humans. Chimpanzee attention is mainly controlled voluntarily, while human social attention is reflexive. It is plausible that this difference is related to the fundamental differences in "theory of mind" abilities between these species. Further investigations will address these speculations.

Acknowledgments

The research presented here, and the preparation of the manuscript, were financially supported by Grants-in-Aid for Scientific Research from the Japanese Ministry of Education, Culture, Sports, Science, and Technology (MEXT), the Japan Society for the Promotion of Science (JSPS; Grant Nos. 04710053, 05206113, 05710050, 06710042, 07102010, 09207105, 10CE2005, 11710035, 12002009, 13610086, 16002001, 16300084), and MEXT/JSPS Grants-in-Aid for the 21st Century COE Program (A14 and D10). I thank Dr. Tomoko Imura, collaborator in most of the studies described. I wish to thank Drs. Tetsuro Matsuzawa, Masayuki Tanaka, and Misato Hayashi for their valuable comments and support for these studies. Thanks are also due to Sumiharu Nagumo, Kiyonori Kumazaki, Norihiko Maeda, Shohei Watanabe, and the staff of the Center for Human Evolution Modeling Research at the Primate Research Institute, Kyoto University (KUPRI) for their technical advice and their care of the chimpanzees, as well as to Tomoko Takashima, Suzuka Hori, and the other staff of the Language and Intelligence Section of KUPRI for their support of the experiments. The care and use of the chimpanzees adhered to the 2002 version of the "Guide for Care and Use of Laboratory Primates" of KUPRI (fully compatible with the guidelines of the National Institute of Health in the United States) and the Regulation on Animal Experimentation at Kyoto University. The research design was approved by the Animal Welfare and Animal Care Committee at KUPRI and the Animal Research Committee at Kyoto University.

Literature Cited

Ando, S. 2002. Luminance-induced shift in the apparent direction of gaze. *Perception* 31, 657–74.

Bard, K. A., M. Myowa-Yamokoshi, M. Tomonaga, M. Tanaka, A. Costal, and T. Matsuzawa. 2005. Group differences in the mutual gaze of chimpanzees (*Pan troglodytes*). *Developmental Psychology* 41:616–24.

Butterworth, G., and N. L. M. Jarrett. 1991. What minds have in common is space: Spatial mechanism serving joint visual attention in infancy. *British Journal of Developmental Psychology* 9:55–72.

Bråten, S., ed. 1998. *Intersubjective Communication and Emotion in Early Ontogeny*. Cambridge, UK: Cambridge University Press.

Call, J., B. A. Hare, and M. Tomasello. 1998. Chimpanzee gaze following in an object-choice task. *Animal Cognition* 1:89–99.

Campbell, R., C. A. Heywood, A. Cowey, M. Regard, and T. Landis. 1990. Sensitivity to eye gaze in prosopagnosic patients and monkeys with superior temporal sulcus ablation. *Neuropsychologia* 28:1123–42.

Deaner, R. O., and M. L. Platt. 2003. Reflexive social attention in monkeys and humans. *Current Biology* 13:1609–13.

Diamond, R., and S. Carey. 1986. Why faces are and are not special: An effect of expertise. *Journal of Experimental Psychology: General* 115:107–17.

Eastwood, J. D., D. Smilek, and P. M. Merikle. 2001. Differential attentional guidance by unattended faces expressing positive and negative emotion. *Perception & Psychophysics* 63:1004–13.

Emery, N. J. 2000. The eyes have it: The neuroethology, function and evolution of social gaze. *Neuroscience and Biobehavioral Reviews* 24:581–604.

Emery, N. J., E. N. Lorincz, D. I. Perrett, M. W. Oram, and C. I. Baker. 1997. Gaze following and joint attention in rhesus monkeys (*Macaca mulatta*). *Journal of Comparative Psychology* 111:286–93.

Fagot, J., and C. Deruelle. 2002. Perception of pictorial eye gaze by baboons (*Papio papio*). *Journal of Experimental Psychology: Animal Behavior Processes* 28:298–308.

Farroni, T., G. Csibra, F. Simion, and M. H. Johnson. 2002. Eye contact detection in humans from birth. *Proceedings of the National Academy of Sciences of the USA* 99:9602–05.

Friesen, C. K., and A. Kingstone. 1998. The eyes have it! Reflexive orienting is triggered by nonpredictive gaze. *Psychonomic Bulletin & Review* 5:490–95.

Frischen, A., A. P. Bayliss, and S. P. Tipper. 2007. Gaze cueing of attention: Visual attention, social cognition, and individual differences. *Psychological Bulletin* 133:694–724.

Hershler, O., and S. Hochstein. 2005. At first sight: A high-level pop out effect for faces. *Vision Research* 45:1707–24.

Itakura, S. 1996. An exploratory study of gaze-monitoring in non-human primates. *Japanese Psychological Research* 38:174–80.

Itakura, S., and M. Tanaka. 1998. Use of experimenter-given cues during object-choice tasks by chimpanzees (*Pan troglodytes*), an orangutan (*Pongo pygmaeus*), and human infants (*Homo sapiens*). *Journal of Comparative Psychology* 112:119–26.

Kobayashi, H., and S. Koshima. 1997. Unique morphology of the human eye. *Nature* 387:767–68.

Langton, S. R. H., and V. Bruce. 1999. Reflexive visual orienting in response to the social attention of others. *Visual Cognition* 6:541–67.

Martin-Malivel, J., and J. Fagot. 2001. Perception of pictorial human faces by baboons: Effects of stimulus orientation on discrimination performance. *Animal Learning & Behavior* 29:10–20.

Matsuzawa, T., M. Tomonaga, and Tanaka, M., eds. 2006. *Cognitive Development in Chimpanzees*. Tokyo: Springer-Verlag.

Mendelson, M. J., M. M. Haith, and P. S. Goldman-Rakic. 1982. Face scanning and responsiveness to social cues in infant rhesus monkeys. *Developmental Psychology* 18:222–28.

Myowa-Yamakoshi, M., and M. Tomonaga. 2001. Perceiving eye gaze in an infant gibbon (*Hylobates agilis*). *Psychologia* 44:24–30.

Myowa-Yamakoshi, M., M. Tomonaga, M. Tanaka, and T. Matsuzawa. 2003. Preference for human direct gaze in infant chimpanzees (*Pan troglodytes*). *Cognition* 89:B53–64.

Müller, H. J., and P. M. A. Rabbitt. 1989. Reflexive and voluntary orienting of visual attention: Time course of activation and resistance to interruption. *Journal of Experimental Psychology: Human Perception and Performance* 15:315–30.

Okamoto, S., M. Tanaka, M., and M. Tomonaga. 2004. Looking back: The "representational mechanism" of joint attention in an infant chimpanzee (*Pan troglodytes*). *Japanese Psychological Research* 46:236–45.

Okamoto, S. M. Tomonaga, K. Ishii, N. Kawai, M. Tanaka, and T. Matsuzawa, T. 2002. An infant chimpanzee (*Pan troglodytes*) follows human gaze. *Animal Cognition* 5:107–14.

Okamoto-Barth, S., J. Call, and M. Tomasello. 2007a. Great apes' understanding of other individuals' line of sight. *Psychological Science* 18:462–68.

Okamoto-Barth, S., N. Kawai, M. Tanaka, and M. Tomonaga. 2007b. Looking compensates for the distance between mother and infant chimpanzee. *Developmental Science* 10:172–82.

Okamoto-Barth, S., M. Tomonaga, M. Tanaka, and T. Matsuzawa. 2008. Development of using experimenter-given cues in infant chimpanzees: Longitudinal changes in behavior and cognitive development. *Developmental Science*, in press.

Overman, W. H., and R. W. Doty. 1982. Hemispheric specialization displayed by man but not macaques for analysis of faces. *Neuropsychologia* 20:113–28.

Parr, L. A., T. Dove, and W. D. Hopkins. 1998. Why faces may be special: Evidence of the inversion effect in chimpanzees. *Journal of Cognitive Neuroscience* 10:615–22.

Phelps, M. T., and W. A. Roberts. 1994. Memory for pictures of upright and inverted primate faces in humans (*Homo sapiens*), squirrel monkeys (*Saimiri sciureus*), and pigeons (*Columbia livia*). *Journal of Comparative Psychology* 108:114–25.

Posner, M. I. 1980. Orienting of attention. *Quarterly Journal of Experimental Psychology* 32A:3–25.

Povinelli, D. J., and T. J. Eddy. 1996. Chimpanzees: Joint visual attention. *Psychological Science* 7:129–35.

Ricciardelli, P., G. Baylis, and J. Driver. 2000. The positive and negative of human expertise in gaze perception. *Cognition* 77:B1–B14.

Senju, A., Y. Tojo, H. Dairoku, and T. Hasegawa. 2003. Reflexive orienting in response to eye gaze and an arrow in children with and without autism. *Journal of Child Psychology and Psychiatry* 44:445–58.

Tanaka, M. 2003. Visual preference by chimpanzees (*Pan troglodytes*) for photos of primates measured by a free choice-order task: Implication for influence of social experience. *Primates* 44:157–65.

Tanaka, M. 2007. Development of visual preference of chimpanzees (*Pan troglodytes*) for photographs of primates: Effect of social experience. *Primates* 48:303–9.

Thomsen, C. E. 1974. Eye contact by non-human primates toward a human observer. *Animal Behaviour* 22:144–49.

Tomasello, M. 1999. *The Cultural Origins of Human Cognition*. Cambridge, MA: Harvard University Press.

Tomasello M., J. Call, and B. Hare. 1997. Five primate species follow the visual gaze of conspecifics. *Animal Behaviour* 55:1063–69.

———. 2003. Chimpanzees understand psychological states: The question is which ones and to what extent. *Trends in Cognitive Science* 7:153–56.

Tomasello, M., B. Hare, and B. Agnetta. 1999. Chimpanzees, *Pan troglodytes*, follow gaze direction geometrically. *Animal Behaviour* 58:769–77.

Tomasello, M., B. Hare, and T. Fogleman. 2001. The ontogeny of gaze following in chimpanzees, *Pan troglodytes*, and rhesus macaques, *Macaca mulatta*. *Animal Behaviour* 61:335–43.

Tomonaga, M. 1994. How laboratory-raised Japanese monkeys (*Macaca fuscata*) perceive rotated photographs of monkeys: Evidence for an inversion effect in face perception. *Primates* 35:155–65.

———. 1997. Precuing the target location in visual searching by a chimpanzee (*Pan troglodytes*): Effects of precue validity. *Japanese Psychological Research* 39:200–211.

———. 1999a. Inversion effect in perception of human faces in a chimpanzee (*Pan troglodytes*). *Primates* 40:417–38.

———. 1999b. Visual search for orientation of faces by a chimpanzee (*Pan troglodytes*). *Primate Research* 15:215–29.

———. 1999c. Attending to the others' attention in macaques: Joint attention or not? *Primate Research* 15:425 (Japanese abstract only).

———. 2001. Investigating visual perception and cognition in chimpanzees (*Pan troglodytes*) through visual search and related tasks: From basic to complex processes. In T. Matsuzawa, ed., *Primate Origin of Human Cognition and Behavior*, 55–86. Tokyo: Springer.

———. 2006. Development of chimpanzee social cognition in the first 2 years of life. In T. Matsuzawa, M. Tomonaga, and M. Tanaka, eds., *Cognitive Development in Chimpanzees*, 182–97. Tokyo: Springer.

———. 2007a. Visual search for orientation of faces by a chimpanzee (*Pan troglodytes*): Face-specific upright superiority and the role of configural properties of faces. *Primates* 48:1–12.

———. 2007b. Is chimpanzee (*Pan troglodytes*) spatial attention reflexively triggered by the gaze cue? *Journal of Comparative Psychology* 121:156–70.

Tomonaga, M., and T. Imura. 2003. Do social gestures cause attentional shift in chimpanzees? *Japanese Journal of Animal Psychology* 53:107 (Japanese abstract only).

———. 2005. Orienting of attention by gaze cues in chimpanzees (*Pan troglodytes*). Paper presented at the 16th annual meeting of the Japanese Society of Developmental Psychology, March 2005, Kobe, Japan.

———. 2006a. Where's the chimp? Efficient search for a face in chimpanzees. *Primate Research* 22:S24–25 (Japanese abstract only).

———. 2006b. Face captures young chimpanzee's visuospatial attention. *Japanese Journal of Animal Psychology* 56:172 (Japanese abstract only).

———. 2007. Where's the chimp 2.0: Efficient search for a face in chimpanzees is based on face recognition. *Primate Research* 23:S28–29 (Japanese abstract only).

———. 2008a. Voluntary orienting of attention by gaze cues in young chimpanzees (*Pan troglodytes*). Paper presented at the 19th annual meeting of the Japanese Society of Developmental Psychology, March 2008, Osaka, Japan.

———. 2008b. Visual search for human gaze direction by the chimpanzee (*Pan troglodytes*). Manuscript submitted for publication.

Tomonaga, M., M. Myowa-Yamakoshi, Y. Mizuno, S. Okamoto, M. K. Yamaguchi, D. Kosugi, K. A. Bard, M. Tanaka, and T. Matsuzawa. 2004. Development of social cognition in infant chimpanzees (*Pan troglodytes*): Face recognition, smiling, gaze and the lack of triadic interactions. *Japanese Psychological Research* 46:227–35.

Valentine, T. 1988. Upside-down faces: A review of the effects of inversion upon face recognition. *British Journal of Psychology* 79:471–91.

Vick, S.-J., I. Toxopeus, and J. R. Anderson. 2006. Pictorial gaze cues do not enhance long-tailed macaques' performance on a computerised object-location task. *Behavioural Processes* 73:308–14.

Von Grünau, M., and C. Anston. 1995. The detection of gaze: A stare in the crowd effect. *Perception* 24:1297–313.

Yin, R. K. 1969. Looking at upside-down faces. *Journal of Experimental Psychology* 81:141–45.

5

Understanding the Expression and Classification of Chimpanzee Facial Expressions

Lisa A. Parr

Imagine two friends twice ending a conflict with the verbal reconciliation "I forgive you." In one instance the statement is given with a smile, slight head tilt, and averted eyes, while in another instance it is given with a wide stare, head tilted slightly back, and a sharp frown. The sincerity of that statement when given in these two different nonverbal scenarios is dramatically different and will produce different social responses by the recipient. How did the use of such nonverbal signals arise—and do they have their origin in the nonverbal behavior of species closely related to humans, such as the chimpanzee?

Although spoken language is the primary modality for social communication in humans, nonverbal signals, such as facial expressions, form a backdrop of nuances that very effectively alter the meaning of even the most direct speech. Understanding the evolution of these nonverbal behaviors can provide important insight into their meaning and social significance. Both humans and chimpanzees have a wide repertoire of facial expressions that are critical for mediating their complex and fluid societies and for maintaining social relationships. Yet for these behaviors to play an important role in communication, individuals have to be able to recognize and discriminate among them, especially when they are used in flexible contexts. In this chapter I describe the facial repertoire of chimpanzees and examine their ability to discriminate among these signals.

Finally, I review a new method for studying comparative social cognition, using standardized facial expressions, that will advance our understanding of these signals' social function with relevance for understanding the evolution of nonverbal communication in humans.

Variability in Primate Facial Expressions

Faces are one of the most important classes of stimuli used in primate social communication. They provide information about individual identity, age, gender, and emotion. Facial expressions enable primates to communicate about ongoing events, such as when they signal status, but they can also provide information about the likelihood of future behavior, such as the motivation to engage in conflict (van Hooff 1967). Moreover, they most often occur within a dynamic interaction between individuals in a range of different social contexts. Thus their specific use must be updated continually, and it requires the sender to take social dynamics into account, in addition to the sender's own internal motivation and desires. In this vein, van Hooff (1967) has described primate facial expressions as a series of behavioral elements that come together within a flexible and changing system to convey information about an individual's ongoing and future motivation.

Numerous ethograms exist that describe the repertoire and proposed social function of facial signals in a variety of nonhuman primates species including chimpanzees, bonobos, rhesus monkeys, and capuchin monkeys (Andrew 1963; Bolwig 1978; de Marco and Visalberghi 2007; de Waal 1988; Flack and de Waal 2007; van Lawick-Goodall 1968; Hinde and Rowell 1962; van Hooff 1962, 1967, 1973; Parr, Cohen, and de Waal 2005; Preuschoft and van Hooff 1997; Waller and Dunbar 2005; Weigel 1979). Although there is considerable continuity in the appearance of many expressions across phylogeny, there are also distinct differences. These differences can take three basic forms. First, a signal may be preserved in both form and function across a range of species and taxonomic groups. One example is the scream, which is similar in appearance across many species and is typically used after some form of agonistic conflict. Other signals also appear to be preserved across a broad taxonomic range. The lip-smack, for example, is common in New World capuchin monkeys and Old World macaques, but it appears to have been lost in the hominids. Chimpanzees, for example—a species whose facial expressions have been described in detail—do not appear to have any form of this behavior. Second, some expressions may appear very similar across a range of species, but take on different meanings depending on the distinct socioecology of each species. The bared-teeth display is a well-studied example (see figure 5.1). In some very despotic species, like the rhesus macaque, this display is highly ritualized, appearing similar in form regardless of the sender. Its use is also very constrained. In the rhesus macaque it is used as a formal signal of dominance, performed only by low-ranking individuals to more dominant ones, but never by dominants towards subordinates (de Waal and Luttrell 1985). However, in closely related macaque species with more flexible, egalitarian hierarchies—such as pigtail macaques and some species of Sulawesi macaques—the bared-teeth display is performed by a variety of individuals and appears to signal both status and appeasement (Flack and de Waal 2007; Petit and Thierry 1992). Among chimpanzees the display takes many different forms—staring bared-teeth, vertical bared-teeth, frowning bared-teeth face (van Hooff 1967, 1973)—and it has a distinct friendly, appeasing function that is common during greetings, play, sex, and feeding (Parr et al. 2005). Finally, some species appear to have evolved unique signals that are distinct in both form and function. The stretch pout whimper in chimpanzees is one such example. It appears as a blend between the bared-teeth display and the pout, sharing features with

Figure 5.1   Bared-teeth display in a human, chimpanzee, and rhesus macaque. It has been proposed that this expression is homologous in these three species (van Hooff 1972).

each expression, and is analogous to human sulking (van Hooff 1973; Parr et al. 2005).

Chimpanzees have a well-characterized repertoire of facial expressions and vocalizations that include screams, bared-teeth displays, pant-hoots, pouts, whimpers, squeaks, pant-grunts, food barks, play faces, and bulging-lip faces (see figure 5.2; Goodall 1968; Parr et al. 2005). Several of the facial expressions are rarely or never accompanied by a vocalization. The pout, for example, is almost always silent, while the bulging-lip face precludes a vocalization, as the lips are pressed together. Other facial expressions, like the stretch-pout whimper (moan), scream (scream), bared-teeth display (low-intensity scream, squeak), and play face (laughter) can occur either silently or with the distinctive vocalization listed here in parentheses. Moreover, as researchers learn more about the acoustic properties of these vocalizations (see chapter 16), evidence is being found for acoustic subtypes within these categories, thus adding to the complexity of the communicative repertoire of this species. By necessity, any vocalization can be said to have an accompanying facial configuration—like the pant-hoot, which consists of a long series of voiced inspiration-expiration sounds ("ooooh-ooooh") that rise in pitch and intensity and end with a type of scream. During the build up for this vocalization, however, the mouth forms a pucker (figure 5.2) and this facial configuration can be maintained in the absence of any detectable vocalization. Therefore, while there is a close relationship between facial expressions and vocalizations, the use of facial expressions as silent nonverbal signals in this species is also pronounced.

Despite considerable complexity in the form and function of nonhuman primate facial expressions, as described above, extremely little is known about how these signals are discriminated at a perceptual level, or whether such perceptual processes are in any way similar to those used by humans to categorize facial expressions. Even among chimpanzees, a species for which considerable information is now known about the perceptual processes un-

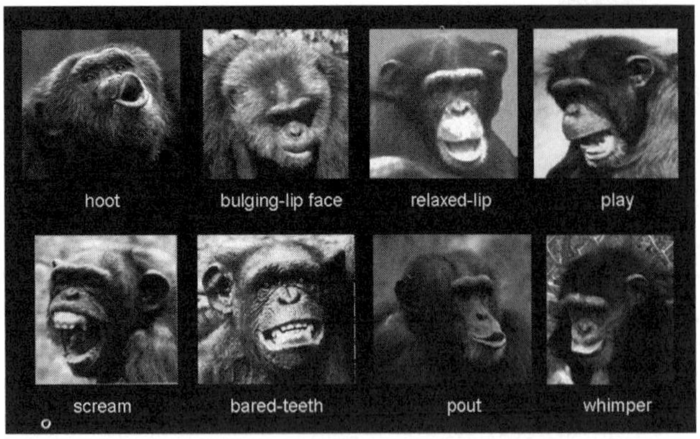

Figure 5.2   Several prototypical facial expressions in chimpanzees.

derlying face recognition, studies of facial expression categorization are rare (Parr et al. 1998; Parr et al. 2000; Parr et al. 2006).

One of the challenges researchers face when preparing computer experiments to assess these issues is that of acquiring high-quality digital images to address the question of interest. This is because of the difficulty in acquiring images of facial expressions from ongoing behavior. Although the advent of digital video and photography has aided this endeavor, most nonhuman primates do not pose for photographs. In the wild, for example, chimpanzees live in densely forested habitats where photographs of even well-habituated individuals are often difficult to obtain. Similarly, in captivity the environment often contains many structures and climbing facilities that are necessary for the physical and mental well-being of the animals, but which inevitably obstruct clear views for photography. Perhaps most problematic for stimulus collection is that facial expressions occur in an ongoing social context that often involves fast action by many different individuals. In such a situation, the likelihood of capturing a full frontal, unobstructed view of an individual's expression is low.

Developing ChimpFACS

Partly to facilitate studies on the perception of facial expressions by chimpanzees, and to describe their exceedingly complex communicative repertoire more accurately, researchers have recently developed an anatomical coding system for measuring facial movement in the chimpanzee, ChimpFACS (Vick et al. 2007). This is based on the well-known Facial Action Coding System, or FACS, developed by Paul Ekman and colleagues for use with humans (Ek-

man and Friesen 1978). The FACS is an objective tool for measuring facial behavior in humans that is anatomically-based, as it describes how the face changes in appearance due to the action of the underlying facial muscles. These changes are referred to by numeric codes called action units, or AUs, which indicate the minimal independent and observable units of facial movement. This eliminates any bias humans may have in interpreting facial expressions using emotional labels, and also provides an objective method for comparing facial expressions across vastly different populations of people, and now species. Because of this precision and objectivity, FACS is now the most widely used technique for measuring facial movement in human facial expression research and provides a standardized language for comparing behavior across individuals (Ekman and Rosenberg 1997).

Table 5.1 provides a list of each action unit identified for the chimpanzee, and its muscular basis (Vick et al. 2007).

To help ensure that ChimpFACS would be a robust tool for comparative studies, it was modeled as closely as possible to the human FACS. This involved three main stages. As the human FACS was based on muscle movements, the first stage in developing a FACS for chimpanzees was to validate the presence and anatomical organization of their

Table 5.1   A summary of action units present in ChimpFACS and their muscular basis (from Vick et al. 2007).

Action unit	Descriptive name	Muscular basis
AU1+2	brow raiser	frontalis
AU4	brow lowerer	procerus, depressor supercilii[*]
AU5	upper lip raiser	orbicularis oculi
AU6	cheek raiser	orbicularis oculi (pars orbitalis)
AU9	nose wrinkler	levator labii superioris (alaeque nasi)
AU10	upper lip raiser	levator labii superioris
AU12	lip corner puller	zygomatic major
AU16	lower lip depressor	depressor labii
AU17	chin raiser	mentalis
AU22	lip funneler	orbicularis oris
AU24	lip pressor	orbicularis oris
AU28	lip suck	orbicularis oris
AU43	eye closure	orbicularis oculi
AU45	blink	orbicularis oculi

[*]In humans this movement lowers and knits the brows. The knitting is attributed to the corrugator muscle, which is present in chimpanzees, but no knitting action has been observed.

facial muscles with direct comparison to humans. The literature on chimpanzee facial muscle organization was extremely limited and often employed a hierarchical, phylogenetic interpretation in which the features of primate evolution ascended linearly towards humans (Huber 1931). To ensure an accurate picture of the organization of chimpanzee facial muscles, new dissections were conducted by Burrows and colleagues (Burrows et al. 2006). Facial muscles are an intricate web of delicate structures, and unlike other skeletal muscles they often attach to soft tissue, like skin, rather than to bone. To examine the dermal attachments and muscle origins at a new level of detail, Burrows removed the facial tissue from the skull and performed dissections from the inside out, a technique referred to as the "facial mask" technique (Burrows et al. 2006). These new dissections were able to confirm the presence and anatomical organization of all 23 mimetic facial muscles in the chimpanzee that have also been described for humans with only a few minor differences (Burrows et al. 2006).

The next step was to consider how the movement of these muscles would function to change the appearance of the chimpanzee face, given the differences in its facial architecture from that of humans. Its morphology, for example, features a greater degree of prognathism, a lack of cheek fat, less nasal cartilage, and a heavy brow ridge—all of which can affect how the facial muscles change the face's appearance. Following the landmark studies of Duchenne (1862), who documented emotional countenance in humans by stimulating configurations of facial muscles using surface electrodes, Waller and colleagues stimulated facial muscles in chimpanzees and humans to document how muscle action changes the appearance of the face (Waller et al. 2006). Thin microelectrodes were directly inserted into the belly of facial muscles in awake humans and anesthetized chimpanzees (the anesthesia coincided with annual health exams) and were then stimulated to achieve contraction. The results confirmed first that movements of the human face were equivalent to human FACS action units, demonstrating that FACS was indeed documenting specific muscle movements. Second, the stimulation of equivalent muscles in the chimpanzees was found to produce very similar appearance changes. Thus, regardless of differences in facial morphology, the same muscular action produced similar appearance changes in chimpanzees and in humans (Waller et al. 2006).

With the anatomical and functional bases for comparative facial movement thus validated, the ChimpFACS was created by identifying these movements in video and photo records of spontaneously occurring behavior in more than 200 chimpanzees. The process of documenting the chimpanzee action units (AUs) relied heavily on detailed observation. Some AUs were never seen in isolation but were inferred by comparing combinations of movement and extrapolating the minimal units in common. To maintain consistency with the human FACS, the same terms and codes for each AU were used. These AUs have been described in detail with reference to the muscular basis, appearance changes on contraction, minimum criteria for coding, subtle differences from similar AUs, and comparisons with equivalent human AUs (Vick et al. 2007). In total, 43 AUs were described for the chimpanzee, 15 of which related to specific facial muscles (see table 5.1). Remaining were miscellaneous action descriptors, such as head and eye movements, similar to those provided by the human FACS. Interestingly, some movements common in humans, such as the knitting of the brow caused by contraction of the corrugator and associated muscles (AU4), were never observed in the chimpanzee. Moreover, the action of the corrugator could not be stimulated, even though its presence and anatomical location were confirmed (Burrows et al. 2006; Waller et al. 2006). In sum, ChimpFACS represents an exciting and significant development in the study of comparative emotional communication, and paves the way for a wealth of evolutionary and comparative investigations.

Animating Facial Movements Using Poser

Using the ChimpFACS system, my colleagues and I began two projects to aid in the classification of chimpanzee facial expressions and facilitate studies on their categorization. The first project was to determine the effectiveness of ChimpFACS in classifying the expression categories present in photographs of naturally occurring facial expressions. As stated earlier, obtaining high quality, unobstructed, full frontal photographs of chimpanzee facial expressions at their peak intensity during ongoing behavior is extremely difficult. As a result, the stimulus library maintained by the author contains many examples of expressions that are less than prototypical, either weak in intensity or blended with other expression types. More than 250 facial expressions were coded using the Chimp-

FACS by colleagues at the University of Portsmouth, UK. These were also coded by the author for expression type, which resulted in eight categories: bared-teeth, play face, pant-hoot, whimper, neutral, pout, scream, and alert face. Around 20 expressions were placed into a ninth category: ambiguous. The action unit codes were then entered as dependent variables into a discriminant functions analysis (DFA) in which the a priori expression categories served as the grouping factor. The DFA provided the best fit for expression categories based on their shared configuration of action units. It also aided in identifying which configuration was most prototypical for each expression category (Parr et al. 2007). The results of this analysis can be seen in table 5.2. Most categories were validated by AU configurations, showing a high percentage of agreement between ChimpFACS coding and the author's subjective classifications. Several expressions, however, were not well classified, including the neutral face, alert face, and pout. The neutral-face pictures did contain some movements, and thus were not completely neutral. The pout expression was most often identified as the pant-hoot, as both those expressions share the puckered lip movement (code AU22, lip funneler). The alert face had previously been undescribed, and may represent a new category of chimpanzee expression (Parr et al. 2007). More important for studies of facial expression perception was the fact that while expressions often shared individual component movements, as in the AU22 shared by the pout and pant-hoot, each facial-expression category was found to have a unique overall configuration of movements.

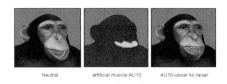

Figure 5.3 Process used to create custom facial animations in Poser 6.0 (efrontiers.com). The left figure shows an open mouth neutral face. The middle figure shows the location of the artificial muscle structure overlaid on a wire grid. This simulated "muscle" is the levator labii superioris, which raises the upper lip, and is Action Unit 10. The right figure show this animated movement on the Poser model.

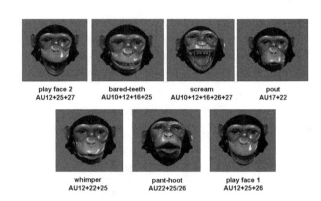

Figure 5.4 Prototypical facial expression configurations animated on a three-dimensional chimpanzee model using Poser 6.0 software, and their action unit codes.

The next project was to try and animate these movements on a virtual model chimpanzee, thereby creating a standardized stimulus and eliminating the problems in acquiring prototypical photographs of chimpanzees from ongoing behavior. A three-dimensional model chimpanzee, for example, could be posed to show prototypical configurations, ambiguous configurations, and even individual facial movements (action units) while at the same time controlling for stimulus size, individual identity, and head and gaze orientation. This model was created using Poser 6.0 digital character animation software (produced by efrontiers.com). Poser 6.0 enables the creation of custom facial morphs using a process whereby the region to be animated is first identified by highlighting specific grids on the face, analogous to overlaying an artificial muscle structure on the skull (figure 5.3). Using these procedures, my colleagues and I created custom morphs for each individual ChimpFACS AU, and then posed the chimpanzee model in each of the prototypical facial expression configurations identified using the DFA, as described above. These prototypical expressions can be seen in figure 5.4 along with their action unit configurations.

Table 5.2 The results of a discriminant functions analysis showing the percentage of agreement between shared action unit configurations and a priori classifications of chimpanzee facial expressions made by the author.

	bt	pf	ho	am	N	sc	al	po	wh
bared-teeth, bt	70.7	17.1	0	2.4	0	9.8	0	0	0
play face, pf	0	87.1	0	0	0	6.5	6.5	0	0
pant-hoot, ho	0	0	94.6	0	0	0	0	5.4	
ambiguous, am	16.7	8.3	20.8	16.7	8.3	8.3	20.8	0	0
neutral, n	0	0	0	6.7	40.0	0	40.0	13.3	0
scream, sc	0	2.9	0	0	0	97.1	0	0	0
alert, al	5.6	16.7	22.2	0	5.6	0	50.0	0	0
pout, po	0	0	72.4	0	0	0	0	24.1	3.4
whimper, wh	9.1	0	18.2	0	9.1	0	0	0	63.6

Facial Expression Categorization by Chimpanzees

Over the last decade my colleagues and I have conducted computer-based studies of social cognition in chimpanzees at the Yerkes National Primate Research Center in Atlanta, Georgia. These studies have focused primarily on facial information important for discriminations based on individual identity (Parr 2008). The main format for these tasks involves a computerized testing system and matching-to-sample (MTS) format. In brief, subjects are given access to a computer, either with a joystick interface in their home cages or with a touchscreen in a dedicated testing room. They are first presented with a sample stimulus, the image to match. They either contact this with the joystick-controlled cursor or touch it on the touchscreen. This action clears the screen, and two comparison images are then presented. One of these images matches the sample in that it is either an identical match, an image of the same general category, or a nonmatching (foil) image. To make a correct response the subject must move the cursor or touch the image that looks most like the sample. The subject is then given a food reward for a correct response, and proceeds to the next trial. Incorrect answers are not food rewarded, and are typically followed by a brief delay before the next trial.

Almost a decade ago, we conducted the first experiment on facial expression categorization by chimpanzees (Parr et al. 1998). This used the MTS format and pictures taken from natural behavior to examine whether chimpanzees were able to categorize their facial expressions and what types of facial features were important. The first task presented two different photographs from the same expression category as the matching pair, and a neutral portrait as the foil image. All three photographs showed different individuals so that individual identity could not guide the discrimination (Parr et al. 1998). Initial results showed accurate matching of faces based on expression type, particularly bared-teeth, scream, and play faces. The expression that required the greatest number of trials to discriminate was the pant-hoot. This is primarily an auditory signal used to communicate over long distances (Mitani et al. 1996), and subjects appear to pay selective attention to its auditory component when the signal is presented in a multimodal format (Parr 2004). In a subsequent task, each expression

was paired with every other expression category, so that every dyadic combination of expressions was presented. We then correlated performance on each dyad with the number of features the two expressions had in common: teeth visible, mouth open, eyes staring, etc. The hypothesis was that if specific facial features were important for expression categorization, then performance should be best for dyads that contained little feature overlap compared to dyads in which the expressions shared many features in common. This hypothesis was supported, but only for some expression types, indicating that the subjects categorized facial expressions using more than simple distinctive features.

Because this initial experiment used naturalistic stimuli, it was possible that some images showed expressions that were more prototypical than others, or that some individuals' facial features were more distinctive than those of others. Therefore, it was the first goal of the studies using the Poser stimuli (see figure 5.4) to replicate these initial findings in the same group of chimpanzees. Such a replication would, in part, validate the use of the animation faces for studies on primate social cognition. Thus, using the Poser stimuli, every dyadic combination of the following expressions was presented to determine the role of distinctive component features versus overall configuration: bared-teeth (BT), pant-hoot (HO), play face 1 (PF1), play face 2 (PF2), scream (SC), pout (PO), and whimper (WH). The previous discriminate function analysis identified two categories of play face that are distinct in terms of their facial features, so both of these were used (Parr et al. 2007). Also, the previous study (Parr et al. 1998) did not include the pout or whimper expressions because these occur infrequently and there were simply not enough photographs available for them to be included, so the use of the Poser stimuli advances this study to include these expressions.

The results of this initial discrimination experiment are described in detail elsewhere (Parr et al. 2008). In brief, only the scream expression was discriminated above chance on the very first presentation, but subjects exceeded chance on the other expression categories by the second testing session for all but the PF1 category. Because these expressions had been created using ChimpFACS, the number of features—that is, individual AU movements—that were similar in each expression dyad could be objectively quantified. Table 5.3 lists the percentage of shared AUs in each expression dyad. This measure of similarity was then correlated with performance on that dyad to

Table 5.3 Percentage of similarity in the individual component movements in each expression dyad coded by ChimpFACS.

	BT	HO	PO	PF1	PF2	SC	WH
Bared-teeth	x	0	0	25	25	75	20
Pant-hoot		x	50	33	25	0	33
Pout			x	0	0	0	50
Play face 1				x	33	20	33
Play face 2					x	50	33
Scream						x	25
Whimper							x

determine how subjects' performance would be affected by overlap in individual component features. If individual features are important, correlations between shared AUs and performance should be negative: the more features shared, the worse the performance. However, each overall facial expression configuration is unique. Therefore, if subjects discriminate expressions using their overall configuration, then the overlap in individual component movements should be minimal. These correlations were performed for each expression category and revealed weak negative correlations for whimper ($r = -0.27$), play face 2 ($r = -0.29$), and pout ($r = -0.41$). Positive correlations were observed for bared-teeth ($r = 0.73$) and pant-hoot ($r = 0.47$). Only very weak positive correlations were observed for play face 1 ($r = 0.12$) and scream ($r = 0.07$). Therefore, single features may be more important in discriminating pouts and whimpers, and the high intensity form of the play face, but they had little to no influence in discriminating screams or the low-intensity form of the play face. Performance in discriminating bared-teeth displays and pant-hoot expressions seemed to be enhanced when the individual features of sample and foil were similar.

These results using the standardized Poser expressions largely replicate our earlier findings (Parr et al. 1998), although the distinctive features in that previous study were identified subjectively. The take-home message from both these studies is that chimpanzees appear to use a combination of distinctive features and configuration when categorizing their facial expressions, and their strategies appear to differ depending on the expression type. Highly salient are the bared-teeth and scream faces (see also Parr 2004), and most difficult are pouts and whimpers. These results are also consistent with findings from humans that implicate an overall bias for configural cues in facial expression processing (Calder et al. 2000), but also an important role for distinctive features in influencing that process. Researchers have shown that a single facial feature—for example, a raised brow—is unable to convey emotion accurately on its own, but the inclusion of an individual feature within an existing configuration can bias the emotional interpretation of the face (Wallbott and Ricci-Bitti 1993). Adding or subtracting non-prototypical movements has not yet been attempted with the Poser stimuli, but it will be one avenue of continued research.

Implications and Future Directions

To conclude, facial expressions are a primary means of visual communication among many species of primates. They may either be conserved across phylogeny in both form and function, they may have a similar form but fulfill different functions depending on the socioecological niche of the species, or they may have unique forms in some species. Although the recognition of facial expressions is an essential means of social communication, very little is known about how nonhuman primates process facial expressions, with how much importance they view distinctive features as opposed to configuration, or whether they process expressions categorically as humans do (Young et al. 1997). This is perhaps due to the difficulty in acquiring high-quality stimuli for conducting these studies, particularly those addressing categorical perception, as human studies typically morph expressions from two categories together, which requires highly standardized images from the same individual. My colleagues and I overcame this problem by creating a three-dimensional chimpanzee model that could be animated into anatomically correct and prototypical facial expression configurations. Using these well controlled, posable stimuli we found that chimpanzees use a combination of overall configuration and distinctive features to discriminate among their major facial expression categories, and that combination appears to differ depending on the expression type. Similar findings have been reported in humans, where expression discrimination appears to be highly sensitive to overall configuration but the manipulation of single component movements within a configuration can alter the emotional interpretation.

Much remains to be learned about how facial expressions are recognized and interpreted within a social context, both in humans and especially in other species. Only a few studies have addressed the social function of facial expressions in chimpanzees, and this work needs to be expanded before a more complete evolutionary picture will emerge. The benefit of introducing ChimpFACS into these studies is that it enables a comparison of the contextual differences in the meaning of facial expression with the actual forms of those expressions. Such microanalyses will enable a detailed assessment of form and function, similar to the advancements that rigorous acoustic analyses have been able to provide in identifying subtypes of vocalizations. Ultimately, these studies will help to give us a better understanding of the socio-emotional function of facial expressions in chimpanzees.

Literature Cited

Andrew, R. J. 1963. The origin and evolution of the calls and facial expressions of the primates. *Behaviour* 20:1–109.

Bolwig, N. 1978. Communicative signals and social behaviour of some African monkeys: A comparative study. *Primates* 19:61–99.

Burrows, A. M., B. Waller, L. A. Parr, and C. J. Bonar. 2006. Muscles of facial expression in the chimpanzee (*Pan troglodytes*): Descriptive, ecological and phylogenetic contexts. *Journal of Anatomy* 208:153–68.

Calder, A. J., A. W. Young, J. Keane, and M. Dean. 2000. Configural information in facial expression perception. *Journal of Experimental Psychology: Human Perception and Performance* 26:527–51.

De Marco, A., and E. Visalberghi. 2007. Facial displays in young tufted capuchin monkeys (*Cebus apella*): Appearance, meaning, context and target. *Folia Primatologica* 78:118–37.

De Waal, F. B. M. 1988. The communicative repertoire of captive bonobos (*Pan paniscus*), compared to that of chimpanzees. *Behaviour* 106:183–251.

De Waal, F. B. M., and L. M. Luttrell. 1985. The formal hierarchy of rhesus macaques: An investigation of the bared-teeth display. *American Journal of Primatology* 9:73–85.

Duchenne de Boulogne, G. B. 1862/1990. *The Mechanism of Human Facial Expression*. R.A. Cuthbertson, ed. New York: Cambridge University Press.

Ekman, P., and W. V. Friesen. 1978. *Facial Action Coding System*. California: Consulting Psychology Press.

Ekman, P., and E. Rosenberg, eds. 1997. *What the Face Reveals: Basic and Applied Studies of Spontaneous Expression Using the Facial Action Coding System (FACS)*. New York: Oxford University Press.

Flack, J., and F. B. M. de Waal. 2007. Context modulates signal meaning in primate communication. *Proceedings of the National Academy of Sciences* 104:1581–86.

Hinde, R. A., and T. E. Rowell. 1962. Communication by postures and facial expressions in the rhesus monkey (*Macaca mulatta*). *Proceedings of the Royal Society of London B* 138:1–21.

Huber, E. 1931. *The Evolution of Facial Musculature and Facial Expression*. Baltimore: Johns Hopkins Press.

Mitani, J. C., J. Gros-Louis, and J. M. Macedonia. 1996. Selection for acoustic individuality within the vocal repertoire of wild chimpanzees. *International Journal of Primatology* 17:569–81.

Parr, L. A. 2004. Perceptual biases for multimodal cues in chimpanzee affect recognition. *Animal Cognition* 7, 171–78.

———. 2008. The primate face as a source of information. In L. Squire, ed., *The New Encyclopedia of Neuroscience*. Oxford, UK: Elsevier.

Parr, L. A., M. Cohen, Mand F. B. M. de Waal. 2005. The influence of social context on the use of blended and graded facial displays in chimpanzees (*Pan troglodytes*). *International Journal of Primatology* 26:73–103.

Parr, L. A., M. Heintz, and U. Akamagwuna. 2006. Three studies of configural face processing by chimpanzees. *Brain and Cognition* 62:30–42.

Parr, L. A., B. M. Waller, and M. Heintz. 2008. Facial expression categorization by chimpanzees using standardized stimuli. *Emotion* 8:216–31.

Parr, L. A., W. D. Hopkins, and F. B. M. de Waal, 1998. The perception of facial expressions in chimpanzees (*Pan troglodytes*). *Evolution of Communication* 2:1–23.

Parr, L. A., B. M. Waller, S. J. Vick, and K. A. Bard. 2007. Classifying chimpanzee facial expressions using muscle action. *Emotion* 7:172–81.

Parr, L. A., J. T. Winslow, W. D. Hopkins, and F. B. M. de Waal. 2000. Recognizing facial cues: Individual recognition in chimpanzees (*Pan troglodytes*) and rhesus monkeys (*Macaca mulatta*). *Journal of Comparative Psychology* 114:47–60.

Petit, O., and B. Thierry. 1992. Affiliative function of the silent bared-teeth display in moor macaques (*Macaca maurus*): Further evidence for the particular status of Sulawesi macaques. *International Journal of Primatology* 13:97–105.

Preuschoft, S., and J. A. R. A. M. van Hooff. 1997. The social function of "smile" and "laughter": Variations across primate species and societies. In U. Segerstrale and P. Mobias, eds., *Nonverbal Communication: Where Nature Meets Culture*, 252–81. New Jersey: Erlbaum.

Slocombe, K., and K. Zuberbuhler. 2005. Agonistic screams in wild chimpanzees (*Pan troglodytes schweinfurthii*) vary as a function of social role. *Journal of Comparative Psychology* 119:67–77.

Van Hooff, J. A. R. A. M. 1962. Facial expressions in higher primates. *Symposia of the Zoological Society of London* 8:97–125.

———. 1967. The facial displays of the Catarrhine monkeys and apes. In D. Morris, ed., *Primate Ethology*, 7–68. Chicago: Aldine.

———. 1973. A structural analysis of the social behavior of a semi-captive group of chimpanzees. In M. von Cranach and C. I. Vine, eds., *Social Communication and Movement*, 75–162. London: Academic Press.

Van Lawick-Goodall, J. 1968. A preliminary report on expressive movements and communication in the Gombe Stream chimpanzees. In P. C. Jay, ed., *Primates: Studies in Adaptation and Variability*, 313–519. New York: Holt, Rinehart, and Winston.

Vick, S. J., B. M. Waller, L. A., Parr, M. Pasqualini-Smith, and K. A. Bard. 2007. A cross species comparison of facial morphology and movement in humans and chimpanzees using FACS. *Journal of Nonverbal Behavior* 31:1–20.

Wallbott, H. G., and P. Ricci-Bitti. 1993. Decoders' processing of emotional facial expression—a top-down or bottom-up mechanism? *European Journal of Social Psychology* 23:427–43.

Waller, B., and R. I. M. Dunbar. 2005. Differential behavioural effects of silent bared teeth display and relaxed open mouth display in chimpanzees (*Pan troglodytes*). *Ethology* 111:129–42.

Waller, B. M., S. J. Vick, L. A. Parr, K. A. Bard, M. Pasqualini-Smith, K. M., Gothard, et al. 2006. Intramuscular electrical stimulation of facial muscles in humans and chimpanzees: Duchenne revisited and extended. *Emotion* 6:367–82.

Weigel, R. M. 1979. The facial expressions of the brown capuchin monkey (*Cebus apella*). *Behaviour* 68:250–76.

Young, A. W., D. Rowland, A. J. Calder, N. L. Etcoff, A. Seth, and D. I. Perrett. 1997. Facial expression megamix: Tests of dimensional and category accounts of emotion recognition. *Cognition* 63:271–313.

6

Behavioral and Brain Asymmetries in Chimpanzees

William D. Hopkins, Jared Taglialatela, David A. Leavens, Jamie L. Russell, and Steven J. Schapiro

Clutching a clipboard, the new lab assistant strides purposefully down the corridor of the great ape wing. As he walks past the row of enclosures, he engages in good-natured banter with the chimpanzees, tossing peanuts as he goes, stopping here and there to engage with some of the more friendly apes. But as he approaches one particular female chimpanzee's cage, his steps falter, his demeanor becomes more guarded and watchful. He slows to a near-crawl, his confident stride reduced to a slinking, furtive posture. And there she is, ever watchful, waiting for him.

Mega is a young adult female chimpanzee. She is not particularly large, nor is she dominant in her group. But where other chimpanzees enjoy throwing the occasional malodorous fistful of feces at people, for Mega it seems a divine calling. With a fistful of excrement clutched in her left hand, Mega becomes a fearsome and terrible menace with the strength of a boxer and the accuracy of a professional baseball pitcher. She even hoards her ammunition, protecting it against the cage-cleaning efforts of the husbandry staff. Nobody here has escaped her prodigious talent. As the lab assistant approaches, she raises her left hand in the air, oscillating with quiet menace. The lab assistant decides he'll make a run for it. He raises his clipboard as a shield and begins to sprint down the corridor, the tails of his pristine white lab coat flapping behind him. With his first step Mega holds her position, waiting for the exact millisecond of opportunity when the V-neck of the human's surgical gown flashes vulnerably open. And by the *second step in his desperate run, she has released her fetid missile, unerringly finding her mark. As the soiled lab assistant trudges past in defeat, he wonders: Why does she always use her left hand when so many of the others use their right?*

Hemispheric specialization refers to perceptual, motor, and cognitive processes that are differentially performed by the left and right cerebral hemispheres. Arguably, the two most robust manifestations of hemispheric specialization in humans are handedness and language. Humans are universally right-handed, although there is some variation between cultures (McManus 2002; Perelle and Ehrman 1994; Porac and Coren 1981; Raymond and Pontier 2004). Archeological data suggest that handedness was present at least two million years ago, as is manifest in depictions from cave art or from examination of the flints from early stone tools (Porac and Coren 1981).

Additionally, most individuals, particularly right-handed individuals, are left-hemisphere dominant for language and speech (Beaton 1997). For example, using both the sodium amytal test (a procedure that temporarily disrupts functions of one half of the brain) as well as more recent Doppler sonography techniques (which measure blood flow into the brains via the carotid artery), it has been found that approximately 96% of right-handed individuals are left-hemisphere dominant for speech in contrast to only 70% of left-handed individuals (Knecht et al. 2000; Rasmussen and Milner 1977). There is also a significant body of evidence showing neuroanatomical differences between the left and right cerebral hemispheres as a function of handedness (Foundas et al. 1995; Foundas et al. 2002) as well as speech production and comprehension (Cooper 2006).

The observed association between handedness, language, and speech in relation to neuroanatomical and neurophysiological asymmetries has stimulated a significant amount of scientific debate over the evolution of hemispheric specialization in relation to handedness and language. Specifically, some have argued that the existence of population-level handedness may have set the stage for the subsequent evolution of left hemisphere specialization in language processing (Bradshaw and Rogers 1993; Corballis 1992). From this perspective, some behaviors in early hominids as well as in our closest living relative, the chimpanzee, may have served as preadaptations for lateralization in language. These behaviors include throwing (Calvin 1983), bimanual feeding (Byrne and Byrne 1991; Hopkins 1994), bipedalism, and tool use (Gibson and Ingold 1993; Greenfield 1991). In contrast, there are others who have suggested that population-level handedness is a consequence of an extant left-hemisphere specialization for the motor control of spoken language (Annett 1985; Annett 2002; Warren 1980) or manual gestures (Corballis 2002). From these theoretical perspectives, right-handedness as a trait is a by-product of early asymmetries in communication in either the vocal or gestural modality. Finally, some have suggested that language lateralization and right-handedness in modern humans may have been a consequence of existing asymmetries in the temporal processing of species-specific signaling, including those forms present in other species, such as vocalization (Bradshaw 1997).

Which of these evolutionary theories best explains the extant data remains unclear, but in all of these different scenarios there are three consistent, underlying themes: (a) the evolution of manual motor skill is fundamentally related to the evolution of oro-facial musculature control potentially related to speech, (b) common neural or cognitive mechanisms might underlie individual and species differences in motor control, and (c) motor control for manual actions and speech are lateralized principally to the left hemisphere.

For the past 15 years we have been studying the relationship between manual skill, communication and organization of the central nervous system with specific reference to lateralization (Hopkins et al. 2007b). Initially, the focus was on determining whether chimpanzees show population-level behavioral and neuroanatomical asymmetries, a trait many have historically argued to be uniquely human (Crow 2004; Warren 1980). We have primarily focused on handedness but other measures of behavioral asymmetries have been quantified, such as oro-facial asymmetries. Hand-use preference data have been collected in chimpanzees from two different facilities, the Yerkes National Primate Research Center (Atlanta, Georgia) and the M. D. Anderson Cancer Center (Bastrop, Texas). For some measures, data have also been collected from the chimpanzees housed at the Alamagordo Primate Facility (Alamagordo, New Mexico). Subsequently, we have been interested in the relationship between lateralization in communicative behaviors and noncommunicative behaviors and how this relationship might contribute to the understanding of evolutionary factors that have shaped the development of lateralization in humans. Lastly, our most recent work has sought to examine the neuroanatomical correlates of lateralized communicative and noncommunicative motor systems using noninvasive imaging technologies including magnetic resonance imaging (MRI) and positron emission tomography (PET). The aim of this chapter is to highlight our findings on neural correlates of communicative and noncommunicative motor functions in chimpanzees.

We begin by providing some descriptive data on behavioral lateralization for communicative and noncommunicative motor functions in chimpanzees. Starting with a discussion of behavioral studies of hemispheric specialization, we focus primarily on handedness, tool use, and aimed throwing. Subsequently, we examine the relationship between oro-facial motor control and handedness. We then shift focus to descriptions of our various brain-imaging studies and results. Lastly, we present data on neuroanatomical correlates of asymmetries in handedness.

Behavioral Studies of Hemispheric Specialization

Handedness

To measure handedness we have used a number of different tasks with varying degrees of complexity, many of which have been described elsewhere (Hopkins 2007). For simple reaching, on each trial a raisin was thrown into the subject's home cage. The chimpanzee had to locomote to get to the raisin, pick it up, and bring it to his or her mouth for consumption. The experimenter recorded the hand used as left or right. For bimanual feeding (figures 6.7b and 6.7c), the chimpanzees were given a cache of food during enrichment activities or during regular feeding times. Often they hold the food with one hand and consume it with the opposite; during these occasions we recorded whether they used their right or left hands to consume the food. Hand use for coordinated bimanual actions was assessed using a measure called the TUBE task (figure 6.7a). For this task, peanut butter is smeared on the inside edges of the ends of polyvinyl chloride (PVC) pipes. The peanut butter is placed far enough down inside the tube so that the subjects cannot lick the contents off with their mouths but rather must use one hand to hold the tube and the other to remove the substrate. The hand used to extract the peanut butter was recorded as either right or left by the experimenter. In the test for manual gestures (figure 6.8b), an experimenter would approach the chimpanzee's home cage and center themselves in front of the chimpanzee at a distance of approximately 1.0 to 1.5 m. If the chimpanzee was not already positioned in front of the experimenter at the onset of the trial, he or she would immediately move towards the front of the cage when the experimenter arrived with food. The experimenter then called the chimpanzee's name and offered a piece of the food until the chimpanzee produced a unimanual gesture. Hand use was recorded as being right or left.

With specific reference to tool use, we have developed several devices for measuring hand use for probing actions in captive chimpanzees that have been designed to mimic the motor demands of termite or ant fishing observed in wild chimpanzees. In these tasks, the apes insert sticks into baited PVC pipes attached to their home cages (figure 6.8a) and the experimenters record hand use as being left or right. Throwing is another form of tool use that has been recorded in the chimpanzees, but it is very situational; thus, data collection has not occurred systematically but rather opportunistically. When apes have been observed to throw, observers have recorded their identity, sex, hand use, posture, and whether they threw overhand or underhand.

During handedness testing, we record the frequency in

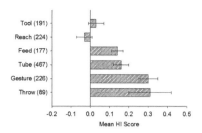

Figure 6.1 Mean handedness index (HI) (+/−s.e.) for the different measures of hand use in captive chimpanzees. HI values were derived for each subjects following the formula [HI = (#R−#L) / (#R + #L)] where #R and #L refer to the number of left- and right-hand responses. Numbers in parentheses indicate the number of chimpanzees tested on that specific task.

left- and right-hand use during each task and derive a handedness index (HI) following the formula [HI = (#R−#L) / (#R + #L)]. Positive HI values indicate right-hand preference, while negative values indicate left-hand preference. The absolute value of the HI reflects the strength of hand preference. Shown in figure 6.1 are the mean HI values for each of the behaviors described so far. Two observations are noteworthy. First, when considering each task, the chimpanzees display varying degrees of right-handedness, with manual gestures and throwing being the most robust while simple reaching and tool use are least robust. Thus, chimpanzee handedness is task-specific. Second, when handedness is averaged across measures, the chimpanzees show a population-level right-hand bias.

Handedness and Tool Use

The issue of handedness and tool use merits some additional analysis because it has been hypothesized to be a potential preadaptation for the evolution of lateralization in motor functions, including those associated with speech and language. Additionally, handedness for tool use has by far been the most frequently studied lateralized behavior in wild chimpanzees, and therefore provides a framework for comparison to our studies in captive chimpanzees. From figure 6.1 it can be seen that the results for handedness and tool use are quite disparate in our captive sample of chimpanzees. Throwing is one of the most pronounced manifestations of right-handedness in our sample, while a task designed to simulate termite fishing yields no evidence of population-level handedness. Handedness for throwing has not been reported in wild apes, but the lack of evidence for population-level hand use for the simulated termite-fishing task stands in strong contrast to reports of population-level left-hand use in wild chimpanzees at Gombe for termite fishing (Lonsdorf and Hopkins 2005). Of the 71 chimpanzees assessed for handedness in termite fishing at Gombe, about two-thirds exhibit a significant left-hand preference while the remaining ones exhibit a right- or no-hand bias. Thus, differences in hand preference between wild apes and our captive apes are evident for this particular task, however, and the reason for the discrepancy in results is not immediately apparent. Studies of hand use for simulated termite fishing in chimpanzees housed in other captive settings have been more consistent with those reported in wild chimpanzees. This leads to the suggestion that either the apes at Yerkes are quite different from other apes in their behavioral laterality, or that aspects of our simulated termite-fishing task are sufficiently different from these types of tasks in other captive settings and in the wild that, in essence, we are not tapping into the same motor and cognitive processes.

To test this latter hypothesis, we retested 66 chimpanzees at the Yerkes Primate Center on a simulated termite-fishing task, but increased its motor and spatial demands by decreasing the size of the hole into which the apes had to dip their sticks. In our original task the hole had been 2.5 cm in diameter; in our revised task we reduced it to 1 cm. The chimpanzees showed a borderline significant left-hand preference for this version of the simulated termite-fishing task $t(54) = -1.69, p < .10$ and these results were particularly strong for the males (*mean* HI $= -.18$) compared to the females (*mean* HI $= -.08$).

Handedness and Tool Use in Wild Chimpanzees

The evidence of population-level left-handedness for termite fishing in wild chimpanzees differs from other reports of handedness in wild settings, particularly for ant dipping, nut cracking, and leaf sponging. Shown in table 6.1 is a summary of the distribution of handedness for tool use in wild chimpanzees. Three general findings emerge. First, chimpanzees show population-level left-handedness for termite fishing and population-level right-handedness for leaf sponging. Wild chimpanzees show a borderline significant right-hand bias for nut cracking and ant dipping. Second, the results from wild chimpanzees are relatively consistent between field sites. For example, for both nut cracking and leaf sponging, the handedness distributions are relatively consistent between Bossou and

Table 6.1   Handedness data on tool use in wild and captive chimpanzees.

	#L	#A	#R
Behavior			
Termite fishing (McGrew and Marchant 1996)	29	10	15
Termite fishing (Lonsdorf and Hopkins 2005)	12	1	4
Total	41	11	19
Nut cracking (Boesch 1991)	36	3	46
Nut cracking (Biro et al. 2004)	7	2	11
Total	43	5	57
Leaf sponging (Boesch 1991)	5	2	10
Leaf sponging (Biro et al. 2006)	4	2	11
Total	9	4	21
Ant dipping (Marchant and McGrew 2007)	1	8	6
Algae dipping (Sugiyama et al. 1995)	2	2	4

NOTE: Data from Boesch (1991) include those reported for the infants. #L = number of left-handed apes; #R = number of right-handed apes; #A = number of ambiguously-handed apes.

the Taï Forest (Biro et al. 2003; Biro et al. 2006; Boesch, 1991; Sugiyama 1995; Sugiyama et al. 1993). Handedness data for termite fishing has only been reported at Gombe, but a significant number of individuals have been studied and the results have been consistent (Marchant and McGrew 2007; McGrew and Marchant 1992; McGrew and Marchant 1996). Finally, it is important to note that statistical power and sample size are important factors in the assessment of handedness in wild chimpanzees. Specifically, for termite fishing, leaf sponging, and nut cracking, analysis of the handedness results for any single study has not revealed evidence of population-level asymmetries; however, combining data across the sites does reveal population-level handedness. This is because the sample sizes are relatively small within a study, and the effect size for handedness in both captive and wild chimpanzees is relatively small—about a 2:1 ratio of dominant to nondominant handedness (Hopkins 2006). Combining the data across study sites results in an increase in sample size which in turn causes an increase in statistical power, thus allowing for statistical significance to be detected. These results are more consistent with the argument that differences between captive and wild chimpanzees are not due to the setting in which they are tested per se (captive versus wild), but are simply an artifact of differences in statistical power between studies of these different populations (Hopkins and Cantalupo 2005).

Aimed Throwing

As noted above, one of the most robust manifestations of handedness for tool use in captive chimpanzees is aimed throwing (Goodall 1986; Hopkins et al. 2005a; Marchant 1983). Anecdotal reports abound of apes in zoos and research facilities throwing feces or other materials at visitors or staff. Goodall (1986) reported that about one-third of the male chimpanzees at Gombe had been observed to engage in aimed-throwing. As far as we know, save a single report of the behavior in capuchin monkeys, there are virtually no reports of aimed throwing in other nonhuman primates (Westergaard et al. 2000).

Throwing in chimpanzees is an interesting behavior from psychological and neurophysiological perspectives. Psychologically it is interesting because the reinforcement that governs the acquisition and maintenance of the behavior is not nutritive. (Indeed, none of us have ever seen anyone give food to a chimpanzee for throwing feces at them!) In many other forms of tool use, chimpanzees seek to obtain food (nut kernels, termites, ants) or water. In contrast, throwing influences the behavior of the recipient (or target), which in turn seems to reinforce the thrower's own behavior. Thus, throwing behavior is maintained by the reinforcing reactions of the recipient. We would note that this is quite different from the report by Westergaard and colleagues (2000) in which capuchin monkeys tried to throw an object into a bucket containing peanut butter, and then had the object—with the peanut butter sticking to it—handed back to them if they were successful. Thus the capuchin could be trained to throw only through direct food reinforcement, which seems different from the manner in which the chimpanzees have acquired and maintained their throwing behavior.

Calvin (1983) has argued that the evolution of throwing would have been selected in primate evolution for several reasons, notably because large prey could be hunted for food with less risk to the individual hunter. Throwing clubs or spears would have allowed the hunters to maintain a greater distance from the large prey that could potentially have inflicted injury. In addition, selection for throwing in the neural organization of the cortex would have increased temporal and spatial summation of large cortical networks in order to coordinate the different muscles as well as calculate the "timing" or "launch window" for the object being thrown relative to the target's distance and movement.

Handedness in Other Apes and Monkeys

The focus of our work has largely been on chimpanzees, but data from other great apes as well as monkeys are available for some measures of handedness (Hopkins 2006; Papademetriou et al. 2005). In particular, data for the previously described TUBE task have been reported for a number of primate species. The mean HI for each species tested to date are presented in figure 6.2. Chimpanzees, gorillas, and baboons show population-level right-handedness while orangutans and De Brazza monkeys show population-level left-handedness. Rhesus monkeys and capuchin monkeys, when averaged across all studies, do not show a bias although some individual studies have reported evidence of population-level right-handedness in each species. The factors that might explain the differences between species are not obvious, however, and additional data would be very useful in trying to isolate the potential role of ecological or biomechanical factors in handedness.

In addition to the TUBE data, one of the more interesting comparative sets of data that has emerged from the literature is the distribution of hand preference for manual gestures in baboons compared to chimpanzees. Meguerditchian and Vauclair (2006) examined hand preference for a manual threat gesture in baboons that was directed toward both conspecifics and humans. For comparison to the gesture data, they also presented data on hand preferences for the TUBE task in their baboon sample. Shown in table 6.2 are the mean HI values for the TUBE task and manual gestures in the baboons. For comparison, we have presented data from chimpanzees. Both species show population-level right-handedness for both measures; in both species, however, the HI values are significantly

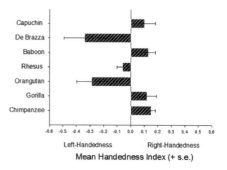

Figure 6.2 Mean handedness index (HI) (+ / − s.e.) for the different measures for primate species on the TUBE task. HI values were derived for each subject following the formula [HI = (#R − #L) / (#R + #L)] where #R and #L refer to the number of left- and right-hand responses.

Table 6.2 Mean HI (+ / − s.e.) for the TUBE task and manual gestures in captive chimpanzees and baboons.

	Gesture Mean HI	s.e.	TUBE Mean HI	s.e.
Baboons	.32	.09	.13	.06
Chimpanzees	.33	.06	.15	.04

higher for gestures than for the TUBE task. These results suggest that chimpanzees and baboons are more strongly lateralized for communicative signaling than for noncommunicative motor tasks, thus supporting the view that selection for gestural communication may have been important for the emergence of hemispheric specialization (see Bradshaw 1997; Corballis 2002).

Motor Control of Oro-Facial Movements and Their Relationship to Handedness

The Relationship between Manual Gestures and Vocal Production

One intriguing observation made during our initial studies on hand use and manual gestures was the concurrence of vocal signals with manual gestures in a subpopulation of the chimpanzees. Specifically, a number of recent studies in our laboratory and others have shown that captive chimpanzees produce manual gestures as well as some vocalizations and sounds to request foods in the presence but not the absence of an audience (e.g., human caretaker) (Hostetter et al. 2001; Leavens and Hopkins 1998; Leavens et al. 1996). Additional studies have shown that they will alter their modality of communication between manual gestures and vocal signals depending on the attentional status of a human (Hostetter et al. 2007; Leavens et al. 2004; Leavens et al. 2005b). When a human with food is oriented toward them or looking at them, they are more inclined to use a visual manual gesture to request the food. In contrast, when a human is oriented away from them or looking at a different individual in the group, the chimpanzees will use vocal signals significantly more often to capture the human's attention. Chimpanzees have also been shown to selectively produce one of two sounds, either a "raspberry" (RASP) or "extended food grunt" (EFG) to capture the attention of an otherwise inattentive human, and to suppress these sounds in the presence of the food or

human alone (Hopkins et al. 2007c). The RASP and EFG are interesting sounds from a phonological perspective because one is voiced (EFG) and one is not (RASP); their functional use indicates that they are under voluntary control, a finding that stands in strong contrast to the prevailing view that primate vocalizations and facial expressions are involuntarily produced, and are not intentional signals (Premack 2004; see also chapter 16).

Although we have not made a distinction regarding the types of sound produced in relation to manual gestures, we have found in both the Yerkes and Bastrop colonies an increased preferential use of the right hand for manual gestures associated with these sounds as compared to gestures made without vocalization (Hopkins et al. 2005b; see figure 6.3). These results suggest that vocal production may be lateralized in the left hemisphere in chimpanzees, and that this facilitates use of the right hand in gesturing.

This pattern of results is not restricted to within-subject variation in the production of manual gestures and either the EFG or RASP sound. It is important to note that not all chimpanzees reliably produce these sounds. Some produce them very reliably while others are less reliable or simply fail to use them at all. The origin of this difference remains unclear, but if a central–left-hemisphere lateralized motor system controls both manual and, potentially, oro-facial movements, then it seems possible that individual differences in the ability to learn to manipulate facial musculature to produce the RASP or EFG sound might also be associated with variation in motor skill and/or hand preference. To test this hypothesis, we compared the HI measures for several of the tasks described above (see section on "Handedness") in a sample of 121 chimpanzees including 52 males and 69 females. Each chimpanzee was classified as being a vocalizer or non-vocalizer based on whether they had been recorded producing an EFG or RASP during our studies (Hostetter et al. 2001; Leavens

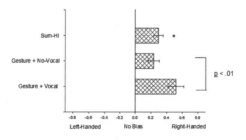

Figure 6.3 Mean handedness index for manual gestures while simultaneously (gesture + vocal) or not simultaneously (gesture + no vocal) producing either an EFG or RASP vocalization.

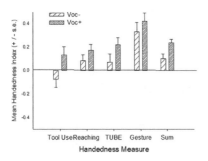

Figure 6.4   Mean HI (+ / − s.e.) values for four handedness tasks for chimpanzees that reliably produce EFG or RASP sounds (voc +) and those who do not (voc −).

and Hopkins 1998; Leavens et al. 2004). Classification of individuals into these groups resulted in 63 vocalizers and 58 non-vocalizers. A mixed-model analysis of variance was used to compare the HI values for each task as a function of sex and the production of attention-getting vocalizations. A significant effect for vocal production was found: $F_{(1, 116)} = 7.414, p < .03$. Chimpanzees that reliably produced attention-getting sounds were significantly more right-handed than those that did not, and these findings were consistent across all four handedness tasks (see figure 6.4).

Oro-Facial Movements

The previous section outlined results suggesting that there are centrally controlled lateralized processes that influence both hand and oro-facial movements. In light of the fact that hand use, particularly for communicative behaviors, is controlled by the left hemisphere, it seems reasonable to ask whether a similar left-hemisphere asymmetry might be present in the control of oro-facial movements, specifically during vocal production. Assessing asymmetries in vocal communication is far more difficult in primates than assessing them in other motor functions such as hand use, but recent studies have attempted to address this area by measuring oro-facial asymmetries when primates make different facial expressions that may or may not have an accompanying vocalization (Fernandez-Carriba et al. 2002; Hauser 1993; Hook-Costigan and Rogers 1998). In these studies, the animals are filmed during social interactions and video analysis allows the scientist to measure which side of the face moves first (Hauser, 1993) or which half of the face expresses an emotion more intensely than the other (Fernandez-Carriba et al. 2002; Hook-Costigan and Rogers 1998).

In previous work in our laboratory, we have reported a left-hemiface bias for several facial expressions of chimpanzees including the scream, play face, hoot, and silent–bared teeth (Fernandez-Carriba et al. 2002). More recently we have examined oro-facial asymmetries in the EFG and RASP sounds ($n = 42$) as well as two more typical chimpanzee vocalizations, the hoot ($n = 22$) and food call ($n = 24$) (Losin et al. 2008) in order to assess more directly the lateralized motor control of the EFG and RASP sounds. For this study, digital video was recorded of the chimpanzees. Video recordings of pant-hoot expressions were made ad libitum in the context of the chimpanzees' normal social interactions. Usable still images of pant-hoot expressions were captured primarily during display behavior because it was most likely that chimpanzees would be facing the experimenter in this context (see Goodall 1986). The hoot, food bark, and raspberry were filmed during the presentation of a food item by a research assistant, and specific facial expressions were selected during frame-by-frame video analysis. Consistent with previous studies in chimpanzees and other nonhuman primates (Fernandez-Carriba et al. 2002; Hauser 1993; Hook-Costigan and Rogers 1998), strict criteria were used to select still images from videotaped sequences. When expressions occurred in bouts, the clearest and most frontal was selected and the frame depicting the point of greatest exaggeration was captured for analysis. For instance, during the food bark and pant-hoot this point occurred when the mouth was open the widest, and during the raspberry it occurred when the bottom lip was extended farthest and the sound was produced.

All images were analyzed using the measurement procedure pioneered by Hook-Costigan and Rogers (1998) and later used by Fernandez-Carriba et al. (2002) in chimpanzees. In this procedure the images were rotated into a symmetrical frame; a line was drawn between the inner corners of the eyes and compared to the horizontal lines on a fixed grid in Adobe Photoshop. Next, the midpoint of the line between the inner eye corners was calculated, and a perpendicular line was drawn at this point to bisect the face.

Once all the lines were drawn, the distance between the outer eye corners and mouth corners to the midline was calculated and each hemi-mouth area was measured for each expression (figure 6.5). Facial asymmetry indices (FAIs) were calculated for the distances to the outer eye corners (eye FAIs) and the mouth area (mouth-area FAIs) by subtracting the left side from the right and dividing that value by the sum of the right- and left-side measurements.

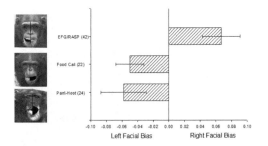

Figure 6.5 Mean facial asymmetry index (FAI) (+ / − s.e.) for three classes of facial expressions that have vocalizations accompanying their production. EFG/RASP = extended food grunts or raspberries. Each image on left panel shows the expression with the lines drawn and areas measured from one side of the face (see text for description). Numbers in parentheses indicate the number of subjects for which data were available.

Negative FAI values indicate leftward asymmetry and positive FAI values indicate rightward asymmetry.

The mouth area FAIs were adjusted for possible asymmetries in the image due to rotation of the face relative to the camera or asymmetries inherent in the subject's face by subtracting the eye FAIs of each image from its mouth-area FAIs. These adjusted FAI values for mouth area were then used in the remaining analyses. The mean FAI values for each expression type are shown in figure 6.5. Significant population left-hemiface biases were found for hoot and food calls. In contrast, population-level rightward asymmetries were found for the RASP and, to a lesser extent, for the EFG sounds. Thus, it seems that oro-facial asymmetries associated with the production of certain sounds and vocalizations are present, and vary according to their different functions.

Structural Magnetic Resonance Imaging the Chimpanzee Brain

Despite the longstanding interest in the evolution of neural substrates associated with lateralization in hand use and communication in nonhuman primates, there has been little or no direct empirical research on this topic. One limitation has no doubt been the lack of available brain tissue to correlate with specific lateralized behavioral phenotypes, such as tool use or gestural communication, in chimpanzees. Moreover, experimental euthanasia or invasive lesion studies in chimpanzees would not be ethical, particularly in modern times. However, over the past 10 years we have been able to capitalize on the use of non-invasive in vivo imaging technologies to examine whether chimpanzees and other great apes show population-level neuroanatomical asymmetries, and whether these asym-

metries are associated with functional asymmetries such as handedness for tool use. Because we can collect neurological data in living apes, the correlation of anatomy and behavior in relatively large samples of apes is feasible and can be performed with minimal risks to the apes.

To date, we have collected structural magnetic resonance images (MRI) in a sample of 88 chimpanzees housed at Yerkes. Some of the brains have been imaged post mortem when the apes died from natural causes (n = 24) but the majority of scans have been collected in vivo (n = 64). Once the MRI images are acquired, multiplanar reformatting software (ANALYZE 6.0 or 7.0) is used to align the brain and virtually slice it in different planes (coronal, sagittal, or axial) depending on the region of interest. To date, from the MRI scans, volumes of the left and right cerebral hemispheres have been measured for a variety of brain regions including the amygdala and hippocampus, ventricles, motor-hand area of the precentral gyrus, inferior frontal gyrus, planum temporale, inferior parietal lobe, and planum parietale (Cantalupo and Hopkins 2001; Cantalupo et al. 2003; Freeman et al. 2004; Hopkins and Pilcher 2001; Hopkins et al. 2000; Taglialatela et al. 2007). In addition to the specific regions of interest, other measures of brain asymmetry and cerebral organization have been obtained in the apes, including torque asymmetries in the cerebrum and cerebellum as well as midsagittal area measures of the corpus callosum (Cantalupo et al. 2008; Dunham and Hopkins 2006; Pilcher et al. 2001). Here we focus on three brain regions including the motor-hand area of the precentral gyrus (KNOB), inferior frontal gyrus (IFG), and planum temporale (PT). A three-dimensional reconstructed chimpanzee brain with the three main cortical regions of interest outlined is shown in figure 6.6.

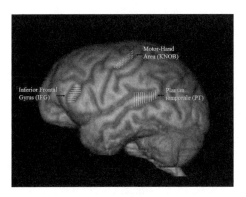

Figure 6.6 Three-dimensional reconstruction of a chimpanzee brain from a magnetic resonance image scan. Different hatched regions reflect regions of interest.

Table 6.3   Distribution of brain asymmetries in three areas of the chimpanzee brain.

	#L	#R	#NB	Mean AQ			
					s.e.	t.	p.
Inferior frontal gyrus (IFG)	51	26	11	−.073	.026	−2.81	.006
Planum temporale (PT)	60	13	12	−.107	.021	−4.99	.001
Motor-hand area (KNOB)	51	35	5	−.001	.051	−0.03	.975

L = number of left lateralized apes; #R = number of right lateralized apes; #NB = number of apes with no bias.

For each region, asymmetry quotients are derived following the formula $[(AQ = (R - L) / (R + L)]$ where R and L refer to the volumes or area measures of the right and left hemispheres respectively.

Shown in table 6.3 are the mean AQ values for each region as well as the distribution of apes classified as being left-, right- or non-dominant. Chimpanzees show a population-level leftward asymmetry for the PT and IFG, but no bias for the motor-hand area of the KNOB. The evidence of leftward asymmetries in the PT and IFG are of particular interest because these brain regions are considered the homologues to Wernicke's and Broca's areas, important cortical regions involved in the comprehension and production of language in the human brain. The evidence of leftward asymmetries in the chimpanzee PT is not restricted to our laboratory. At least two other laboratories have also reported such evidence (Gannon et al. 1998; Gilissen 2001), thus suggesting that the results are consistent across different labs employing different methods of measurement. In short, apes show population-level leftward asymmetries in PT and IFG, regions considered the homologues to the human language brain regions.

Neuroanatomical Correlates of Lateralized Hand Use in Captive Chimpanzees

We have provided descriptive information on behavioral and brain asymmetries in chimpanzees. The natural question that arises from these results is whether variation in behavioral asymmetries correlates with lateralization in any of the neuroanatomical regions that have been described. The general results of our analyses of the associations between hand use and brain asymmetries are summarized in figures 6.7 and 6.8. When considering manual actions such as simple reaching, bimanual feeding, or coordinated bimanual actions, the best and only significant neuroana-

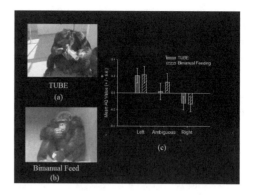

Figure 6.7   (a) Chimpanzees engaged in a task (referred to as the TUBE task) designed to measure coordinated bimanual actions; (b) chimpanzee exhibiting bimanual feeding; (c) mean AQ values for the motor-hand area (or KNOB) as a function of handedness for the TUBE and bimanual feeding tasks.

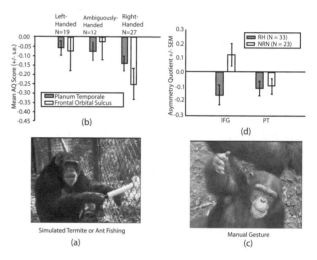

Figure 6.8   (a) Chimpanzees engage in the simulated termite fishing task; (b) mean AQ values for the planum temporale and inferior frontal gyrus in left-handed, ambiguously-handed, and right-handed chimpanzees classified on the basis of their hand use in the tool-use task (data taken from Hopkins et al. 2007); (c) chimpanzees manually gesture for food from a human; (d) mean AQ values for the inferior frontal gyrus (IFG) and planum temporale (PT) for right- and non-right-handed chimpanzees when classified for hand use during gesturing (data from Taglialatela et al. 2006).

tomical correlate is the motor-hand area of the precentral gyrus (KNOB) (Hopkins and Cantalupo 2004). Right-handed chimpanzees have a larger KNOB in the left hemisphere, whereas left-handed chimpanzees have a larger one in the right hemisphere. No significant differences in asymmetry of the KNOB were found in ambiguously-handed chimpanzees. With respect to handedness for the simulated termite-fishing task and manual gestures, the single best correlate is the inferior frontal gyrus (Hopkins et al. 2007a; Taglialatela et al. 2006). Right-handed chimpanzees have a larger left IFG compared to non-right-handed individuals for both tool use (the simulated termite-fishing task) and manual gestures.

Discussion

The results of our longitudinal studies thus far can be summarized as follows. First, captive chimpanzees show population-level asymmetries in communicative signaling, notably a left-hemisphere asymmetry in manual gesture and in oro-facial movements associated with the production of "raspberries" and extended food grunts (but not other calls). Right-hemisphere biases in oro-facial asymmetries were present for play and silent-bared-teeth expressions and for hoots, screams, and food calls. Second, analyses of MRI scans have revealed population-level asymmetries in several brain regions and particularly in the homologues to the human cortical language centers, including the frontal operculum and planum temporale. Third, while hand preferences in chimpanzees correlate with asymmetries in their brains, different tasks correlate with different brain regions, thus revealing a more complicated picture than has been reported in humans.

Motor Control of the Hand, Mouth, and Face

The evidence we have presented here suggests a link between motor control of the hand and mouth in chimpanzees. Specifically, we have found that chimpanzees that produce the EFG and RASP sounds show a rightward oro-facial asymmetry and that the use of the right hand for manual gestures is enhanced in apes that simultaneously produce EFG or RASP sounds. Additionally, chimpanzees that reliably produce EFG or RASP vocalizations are significantly more right-handed in several motor tasks, reinforcing the view that the left hemisphere may be involved in both the motor control of manual and oro-facial movements.

Generally speaking, there has been a tendency to consider the evolution of language and speech as separate, competing views with little overlap. For instance, there have been traditional linguistic theories of language and speech that question the role of gestures in language evolution, yet primates gesture preferentially with the right hand—as do humans, particularly while they speak—and this has not been not adequately explained in any purely linguistic evolutionary framework. Gestural origin theories of language fit more neatly with existing behavioral data from human and nonhuman primates, yet these approaches have difficulty explaining different phonological

phenomena unique to speech, and the exact stages that took place in the transformation from gesture to speech are unclear (Arbib 2005). McNeil (1992) has argued that perhaps it might be better to consider the notion that speech and gesture evolved together rather than independently, and our results are somewhat consistent with this view. We believe our data fit most closely with this perspective because of the following considerations. First, captive chimpanzees functionally use their manual gestures and some vocal signals to indicate a request, but can modify the modality of their signaling depending on the attentional status of the receiver. This implies that chimpanzees have at least some degree of choice over the decision to emit a vocalization, and suggests that our common ancestors were therefore preadapted for intentional communication in both manual and vocal modes. Second, lateralization in manual gestures is enhanced when accompanied by a vocalization. These findings suggest that a common neural substrate rooted in the left hemisphere may explain the evolution of both manual and vocal communication. If this is true, then speculation about the adaptive origins of these specific behavioral and anatomical asymmetries must shift from the Pleistocene (12,000 to 1.8 million years ago) or Plio-pleistocene (1.5 to 2.5 million years ago) epochs to at least Miocene paleoecological contexts (in excess of 8 million years ago). Thus, with respect to the evolution of language, we believe that these findings warrant a shift in focus away from human- or hominid-unique adaptations to the shared adaptive traits of hominoids.

Cognitive Parallels in Tool Use and Gesture

The results of our studies indicate that hand preferences in both manual gestures and tool use correlate with the homologues to the human cortical language areas, notably the frontal operculum. Why would these two apparently motorically different tasks correlate with a common brain area? One argument might be that the evidence of an association between asymmetries in the inferior frontal gyrus and both tool use and manual gestures may reflect similar cognitive mechanisms used in the execution of these behaviors, despite the obvious differences in motor function.

As we have argued elsewhere (Leavens et al. 2005a), the use of manual gestures by captive chimpanzees is essentially a form of social tool use. It is observed primarily when a chimpanzee attempts to draw a human's attention toward an otherwise unattainable food item. The chimpanzee must therefore engage in communicative behavior to obtain the food. Material tool use operates in a similar manner. In the simulated termite-fishing and anvil-use tasks, a food item that is otherwise unattainable to the chimpanzee can be obtained by use of a stick or by the use of substrates as a surface on which to pound. For both tool use and manual gestures, there is an extrinsic means/ends relationship with the food representing the goal and the tool or human (depending on the task) representing the means to achieving it (obtaining the food item). Thus, in captive chimpanzees, the cognitive components of tool use and manual gesturing as assessed in captive settings may tap similar neural systems, notably those that are homologous to linguistic and tool-use abilities in humans.

Alternatively, the association between tool use and neuroanatomical asymmetries in the language homologues may suggest that tool use served as a preadaptation for left-hemisphere specialization in praxic or motor functions, from which language lateralization subsequently evolved. This would suggest that lateralization for motor functions preceded the evolution of specialization associated with communicative functions—a finding at odds with the recent gestural origins theory proposed by Corballis (2002). This interpretation is also at odds with recent studies showing evidence of population-level handedness for manual gestures in baboons, a species that has not been reported to use tools in its natural habitat (Meguerditchian and Vauclair 2006). Indeed, these findings are more consistent with the view that lateralization in species-specific communicative behavior were selected for in evolution, and that handedness for tool use and other types of manual actions postdate these early asymmetries (Bradshaw, 1997).

Evolution of Hemispheric Specialization

We would be remiss if we did not touch at least briefly on mechanisms that selected for the evolution of hemispheric specialization in primates. It should be clear now that the older, more traditional view that hemispheric specialization is uniquely human is simply no longer a tenable thesis, and it is time to consider other evolutionary theories and scenarios (Hopkins and Cantalupo 2008). At the heart of the matter is what the advantages and disadvantages are for hemispheric specialization. It is generally viewed that hemispheric specialization provided an opportunity to double the information processing and cognitive processes of the brain by dividing rather than duplicating function between the left and right cerebral hemispheres, and there is now ample evidence to support this assumption (Rogers 2007). However, as pointed out by Vallortiagara and Rogers (2005), division of sensory and cognitive functions between the two hemispheres can be achieved without all individuals being lateralized in the same direction. Thus it is not readily apparent why a significant majority of individuals conforms to a specific lateralized pattern of behavior. Vallortiagara and Rogers (2005) have suggested that lateralized conformity in socially relevant behavior creates an evolutionary stable strategy (ESS) through group selection, and, at least in lower vertebrates, there seems to be evidence in support of this view. How the ESS theory applies to primate social systems is less clear, but it certainly provides a solid foundation for empirical investigation. In contrast to the ESS theory, others have suggested that hemispheric specialization is simply an artifact of increasing brain size in primate evolution (Rilling and Insel 1999; Ringo et al. 1994). Comparative studies in primates have shown that as brain size increased, the corpus callosum (CC), a set of homotopic fibers connecting the two hemispheres, did not keep pace. Thus, humans have a relatively small CC for an animal of their brain size compared to apes and monkeys (Rilling and Insel 1999). The argument is that as brain size increased, each hemisphere became increasingly independent in function and thus lateralized. This theory is certainly plausible, but the reason why a majority of individuals show the same pattern of lateralized functions is not readily apparent.

Implications and Future Directions

Our results clearly show that captive chimpanzees show population-level handedness for certain tasks. In our view the same is shown by data from wild chimpanzees, even though much has been written about the apparent disparity between findings from these settings. Rather than continue to argue whether the apparent or alleged differences found between wild and captive chimpanzees reflect rearing effects or task differences, we would suggest that two important issues need to be addressed in future studies. First, more data are needed from wild chimpanzees—

and this should include data for tasks other than tool use, which has been the predominant measure studied to date. Some researchers have measured hand use for spontaneous activities in wild and captive chimpanzees with no apparent evidence of asymmetries in hand use at either the individual or group level of analysis (Fletcher and Weghorst 2005; Marchant and McGrew 1996; McGrew 2001). The issue of individual hand preference is not trivial, as it suggests either that the measures are insensitive to the construct of hemispheric specialization or that there are insufficient observations to detect individual hand preference when using the very conservative z-score as an indicator of hand preference. In short, obtaining hand-preference data from wild chimpanzees just for the sake of measuring their hand use is not constructive. Instead, it seems critical to identify behaviors that reliably induce individual asymmetries before deciding which behaviors to focus on within a study. We believe there are several candidate behaviors that should be studied extensively in wild chimpanzees, including manual gestures, grooming, scratching, and certain forms of locomotion and posture. Second, clearly in captive settings there need to be more tasks developed that model the natural behavior of chimpanzees. The fact that there are so few data on tool use from different laboratories or zoo settings is striking, in light of the long-standing interest in the topic and the availability of subjects. In our opinion, unless these two research goals can be met and more common ground can be found between studies in wild and captive chimpanzees, there will be further confusion as to whether population-level handedness represents an endogenous trait of the chimpanzee.

Another interesting aspect of handedness in wild and captive chimpanzees involves the distribution of right- and left-handed individuals within the population. When population-level handedness has been reported in apes (and monkeys), the ratio of dominant to nondominant individuals has been reported to be ∼ 2:1 and at best 3:1. This is quite different from what has typically been reported in humans, where estimates of the ratio range from 6:1 up to 9:1. It should also be noted that in some birds and other vertebrates, ratios of dominant to nondominant individuals approach 8:1 or 9:1 (Rogers 2007). Why nonhuman primates show lower ratios of dominant to nondominant individuals remains unclear but warrants further investigation.

A second overarching issue from our behavioral studies is the individual differences seen in the acquisition and use of vocal sounds. Approximately 50% of the Yerkes chimpanzees reliably produce either an EFG or RASP sound, which of course means that 50% do not. Individual differences in the use of these sounds are not associated with sex or rearing differences of the subjects. Similarly, throwing is not routinely observed in a majority of the chimpanzees we have studied and, like vocal production, individual differences are not attributable to some of the more obvious subject variables such as sex and rearing history. Throwing and the production of EFG and RASP are not the only behaviors that show this type of variability in use. There are other chimpanzee behaviors that are not seen in all individuals and are rarely if ever seen in other nonhuman primates. These include clapping, spitting, and cage banging or door knocking, when used as a means of getting the attention of another chimpanzee or human. As noted above, one possible explanation for the behavioral differences seen in individual chimpanzees may be their differing social cognition abilities, but this remains to be tested.

Related to the issue of the production of the EFG and RASP sounds is the question of volitional control. There is significant resistance to the notion that nonhuman primates have volitional control of their facial musculature and vocalizations (Fitch 2000; Premack 2004); we believe our results challenge these views for several reasons. First, RASP and EFG are produced selectively to capture the attention of an otherwise inattentive audience (Hopkins et al., 2007b), and they are not bound to a specific emotional context. For example, chimpanzees will use these sounds to draw a human's attention not only to food but also to a tool (Russell et al. 2005). Thus, the presence of food is not necessary to elicit the sounds. Second, the sounds are produced almost exclusively in the presence of an audience, which is quite the opposite pattern from that seen for other chimpanzee vocalizations, such as food calls.

This chapter has focused on behavioral correlates of brain asymmetries as reflected in gross morphology. In humans, very little is understood regarding microstructural asymmetries and how they might relate to variability in gross morphological landmarks used to define specific regions of interest, such as Broca's or Wernicke's areas (Schenker et al. 2007). Even less is known about such relations in nonhuman primates, notably the great apes, but progress is now being made in this domain. Several recent studies have attempted to examine asymmetries in minicolumn organization of the planum temporale (Buxhoeveden and Casanova 2000), and Sherwood et al. (2007) have recently examined several histological parameters of the motor-hand area of chimpanzees for which MRI and behavioral data were available. Sherwood et al. (2007) found a population-level leftward asymmetry in layer II/III neuron density. Moreover, asymmetries in the density of layer II/III parvalbumin-immunoreactive interneurons were a significant predictor of handedness. Thus, correlates of behavioral asymmetries are found even at the neuronal level of analysis.

Finally, our studies to date have largely focused on structural aspects of the brain. Functional imaging studies are needed to establish whether the brain areas putatively implicated in our morphology studies are involved in motor and cognitive functions associated with the behavior described in this chapter. There is little doubt that modern imaging technologies have made possible the development of sound studies that allow for the assessment of brain-behavior relationships in chimpanzees and other nonhuman primates. Magnetic resonance imaging also offers many new advantages to the study of the chimpanzee brain.

It seems to us that this technology will only improve with time, which will in turn offer even greater opportunities for understanding the chimpanzee mind from the standpoint of the brain's anatomical and functional properties.

Acknowledgments

This work was supported in part by NIH grants RR-00165, NS-42867, NS-36605, HD-38051, and HD-56232. Some of the MRI scans were collected during many hours of dedication by Drs. Jim Rilling and Tom Insel. Special thanks are also directed to Drs. Elizabeth Strobert, Jack Orkin, and Brent Swenson and the rest of the veterinarian staff for assisting in the care of the animals during scanning. We also thank the many students and staff who have assisted in data collection over the years, notably Dawn Pilcher, Hani Freeman, and Stephanie Braccini. Correspondence and reprint requests should be addressed to Dr. William D. Hopkins, Division of Psychobiology, Yerkes National Primate Research Center, Emory University, Atlanta, Georgia 30322. E-mail: whopkin@emory.edu or whopkins@agnesscott.edu.

Literature Cited

Annett, M. 1985. *Left, Right, Hand, and Brain: The Right-Shift Theory*. London: Lawrence Erlbaum Associates.

———. 2002. *Handedness and Brain Asymmetry: The Right Shift Theory*. Hove: Psychology Press.

Arbib, M. 2005. From monkey-like action recognition to human language: An evolutionary framework for neurolinguistics. *Behavioral and Brain Sciences* :105–67.

Beaton, A.A. 1997. The relation of planum temporale asymmetry and morphology of the corpus callosum to handedness, gender and dyslexia: A review of the evidence. *Brain and Language* 60:255–322.

Biro, D., N. Inoue-Nakamura, R. Tonooka, G. Yamakoshi, C. Sousa, and T. Matsuzawa. 2003. Cultural innovation and transmission of tool use in wild chimpanzees: Evidence from field experiments. *Animal Cognition* 6(4): 213–23.

Biro, D., C. Sousa, and T. Matsuzawa. 2006. Ontogeny and cultural propagation of tool use by wild chimpanzees at Bossou, Guinea: Case studies in nut cracking and leaf folding. In T. Matsuzawa, T. Tomonaga, and M. Tanaka, eds., *Cognitive Development of Chimpanzees*. New York: Springer, 476–507.

Boesch, C. 1991. Handedness in wild chimpanzees. *International Journal of Primatology* 6:541–58.

Bradshaw, B., and L. Rogers. 1993. *The Evolution of Lateral Asymmetries, Language, Tool-use and Intellect*. San Diego: Academic Press.

Bradshaw, J. L. 1997. *Human Evolution: A Neuropsychological Perspective*. Hove, UK: Psychology Press.

Buxhoeveden, D., and M. Casanova. 2000. Comparative lateralisation patterns in the language area of human, chimpanzee, and rhesus monkey brains. *Laterality* 4:315–30.

Byrne, R.W., and J. M. Byrne. 1991. Hand preferences in the skilled gathering tasks of mountain gorillas (*Gorilla gorilla berengei*). *Cortex* 27:521–36.

Calvin, W. H. 1983. *The Throwing Madonna: Essays on the Brain*. New York: MacGraw-Hill.

Cantalupo, C., H. D. Freeman, and W. D. Hopkins. 2008. Handedness for tool use correlates with cerebellar asymmetries in chimpanzees (*Pan troglodytes*). *Behavioral Neuroscience* 122:191–98.

Cantalupo, C., and W. D. Hopkins. 2001. Asymmetric Broca's area in great apes. *Nature* 414:505.

Cantalupo, C., D. Pilcher, and W. D. Hopkins. 2003. Are planum temporale and sylvian fissure asymmetries directly related? A MRI study in great apes. *Neuropsychologia* 41:1975–81.

Cooper, D. L. 2006. Broca's arrow: Evolution, prediction and language in the brain. *The Anatomical Record* (Part B: New Anatomy) 289B:9–24.

Corballis, M. C. 1992. *The Lopsided Brain: Evolution of the Generative Mind*. New York: Oxford University Press.

———. 2002. *From Hand to Mouth: The Origins of Language*. Princeton, NJ: Princeton University Press.

Crow, T. 2004. Directional asymmetry is the key to the origin of modern *Homo sapiens* (the Broca-Annett axiom): A reply to Rogers' review of *The Speciation of Modern Homo Sapiens*. *Laterality: Asymmetries of Body, Brain and Cognition* 9:233–42.

Dunham, L. A., and W. Hopkins. 2006. Sex and handedness effects on corpus callosum morphology in chimpanzees (*Pan troglodytes*). *Behavioral Neuroscience* 120(5): 1025–32.

Fernandez-Carriba, S., A. Loeches, A. Morcillo, and W.D. Hopkins. 2002. Asymmetry in facial expression of emotions by chimpanzees. *Neuropsychologia* 40(9): 1523–33.

Fitch, W. T. 2000. The evolution of speech: A comparative review. *Trends in Cognitive Sciences* 4:258–67.

Fletcher, A. W., and J. A. Weghorst. 2005. Laterality of hand function in naturalistically housed chimpanzees (*Pan troglodytes*). *Laterality* 10:219–42.

Foundas, A., C. Leonard, and K. Heilman. 1995. Morphological cerebral asymmetries and handedness: The pars triangularis and planum temporale. *Archives of Neurology* 52:501–8.

Foundas, A. L., C. M. Leonard, and B. Hanna-Pladdy. 2002. Variability in the anatomy of the planum temporale and posterior ascending ramus: Do right- and left-handers differ? *Brain and Language* 83:403–24.

Freeman, H. D., C. Cantalupo, and W. D. Hopkins. 2004. Asymmetries in the Hippocampus and Amygdala of chimpanzees (*Pan troglodytes*). *Behavioral Neuroscience* 118(6): 1460–65.

Gannon, P. J., R. L. Holloway, D. C. Broadfield, and A. R. Braun. 1998. Asymmetry of chimpanzee Planum Temporale: Humanlike pattern of Wernicke's language area homolog. *Science* 279:220–22.

Gibson, K. R., and T. Ingold. 1993. *Tools, Language and Cognition in Human Evolution*. Cambridge: Cambridge University Press.

Gilissen, E. 2001. Structural symmetries and asymmetries in human and chimpanzee brains. In D. Falk and K. R. Gibson, eds., *Evolutionary Anatomy of the Primate Cerebral Cortex*. Cambridge: Cambridge University, 187–215.

Goodall, J. 1986. *The Chimpanzees of Gombe: Patterns of Behavior*. Cambridge, MA: Harvard University Press.

Greenfield, P. M. 1991. Language, tools, and brain: The ontogeny and phylogeny of hierarchically organized sequential behavior. *Behavioral and Brain Sciences* 14(4): 531–50.

Hauser, M. C. 1993. Right hemisphere dominance in the production of facial expression in monkeys. *Science* 261:475–77.

Hook-Costigan, M. A,, and L. J. Rogers. 1998. Lateralized use of the mouth in production of vocalizations by marmosets. *Neuropsychologia* 36(12): 1265–73.

Hopkins, W. D. 1994. Hand preferences for bimanual feeding in 140 captive chimpanzees (*Pan troglodytes*): Rearing and ontogenetic factors. *Developmental Psychobiology* 27:395–407.

———. 2006. Comparative and familial analysis of handedness in great apes. *Psychological Bulletin* 132(4): 538–59.

———. 2007. Hemispheric specialization in chimpanzees: Evolution of hand and brain. In T. Shackelford, J. P. Keenan, and S. M. Platek, eds., *Evolutionary Cognitive Neuroscience*. Boston: MIT Press. 99–120.

Hopkins, W. D., and C. Cantalupo. 2004. Handedness in chimpanzees is associated with asymmetries in the primary motor but not with homologous language areas. *Behavioral Neuroscience* 118:1176–83.

———. 2005. Individual and setting differences in the hand preferences of chimpanzees (*Pan troglodytes*): A critical analysis and some alternative explanations. *Laterality* 10:65–80.

———. 2008. Theoretical speculations on the evolutionary origins of hemispheric specialization. *Current Directions in Psychological Science*.

Hopkins, W. D., and D. L. Pilcher. 2001. Neuroanatomical localization of the motor hand area with magnetic resonance imaging: The left hemisphere is larger in great apes. *Behavioral Neuroscience* 115(5): 1159–64.

Hopkins, W. D., D. L. Pilcher, and L. MacGregor. 2000. Sylvian fissure length asymmetries in primates revisited: A comparative MRI study. *Brain, Behavior and Evolution* 56:293–99.

Hopkins, W. D., J. Russell, C. Cantalupo, H. Freeman, and S. Schapiro. 2005a. Factors influencing the prevalence and handedness for throwing in captive chimpanzees (*Pan troglodytes*). *Journal of Comparative Psychology* 119(4): 363–70.

Hopkins, W. D., J. Russell, H. Freeman, N. Buehler, E. Reynolds, and S. Schapiro. 2005b. The distribution and development of handedness for manual gestures in captive chimpanzees (*Pan troglodytes*). *Psychological Science* 16(6): 487.

Hopkins, W. D., J. L. Russell, and C. Cantalupo. 2007a. Neuroanatomical correlates of handedness for tool use in chimpanzees (*Pan troglodytes*): Implication for theories on the evolution of language. *Psychological Science* 18(11): 971–77.

Hopkins, W. D., J. L. Russell, S. Lambeth, and S. J. Schapiro. 2007b. Handedness and neuroanatomical asymmetries in captive chimpanzees: A summary of 15 years of research. In W. D. Hopkins, ed., *Evolution of Hemispheric Specialization in Primates*. London: Academic Press, 112–35.

Hopkins, W. D., J. P. Taglialatela, and D. A. Leavens. 2007c. Chimpanzees differentially produce novel vocalizations to capture the attention of a human. *Animal Behaviour* 73:281–86.

Hostetter, A. B., M. Cantero, W. D. Hopkins. 2001. Differential use of vocal and gestural communication by chimpanzees (*Pan troglodytes*) in response to the attentional status of a human (*Homo sapiens*). *Journal of Comparative Psychology* 115(4): 337–43.

Hostetter, A. B., J. L. Russell, H. Freeman, W. D. Hopkins. 2007. Now you see me, now you don't: Evidence that chimpanzees understand the role of the eyes in attention. *Animal Cognition* 10:55–62.

Knecht, S., B. Drager, M. Deppe, L. Bobe, H. Lohmann, A. Floel, E. B. Ringelstein, and H. Henningsen. 2000. Handedness and hemispheric language dominance in healthy humans. *Brain* 123(12): 2512–18.

Leavens, D. A., and W. D. Hopkins. 1998. Intentional communication by chimpanzee (*Pan troglodytes*): A cross-sectional study of the use of referential gestures. *Developmental Psychology* 34:813–22.

Leavens, D. A., W. D. Hopkins, and K. A. Bard. 1996. Indexical and referential pointing in chimpanzees (*Pan troglodytes*). *Journal of Comparative Psychology* 110(4): 346–53.

———, 2005a. Understanding the point of chimpanzee pointing: Epigenesis and ecological validity. *Current Directions in Psychological Science* 14(4): 185–89.

Leavens, D. A., A. B. Hostetter, M. J. Wesley, and W. D. Hopkins. 2004. Tactical use of unimodal and bimodal communication by chimpanzees, *Pan troglodytes*. *Animal Behaviour* 67:467–76.

Leavens, D. A., J. L. Russell, and W. D. Hopkins. 2005b. Intentionality as measured in the persistence and elaboration of communication by chimpanzees (*Pan troglodytes*). *Child Development* 76(1): 291–306.

Lonsdorf, E. V., and W. D. Hopkins. 2005. Wild chimpanzees show population level handedness for tool use. *Proceedings of the National Academy of Sciences* 102:12634–38.

Losin, E. R., H. Freeman, J. L. Russell, A. Meguerditchian, and W. D. Hopkins. 2008. Left hemisphere specialziation for oro-facial movements of learned vocal signals by captive chimpanzees. *PlosOne* 3:1–7.

Marchant, L. F. 1983. Hand preferences among captive island groups of chimpanzees. Unpublished doctoral dissertation. Rutgers, the State University of New Jersey, New Brunswick.

Marchant, L. F., and W. C. McGrew. 1996. Laterality of limb function in wild chimpanzees of Gombe National Park: Comprehensive study of spontaneous activities. *Journal of Human Evolution* 30:427–43.

———. 2007. Ant fishing by wild chimpanzees is not lateralised. *Primates* 48:22–26.

McGrew, W. C., and L. F. Marchant. 1992. Chimpanzees, tools, and termites: Hand preference or handedness? *Current Anthropology* 33:114–19.

———. 1996. In W. C. McGrew, L. F. Marchant, and T. Nishida, eds., *Great Ape Societies.* Cambridge: Cambridge University Press, 255–72.

———. 2001. Ethological study of manual laterality in the chimpanzees of the Mahale mountains, Tanzania. *Behaviour* 138:329–58.

McManus, C. 2002. *Right Hand, Left Hand: The Origins of Asymmetry in Brains, Bodies, Atoms, and Cultures.* London: Weidenfeld & Nicolson.

McNeil, D. 1992. *Hand and Mind: What Gestures Reveal about Thought.* Chicago: University of Chicago Press.

Meguerditchian, A., and J. Vauclair. 2006. Baboons communicate with their right hand. *Behavioural Brain Research* 171:170–74.

Papademetriou, E., C. F. Sheu, and G. F. Michel. 2005. A meta-analysis of primate hand preferences for reaching and other hand-use measures. *Journal of Comparative Psychology* 119:33–48.

Perelle, I. B., and L. Ehrman. 1994. An international study of human handedness: The data. *Behavior Genetics* 24:217–27.

Pilcher, D., L. Hammock, and W. D. Hopkins. 2001. Cerebral volume asymmetries in non-human primates as revealed by magnetic resonance imaging. *Laterality* 6:165–80.

Porac, C., and S. Coren. 1981. *Lateral Preferences and Human Behavior.* New York: Springer.

Premack, D. 2004. Is language the key to human intelligence? *Science* 303:318–20.

Rasmussen, T., and B. Milner. 1977. The role of early left-brain injury in determining lateralization of cerebral speech function. *Annals of the New York Academy of Sciences* 299:355–69.

Raymond, M., and D. Pontier. 2004. Is there geographical variation in human handedness? *Laterality* 9:35–51.

Rilling, J. K., and T. R. Insel. 1999. The primate neocortex in comparative perspective using magnetic resonance imaging. *Journal of Human Evolution* 37:191–223.

Ringo, J., R. Doty, S. Demeter, and P. Simard. 1994. Timing is of essence: A conjecture that hemispheric specialization arises from inter-hemispheric conduction delay. *Cerebral Cortex* 4:331–43.

Rogers, L. J. 2007. Lateralization in its many forms and its evolution and development In W. D. Hopkins, ed., *Evolution of Hemispheric Specialization in Primates.* London: Academic Press.

Russell, J. L., S. Braccini, N. Buehler, M. J. Kachin, S. J. Schapiro, and W. D. Hopkins. 2005. Chimpanzees (*Pan troglodytes*) intentional communication is not contingent upon food. *Animal Cognition* 8(4): 263–72.

Schenker, N. M., C. C. Sherwood, P. R. Hof, and K. Semendeferi. 2007. Microstructural asymmetries of the cerebral cortex in humans and other mammals. In W. D. Hopkins, ed., *Evolution of Hemispheric Specialization in Primates.* London: Academic Press.

Sherwood, C. C., E. Wahl, J. M. Erwin, P. R. Hof, and W. D. Hopkins. 2007. Histological asymmetries of primary motor cortex predicts handedness in chimpanzees (*Pan troglodytes*). *Journal of Comparative Neurology* 503:525–37.

Sugiyama Y. 1995. Tool-use for catching ants by chimpanzees at Bossou and Monts Nimba, West Africa. *Primates* 36:193–205.

Sugiyama, Y., T. Fushimi, O. Sakura, and T. Matsuzawa. 1993. Hand preference and tool use in wild chimpanzees. *Primates* 34:151–59.

Taglialatela, J. P., C. Cantalupo, W. D. Hopkins. 2006. Gesture handedness predicts asymmetry in the chimpanzee inferior frontal gyrus. *NeuroReport* 17(9): 923–27.

Taglialatela, J. P., M. Dadda, and W. D. Hopkins. 2007. Sex differences in asymmetry of the planum parietale in chimpanzees (*Pan troglodytes*) *Behavioural Brain Research* 184(2): 185–91.

Vallortigara, G., and L. J. Rogers. 2005. Survival with an asymmetrical brain: Advantages and disadvantages of cerebral lateralization. *Behavioral and Brain Sciences* 28:574.

Warren, J. M. 1980. Handedness and laterality in humans and other animals. *Physiological Psychology* 8:351–59.

Westergaard, G. C., C. Liv, M. K. Haynie, and S. J. Suomi. 2000. A comparative study of aimed throwing by monkeys and humans. *Neuropsychologia* 38(11): 1511–17.

7

Trapping the Minds of Apes: Causal Knowledge and Inferential Reasoning about Object-Object Interactions

Josep Call

A six-year-old child is presented with the following task: A pair of identical strips of cloth is placed on a table facing the child such that he can reach the near but not the far end of each cloth. A small toy is placed on the far end of the left cloth and another identical toy is placed to the side of the far end of the right cloth. So both toys are at the same distance from the child and out of his reach. The child is then told that he can keep the toy if he can get it without getting up. The child smiles and pulls the cloth with the toy on it. Why does he do this? It is conceivable that he learned in the past that certain perceptual configurations (i.e., toy on cloth) are more likely to produce desirable outcomes than others (i.e., toy off the cloth). It is also possible that he can attribute some causal reason for this outcome (i.e., the cloth can cause the toy to change location). So we ask him, and this is what he says: "I pulled this cloth because the car is on top of it." Then we ask why it is important that the car is on top of it and he says: "Because when I pull the cloth the car will get closer."

Studying the minds of nonhuman animals, just like studying those of infants, is a fascinating and challenging enterprise. We seek to capture the elusive cognitive mechanisms that regulate the behavior of the subjects under study, so that we can better understand how they process information and how they respond to new challenges. Since we cannot directly ask nonverbal subjects why they do certain things, empirically oriented scientists have to find indirect

means to ask those questions. The crux of this endeavor involves designing tasks that can accurately "translate" our verbal question into nonverbal means.

In recent years the study of causal knowledge in nonhuman animals has received considerable research attention that has sparked a lively debate regarding the processes underlying complex problem-solving behaviors such as tool use. Whereas some authors have suggested that subjects may solve problems by responding to the presence of certain observable cues but without any real causal understanding, others have suggested that some level of comprehension of object-object relations may underlie tool use (see Call 2006a; Hauser and Spaulding 2007; Povinelli 2000; Visalberghi and Limongelli 1994). In general, the controversy in this area, just like other controversies in the field of animal cognition (e.g., mind reading; see chapter 19) and psychology as a whole, has been polarized between those who propose that associative processes can explain the observed behavior and those who favor the involvement of non-associative processes (e.g., inferential reasoning).

A key step towards advancing our knowledge in this area is to empirically distinguish between the various processes that subjects may use to solve a particular problem. Before we proceed any further, I will provide some working definitions. *Knowledge* is the broadest term used in this chapter; it refers to information that subjects have encoded and stored in their memories. *Causal knowledge* is the subset of encoded information that refers to object-object relations and their interactions, such as those typically found in tool-use behavior. *Inferential reasoning* is the mechanism that allows subjects to use and combine old and new information in new ways to find solutions to novel problems. *Inference* can be distinguished from association because in the latter, subjects have already experienced a particular configuration or the solution simply represents a generalization based on extending their knowledge about a particular stimulus along a particular perceptual dimension (e.g., color hue). Finally, *causal inferences* are inferences made on the basis of information about object-object relations. Solutions based on this mechanism typically appear suddenly, without trial-and-error learning.

The first goal of this chapter is to explain how scientists have investigated the causal knowledge and inference of object-object interactions in the great apes (and other nonhuman animals) and to present the current state of the art. It is not my intent to produce a complete review of the available findings. I will be selective and concentrate on the use of tools to get rewards to which direct access is blocked by the presence of a trap. Readers interested in a more thorough coverage of the literature are directed to other sources (e.g., Antinucci 1989; Povinelli 2000; Tomasello and Call 1997; Call and Tomasello 2005).

Although part of the controversy on causal reasoning is theoretical in nature, there are also methodological aspects that have contributed to the current debate on the level of causal knowledge that chimpanzees and other animals possess. Designing experiments that measure exactly what we want, not more but not less, is not so simple. More often than we would like to admit it, we, the scientists, fall prey of our own assumptions when designing tasks, and it is fair to say that we become trapped in the traps that we have designed to trap our subjects' minds. Discussing this more methodologically based issue is important enough to constitute the second goal of this chapter.

The chapter is organized as follows. The first section provides some historical background to the study of apes' causal knowledge about objects, and a description of some of the most prominent tasks used in that study. Here the results and caveats for each of the tasks are presented in some detail. The next section takes a brief methodological detour to emphasize the importance of taking negative results seriously. The following three sections present ways to test causal comprehension: transferring tests to functionally equivalent tasks, removing the potential confounding effect of complex manipulation, and directly assessing whether perceptual or associative processes can explain particular results. The final section presents the implications of this work and outlines current challenges and potential future directions.

Traps, Tools, and Minds

It is fair to say that the discovery of chimpanzees using tools in the wild (van Lawick-Goodall 1968) was one of the most significant discoveries in field primatology and perhaps in animal behavior as a whole. This finding not only questioned the long-held assumption that using and manufacturing tools were the hallmark of humanity, but also reinvigorated the study of cognitive processes in apes. I say reinvigorated because tool use had been used by psychologists to investigate ape cognition long before its discovery in the wild.

Both Köhler's and Yerkes' pioneering work on ape problem solving drew heavily on tool use (Köhler 1925; Yerkes and Yerkes 1929). Yerkes saw tool use as an example of ideation, thinking in the absence of the actual stimuli, whereas Köhler saw it as an example of finding indirect solutions when the direct route was not possible. Köhler emphasized the insightful nature of several tool-use episodes, and he contrasted such a sudden mode of solution with the gradual trial-and-error learning that Thorndike had described with his puzzle boxes. Historically, this is the forerunner of a debate between association and reasoning that continues unresolved today, and to which we have alluded to earlier. Psychologists studying apes were not the only ones who noted the importance of tool use. Piaget (1952) also explored its development in children in great detail. For him, tool use was important because it involved the coordination of means and ends. It was one of the first manifestations of causal knowledge understood as knowledge about interactions between objects, not just knowledge about actions on the self or on objects—something that Piaget called tertiary schemata.

Given these historical precedents, it is therefore not surprising that tool use has continued to play a major role in the investigation of causal knowledge in nonhuman primates. Researchers have used relatively simple tasks—such as the cloth problem alluded to at the beginning of this chapter, in which subjects must retrieve a reward that is out of reach by pulling the cloth toward themselves—or more complex tasks in which subjects must use a stick to bring the reward within reach (Natale et al. 1988; Piaget 1952; Spinozzi and Potì 1989). Even more complex are those tasks in which subjects must overcome some obstacle (not just distance) on the way to getting the reward with a tool.

One of the most well-known tasks of this kind is the trap-tube (Visalberghi and Limongelli 1994; Limongelli et al. 1995; see figure 7.1a). Subjects are faced with a transparent tube that has a trap in its center and a reward inside the tube next to the trap, outside their direct reach. They are provided with a stick whose diameter is slightly smaller than the inner diameter of the tube. The solution is to insert the stick into the tube, and then push the reward away from the trap and out of the tube. This is a hard task for capuchin monkeys and chimpanzees, as only a minority of subjects solved it even after dozens of trials. Moreover, when successful subjects are presented with a condition in which the trap is inverted 180 degrees, thus making it nonfunctional, they still continue to avoid the trap (Visalberghi and Limongelli 1994; Povinelli 2000). This has led several authors to conclude that subjects may have used a perceptual strategy based on using the position of the trap to determine the appropriate insertion point, but without understanding that the position of the reward in relation to the trap hole is the critical feature in this task.

When researchers find no evidence that subjects are capable of solving a task, they wonder whether there may be a specific superfluous feature of the task, not directly linked to the issue under scrutiny, that may have prevented them from solving it. Mulcahy and Call (2006) presented

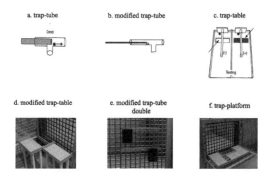

Figure 7.1 Experimental setup for various trap tasks. Tube: Visalberghi and Limongelli, 1994, Povinelli, 2000. Modified tube: Mulcahy and Call 2006; Tebbich and Bshary 2004. Table: Povinelli 2000; Girndt et al. 2008. Modified table: Girndt et al. 2008. Modified tube double: Martin-Ordas et al. 2008. Platform: Martin-Ordas et al. 2008.

a modified trap-tube task to great apes in which the subjects could choose between raking the reward in or pushing it out (see also Tebbich and Bshary 2004). The authors did this by enlarging the diameter of the tube and reducing that of the tool so that both raking and pushing the reward were possible (figure 7.1b). Mulcahy and Call (2006) reported three interesting results. First, apes preferred to rake rather than push the reward out of the tube. Second, although only three of the ten apes tested performed above chance, those three solved the task much faster than any animal tested before. Third, when the trap was inverted, subjects did not avoid it, thus ruling out the possibility that they were simply using the trap's position as a cue to decide where to insert the tool. Taken together, this study showed the best results on the trap tube to date, and suggested that some subjects, at least, seemed to understand the trap's functional properties. Mulcahy and Call (2006) suggested that forcing subjects to get the reward by pushing it out (and thus away from themselves) may have contributed to the poor performance found in previous studies, thus effectively masking the knowledge that subjects may have had about the trap.

Povinelli (2000) tried to simplify the trap problem in a different way: with the trap table. He presented chimpanzees with a table divided into two sections (figure 7.1c). One section had a hole cut in it (i.e., the trap) and the other had a blue rectangle painted on it (i.e., a fake trap whose dimensions were identical to those of the real trap). Rewards were placed in front of the trap and the rectangle, and the subject could pull one of two rakes that had been placed on the table right behind each of the rewards. Thus, the only thing that subjects had to do was to pull the rake placed behind the fake trap in the first place. Only one ape did so consistently. Povinelli (2000) concluded that apes had a limited understanding of the physical properties of the trap. Subsequent studies carried out with capuchins and gibbons using a similar paradigm also concluded that subjects did not have a total comprehension of the elements of the problem, but that they might have learned certain associative rules to solve it (Fujita et al. 2003; Cunningham et al. 2006).

Girndt et al. (2008) investigated which features of the trap table made it such a difficult task for chimpanzees. The authors compared the original setup used by Povinelli (2000) with one in which they varied the type of trap, the positions and number of tools, the type of tools, and the reinforcement regime. In a series of five experiments, they demonstrated that great apes consistently avoided the trap (including in the first trial) when they were allowed to insert the tool in one of the two sections, but not when the tools were pre-positioned on the table (figure 7.1d). In contrast, when apes were required to pull one of the pre-positioned tools they failed the task, thus replicating Povinelli's (2000) original results. The authors concluded that apes were indeed capable of spontaneously solving the table task if they were allowed to insert the tool to get the reward. Although this is consistent with the idea that apes possess causal knowledge about the effect the trap can have on a reward that is moving on it, it does not conclusively prove that they possess that knowledge, as I will discuss later.

Martin-Ordas et al. (2008) presented apes with the Mulcahy and Call (2006) modified version of the trap tube, in which they were able both to rake and to push the reward out. Additionally, the authors attached two apparatuses to the fence, one above the other (figure 7.1e). One apparatus had the trap on the left side while the other had it on the right side. In each trial the subject saw the experimenter bait one of the apparatuses and then offer a tool to get the reward out. Thus, unlike in previous trap-tube studies, subjects experienced two apparatuses simultaneously: one baited, the other one not. Forty percent of the subjects reached above chance performance after 36 trials in this setup. This percentage of success is the highest ever reported in any of the trap-tube studies conducted to date after so few trials.

One possible explanation for this result is that seeing two apparatuses at the same time facilitated the task because it constantly highlighted the two possible trap positions. Alternatively, this task may have been easier because subjects may have learned to use a set of actions with one tube (e.g., raking from the left side) and the opposite set of actions with the other tube (e.g., raking from the right side). This means that they were not required to apply two different actions to the same apparatus depending on the trap position. Currently it is unclear which of these two alternatives best explains the data. Moreover, future studies should directly compare the performance of subjects with a single apparatus with those that have experienced two apparatuses simultaneously.

Martin-Ordas et al. (2008) also presented another task functionally equivalent to the trap tube but based on a platform setup (figure 7.1f). Here apes faced an inverted

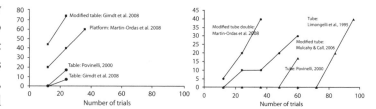

Figure 7.2a Percentage of subjects who performed above chance as a function of the trial number for table/platform tasks. Squares denote tasks in which subjects were handed one tool, while circles denote tasks in which subjects were presented with two pre-positioned tools.

Figure 7.2b Percentage of subjects who performed above chance as a function of the trial number for tube tasks. Diamonds denote tasks in which subjects could either push or rake the reward, while triangles denote tasks in which subjects could only push the reward out of the tube.

U-shaped platform with a reward placed on its center and a trap to one side. The solution consisted of using the tool to move the reward towards the side of the platform without the trap in order to bring it within reach. This task shares some features with the trap tube. For instance, there is a single reward that must be moved to the side to get access to it while avoiding the trap. Also, there is only one tool, not two as with the trap-table. But, like the trap table, it presents the problem on a flat surface. Subjects also solved this task at high levels (60% did so within 36 trials) although not as high as in the table task.

In summary, when only one tool is provided so that they can decide where to insert it, a substantial proportion of subjects can solve the table and platform tasks quite quickly (figure 7.2). Furthermore, when subjects can choose between raking and pushing the reward out of the tube, they perform better in that task than when they can only push the reward out (figure 7.2b). This means that implementing quite small procedural modifications that give the subjects more choices can substantially improve their performance and reveal capabilities not previously observed. Confronted with such a mixed bag of positive and negative results, we face a couple of dilemmas. First, do the positive results constitute evidence of causal knowledge? Second, do negative results have any value? Allow me to make a methodological detour to briefly tackle this last question first, before discussing what may constitute evidence of causal knowledge.

Negative Results: A Blessing, a Curse, or Both?

Several studies found no evidence that chimpanzees spontaneously avoided knocking the reward into the trap when attempting to retrieve it, leading some authors to suggest that their causal understanding of the task was quite limited. Yet we have also seen that certain slight procedural changes dramatically improved the apes' performance. Although one may be tempted to view negative results as valueless, that could lead to serious oversimplification. It is true that negative results are hard to interpret because the absence of evidence is not evidence of absence. Indeed, negative results are not useful if they are not a faithful representation of reality—for instance, if they have arisen due to some methodological flaw such as a poorly executed procedure, a coding mistake, an analytical error, or a lack of statistical power. Ignoring this kind of "artificial" negative result has no detrimental consequences for our conclusions.

In contrast, negative results that are not the product of methodological glitches—and to this category belong the negative results on the trap tasks presented above—are valuable and can play a major role in shaping our conclusions. Negative results are as important to the experimental method as positive results. In fact, the only reason positive results are meaningful is that we can contrast them to "real" negative results. If all studies produced positive results in all their conditions, we would learn nothing. The scientist who could only produce positive results would as unhappy as King Midas, who, wishing to be rich beyond belief, was granted his wish for everything that he touched to turn to gold—including the food he had to eat.

There is another reason why it is important to consider negative results carefully. Negative results often highlight the circumstances under which an individual may not display certain abilities that she possesses. For instance, many subjects failed the table task with pre-positioned tools, but not with a single non-pre-positioned tool. This was quite surprising, given that pre-positioning the tools was conceived as a way to simplify the problem. In reality this manipulation just did the opposite: it made the problem harder for the apes to solve. Thus, while new designs do not necessarily invalidate old results, they surely make our conclusions more relativistic, which may not be a bad thing after all. Thus, faced with a set of negative results, caution and careful scrutiny seems more fruitful than dismissal. This should not be taken to mean that positive results are free of ambiguity. They are not, at least not when it comes to investigating the cognitive processes underlying behavior. There are always multiple explanations for a positive

result, not to mention the fact that positive results may have arisen by chance. The point here is that any strong theory must be able to adequately explain both positive and negative results (see chapter 23).

Transfer Tasks: One Road to Causal Understanding

Although the positive results in the modified trap table are consistent with the idea that apes possess causal knowledge, they do not constitute proof because success can arise via different cognitive processes. Thus, chimpanzees may solve the modified table task due to the presence of some perceptual feature, not because they understand the tertiary relations between objects. For instance, they may like the painted surface in the table task (Povinelli 2000), or they may find discontinuous surfaces aversive independently of the effects they may have on a moving reward. Nevertheless, showing positive results constitutes the first important step towards determining the cognitive processes underlying successful performance.

The next step is to find out the cognitive mechanism responsible for the observed behavior. The key is to combine the results of multiple experiments into a coherent whole. To this end, one should follow complementary strategies. One can test whether specific features control the subjects' responses in a particular task. For instance, we could change the color of the painted rectangle or the type of trap, and observe whether this has any effect on the subjects' performance. Manipulating one variable at a time and observing its effects on performance can help researchers to rule out some of the most obvious alternative explanations. For instance, Hauser and colleagues (Hauser 1997; Santos et al. 1999) used this approach when studying the tool features that various monkey species attend to in a raking task. They found that monkeys ignored irrelevant features, such as the tool's color or size, while not ignoring key features such as its shape or rigidity. Similarly, Girndt et al. (2008) varied the type of trap that subjects faced—all of them functional—and found that subjects succeeded regardless of the type of task that was used. However, one cannot manipulate every variable in a given task; that would require conducting an infinite number of manipulations.

One solution to this problem consists of presenting a task with different perceptual features but which is functionally equivalent to the original task. For instance, the trap-tube task and the trap-platform task are functionally equivalent but they differ in their perceptual features. This strategy has two desirable outcomes. First, it rules out many of the potential (and superficial) explanations that have applied to the original task. This is especially important if certain actions may not be appropriate for certain subjects or species. Second, this additional task can be used as a transfer task. Indeed, the use of transfer tasks is one of the most powerful methods used to investigate the cognitive processes that underlie behavior. Interestingly, when it comes to trap tasks, few studies have shown robust evidence of transfer within and especially between tasks. With regard to within-task transfer, only a minority of birds and primates have shown evidence of transfer from the original to the modified tasks (Seed et al. 2006; Tebbich and Bshary 2004; Visalberghi and Limongelli 1994; Povinelli 2000). Testing transfer across different tasks has not produced better results. Martin-Ordas et al. (2008) found that although apes could solve both the trap-tube and the platform-trap tasks, there was no correlation between the two (but see Martin-Ordas and Call 2009). Since the evidence of transfer is so weak, one could conclude that there is no evidence of causal knowledge in trap tasks.

There are, however, two potential problems that may make this conclusion premature. First, when tasks require different actions the problem may reside with the actions themselves, not with the knowledge subjects may have about the task. Recently, Seed et al. (2009) found evidence of positive transfer between two versions of the trap tube, but only when chimpanzees were not required to use a tool to get the reward and could simply move it away from the trap with their fingers. As noted above, this is a non-trivial problem that is easily overlooked. This issue does not exist when the transfer takes place within the same task rather than between tasks. However, here the perceptual similarity between the original and the transfer task may increase the difficulty considerably, thus overriding the use of the appropriate responses.

Second, and more important, although transfer tasks are commonly used to assess causal knowledge, a transfer task may be much more than that. Successful transfer may require that subjects not only know about the relation between elements of a particular task, but also that they perceive equivalence between the relation of elements in one task and the relations in the other. Such an operation can be better described as analogical in nature (relations between the relations between elements) rather than causal (relations between elements themselves). Thus, transfer tasks may not be the best way, and certainly are not the only way, to assess causal knowledge. One complementary way to do so is to directly assess the predispositions of the subjects and assess how much they can learn when causal relations are not a key component of the task. Would subjects, for instance, learn arbitrary relations with the same ease as causal relations? To answer this question we must switch now to a completely different paradigm based on inferential reasoning, to assess the likelihood that subjects' responses in trap tasks are based solely on learned association and/or innate predisposition.

Causal Reasoning without Tools: When Inference Meets Causation

Tool use has been the main avenue for investigating causal knowledge in nonhuman primates—most likely a legacy of the Piagetian approach. This means that the use of actions has been a key component in all these studies. An excessive reliance on action, however, can be problematic since it has been shown that apes can be very sensitive to the way the task is presented (pre-positioned tools or not) and the actions that subjects have at their disposal to solve the task (rake versus push out). Since the presence of these factors can mask the knowledge that subjects may have about the problem, it is necessary to search for alternative paradigms that are free of those constraints.

One complementary way to tackle the question of causal knowledge in nonhuman animals is to present a task in which action does not occupy such a prominent role. Here is the basic procedure for one such task. A reward is hidden in one of two containers, and subjects witness some event involving a physical interaction between the reward and the containers. Using this information, they must determine the location of the reward. For instance, the subject witnesses the experimenter placing two boards flat on a platform. After raising a screen in front of the platform, the experimenter places a reward under one of the boards, (out of the subject's view) so that it acquires a sloping orientation. Once the screen is removed, subjects must find the reward on their first attempt. Thus, whereas tool-use tasks can be construed as requiring the subject to obtain a visible reward after bypassing an obstacle, inferential tasks can be construed as requiring the subject to find a hidden reward. Thus, unlike tool-use tasks, inferential tasks do not require complex actions. Subjects can make their choice by simply touching one of the two alternatives.

Inferential abilities have been studied in various domains. Monkeys and apes can infer the location of hidden objects based on their past association with certain landmarks, the geometric disposition of other target objects, or the successive displacements behind several barriers (Menzel 1996; Hemmi and Menzel 1995; Call 2001). There are also studies on transitive inference (Boysen et al. 1993; Gillan et al. 1981; Yamamoto and Asano 1995) and various studies that have found evidence of inference by exclusion (Premack and Premack 1994; Hashiya and Kojima 2001; Tomonaga 1993; Call 2006b; Call and Carpenter 2001). Chimpanzees can also attribute certain effects to certain objects (Premack and Premack 1983; see also Hauser 2006, Hauser and Spaulding 2007 for data on rhesus macaques).

Much less is known about inferences involving tertiary object relations (i.e., object-object interactions). Recently, this approach has been successfully used in several tasks. In addition to the aforementioned task with the reward placed under one of the boards, others have included investigations of the concepts of solidity and gravity. Thus, apes can infer the location of food based on the noise that it makes or, more important, by the noise it should have made if it had been there (Call 2004). Chimpanzees can also infer the location of food by the effect its weight has on the environment (Hanus and Call 2008). In this study, chimpanzees faced a beam resting on a pivot at equilibrium with an opaque cup mounted at each end. The experimenter showed the subject a reward, moved his hand behind both cups, and deposited the reward into one of them. He then took his hands away from the beam, letting it tilt under the weight of the reward until it reached its new equilibrium point when its loaded end hit the substrate.

One key feature of these causal inference experiments is that they are always composed of an experimental condition and a control condition (figure 7.3). The control condition is perceptually as similar as possible to the experimental condition but, crucially, it differs in the underlying causal relations between its elements. Thus, subjects in the control condition of the inclined board experiment are confronted with a wedged board and a flat board placed

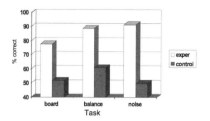

Figure 7.3   Percentage of correct trials for three inference tasks as a function of condition

side by side, whose appearance closely mirrors that of the stimuli presented in the experimental condition. Each of these two stimuli covers a hole in the platform where the reward can be hidden. This means that the reward cannot cause the board to slope, yet its appearance is very similar to that of the reward in the experimental condition, which the corresponding board does slope. In contrast to the experimental condition, subjects in the control condition showed no preference for the sloped surface (belonging to the wedge) over a flat one. Similarly, chimpanzees did not show a preference for the bottom cup in the balance experiment when the cups were mounted on a fixed sloping surface. Finally, apes did not preferentially select cups that were not shaken but were simply paired with the noise that food makes when shaken inside a cup.

The point here is that innate or acquired predispositions for certain perceptual stimuli or configurations of stimuli cannot explain the success in the experimental conditions; otherwise such success would be also present in the control conditions. Note how crucial the negative results in the control conditions are to interpret the positive results. Taken together, these results suggest that apes attribute a special status to the experimental conditions. I have hypothesized that this special status is based on some grasp of the causal relations between objects—that is, of some level of causal knowledge that apes possess about their worlds (Call 2006a).

The skeptic will say that each of the results presented in this section represents a different phenomenon. Such a collection of results does not create a coherent picture, and therefore it is open to alternative accounts. It is always possible to come up with a specific explanation for each set of results while ignoring the data set as a whole, in what we could call the scientific version of "divide and conquer." Although there is no question that a detailed analysis of each experiment is a valid analytical tool, when taken to the extreme it may produce an infinite series of particular post-hoc explanations for each single empirical result. And this is not as parsimonious as it seems (for more on this topic see Tomasello and Call 2006, Call 2007). Nevertheless, requiring a transfer across tasks that share functionally equivalent properties is fair. This is the issue that we address in the next section.

Transfer to Non–Tool-Using Tasks

It is conceivable that the use that we make of transfer tests is overrated. Transfer tests may not be so conclusive, given that problems can taint its interpretation. Moreover, they are not the only way to assess causal knowledge. In the previous section I have already outlined an alternative complementary method which consists of assessing the innate and learned predispositions that subjects may have. Nevertheless, the use of transfer tasks is still useful, and some scholars may not be convinced by anything other than transfer data. So let me use my last example to illustrate some evidence of transfer across trap tasks (Martin-Ordas and Call 2009).

Chimpanzees and other great apes were presented with a platform with two square holes cut in it to create three solid areas on its surface: one central area and two smaller ones on either side (figure 7.4). One hole was covered with transparent plastic and the other was left uncovered. Two opaque plastic cups were placed upside down, side by side, on the central area of the platform. The experimenter showed a reward to the subject and, from behind a screen and out of the subject's sight, placed it under one of the cups. The experimenter then removed the screen and moved each cup from the central area to one side so that it crossed over one of the holes (figure 7.4a). After both cups had been moved, the ape could select one by touching it.

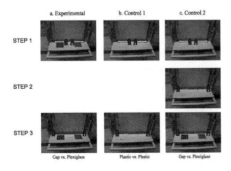

Figure 7.4 Experimental setup and procedure for the gap task (Call, unpublished data)

Martin-Ordas and Call (2009) found that apes selected the baited cup above chance levels on the first trial, and overall, but failed to do so if both holes were covered with opaque or transparent plastic (figure 7.4b). This ruled out the possibility that subjects used inadvertent cues left by the reward or the experimenter to solve the problem. Apes also failed to select the baited cup if it was moved while both holes were covered but later one hole was covered with a transparent piece of plastic and the other left uncovered (figure 7.4c). Since this is the same perceptual configuration that subjects encountered at the time of choice in the experimental condition, we can rule out the possibility that subjects had a predisposition for avoiding uncovered holes regardless of how the reward was moved.

There is another finding that makes this study interesting. Apes' performance in the gap task is positively correlated with their performance in the modified table task to which I alluded earlier (Girndt et al. 2008). This means that subjects transferred their performance across tasks that were functionally equivalent but which required different response modes. Moreover, it demonstrated that subjects did not have a predisposition to avoid stimuli that had been associated with holes in the absence of causal relations between the hole, the cup, and the reward.

Although there was a positive transfer between the modified table task and the gap task, other studies (e.g., Martin-Ordas et al. 2008) have not found any evidence of transfer across different trap tasks. Is this lack of transfer an indication of lack of causal knowledge? Not necessarily. Martin-Ordas et al. (2008) suggested that the lack of transfer may be related to a failure in analogical rather than causal knowledge. Although subjects may have knowledge about the causal relations within each task, they may not establish a connection between those relations across tasks. In other words, it is not a problem with relations between elements of the task, but with relations between relations (Gentner 2002). In this sense, subjects may understand each task as an isolated entity, but not a family of tasks that are functionally equivalent. Such a mixed picture is not totally surprising if one considers the literature in cognitive development in children. For instance, two of the key false-belief tasks (location change and deceptive box) do not correlate either (Naito 2003)—and when asked why, experts say that it is because in addition to false belief reasoning, each task also has different demands. Given the deep impact, highlighted in this chapter, that action can have on performance, the lack of correlation between trap tasks is not entirely unexpected. Indeed, the surprising thing is that some trap tasks correlate at all!

Finally, even if we accept the mixed data on transfer as adequate, one could argue that such a transfer was done at the perceptual level, not at the conceptual level. Note, however, that subjects treat holes as they treat other obstacles, even though they are perceptually very different. Unless we are willing to accept this sort of evidence as going beyond the information given, this is a problem that is hard to resolve, because studying subjects that lack language limits the questions that we can pose and the answers that we can get in return. The use of language may also be the reason why developmental psychologists are not deeply troubled with the lack of correlation between key false-belief tasks. Being able to ask subjects questions and, more importantly, listen to them give reasons for their behavior, just as our child did at the beginning of this chapter, is what is really convincing, some will argue. For those who think that way, I suspect that nonverbal measures will always remain unsatisfactory and inadequate to assess whether apes, or any other nonverbal organisms, show causal knowledge. Those scholars are probably correct in pointing out that verbalizing reasons requires a deeper knowledge than not doing so. However, this does not mean that knowledge is absent if subjects cannot talk about it.

Implications and Future Directions

It is very unlikely that we will ever be able to ask a chimpanzee why he pulled the cloth with the food on top of it. If receiving an answer to this question is the key for finding out whether chimpanzees possess causal knowledge about object-object interactions, as some scholars have suggested, then I venture that we will never know the answer. However, such a conclusion is in my view unsatisfactory. The evidence that has begun to accumulate casts serious doubt on the "externalist" view that chimpanzees solve the cloth task and other problems by attending only to perceptual features of objects without any insight about the interactions between objects and the causes of external events. Currently, the most promising hypothesis to explain the apes' behavior in situations such as those reviewed in this chapter is that they are sensitive to the causal relations between objects and are capable of using that knowledge to

avoid obstacles to obtain food and make inferences to locate hidden food.

This does not mean that associative processes play no role in shaping the observed responses. Indeed, associative processes are very likely implicated in some aspects of problem solving, but I doubt that they are the only process contributing to the observed behavior. My proposal is that various animal species have at their disposal both associative and nonassociative processes, and that they engage them depending on the task at hand and the specific aspect of the task. Thus, observable responses are a mixture of processes, not the product of a single process. This so-called "composite hypothesis" represents a clear departure from a Cartesian view that postulates a clear differentiation between the cognitive processes of humans and nonhuman animals (humans reason, whereas nonhuman animals are merely stimulus response machines). The composite hypothesis is also a departure from the view that the problem-solving behavior of certain organisms can be explained by invoking an intermediate process that is different from both associative and "higher-level" reasoning. Without denying the existence of processes that are neither association nor reasoning, the composite hypothesis considers the joint contribution of all of these processes in finding solutions to a variety of problems.

Future research is needed to identify the processes implicated in the production of these solutions. To this end, it is crucial that we use paradigms that capitalize on the response mode individuals use more readily without requiring a lengthy training regime. Assessing and isolating the contribution of both the motor and perceptual components implicated in the observed responses is one key aspect of this research. It is equally important to consider transfer tasks and assess the likelihood that certain responses reflect innate or learned predispositions rather than causal knowledge. It is also particularly important to assess the likelihood that subjects can learn to solve a task by association alone, without any participation of causal knowledge, or that they have a predisposition to respond in particular ways to certain configurations of stimuli. Finally, interpreting the results (positive and negative) in a balanced way is critical. A theory that can predict both positive and negative results will always be more successful than one that only attends to positive results and sweeps negative results under the rug.

Literature Cited

Antinucci, F. 1989. *Cognitive Structure and Development in Nonhuman Primates*. Hillsdale, NJ: Lawrence Earlbaum Associates.

Boysen, S. T., G. G. Berntson, T. A. Shreyer, and K. S. Quigley. 1993. Processing of ordinality and transitivity by chimpanzees (*Pan troglodytes*). *Journal of Comparative Psychology* 107:208–15.

Call, J. 2001. Object permanence in orangutans (*Pongo pygmaeus*), chimpanzees (*Pan troglodytes*), and children (*Homo sapiens*). *Journal of Comparative Psychology* 115:159–71.

———. 2004. Inferences about the location of food in the great apes (*Pan paniscus, Pan troglodytes, Gorilla gorilla, Pongo pygmaeus*). *Journal of Comparative Psychology* 118:232–41.

———. 2006a. Inferences by exclusion in the great apes: The effect of age and species. *Animal Cognition* 9:393–403.

———. 2006b. Descartes' two errors: Reasoning and reflection from a comparative perspective. In S. Hurley and M. Nudds, eds., *Rational Animals*, 219–34. Oxford: Oxford University Press.

———. 2007. Past and present challenges in theory of mind research in primates. In C. van Hofsten and K. Rosander, eds., *Progress in Brain Research*, vol. 164, 341–54. Amsterdam: Elsevier.

Call, J., and M. Carpenter. 2001. Do chimpanzees and children know what they have seen? *Animal Cognition* 4:207–20.

Call, J., and M. Tomasello. 2005. Reasoning and thinking in nonhuman primates. In K.J. Holyoak and R.G. Morrison, eds., *Cambridge Handbook on Thinking and Reasoning*, 607–32. Cambridge: Cambridge University Press.

Cunningham, C. L., J. R. Anderson, and A. R. Mootnick. 2006. Object manipulation to obtain a food reward in hoolock gibbons, *Bunopithecus hoolock*. *Animal Behaviour* 71:621–29.

Fujita, K., H. Kuroshima, and S. Asai. 2003. How do tufted capuchin monkeys (*Cebus apella*) understand causality involved in tool use? *Journal of Experimental Psychology: Animal Behavior Processes* 29:233–42.

Gentner, D. 2002. Analogical reasoning, psychology of. *Encyclopedia of Cognitive Science*. London: Nature Publishing Group.

Gillan, D. J. 1981. Reasoning in the chimpanzee: II. Transitive inference. *Journal of Experimental Psychology: Animal Behavior Processes* 7:150–64.

Girndt, A., T. Meier, and J. Call. 2008. Task constraints mask great apes' proficiency in the trap-table task. *Journal of Experimental Psychology: Animal Behavior Processes* 34:54–62.

Hanus, D., and J. Call. 2008. Chimpanzees infer the location of a reward based on the effect of its weight. *Current Biology* 18:R370–372.

Hashiya, K., and S. Kojima. 2001. Hearing and auditory-visual intermodal recognition in the chimpanzee. In T. Matsuzawa, ed., *Primate Origins of Human Cognition and Behavior*, 155–89. Berlin: Springer-Verlag.

Hauser, M., and B. Spaulding. 2007. Wild rhesus monkeys generate causal inferences about possible and impossible physical transformations in the absence of experience. *Proceedings of the National Academy of Sciences, USA* 103:7181–85.

Hauser, M. D. 1997. Artifactual kinds and functional design features: What a primate understands without language. *Cognition* 64:285–308.

Hemmi, J. M., and C. R. Menzel. 1995. Foraging strategies of long-tailed macaques, *Macaca fascicularis*: Directional extrapolation. *Animal Behaviour* 49:457–63.

Köhler, W. 1925. *The Mentality of Apes*. London: Routledge and Kegan Paul.

Limongelli, L., S. T. Boysen, and E. Visalberghi. 1995. Comprehension of cause-effect relations in a tool-using task by chimpanzees (*Pan troglodytes*). *Journal of Comparative Psychology* 109:18–26.

Martin-Ordas, G., and J. Call. 2009. Assessing generalization within and between trap tasks in the great apes. *International Journal of Comparative Psychology* 22:43–60.

Martin-Ordas, G., J. Call, and F. Colmenares. 2008. Apes solve functionally equivalent trap-tube and trap-table tasks. *Animal Cognition* 11:423–30.

Menzel, C. R. 1996. Spontaneous use of matching visual cues during foraging by long-tailed macaques (*Macaca fascicularis*). *Journal of Comparative Psychology* 110:370–76.

Mulcahy, N. J., J. and Call. 2006. How great apes perform on a modified trap-tube task. *Animal Cognition* 9:193–99.

Naito, M. 2003. The relationship between theory of mind and episodic memory: Evidence for the development of autonoetic consciousness. *Journal of Experimental Child Psychology* 85:312–36.

Natale, F., P. Poti, and G. Spinozzi. 1988. Development of tool use in a macaque and a gorilla. *Primates* 29:413–16.

Piaget, J. 1952. *The Origins of Intelligence in Children*. New York: Norton.

Povinelli, D. 2000. *Folk Physics for Apes: A Chimpanzee's Theory of How the World Works*. Oxford: Oxford University Press.

Premack, A. J., and D. Premack. 1983. *The Mind of an Ape*. New York: Norton.

Premack, D. and A. J. Premack. 1994. Levels of causal understanding in chimpanzees and children. *Cognition* 50:347–62.

Seed, A. M., J. Call, N. J. Emery, and N. S. Clayton. 2009. Chimpanzees solve the trap problem when the confound of tool-use is removed. *Journal of Experimental Psychology: Animal Behavior Processes* 35:23–34.

Seed, A. M., S. Tebbich, N. J. Emery, and N. S. Clayton. 2006. Investigating physical cognition in rooks, *Corvus frugilegus*. *Current Biology* 16:697–701.

Spinozzi, G., and P. Potí. 1989. Causality I: The support problem. In F. Antinucci, ed., *Cognitive Structure and Development in Nonhuman Primates*, 113–19. Hilldsale, NJ: Lawrence Erlbaum Associates.

Tebbich, S., and R. Bshary. 2004. Cognitive abilities related to tool use in the woodpecker finch, *Cactospiza pallida*. *Animal Behaviour* 67:689–97.

Tebbich, S., A. M. Seed, N. J. Emery, and N. S. Clayton. 2007. Non-tool-using rooks (*Corvus frugilegus*) solve the trap-tube task. *Animal Cognition* 10:225–31.

Tomasello, N., and J. Call. 1997. *Primate Cognition*. New York: Oxford University Press.

———. 2006. Do chimpanzees know what others see—or only what they are looking at? In S. Hurley and M. Nudds, eds., *Rational Animals*, 371–84. Oxford: Oxford University Press.

Tomonaga, M. 1993. Tests for control by exclusion and negative stimulus relations of arbitrary matching to sample in a "symmetry-emergent" chimpanzee. *Journal of the Experimental Analysis of Behavior* 59:215–29.

Van Lawick-Goodall, J. 1968. The behaviour of free-living chimpanzees in the Gombe Stream reserve. *Animal Behaviour Monographs* 1:161–311.

Visalberghi, E., and L. Limongelli. 1994. Lack of comprehension of cause-effect relations in tool-using capuchin monkeys (*Cebus apella*). *Journal of Comparative Psychology* 108:15–22.

Yamamoto, J., and T. Asano. 1995. Stimulus equivalence in a chimpanzee (*Pan troglodytes*). *Psychological Record* 45:3–21.

Yerkes, R. M., and A. W. Yerkes. 1929. *The Great Apes: A Study of Anthropoid Life*. New Haven: Yale University Press.

PART II

Tool Use
and Culture

8

A Coming of Age for Cultural Panthropology

Andrew Whiten

A chimpanzee clambers somewhat awkwardly into the top of a palm tree, carefully negotiating the stiff, sharp vegetation. She briefly examines the central growing "bud" (the apical meristem) of the tree, and turns to select a big leaf frond already detached from the tree. Then we see an extraordinary sight. She balances bipedally and, holding the stiff central rib of the frond vertically—it is taller than she is—smashes one end directly down into the bud, several times. Because this is the tree's growing point, the mashed pulp that results is highly nutritious, and the chimpanzee is soon reaching down to scoop it up before another bout of pounding. To anyone who has watched African people wielding a similarly massive pestle and mortar to grind food crops, the resemblance is striking. And so the researchers who discovered this foraging technique at their field site, Bossou in Guinea, called it "pestle pounding."

For the scientists who study chimpanzees at sites other than Bossou in Guinea, the absence of pestle-pounding behavior, even when the right kind of palm trees are present, provides a remarkable contrast. This is the case even at the nearest other field site, in the Taï Forest, less than 250 km distant from Bossou. At Taï, however, the chimpanzees employ other forms of tool use that are absent at Bossou. Any fieldworker traveling between these chimpanzee communities thus experiences something of the same surprise that is familiar to us when we exchange alien human cultures for our own. Could this be exactly what

we are seeing in the case of chimpanzees' pestle pounding: a cultural behavior pattern handed down from generation to generation in one small region of Africa but not elsewhere? Certainly, pestle pounding seems such an elaborate and "non-obvious" technique that experienced chimpanzee fieldworkers find it difficult to imagine that each chimpanzee invents it independently. By contrast, youngsters traveling with their mothers have ample opportunity to learn the skill through observation and apprenticeship. Could it be that pestle-pounding not only physically resembles the human act of wielding a pestle and mortar, but is a similarly cultural phenomenon?

This "question of animal culture," as Galef (1992) put it, has become an increasingly heated and controversial one and looks set to be so for a while yet (de Waal 2001; Laland and Galef 2009; see also chapter 14). Chimpanzees have been at center stage in these debates about the nature, or even the existence, of culture in nonhuman animals. In large part this is because evidence has accumulated that among nonhuman creatures, our nearest living relative is also closest to us in the richness of its claims to be an interestingly cultural being. However, we should also recognize that another contributing factor may be that chimpanzees' evolutionary proximity to us has generated a disproportionate number of long-term studies, so that we know more about matters relevant to inferring culture in them than in other species.

Either way, chimpanzee research has played a notably

prominent part in the science of social learning, tradition, and culture. Thus, whilst this book can be seen most obviously as a celebration of all we have come to know about this inherently fascinating genus, it is also appropriate to highlight the pioneering role chimpanzee research has played in advancing our understanding of the nature and evolution of culture and kindred topics more broadly in the animal kingdom, as well as in the special case of human cultural evolution. The impact of such research is due not only to the nature of chimpanzees, of course: it is also the result of work by a considerable number of gifted, innovative, and industrious scientists who in recent decades had the privilege of studying our sister species.

The bulk of this chapter will focus on the newest discoveries made in recent years and how they have taken our understanding of the cultural nature of the chimpanzee mind to new levels, reshaping perspectives that were in common currency just a decade ago. Cultural "*pan*thropology" has to this extent "come of age" (deWaal 1999). A predecessor to the present volume (Wrangham et al. 1994) was aspirationally called *Chimpanzee Cultures* but, as more than one reviewer noted, it was rarely able to address this topic explicitly. Ironically the one chapter that did so, which has since been much cited (Tomasello 1994), echoed Galef both in its title—"The question of chimpanzee culture"—and its skepticism, charging that to a hard-nosed experimentalist surveying the evidence, the case for chimpanzee culture was not yet compelling.

In this chapter I will argue that more rigorous analyses of field data and new social transmission experiments have since converged on a picture that is both more compelling and significantly richer in scope than what was available just a few years ago. However, this work deserves to be set in a proper historical context. To this end, the following section begins with a brief overview of how developments in the last century provided the foundations for the picture we now have.

Discovering Wild Chimpanzee Cultures: The Twentieth Century

Early field research at Gombe and Mahale in the 1960s soon led to suspicions that cultural transmission was at work, particularly in the wake of the famous discovery of skilled tool use (Goodall 1964). The sophistication implied in such behavior, and the way in which it is avidly watched by youngsters in the course of mastering the

relevant techniques, was readily interpreted as evidence of social learning.

Skeptics (e.g., Tomasello 1994) later argued that this conclusion was unwarranted because confirming that social learning is taking place requires a controlled experimental contrast in which some individuals witness a model and others do not. Only if learning occurs in the first of these conditions can we know that it is *social* learning that permits the later skill to emerge. However, as field studies proliferated, by 1973 Goodall felt able to write of "cultural elements in a chimpanzee community." In this paper she began to marshal the convergent evidence from other field sites that the skills of interest varied in their expression in ways consistent with a cultural interpretation. Goodall assembled some of the then sparse information from five main field locations. She noted putative cultural variations such as the use of a fly-whisk, recorded at Budongo in Uganda yet not recorded in 12 years of observations at Gombe despite obvious situations in which a custom of this kind would have been adaptive. In this pioneering analysis Goodall acknowledged that "much of the material is speculative," but added that "it will, I hope, draw attention to areas that with increasing sophistication of methodology and increasing cooperation between investigators could lead to major advances in the understanding of an extremely complex subject" (Goodall 1973, p. 180). This is precisely what the following quarter century has witnessed.

Goodall (1986) herself extended the comparative approach, tabulating the varying occurrence of numerous forms of tool use as reported from several field sites. McGrew later offered the first book-length analysis of the accumulating picture (McGrew 1992), including a landmark table charting the distributions of 19 forms of tool use across seven principal field sites and five lesser-studied ones. Boesch and Tomasello (1998) took a further step, applying this approach to a broader set of 25 variants that included forms of communication and other social behavior.

Invited to review these achievements (Whiten 2000), I was struck by the excitement of what the emerging picture apparently implied about the complexity of chimpanzee culture. By contrast I found the evidential base for these integrative studies unsatisfying, principally because they were based on the published literature, supplemented by personal knowledge. Relying on published reports to build a definitive chart of the distribution of behavioral variants is inadequate in a number of respects, principally in that

(1) not all behavior patterns occurring at a site are necessarily published; (2) when they are, it is not necessarily reported whether they are widespread in the community, as one might expect for a local custom; and (3) the most critical information, on absence of behavior patterns that are customary elsewhere, may well never have been published.

To overcome these limitations I was able to persuade the leaders of the principal long-term study sites to pool their local knowledge in a systematic and thorough collaborative effort. Phase 1 of this effort drew up a list of 65 behavior patterns that participants suggested might vary culturally across Africa. These were consensually defined, allowing the senior researchers at each site to then allocate them to classes of which the key ones were *customary* (common in at least one age-sex class), *habitual* (occurring at least repeatedly in several individuals, consistent with social learning), or *absent*. A crucial distinction where absence was recorded was whether there was a likely environmental explanation (the relevant food items or materials were not available, for example) or not. We classified the behavior pattern as a putative cultural variant if it was absent in at least one community without such environmental explanation being apparent, yet was customary or habitual elsewhere. Genetic explanations for these variations were thought even less likely than environmental

explanations, partly because many differences occurred between geographically close communities and partly because many variants, notably tool use, are known from experimental studies to be acquired by chimpanzees through learning (Tomasello and Call 1997).

Two core findings of these collaborative studies (Whiten et al. 1999, 2001) deserve to be emphasized. First, as many as 39 cultural variants were identified—an unprecedented figure in the literature on traditions among non-human animals, which in the majority of cases concerns only a single behavior pattern, such as bird-song dialect. A very close replication of these methods applied to wild orangutans has yielded a strikingly similar story, initially identifying 19 clear variants and five more tentative ones (van Schaik et al. 2003), more recently updated to a figure of 33 (van Schaik 2009). However, although other studies have since documented such multiple-tradition cultures in other primate and cetacean species, a recent review concluded that none of these repertoires of variants convincingly exceeded single figures (Whiten and van Schaik 2007). At present, the great apes thus appear to be rather distinctive in their cultural complexity.

The second important finding, illustrated in figure 8.1, is for two communities that are only about 160 km apart. Each community can be distinguished by its own unique

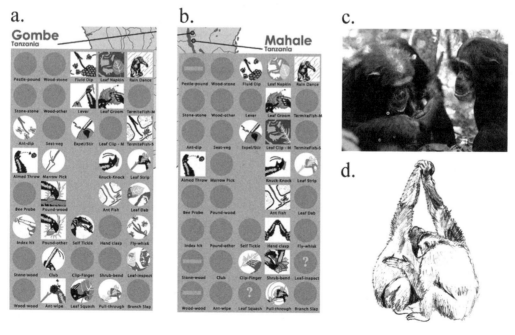

Figure 8.1 Cultural variants in wild chimpanzees: (a, b) panels contrasting two communities of the nine analyzed by Whiten et al. (2001), Gombe and Mahale, which are just 160 km apart. Square icons = customary; disc = habitual; barred disc = absent with no apparent environmental explanation; faded disc = present; blank disc = absent with no environmental explanation. Note that each community displays a unique culture defined by its particular patchwork of customary and habitual traditions. (c) Leaf-grooming: note the close visual attention. Photo by David Bygott. (d) Hand-clasp grooming at Mahale. Drawing by David Bygott.

array of customary and habitual variants (figures 8.1a, 8.1b). If we call these variants "traditions" and the local array of these traditions a "culture" (see Whiten 2005, 2009 for more extended discussion of such terminology in animal culture studies), then we see here a simple form of exactly what we usually mean when we talk of human cultures. Chinese culture is distinguished from Scottish culture, for instance, on the basis of multiple specific traditions concerning preferred foods, tools, communication patterns, and so on.

This body of work has had considerable impact in the field of animal social learning generally, but also beyond, with the full chart of which figure 8.1 is a part being replicated in secondary reports ranging from serious journalism (Angier 1999) to human evolution textbooks (Lewin and Foley 2004). The approach developed has been adopted as a template for investigating other species, including van Schaik et al. (2003) for orangutans, Stoinski (2005) for zoo-living gorillas, and Furuichi and colleagues (in progress) for bonobos (see also Hohman and Fruth 2003). Others have completed numerical analyses of the original database in ways that reinforce and elaborate on our original conclusions (van Schaik et al. 2003; Lycett et al. 2007). In sum, not only has this Collaborative Chimpanzee Cultures Project (CCCP) provided us with the most definitive picture of chimpanzee culture to date, but through it, chimpanzee research has had numerous ramifications for both closely and distantly related disciplines. Round two of the CCCP (CCCP2) is ongoing, led by Whiten, McGrew, and Boesch and now extending to 12 research sites across Africa.

But all is not as rosy as this last celebratory paragraph would suggest. This approach to studying culture in the wild has some obvious weaknesses. One is that by focusing only on cultural variants, it neglects behavior patterns that may be culturally transmitted yet important enough to occur in all communities, like such aspects of human culture as knife use. A likely chimpanzee equivalent may be using a leaf sponge for drinking otherwise inaccessible water, which was recorded at all sites and so was not included in our list of 39 cultural *variants*. However, given all we know of the way in which chimpanzees learn tool use observationally (Tomasello and Call 1997; Whiten et al. 2004, 2005, 2007), it would be surprising if leaf-sponging was not acquired through social learning. If it is, it and perhaps some other "universal" behavior patterns may be cultural

phenomena our method misses. A second weakness is that by rejecting common-versus-absent contrasts that are environmentally explicable, we may sometimes neglect cases in which culture is providing an *adaptation* to the local ecology, as it does so importantly in such human cultural phenomena as the igloo. Since we also excluded analysis of vocalizations (because analysis requires specific technical recording methods) and diet (because it is particularly difficult to exclude the influence of unknown differences in composition, etc.) there are thus numerous ways in which our analysis may underrepresent the complexity of chimpanzee culture. It is thus possible that we have so far described only the tip of a chimpanzee "cultural iceberg."

There is a converse weakness that may mean we have instead *over*represented the scope of chimpanzee culture through failing to recognize alternative genetic or environmental explanations. In short, it is inherently difficult to demonstrate unequivocally that wild chimpanzees are indeed learning, through observation, the behavior patterns we identified as putative cultural variants. Above all others, this is a weakness that commentators from outside our project have critiqued (Fragaszy and Perry 2003; Byrne 2007; Laland and Galef 2008). Careful research on one of the putative cultural variants, concerning techniques for gathering ants from their swarming columns and nests by dipping sticks and stems into the swarm, has shown that environment shapes these behaviors more than had been earlier inferred (Humle and Matsuzawa 2002). However, this picture has now been discovered to be yet more complex, with some features apparently shaped by environment and others representing cultural variants (Humle 2006; see also chapter 10). Field primatologists appear to vary in the extent to which they acknowledge and share these concerns, which I suspect is in part because they know wild chimpanzees so well that the confidence of their own judgments can be difficult to convert into objective scientific terms necessary to convince outside commentators. Yet, it is in the nature—and the strength—of good science to have to achieve exactly that. How can we do better?

In theory, one way to clinch the matter would be through field experiments. One such approach would be to translocate a chimpanzee proficient in some putative cultural variant, such as nut cracking, to a naïve community and record whether the behavior spreads. The ethical and logistic difficulties this poses probably mean it will not happen. However, a variant on this has been achieved, through the work of Matsuzawa (1994 and chapter 1) and colleagues introducing novel nut species into a nut-cracking community. Biro et al. (2003) provide an important and detailed analysis of the subsequent spread of cracking the new nuts. However, this still suffers the weakness that we cannot be sure how much of this was due to progressive individual experimentation over the years involved, rather than social learning. A different approach would be to introduce into a wild community one of the "artificial fruits" used in our studies with captive chimpanzees (e.g., Whiten et al. 2000), an avenue we ourselves are exploring (but haltingly, given the neophobia of wild chimpanzees).

Given the constraints on such efforts in the field, I have instead sought to complement the CCCP2 project with large-scale social transmission experiments with captive chimpanzees. These have now become quite extensive, and it is to these we now turn.

Cultural Diffusion Experiments

What kind of experiments does one need to establish chimpanzees' capacity to sustain cultural variation in behavior? Strangely, although more than 40 chimpanzee social learning experiments have been conducted over the last century (Tomasello and Call 1997; Whiten et al. 2004), often with relevance to cultural transmission being an implicit or explicit rationale, they remain of limited significance because they were typically merely dyadic, asking only what an observer subject learned from watching a model (who was often a human, not a chimpanzee).

A different and more relevant paradigm has existed since the time of Sir Frederick Bartlett's (1932) pioneering memory experiments, in which changes in behavior were tracked along chains of individuals, each individual learning from the last in the chain. This embodies the essence of true cultural transmission, and in recent decades this paradigm has been adopted by a handful of researchers in animal behavior, most studying rats and birds. Whiten and Mesoudi (2008) provide a comprehensive survey of this little corpus of just over 30 studies. Interestingly, we believe the first such animal study was of chimpanzees. Emil Menzel, ever ahead of his time, exposed three juvenile chimpanzees to two novel objects, which they tended to avoid; then one youngster was removed and another added, a process that was repeated 16 times, thus involving 19 individuals in total (Menzel et al. 1972). Over the first

half-dozen "generations" habituation grew and peaked, after which naïve immigrants quickly followed suit. A culture of acceptance of the objects had been established.

More powerful designs were later developed to contrast an experimental condition, in which some behavior was seeded using a trained model, with a control condition that had no model. In this way, Laland and Plotkin (1990), for example, demonstrated the transmission of digging for buried pieces of carrot along chains of eight rats. A more elaborate experimental design of this type particularly caught my eye: Galef and Allen (1995) seeded *two* experimental chains of rats with preferences for differently flavored foods. They then tracked the continuation of these preferences in the two respective traditions as they passed along transmission chains, with a new, naive rat replacing the longest-serving one in a group of four at each step. It occurred to me that if we replaced simple dietary preferences with the more elaborate behavioral techniques of which chimpanzees are capable, here would be the optimal design for an experimental study that sought to test the capacity of the species to sustain differential traditions of the kinds inferred in the wild.

This essential idea has now been implemented in a suite of studies completed in collaboration with colleagues at the Yerkes National Primate Research Center (Atlanta, Georgia,) and at the University of Texas M.D. Anderson Cancer Center (Bastrop, Texas), where appropriately sized communities of chimpanzees live in large enclosures. The first three of these studies, completed at Yerkes, are described in detail in this volume by Victoria Horner (see chapter 9). Here I focus more on later studies conducted at Bastrop by my colleague Antoine Spiteri.

Each of our first three experiments at Yerkes showed that two groups of chimpanzees, seeded with different techniques for achieving the same valued outcome, would perpetuate them to generate two different traditions (Whiten et al. 2007; Horner et al. 2006; Bonnie et al. 2007). The next outstanding question was whether the same would be true of transmission between groups. In the wild, the regional distribution of such behaviors as nut cracking and leaf grooming are thought to be due to social diffusion (Boesch et al. 1994; McGrew et al. 1997; Whiten et al. 2001), but this would require repeated intergroup transmission. The fact that varying degrees of corruption occurred in each of our Yerkes studies meant that whether techniques would spread with significant fidelity to neighboring groups, where a number of factors might make transmission a more fragile phenomenon, remained unknown. Bastrop offers excellent opportunities to investigate this issue because its chimpanzees live in communities numbering nine to eleven individuals in eight adjacent enclosures that have large barred windows, allowing the inhabitants to watch their neighbors. Here we ran intergroup diffusion studies using two artificial foraging tasks: a "probe task" that could be tackled using either of two techniques, stab or slide (figures 8.2a, 8.2b) and the "turn-ip," where the alternatives were turn-and-press or ratchet-and-slide (figures 8.2c, 8.2d; Whiten et al. 2007). Each task was challenging, requiring a sequence of two different actions to gain food, and in baseline tests chimpanzees explored them extensively yet never extracted the food.

For each task, the experiment began with a chimpanzee in one group being trained to perform one technique, and a chimpanzee in a second group being trained to perform the other, each individual out of sight of the rest of her own group. Each expert was then reunited with her group. As in the Yerkes studies, we found the predicted differen-

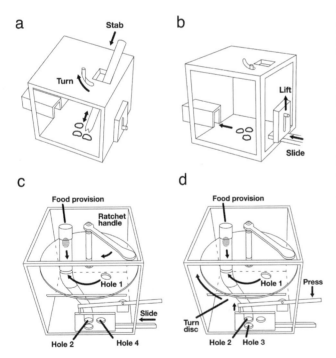

Figure 8.2   Foraging devices used in social diffusion studies: (a) Stab technique used to extract food items from probe task by first pushing button to open hole, then using tool to stab food. (b) Slide technique, in which hatch is opened and flat tool used to push food out. (c) Turn-ip, ratchet-then-slide technique, in which the ratchet arm is used to turn the disc so food falls through the tube and is then extracted by sliding the lever to align holes to release food. (d) Turn-ip, turn-then-press technique in which the disc is directly turned and a second lever is pressed down to align holes and release the food.

tial spread of the alternative traditions, together with some corruption (i.e., some individuals in each group discovered some of the alternative approaches employed by the other group). Once half the individuals in a group were operating successfully, the foraging devices were moved to a position where the neighboring group could watch their use, and after three two-hour observation sessions, the neighboring group was given the task. This procedure was then repeated for a third group. Thus the experiment eventually extended to two trios of groups for each task.

Each technique was found to spread with rather impressive fidelity. The third of each trio of groups typically displayed no more corruption than had occurred in the earlier groups. In one case (the slide technique for the probe task) the final group showed 100% fidelity to the technique seeded in the first group, and in the turn-ip task each respective technique was completed only in the final group using only the technique seeded in the original seeding event. We suspect that this degree of fidelity was due to the very complexity of these two-step tasks, which—particularly in the case of the turn-ip—exceeds that of the tasks in our earlier studies. Since control subjects did not succeed in the tasks, it was clear that the experimental subjects learned a quite novel skill through observation.

These results offer significant support to the inference arising from field research that some chimpanzee traditions have spread regionally through repeated inter-community transmission events. However, it is important to acknowledge that in the wild, the particular kind of group-to-group observation opportunities we engineered at Bastrop would not be expected. Parties of chimpanzees encountering neighboring communities are antagonistic (Goodall 1986). The likely route to inter-community transmission would be through immigration of skilled individuals into naïve communities, a scenario entirely consistent with the results of the field experiments of Matsuzawa and colleagues outlined earlier (Matsuzawa 1994, 2001; Biro et al. 2003; see also chapters 1, 12). Simulating this with captive chimpanzees would be an undertaking more challenging than our experiments outlined here. One implication of our results is thus to justify the additional care (mindful of animal welfare) and effort that will be required to conduct any such immigration (translocation) experiment.

We can now take stock of this suite of diffusion studies as a whole. Above, I noted three experiments at Yerkes and two at Bastrop. A sixth experiment completed at Bastrop (Hopper et al. 2007) is the odd one out in that although there was a significant social learning effect favoring the seeded technique, this produced no clear differential traditions, despite the experiment being conducted with one of the tasks (the "panpipes") that generated differential traditions at Yerkes. We suspect this resulted from enrichment routines at Bastrop producing a significant bias favoring one of the two techniques, and we are running further experiments to investigate this. Setting this result aside, the other five experiments produced clear differential traditions, summarized in figure 8.3. Note that figure 8.3 displays an important resemblance to figure 8.1, insofar as in both cases, each community differs from another in a profile that spans more than one behavior pattern. In the case of figure 8.1, the hypothesis that this is due to social learning is supported by strong circumstantial evidence—but as we have seen above, not all commentators find this compelling. In the case of figure 8.3, the role of social learning is confirmed by experimental procedures that leave much less room for doubt.

The consistent and coherent picture generated by the complementary field and experimental diffusion data is thus of a species capable of maintaining unique local cultures, each constituted by multiple traditions (figure 8.1 shows two of the six such cultures illustrated in figure 1 of Whiten et al. 1999; figure 8.3 here illustrates four cultures defined in this way). At present, cultural panthropology is distinctive in this respect. Aside from the complexity illustrated in figure 8.3, the only primate diffusion experiments completed are the pioneering exploration of Menzel et al. (1972) and a partial replication of the Horner et al. (2006) study with capuchin monkeys (Dindo et al. 2008). This paucity of primate social diffusion experiments may seem surprising, given the power of the two-action, three-group paradigm we have developed for addressing questions about the transmission of culture. Part of the answer may lie in the numerous practical and socio-dynamic matters that need to be negotiated in completing such work with socially sensitive animals, especially in the case of diffusion chains, as Horner makes clear in chapter 9. Diffusion experiments in other species amount to a total of approximately two dozen, each concerned with the spread of a single potential tradition; this corpus is reviewed in detail by Whiten and Mesoudi (2008).

Social Learning Mechanisms

The diffusion experiments are limited in what they can tell us about potential underlying learning mechanisms. The point of these experiments is to investigate fidelity of transmission, irrespective of the mechanisms involved. However, the three-group comparison does reveal more than two-group studies can. If, in diffusion studies that compare only two groups, spread of a seeded behavior in the experimental group occurs with little or no such spread in a no-model control group, then we know that social learning has occurred, but not what kind of learning it is. If in our own studies we had compared only one experimental group to controls and found the differences we did, we would have to acknowledge that possibly the group able to watch a model simply learned that food could be obtained from the foraging device, and then persisted in achieving its own solution. However, when we find the differences illustrated in figure 8.3 between the groups seeded with the two different techniques, we know that in some way those techniques are being differentially "copied" or "replicated" in the two groups.

Still, it remains the case that this replication could have been generated by various social learning mechanisms. The most commonly considered of these mechanisms are stimulus/local enhancement (merely having one's attention drawn to some relevant object or locality: Whiten and Ham 1992), emulation ("learning things about the environment by observing the manipulations of others": Tomasello 1996, p. 321) and imitation ("copying some part of the form of an action": Whiten and Ham 1992). Often these three will represent a continuum of potential fidelity of copying, from the minimum involved in mere enhancement to the maximum in faithful imitation. Accordingly it has often been assumed that imitation would be crucial for cultural transmission (Tomasello 1994), and that provides one reason to investigate the mechanism underlying the diffusion effects we have documented. However, this means it would be no less interesting if the processes were found to be as simple as enhancement, for this would challenge the basic assumption that such diffusions must rest on the more sophisticated processes.

I shall focus on two of a suite of recent experiments we have completed on this topic. The first is aimed rather directly at testing for emulation learning—an important is-

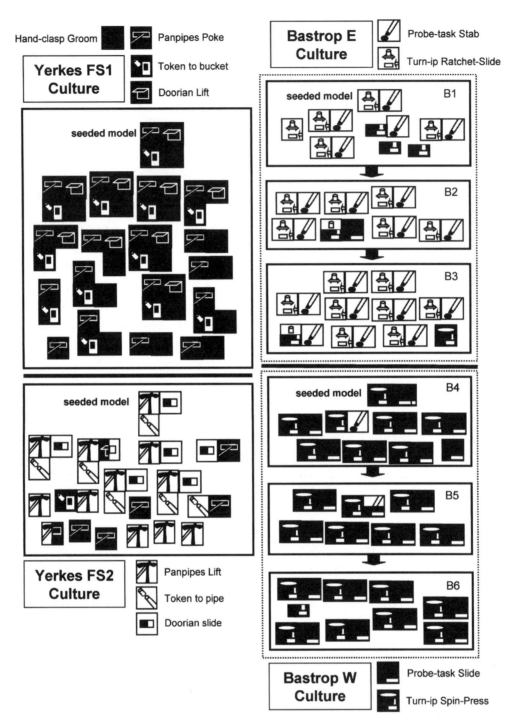

Figure 8.3   Overview of results of diffusion experiments, creating four different cultures. Each cluster of icons represents the behavioural profile of one chimpanzee, made up of up to four behavior patterns as indicated in the key. In separate experiments the Yerkes FS1 group was seeded with a model displaying three different foraging methods (dark symbols) and FS2 was seeded with alternative methods (light symbols). These spread with the fidelity indicated in each group. The additional difference between FS1 and FS2 in handclasp grooming arose spontaneously. At Bastrop two-seeded techniques spread from group B1 via B2 to B3, while the alternative techniques seeded in B4 spread via B5 to B6. In both cases fidelity was high but not perfect, as indicated. For further explanation see Horner (chapter 9), and the more detailed results and figures in Whiten et al. (2007).

sue because of influential conclusions drawn by Tomasello and Call (Tomasello and Call 1997; Call et al. 2005), on the basis of several studies, that chimpanzees (apart from those "enculturated" by intimate association with humans) are essentially emulators—contrasting with children, who are true imitators. If this is true, it likely has big implications for the cultural capacities of the two species (Tomasello 1999; Whiten et al. 2009).

We have aimed to test for emulation learning using the procedure called a "ghost condition," advocated for this purpose also by Tennie et al. (2006). In a ghost condition, the manipulandum normally moved by a model in a full social learning condition is instead made to move by some remote means, with no model responsible. The rationale is that this will provide precisely the information which for an emulator is hypothesized to be sufficient to shape a matching response. Just a handful of such studies have been completed to date for nonhuman animals and children. These are reviewed in the only two chimpanzee reports I am aware of, by Tennie et al (2006) and ourselves (Hopper et al. 2007). Tennie's pioneering study included a full social learning condition in which a chimpanzee model either pulled or pushed open a door flap to obtain food, a control condition with no model, and a ghost condition in which fishing line was used to move the flap, with no model present. Unfortunately, since there was no matching in the full model condition, the question of whether emulation might account for any such effect became rather redundant in this study.

We instead looked at a task that we had already shown to be socially learned—indeed, sufficiently well to generate differential traditions—the "panpipes" (Whiten et al. 2005; see also chapter 9). Since the evidence for social learning is strongest for the "lift technique," in which a stick-tool is inserted under the T-bar of this device and pulled upwards to remove a blockage and release food, we focused only on that technique and arranged two different ghost conditions in which fishing line was used to either (1) raise the T-bar to release the food, or (2) raise both tool and T-bar, as happens when a model performs the action normally. Observer chimpanzees watched either of these displays more than 200 times, corresponding to what happened in actual diffusion studies. None, however, learned the lift technique. To a human who has watched the ghost-with-tool condition in particular (the reader can probably

imagine what they themselves would expect to learn from seeing it), it is quite striking, even surprising, that this was of so little help to chimpanzees.

Thus, if the notion that chimpanzees are emulators implies that they should have learned from our ghost condition (which Tomasello and Call's collaboration in the Tennie et al. study appears to imply) then our results must be seen as demonstrating rather dramatically that this description of chimpanzees is incorrect. Thorndike (1898) defined imitation, by contrast, as "doing an act from seeing it done," and our results suggest that this is exactly what chimpanzees need to witness in order to replicate what they see.

This, however, still leaves open the question of how much chimpanzees may also be learning from changes in the environment in the common situation where they are watching a chimpanzee make the events happen. Our research suggests they are learning significant things. For example, Horner and Whiten (2005) showed that when young chimpanzees witnessed a model performing a series of novel tool-based actions to extract food from a box, they would later neglect to copy components of the sequence that could be seen to be ineffectual in a transparent version of the task, but would include them in a fuller copy of what they saw when the box was instead opaque. We interpreted these results by conceptualizing emulation and imitation not simply as a dichotomy but as a continuum of fidelity, with the opaque box results thus seen as a relatively faithful imitation of the whole sequential performance, and the more selective match in the transparent box condition as being more emulative.

Taking into account this new range of findings, other findings over the last twenty years (reviewed by Whiten et al. 2004), and several of our recent and related studies yet to be published, a picture is emerging of chimpanzees as being in some ways more restricted (noting the ghost condition results), yet in many other ways more flexible social learners (noting the Horner and Whiten results, for example), than has been appreciated in former times. It is less appropriate to try to pigeonhole them as emulators or imitators than to recognize that they possess a "tool kit" of social (and nonsocial) learning strategies, and will deploy them in ways that can be adaptive in the particular circumstances of any one learning opportunity, depending on such factors as the nature of their preexisting knowledge and the nature of the task at hand (Hopper et al. 2008).

Back to the Wild

The new findings from the experimental studies give us fresh confidence that chimpanzees have cultural capacities consistent with the inference drawn—on the basis of very different kinds of evidence—that aspects of behavioral diversity in wild chimpanzees are cultural in nature. Armed with this knowledge, we now return to a focus on the wild and survey some more recent developments.

Following the discoveries resulting from the systematic analyses made possible by the CCCP, a second round in this program (CCCP2) has been initiated in the present century. CCCP2 is more ambitious than its predecessor in a number of respects, notably inclusion of more study sites (12), refinements in methodology, and analysis of a much broader range of candidate behavior patterns. In part because of this, CCCP2 remains in progress at the time of this writing, and a finished set of conclusions remains for the future.

In the meantime we can note that evidence has continued to accumulate in the literature, suggesting that more cultural variants exist. Indeed, each year of this century has seen at least one such report, and often several. The new studies can be loosely allocated to either of two main categories: (1) new subtleties described for long-studied populations; and (2) discoveries at relatively new study sites. Both of these would be predicted by our earlier work—the first because the CCCP1 reports have since focused field researchers' keen eyes on spotting previously undocumented differences between sites, and the second because if chimpanzees are substantially cultural animals, investigation of each new locality can be expected to reveal new behavioral variants, as has occurred when explorers contact new and far-flung pockets of humanity.

Subtle Variations

An instructive example of what is meant by this arose soon after the CCCP reports. Assembly of video clips to illustrate the behavior categories we analyzed happened to include leaf-sponging, which as noted earlier had been found at all sites and therefore was not included in the original list of 39 cultural variants. Gombe chimpanzees, however, were seen to leaf sponge in streams, and this was immediately recognized as a variant absent at Taï. This relative chance discovery encouraged an expectation that examining behavioral subtleties with a finer-tooth comb might reveal variants not hitherto recognized.

Such has proved the case. Discoveries have been facilitated most usefully when a researcher with extensive knowledge of chimpanzees at one site has visited another. Some such differences may be recognized quite quickly, as happens to all of us when visiting foreign lands. For example, inter-site visits led to the recognition at Mahale of a behavior called "social scratch," that was not known at Gombe (Nakamura et al. 2000); later it took Nishida, who had decades of experience at Mahale, only a month to see that at a third site, Ngogo, social scratch was in place yet took a different form (Nishida et al. 2004). At Mahale social scratch is typically a long, raking movement, whereas at Ngogo it involves short jabs.

Accordingly Nakamura and Nishida (2006) have advocated a project of "culture hunting" in which experienced field researchers shift sites and actively scan behavioral repertoires for patterns different from those familiar in the original chimpanzee community. This may uncover more subtle differences in behavior, like those we are barely aware of in crossing between human cultures but which can be documented by the attentive anthropologist. With this aim, Nakamura visited Bossou and identified four subtle contrasts with behavior familiar to him at Mahale: a greeting involving a *mutual genital touch* (common at Bossou, absent at Mahale); a courtship display involving a *heel tap* (common among males at Bossou, absent at Mahale where males instead foot-stamp); an *index-finger-to-palm* action in grooming, probably dealing with tiny ectoparasites (habitual at Bossou, absent at Mahale where a variant incorporating leaves occurs: see figure 8.1c); and a *sputter* sound during grooming (habitual in some chimpanzees at Bossou, absent at Mahale: Nakamura and Nishida 2006).

Comparison of neighboring communities obviously offers similar opportunities for comparison by experienced observers, but achieving this represents a major practical challenge and still remains relatively rare. A nice example concerns the famous grooming handclasp (figure 8.1d), originally described in K-group at Mahale but more recently discovered to take a subtly different form in neighboring M-group, where groomers tend instead to use a wrist-to-wrist configuration (McGrew et al. 2001).

An adult female, Gwekulo, who transferred from K-group to M-group, adopted the preferred wrist-wrist convention of her new community some of the time, but also managed to get them sometimes to make palm-to-palm contact (Nakamura and Uehara 2004). To do this, however, she had to make delicate adjustments, flexing her elbow in the local customary fashion rather than keeping it straight, as occurs in the classic A-frame configuration (figure 8.1d). Observation of female inter-community transfers and their assumed crucial role as culture-bearers remains an uncommon achievement. This particular example offers perhaps the first nonhuman instance of a kind of cultural blending, at least in this one individual—thus addressing a controversy current in the human cultural literature, of whether cultural evolution conforms largely to a branching phylogenetic tree structure, analogous to biological evolution (Mesoudi et al. 2004), or instead exhibits blending (Tehrani and Collard 2002).

Discoveries at Younger Research Sites

If chimpanzees are as cultural as this chapter suggests, we shall not be surprised to hear that studies of different and more distant communities reveal novel forms of behavior. Recent publications support this.

One set of impressive examples comes from studies in the Goualougo triangle of the Congo region, where hidden cameras have been used to good effect to obtain early high quality information (Sanz et al. 2004). These chimpanzees display an impressive range of tool use which includes a tool set for termite fishing, described in detail by Sanz and Morgan in chapter 11 of this volume. This and other techniques such as beehive pounding have not been observed at other sites. At Fongoli in Senegal, another suite of distinctive behavior patterns has been discovered, including use of cave shelters (Pruetz 2001), bathing, and weapons for stabbing prey within tree holes (Pruetz and Bertolani 2007). These patterns have yet to be assimilated into the CCCP2 analysis, and the viability of alternative environmental and cultural explanations has yet to be fully evaluated. Bathing is one obvious case that is likely to reflect adaptation to local, extremely hot seasonal conditions. As noted earlier, this does not negate the possibility that like other cases it may be a culturally transmitted adaptation, but it does mean that the conservative scythe of our exclusion methodology may exclude it from CCCP2's list

of putative cultural variants. The subtle variants discussed in the previous subsection are less prone to this concern because of their inherently social nature.

In summary, the CCCP phase of cultural panthropology has ushered in a cluster of more systematic and focused efforts to delineate the scope of culture in chimpanzees, and a burgeoning corpus of exciting and high-quality new discoveries is accumulating. Other components of this growing corpus and their cultural status are reviewed in chapters 10, 11, 13, and 14. It is also important to note that approaches other than inter-site comparison can make powerful contributions to our understanding of cultural phenomena in chimpanzees. These include studies that test specific predictions at a single site, as in the work of Lonsdorf (2005) showing that differences in the involvement of young males and females with expert termite fishers predicts differences in their later competence and style, and the use of field experiments pioneered by the Matsuzawa group, discussed earlier.

Implications and Future Directions

One of the most exciting aspects of the recent findings reviewed above is their convergence on a coherent picture of the "cultural mind" of the chimpanzee, which is especially the case in examining data from the wild (figure 8.1) and our diffusion experiments (figure 8.3): If there is a particular theme running through the discoveries of the last decade, we might label it "richness" or "complexity": from the perspective of comparison with other nonhuman taxa, the cultures of chimpanzees (and at least some other apes) are rich in traditions, both locally and across Africa as a whole; diffusion experiments have likewise generated rich outcomes (figure 8.3); underlying mechanisms increasingly defy simple classifications; and as work expands across Africa and with captive populations, the story gets only richer. Fifty years ago, this richness in the cultural mind and achievements of chimpanzees could scarcely have been dreamt of. These conclusions have a number of implications.

The Cultural Mind and Brain of the Chimpanzee

Theories abound for why humans have the enormous brains we do, but a hypothesis little considered is that it may be at least partly because our brains must store in accessible form all the vast information we acquire culturally. Could the same be true for a nonhuman animal who appears also to rely considerably on culture (Whiten 2006), even if in a much lesser way than we do? Whiten and van Schaik (2007) addressed this question in some depth and concluded that there is supporting evidence for two complementary hypotheses. The first is that an ape needs to be smart to learn socially all that he or she acquires. Evidence consistent with this includes a variety of ways in which the social learning capacities of great apes appear sophisticated compared to those of other animal taxa, as reviewed by Whiten et al. (2004) and Whiten and van Schaik (2007). The second hypothesis is that, in van Schaik's words, "culture makes you smart"—that is, inheriting significant expertise socially makes you a significantly more competent operator. Whiten and van Schaik offer evidence that for both chimpanzees and orangutans, greater availability of models predicts a greater repertoire of the more complex types of cultural behavior patterns. In sum, there is now increasing evidence supporting a "cultural intelligence hypothesis" that explains the large relative and absolute size of great ape brains as, at least in part, adaptations for the levels of cultural complexity they sustain.

The Evolution of Culture

Probably all of the principal features of the cultural mind of the chimpanzee are shared with humans (which remains consistent with the fact that in many respects they are also surpassed by those of humans); parsimony thus suggests that these features were shared also with our common ancestor of five to six million years ago. This issue is discussed at greater length in Whiten et al. (2003) and Whiten (2009). It implies that our own distinctive cultural trajectory did not emerge out of the blue, but was built on rich foundations of the kind we can now glimpse in chimpanzees—although we must remember that they too have been evolving biologically and culturally from that common ancestor over the same period.

Welfare

The cultural capacity of chimpanzees is one more factor to take seriously in their captive care, whether in zoos or at research institutions (see chapter 24). Both these environments could enrich their lives by giving them the op-

portunity to acquire and perhaps modify their own local cultures. If implemented in zoos, this could simultaneously offer the prospect of a major initiative in public engagement with science, as visitors would directly perceive chimpanzees' cultural propensities.

Conservation

A message of the CCCP is that not only are we losing chimpanzees but, likely at a faster rate, we are losing chimpanzee cultural diversity (Whiten 2005a, b). Perhaps, however, the discoveries reviewed above and in the rest of this volume can be used more effectively to help protect them. I recall an African accustomed to eating chimpanzee meat who, upon viewing a video of chimpanzees using sign language, commented to the effect that "I can no longer eat this creature; he is too much like me." Learning more of the cultural nature of chimpanzees could surely have a similar effect.

Future Research

The work reviewed here suggests numerous future research directions. I end by briefly highlighting four of them.

DIFFUSION STUDIES. We hope we have laid the foundations for a wider application of diffusion experiments. To give but one example of an issue not yet addressed, diffusion studies offer the prospect of more systematically exploring chimpanzees' capacity for cumulative cultural evolution, and of learning what holds them back in comparison to ourselves (see Marshall-Pescini and Whiten 2008).

SOCIAL LEARNING MECHANISMS. Many experiments on social learning mechanisms have perhaps not started in the most productive place. In the context of our research program as a whole, I would now suggest that an optimal strategy is to first discover what behavior patterns do diffuse across communities with appreciable fidelity, and only then address the underlying mechanisms. Research from this perspective has only just begun.

SOCIO-DYNAMICS. The diffusions we have documented raise many unanswered question about the underlying social dynamics, such as, for example, who watches and learns from whom. Chapter 9 discusses this issue further.

COMPARATIVE RESEARCH. After more than a century of research on social learning, it is a striking fact that our social diffusion studies remain virtually the only ones in primatology. We hope the various methods we have developed will be applied by other researchers not only to primates but also to other animals such as birds, where excellent research foundations already exist. This offers the exciting prospect of an era of comparative culture studies for which the chimpanzee research reviewed here would provide a substantial foundation.

Acknowledgments

Thanks go to the Biotechnology and Biological Sciences Research Council, Leverhulme Trust, Max Planck Institute, National Science Foundation, Royal Society, University of St. Andrews, and Emory University for supporting various parts of the research program outlined here. The author was supported by a Royal Society Leverhulme Trust Senior Research Fellowship during the writing of this chapter.

Literature Cited

Angier, N. 1999. Doin' what comes culturally. *New York Times*, June 17, 1999, p. 1, A28.

Bartlett, F. C. 1932. *Remembering*. Oxford: Macmillan.

Biro, D., N. Inoue-Nakamura, R. Tonooka, G. Yamakoshi, C. Sousa, and T. Matasuzawa. 2003. Cultural innovation and transmission of tool use in wild chimpanzees: Evidence from field experiments. *Animal Cognition* 6:213–23.

Boesch, C., and M. Tomasello. 1998. Chimpanzee and human cultures. *Current Anthropology* 39:591–694.

Boesch, C., P. Marchesi, N. Marchesi, B. Fruth, and F. Jonlian. 1994. Is nut-cracking in wild chimpanzees a cultural behaviour? *Journal of Human Evolution* 26:325–38.

Bonnie, K., V. Horner, A. Whiten, and F. B. M. de Waal. 2007. Spread of arbitrary customs among chimpanzees: A controlled experiment. *Proceedings of the Royal Society of London B* 274:367–72.

Bonnie, K. E., and F. B. M. de Waal. 2006. Affiliation promotes the transmission of a social custom: Handclasp grooming among captive chimpanzees. *Primates* 47:27–34.

Byrne, R. W. 2007. Culture in great apes: Using intricate complexity in feeding skills to trace the origins of human technical prowess. *Philosophical Transactions of the Royal Society B* 362:577–85.

Call, J., M. Carpenter, and M. Tomasello. 2005. Copying results and copying actions in the process of social learning: Chimpanzees (*Pan troglodytes*) and human children (*Homo sapiens*). *Animal Cognition* 8:151–63.

DeWaal, F. B. M. 1999. Cultural primatology comes of age. *Nature* 399:635–36.

———. 2001. *The Ape and the Sushi Master: Cultural Reflections of a Primatologist*. New York: Basic Books.

Dindo, M., B. Thierry, and A. Whiten. (2008) Social diffusion of novel foraging methods in brown capuchin monkeys (*Cebus apella*). *Proceedings of the Royal Society of London B*, 275:187–93.

Fragaszy, D. M., and S. Perry, eds. 2003. *The Biology of Traditions: Models and Evidence*. Cambridge: Cambridge University Press.

Galef, B. G., Jr. 2004. Approaches to the study of traditional behaviours of free-living animals. *Learning and Behavior* 32:53–61.

———. 1992. The question of animal culture. *Human Nature* 3:157–78.

Galef, B. G., Jr., and C. Allen. 1995. A new model system for studying animal traditions. *Animal Behaviour* 50:705–17.

Goodall, J. 1964. Tool-use and aimed throwing in a community of free-living chimpanzees. *Nature* 201:1264–66.

———. 1986. *The Chimpanzees of Gombe: Patterns of Behaviour*. Cambridge, MA: Harvard University Press.

Hohmann, G. and B. Fruth. 2003. Culture in bonobos? Between-species and within-species variation in behaviour. *Current Anthropology* 44:563–71.

Hopper, L., A. Spiteri, S. P. Lambeth, S. J. Schapiro, V. Horner, and A. Whiten. 2007. Experimental studies of traditions and underlying transmission processes in chimpanzees. *Animal Behaviour* 73:1021–32.

Hopper, L. M., S. P. Lambeth, S. J. Schapiro, and A. Whiten. 2008. Observational learning in chimpanzees and children studied through "ghost" conditions. *Proceedings of the Royal Society B* 275:835–40.

Horner, V. and A. Whiten. 2005. Causal knowledge and imitation/emulation switching in chimpanzees (*Pan troglodytes*) and children (*Homo sapiens*). *Animal Cognition* 8:164–81.

Horner, V., A. Whiten, and F. B. M. de Waal. 2006. Faithful replication of foraging techniques along cultural transmission chains by chimpanzees and children. *Proceedings of the National Academy of Sciences of the USA* 103:13878–83.

Humle, T. 2006. Ant-dipping in chimpanzees: An example of how microecological variables, tool use, and culture reflect the cognitive abilities of chimpanzees. In *Cognitive Development in Chimpanzees*, ed. T. Matsuzawa, M. Tomonaga, and M. Tanaka, 152–75. Tokyo: Springer-Verlag.

Humle, T., and T. Matsuzawa. 2002. Ant-dipping among the chimpanzees of Bossou, Guinea, and some comparisons with other sites. *American Journal of Primatology* 58:133–48.

Laland, K. N., and B. G. Galef, Jr., eds. 2008. *The Question of Animal Culture*. Cambridge, MA: Harvard University Press.

Laland, K. N., and H. C. Plotkin. 1990. Social learning and social transmission of foraging information in norway rats (*Rattus noregicus*). *Animal Learning and Behavior* 18:246–51.

Lewin, R., and R. Foley. 2004. Principles of human evolution. Oxford: Blackwell Science.

Lonsdorf, E. V. 2005. Sex differences in the development of termite-fishing skills in the wild chimpanzees, *Pan troglodytes schewinfurthii*, of Gombe National Park, Tanzania. *Animal Behaviour* 70:673–83.

Lycett, S. J., M. Collard, and W. C. McGrew. 2007. Phylogenetic analyses of behavior support existence of culture among wild chimpanzees. *Proceedings of the National Academy of Sciences of the USA* 104:17588–93.

Marshall-Pescini, S., and A. Whiten. 2008. Social learning of nut-cracking behaviour in East African sanctuary-living chimpanzees (*Pan troglodytes schweinfurthii*). *Journal of Comparative Psychology* 122:186–94.

McGrew, W. C. 1992. *Chimpanzee Material culture: Implications for human evolution*. Cambridge: Cambridge University Press.

McGrew, W. C., R. M. Ham, L. T. J. White, C. E. G. Tutin, and M. Fernandez. 1997. Why don't chimpanzees in Gabon crack nuts? *International Journal of Primatology* 18:353–74.

McGrew, W. C., L. F. Marchant, S. E. Scott, and C. E. G. Tutin. 2001. Intergroup differences in a social custom in wild chimpanzees: The grooming handclasp of the Mahale Mountains. *Current Anthropology* 42:148–53.

Matsuzawa, T. 1994. Field experiments on use of stone tools by chimpanzees in the wild. In *Chimpanzee Cultures*, ed. R. W. Wrangham, W. C. McGrew, F. B. M. de Waal, and P. Heltne, 351–70. Cambridge, MA: Harvard University Press.

———. 2001. *Primate Origins of Human Cognition and Behavior*. New York: US Publishing.

Menzel, E. W., R. K. Devenport, and C. M. Rogers. 1972. Proto-cultural aspects of chimpanzees' responsiveness to novel objects. *Folia Primatologica* 17:161–70.

Mesoudi, A., A. Whiten, and K. N. Laland. 2004. Is human cultural evolution Darwinian? Evidence reviewed from the perspective of "The Origin of Species." *Evolution* 58:1–11.

Nakamura, M., W. C. McGrew, L. F. Marchant, and T. Nishida. 2000. Social scratch: Another custom in wild chimpanzees? *Primates* 41:237–48.

Nakamura, M., and T. Nishida. 2006. Subtle behavioral variation in wild chimpanzees, with special reference to Imanishi's concept of kaluchia. *Primates* 47:35–42.

Nishida, T., J. Mitani, and D. Watts. 2004. Variable grooming behaviours in wild chimpanzees. *Folia Primatologica* 75:31–36.

Pruetz, J. D. 2001. Use of caves by savanna chimpanzees (*Pan troglodytes verus*) in the Tomboronkoto region of southeastern Senegal. *Pan Africa News* 8:26–28.

Pruetz, J. D., and Bertolani, P. 2007. Savanna chimpanzees, *Pan troglodytes verus*, hunt with tools. *Current Biology* 17:1–6.

Sanz, C., D. Morgan, and S. Glick. 2004. New insights into chimpanzees, tools and termites from the Congo Basin. *American Naturalist* 164:567–81.

Stoinski, T. 2005. Culture in gorillas: Evidence from captive populations. Paper presented at the St. Andrews International Conference on Animal Social Learning, St. Andrews, U.K., June 2005.

Tehrani, J. J., and M. Collard. 2002. Investigating cultural evolution through biological phylogenetic analyses of Turkman textiles. *Journal of Anthropological Anthropology* 21:443–63.

Tennie, C., J. Call, and M. Tomasello. 2006. Push or pull: Imitation vs. emulation in great apes and human children. *Ethology* 112:1159–69.

Thorndike, E. L. 1898. Animal intelligence: An experimental study of the associative processes in animals. *Psychological Reviews Monograph* 2:551–53.

Tomasello, M. 1994. The question of chimpanzee culture. In *Chimpanzee Cultures*, ed. R. W. Wrangham, W. C. McGrew, F. B. M. de Waal, and P. Heltne, 301–17. Cambridge, MA: Harvard University Press.

———. 1996. Do apes ape? In *Social Learning in Animals: The Roots of Culture*, ed. C. M. Heyes, and B. G. Galef Jr., 319–46. London: Academic Press.

———. 1999. *The Cultural Origins of Human Cognition*. Cambridge, MA: Harvard University Press.

Tomasello, M., and J. Call. 1997. *Primate Cognition*. Oxford: Oxford University Press.

Van Lawick-Goodall, J. 1973. Cultural elements in a chimpanzee community. In *Precultural Primate Behaviour*, ed. E. W. Menzel, 144–84. Basel: Karger.

Van Schaik, C. P., in press. Geographic variation in the behaviour of wild great apes: Is it really cultural? In *The Question of Animal Culture*, ed. K. N. Laland and B. G. Galef Jr. Cambridge, MA: Harvard University Press.

Van Schaik, C. P., M. Ancrenaz, G. Borgen, B. F. M. Galdikas, C. D. Knott, I. Singleton, A. Suzuki, S. S. Utami, and M. Merrill. 2003. Orangutan cultures and the evolution of material culture. *Science* 299:102–5.

Whiten, A. 2000. Primate culture and social learning. *Cognitive Science* 24:477–508.

———. 2005a. The second inheritance system of chimpanzees and humans. *Nature* 437:52–55.

———. 2005b. Chimpanzee cultures. In *The World Atlas of Great Apes and their Conservation*, ed. J. Caldecott. and L. Miles, 66–67. United Nations Environment Programme. Berkeley and Los Angeles: University of California Press.

———. 2006. The significance of socially transmitted information for nutrition and health in the great ape clade. In *Social Information Transmission and Human Biology*, ed. J. C. K. Wells, K. Laland, and S. S. Strickland, 118–34. London: CRC Press.

———. 2008. The identification of culture in chimpanzees and other animals: From natural history to diffusion experiments. In *The Question of Animal Culture*, ed. K. N. Laland and B. G. Galef, 99–124. Cambridge, MA: Harvard University Press.

———. 2009. Ape behaviour and the origins of human culture. In *Mind the Gap: Tracing the Origins of Human Universals*, ed. P. Kappeler and J. Silk. Berlin: Springer-Verlag.

Whiten, A., J. Goodall, W. C. McGrew, T. Nishida, V. Reynolds, Y. Sugiyama, C. E. G. Tutin, R. W. Wrangham, and C. Boesch. 1999. Cultures in chimpanzees. *Nature* 399:682–85.

———. 2001. Charting cultural variation in chimpanzees. *Behaviour* 138:1481–16.

Whiten, A., and R. Ham. 1992. On the nature and evolution of imitation in the animal kingdom: reappraisal of a century of research. *Advances in the Study of Behaviour* 21:239–83.

Whiten, A., V. Horner, and F. B. M. de Waal. 2005. Conformity to cultural norms of tool use in chimpanzees. *Nature* 437:737–40.

Whiten, A., V. Horner, C. A. Litchfield, and S. Marshall-Pescini. 2004. How do apes ape? *Learning and Behavior* 32: 36–52.

Whiten, A., V. Horner, and S. Marshall-Pescini. 2003. Cultural panthropology. *Evolutionary Anthropology* 12:92–105.

Whiten, A., N. McGuigan, S. Marshall-Pescini, and L. M. Hopper. 2009. Emulation, imitation, overimitation, and the scope of culture for child and chimpanzee. *Philosophical Transactions of the Royal Society B*, in press.

Whiten, A., and A. Mesoudi. 2008. Establishing an experimental science of culture: Animal social diffusion experiments. *Philosophical Transactions of the Royal Society B* 363:3477–88.

Whiten, A., and C. P. van Schaik. 2007. The evolution of animal "cultures" and social intelligence. *Philosophical Transactions of the Royal Society B* 362:603–20.

Whiten, A., A. Spiteri, V. Horner, K. E. Bonnie, S. P. Lambeth, S. J. Schapiro, and F. B. M. de Waal. 2007. Transmission of multiple traditions within and between chimpanzee groups. *Current Biology* 17:1038–43.

Wrangham, R. W., W. C. McGrew, F. B. M. de Waal, and P. G. Heltne, eds. 1994. *Chimpanzee Cultures*. Cambridge, MA: Harvard University Press.

Yamakoshi, G., and Y. Sugiyama. 1995. Pestlepounding behavior of wild chimpanzees at Bossou, Guinea: A newly observed tool-using behavior. *Primates* 36:489–500.

9

The Cultural Mind of Chimpanzees: How Social Tolerance Can Shape the Transmission of Culture

Victoria Horner

As a new day starts at Yerkes National Primate Research Center, excited food calls rise from one group of chimpanzees (FS1). The chimpanzees are running back and forth embracing each other, a sign of excitement and anticipation as they watch a favorite food puzzle, called the panpipe, being pushed up the hill on a cart by a familiar researcher. As soon as the panpipe has been maneuvered into position, Socrates, the alpha male, seizes a plastic stick the researcher has given him, inserts it into the front of the apparatus, pushes a small plastic block backwards. A grape rolls out into his waiting hand. Socrates is closely watched by several other chimpanzees who are clustered around him. Every time Socrates pushes the block, another grape is released. Two hundred meters away and out of sight of FS1, a second group of chimpanzees has its own tradition. Steward, the alpha male in FS2, uses the stick to lift the block, thus freeing a grape. Both Steward and Socrates have access to the same panpipe, yet one lifts the block while the other pushes it. Both work equally well. These distinct methods are shared within their respective groups: members of FS1 push, while FS2 lifts. Their behavior is not the result of training by the researcher, nor did they figure out the solution on their own. Both groups learned the behavior from other chimpanzees. Much like the use of chopsticks or cutlery in humans, the ability to accurately pass behaviors on from one individual to another, despite more than one possible solution, opens the door to cultural differences in chimpanzees.

Following the first long-term field studies of chimpanzees at Gombe and Mahale in the 1960s (Goodall 1963; Nishida 1968), researchers began to suspect that some behaviors might be acquired culturally, particularly those that involved complex tool use (Goodall 1973). In the decades that have followed, there has been an accumulation of reports of behaviors that are thought to be cultural as they differ between populations without obvious ecological or genetic explanation. A compilation of this data from six major field sites was published in 1999, reporting 39 behavioral variants of courtship, grooming, and tool use which are believed to represent chimpanzee cultures (Whiten et al. 1999). Although reports of culture in other species are still growing, chimpanzees currently exhibit an unprecedented level of cultural complexity within the animal kingdom (see chapter 8). Nevertheless, without experimental manipulations (which can be ethically and logistically challenging) it is difficult to demonstrate conclusively that these putative cultures result from differential invention and social transmission of behavior, rather than from differences in ecology or genetics (Galef 2003; Laland and Janik 2006; but see Lycett et al. 2007). Before a behavior can be classified as cultural, it must therefore be shown that it has been learned socially from others, and hence the question of chimpanzee culture should also be addressed from a cognitive perspective: Can chimpanzees learn behaviors accurately from one another in a manner that could lead to the cultural differences reported in the wild?

To date, cognitive questions about chimpanzees' ability to accurately learn new, potentially cultural behaviors have been addressed from two different perspectives: (1) observational studies and semi-controlled experiments with wild chimpanzees (see chapters 10, 11, and 12), and (2) captive studies of chimpanzee social learning under controlled experimental conditions (reviewed in Whiten et al. 2004, Boesch 2007). While both methods represent complementary and necessary approaches, neither is independently sufficient to fully investigate the cultural minds of chimpanzees. Observational studies provide essential information about the social context in which learning occurs and the types of behaviors that might be candidates for culture (Boesch and Tomasello 1998; Matsuzawa 2001). However, the logistical constraints of fieldwork typically limit the application of stringent control conditions that are necessary to determine whether behavioral differences among populations result from social learning or from individual discovery. On the other hand, cognitive studies in captivity often struggle to reconstruct the complex social and environmental conditions under which learning occurs in the wild. Typically, captive studies have been restricted to investigations of social learning between pairs of individuals, with a human acting as a model for a "naive" chimpanzee observer. Chimpanzees are often physically separated from the human model, preventing close observation and social interaction (see Boesch 2007 for some exceptions). Many of these studies are conducted from a comparative perspective to determine whether chimpanzees share the same cognitive abilities as humans when tested under the same conditions. Because such studies are intrinsically human-centric, they tend to use methodologies that may inadvertently favor human participants while posing handicaps for the apes (Boesch 2007; de Waal et al. 2008). For example, in contrast to chimpanzees, children are typically tested without physical separation, in close proximity to a parent, and by a member of their own species. The poor performance of chimpanzees and other apes in comparison to children has been used to argue that they are incapable of learning with the accuracy required for culture (Richerson and Boyd 2005). However, the environmental and social conditions under which learning is assessed make the relevance of such studies to chimpanzee culture unclear. As illustrated by Nakamura (see chapter 13), it is important to view cognition in the framework of the so-

ciety in which it operates. It therefore follows that chimpanzees are likely to demonstrate learning that is most representative of their natural cognition when learning from conspecifics, without physical separation, and in a relaxed social context. In this chapter I will present a series of studies designed specifically to investigate chimpanzee cognition under these conditions. Chimpanzees were given opportunities to learn from one another in their normal social environment without physical separation between individuals. This naturalistic methodology enabled interpersonal relationships such as rank, affiliation, and social tolerance (see below) to influence the transmission of behaviors in a manner more representative of the learning environment chimpanzees might experience in the wild, while also maintaining the controlled experimental conditions afforded by a captive setting.

Each study was conducted in collaboration with Andrew Whiten (University of St. Andrew, UK), Frans B. M de Waal, and Kristin E. Bonnie (both of Emory University, USA). The studies were conducted with chimpanzees from the FS1 and FS2 social groups at Yerkes National Primate Research Center and were designed to explore (1) the differential spread of alternative technologies within each group, (2) how transmission may be related to social tolerance, and (3) the longevity and sustained fidelity of cultural behavior over multiple simulated "generations." The results are discussed in relation to the cultural minds of chimpanzees and the cognitive components that may be crucial to chimpanzee culture.

General Approach

Participants and Study Site

Each study was conducted at the Field Station of Yerkes Primate Center, at Emory University, USA. The Field Station is home to two social groups of chimpanzees, known as FS1 and FS2. Each mixed-gender group has 17 individuals of varying ages (ages differed at time of each study; see table 9.1). Both groups live in large outdoor enclosures (FS1 = 697m²; FS2 = 520m²) with grass, wooden climbing structures, and enrichment "toys." Each enclosure is attached to an indoor building with five interconnected rooms known as "the sleeping quarters." FS1 and FS2 can hear, but not see each other because their enclosures are approximately 200m apart and are separated by a small hill.

Table 9.1 Background history and ID codes for chimpanzees in the FS1 and FS2 groups.

			FS1					FS2	
Name		Sex	Birth year	upbringing	Name		Sex	Birth year	Upbringing
Georgia	GG	F	1980	mother	Ericka	ER	F	1973	human
Anja	AJ	F	1980	mother	Amos	AM	M	1981	mother
Azalea	AZ	F	1997	mother	Barbie	BB	F	1976	mother
Bjorn	BJ	M	1988	mother	Chip	CP	M	1989	nursery
Borie	BO	F	1964	unknown	Cynthia	CY	F	1980	mother
Claus	CL	M	1992	mother	Daisy	DA	F	1989	mother
Donna	DN	F	1990	mother	Jaimie	JA	F	1995	mother
Katie	KT	F	1989	mother	Julianne	JL	F	1998	mother
Liza	LZ	F	1994	mother	Kerri	KE	F	1995	mother
Mai	MA	F	1964	unknown	Magnum	MG	M	1989	nursery
Missy	MS	F	1993	mother	Reid	RD	M	1993	mother
Peony	PE	F	1968	wild-born	Sean	SN	M	1992	mother
Reinette	RN	F	1987	mother	Steward	ST	M	1993	mother
Rhett	RT	M	1989	mother	Taï	TI	F	1967	unknown
Rita	RI	F	1987	mother	Virginia	VR	F	1991	mother
Socko	SK	M	1987	mother	Vivienne	VV	F	1974	nursery
Tara	TA	F	1995	mother	Waga	WG	F	1982	mother

Starting a Potential Culture

Like human cultures, chimpanzee cultures are likely to arise when new behaviors are introduced to a population either by immigrating individuals (Matsuzawa and Yamakoshi 1998; Nakamura and Uehara 2004), or through invention by existing group members (Goodall 1986; McGrew 2004). These new behaviors are then transmitted within the group by social learning. In each of the three studies presented here we wanted to simulate invention, because successfully introducing an adult chimpanzee to a new group can be dangerous and unpredictable. Innovation was simulated by training a chimpanzee model from FS1 and FS2 to use one of two alternative yet equally difficult behaviors to gain food from the same apparatus. By seeding each group with a different variant of the new behavior, we could determine whether and how each variant spread and was maintained to potentially become a cultural tradition. As detailed below, the choice of model in each group was informed by theoretical models and field data and was based upon social status, kinship, age, and gender. Once trained, the models could then perform their new behavior while being observed by other chim-

panzees in their respective social groups. Social learning, and hence cultural transmission, was inferred by tracking the differential spread of each behavior in the FS1 and FS2 groups and comparing their behavior to a third control group not exposed to either solution.

Social Tolerance and Social Models

In a departure from many previous captive studies which have used humans to model novel behavior, we determined that the conditions most likely to reflect cognitive abilities in the wild were those in which chimpanzees learn spontaneously from other chimpanzees. The identity of the trained model was therefore of critical importance to the potential transmission of behavior. Theoretical models predict that observers should be choosy about who they learn from (Boyd and Richerson 1985; Laland 2004) with several factors such as skill competence, rank, and social tolerance directly influencing the likelihood that the actions of an individual will be copied (Coussi-Korbel and Fragaszy 1995; Boesch and Tomasello 1998; Russon 2003). Of these factors, social tolerance between the model and the observer is believed to be particularly important because it (1) allows close observation of complex behaviors (e.g. tool use), (2) creates a relaxed social atmosphere which encourages attention to the task without fear of aggression, and (3) allows proficient individuals to perform without fear of excessive scrounging (van Schaik et al. 1999; van Schaik 2003). On this basis, we thought the best model for each group would be a highly tolerant chimpanzee with sufficient rank to retain control of the apparatus while being observed by the rest of the group. Additionally, reports from the wild indicate that chimpanzees tend to learn complex foraging skills from older or similar-aged peers (Biro et al. 2003), and that due to social and reproductive constraints, females typically engage in tool use more frequently than males (Goodall 1986; Lonsdorf et al. 2004), and so we thought the ideal model should also be an adult female. We reviewed observational data recording the social relationships among group members collected between 2002 and 2004, and used this to pick two promising high-ranking adult female models. In FS1 we chose Georgia, a chimpanzee with many affiliative relationships within her group (and therefore potentially tolerant of observation by others) and who is believed to have initiated the handclasp grooming tradition in FS1 (Bonnie and de Waal 2006). In FS2 we chose Ericka, who also has numerous affiliative relationships within her group. Ericka spent the first eight years of her life living with a human family; however, we do not consider her to be human-enculturated (cf. Tomasello et al. 1993) as she has been living with members of the FS2 group for more than 20 years and displays typical chimpanzee behavior. Nevertheless, by training her as a model we removed her from the pool of potential learners, and so circumvented any possibility that her learning skills might be unrepresentative. Georgia and Ericka acted as social models for each of the three studies described below.

Study 1: Group Diffusion—The Panpipe

Approach

APPARATUS. The panpipe apparatus was designed to simulate a tool-use foraging task. When the apparatus was baited, a food reward was trapped behind a movable square block in the top pipe. Chimpanzees could retrieve food from the panpipe by using a polycarbonate "stick" to remove the block so that the food rolled down into the

bottom pipe and out towards the operator. There were two methods available to remove the block, and so by training each model to perform a different solution, alternative "inventions" could be introduced to each group. In FS1, Georgia was trained to poke the block backward, making the food drop into the lower pipe (figure 9.1a), while in FS2 Ericka was trained to lift the block out of the way (figure 9.1b).

MODEL TRAINING. Model training was conducted out of sight of other chimpanzees. The Yerkes chimpanzees know their names and can be called in from their outside enclosure to voluntarily participate in research. Georgia and Ericka voluntarily separated from their group to participate in daily 20-minute training sessions in the indoor sleeping quarters. Training involved a period of habituation to the apparatus followed by a combination of human demonstrations and positive reinforcement training (see chapter 25). Models were judged to be proficient when they could perform the trained behavior accurately on 20 consecutive trials. This occurred after approximately four training sessions for each model.

OBSERVATION AND TRANSMISSION. Once Georgia and Ericka were proficient, the panpipe was moved to the outdoor enclosure so that each model could demonstrate their trained behavior while being observed by their respective group mates. Initially, only the models were allowed to interact with the apparatus during a seven-day observation phase of 20-minute daily sessions. This was achieved by pulling the apparatus out of reach if a chimpanzee other than the model picked up the plastic tool that was needed to operate the panpipe (figure 9.1). The observation phase was included to ensure that every chimpanzee had an opportunity to observe the demonstrations before the apparatus was made available to the entire group. Following this, a transmission phase was run for 36 hours in each group spread over 10 days, during which time all chimpanzees had access to the panpipe.

Results

During the observation phase, Georgia and Ericka consistently used their trained method to retrieve food from the panpipe. Their actions generated great interest from the rest of their group, with each demonstration being closely

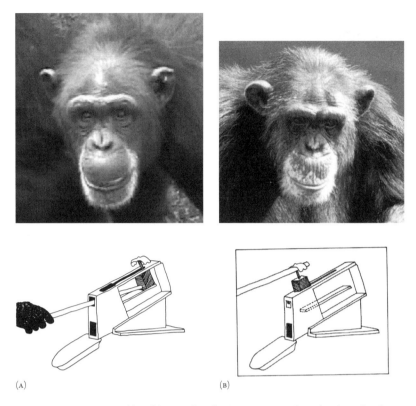

Figure 9.1 Panpipe apparatus: (a) model Georgia from the FS1 group was trained to exclusively use the poke method to retrieve food from the Panpipe; (b) model Ericka from FS2 was trained to exclusively use the alternative lift method. Both methods were always possible. The front panel and front third of the side panel where opaque.

observed (within 1 m) by up to 10 individuals. The task was then made available to all chimpanzees in the transmission phase, during which 15 chimpanzees successfully retrieved food from the panpipe in each group. In both FS1 and FS2, the method that had been introduced by the trained model was used significantly more than the alternative, such that alternative Lift and Poke cultures were created in each group (Whiten et al. 2005).

Six chimpanzees from a third control group who received no training and had no opportunity to observe a knowledgeable conspecific failed to retrieve food from the panpipe despite several hours of interaction. Their attempts involved hitting the top and side of the apparatus with the tool, inserting the tool into the bottom pipe, and biting and touching the food chute which protruded into their enclosure.

Discussion

This study demonstrates that chimpanzees have the cognitive ability to accurately learn from one another in such

a way that two groups can exhibit alternative traditions if different behaviors are introduced by an "inventor." The role of social learning is further supported by the poor performance of the control group. The panpipe study thus provides support for the assertion that group-typical behavior patterns of wild chimpanzees can result from differential invention and transmission of behavior.

THE IMPORTANCE OF SOCIAL TOLERANCE. We believe that the successful, differential transmission of panpipe behavior was greatly influenced by the natural learning environment that we exploited, and by the high degree of social tolerance exhibited by the Yerkes chimpanzees. Both Georgia and Ericka were able to retain control of the panpipe apparatus during the observation phase despite being observed by up to 10 conspecifics closely packed around them. Although some scrounging was seen, models were able to keep the majority of food that they retrieved. Tolerance continued into the transmission phase when the apparatus was made available to everyone. The first observer in each group to succeed at the task was a high-ranking male (in FS1, Socrates, α-male; in FS2, Amos, β-male). Yet despite the high rank of these individuals, they continued to tolerate close observation by the rest of the group—a characteristic that distinguishes chimpanzee and bonobo males from males of other great ape species (Russon 2003; van Schaik 2003). In addition, we witnessed 106 instances of co-action involving 38 different chimpanzee pairs, only 7 of which occurred between kin. Co-action is said to occur when models allow observers to participate intimately in their behavior (Visalberghi and Fragaszy 1990). This was seen when a chimpanzee who was using the panpipe allowed an observer to touch either their hand or part of the tool as they worked. Co-action between mothers and offspring has been observed in captive tufted capuchins using sticks to dip for syrup (Westergaard and Fragaszy 1987) and in the wild chimpanzees during termite fishing (McGrew 1977). Co-action is distinct from scrounging, in which the observer exploits the actions of the model by stealing the food they have worked for, and which may actually impede social learning in some situations (Giraldeau and Lefebvre 1987). Although some limited scrounging was seen during the panpipe experiment, this was not observed during bouts of co-action, where models kept 100% of their food rewards. A higher degree of social tolerance is expected between kin; hence the high number of non-kin co-actors illustrates the level of tolerance that was exhibited in the panpipe study. These results indicate that social tolerance may be a key component of chimpanzee culture.

In addition to visual information about the task, the ability to participate in co-action may provide observers with information about the motor actions required to successfully operate the panpipe, such as amplitude, force, and direction of movement. In a follow-up study conducted with captive chimpanzees at the University of Texas (Bastrop, Texas), researchers showed chimpanzees a "ghost" demonstration of the panpipe solution, in which the tool was attached to clear fishing line and maneuvered to successfully operate the panpipe as if being used by an invisible chimpanzee. They found that seeing the correct solution in the absence of an appropriate social context was not sufficient for the chimpanzees to learn (Hopper et al. 2007). Not surprisingly, it seems that effective social learning must incorporate observational, social, and perhaps physical elements in order to appropriately recreate the effective learning environment that chimpanzees experi-

ence in the wild. Studies of cognition that involve physical separation between observers and models, and which are not sympathetic to inter-individual relationships such as social tolerance, are less likely to elicit social learning that is representative of chimpanzee cultural transmission.

CONFORMITY. In both FS1 and FS2, chimpanzees used the method introduced by the original model significantly more than the alternative. However, in both groups there were a small number of individuals who used the alternative solution to varying degrees. We re-tested each group two months after completion of the original study to see if the inter-group difference had eroded. To our surprise, although they had had no contact with the panpipe for two months, we found that each group performed the original behavior with even greater consistency than before. This effect was most striking in the FS2 group, where Ericka originally introduced the lift method. During the initial study, eight individuals had, at some point, successfully used the alternative poke method. When re-tested two months later, three of the eight had completely reverted to the lift method, and three performed a greater proportion of lift than before. This finding suggests a type of conformity, defined in anthropology and social psychology as the tendency for members of a group to discount personal experience (poke was equally successful) in favor of the behavior most commonly performed by others (lift) (Henrich and Boyd 1998). Boyd and Richerson (1985) propose that in human culture, conforming to the majority behavior conveys a selective advantage by increasing the probability that conformers will adopt behaviors that are adaptive for their environment. The conformity indicated in the panpipe study strongly implies that similar learning biases may operate in chimpanzees, as suggested by theoretical discussions of non-human learning (Boesch and Tomasello 1998; Laland 2004).

TRANSMISSION PATHWAYS. We recorded the order in which chimpanzees first successfully operated the panpipe in each group. Nevertheless, new practitioners typically observed several panpipe users before their own first success, and so our ability to determine who learned from whom is limited. Access to the apparatus was loosely related to rank, with high-ranking individuals being able to assume control of the panpipe sooner than subordinates, making it impossible to determine what each chimpanzee knew

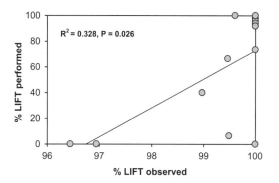

Figure 9.2 Relationship between the percentage of lift performed by chimpanzees from the FS2 group during their first 30 successes with the Panpipe and the percentage of lift that they had seen others use beforehand.

before his or her first turn. One possibility is that all group members learned from the original trained model during the observation phase, but their ability to demonstrate this knowledge was constrained by social structure. If all observers learned from the same model, this represents a star pattern of transmission, with information radiating from a single individual.

Although chimpanzees in each group were influenced by the original model, they also seem to have been influenced to some degree by each other. This was seen most obviously in FS2, where there were a small number of pokers among the lifters. Although as a group FS2 performed the original lift behavior more than the alternative, there was a significantly positive correlation between the percentage of poke performed by each chimpanzee during their first 30 successes with the panpipe and the percentage of poke that they had witnessed others using beforehand (figure 9.2). Another possibility is, therefore, that each successful chimpanzee learned from the preceding performer. This represents a more linear pattern of transmission in which a technique is transmitted from the original model, to a second individual, from that individual to a third, and so forth throughout the group. In fact, transmission of the panpipe solution (and behavior in the wild) is likely to be a complex mixture of both star and linear patterns. While star transmission would be sufficient to spread a novel behavior throughout a group starting from a single source, linear transmission would potentially allow behaviors to spread horizontally and hence persist over more than one generation. Longitudinal observations of wild chimpanzees indicate that infants closely observe the behavior of adults in their group and grow up to use similar technologies and techniques, indicative

of multigenerational transmission (Lonsdorf et al. 2004). However, this assertion has yet to be verified experimentally. We therefore designed study 2 to investigate whether chimpanzees can learn from one another in a linear sequence that could potentially support multigenerational transmission.

Study 2: Chain Diffusion: The Doorian Fruit

Approach

Chain diffusion was first employed to study human memory by exploring how narrative stories altered as they were passed from one person to the next in a chain (Bartlett 1932). This design was later adapted to study social transmission in other animals by substituting stories for behavior patterns. Menzel and colleagues (1972) pioneered the first nonhuman study in which habituation to a novel play object was transmitted successfully down a chain of 19 chimpanzees. The paradigm was later used to investigate the transmission of predator avoidance in birds (Curio et al. 1978), food and foraging preferences in rats (Laland and Plotkin 1993), and foraging pathways in fish (Laland and Williams 1998). These studies have typically investigated transmission of behavior down a single chain, and hence inferences about the transmission of alternative behavioral variants are limited. Study 2 investigated linear transmission along two separate diffusion chains, one in FS1 and the other in FS2. As with the panpipe, each chain was compared to a control group of six chimpanzees who did not observe a successful solution.

APPARATUS AND MODELS. The apparatus was a foraging box adapted from the bottom section of a box used in a previous study of social learning with chimpanzees at Ngamba Island, Uganda (Horner and Whiten 2005). Since the alternative behaviors involved differentially manipulating a door, the foraging box was named the doorian fruit (figure 9.3a). The door could be opened by either a lift or slide technique. Using the lift method, the door frame remained in place while the door was lifted up (figures 9.3b and 9.3d). Using the slide method, the door remained closed while the frame slid to the right (figure 9.3c). The study was conducted inside the sleeping quarters for each group, with the apparatus bolted to the floor of one room. The experimenter could bait the doorian fruit by dropping

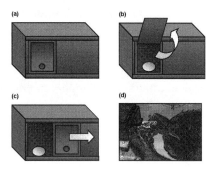

Figure 9.3 Doorian fruit apparatus: (a) the starting position with door closed, (b) lift door method, (c) slide door method, and (d) outlined photograph of model Georgia performing the lift method at the start of the diffusion chain in the FS1 group (Horner et al. 2006).

food rewards down a metal pipe which connected the side of the apparatus to the human area.

Ericka and Georgia were again selected to start each chain, due to the success with which they had introduced the panpipe cultures in study 1. Each model was trained in the sleeping quarters with positive reinforcement training to use one of the two alternative, yet equally viable, methods to open the door: Georgia the lift, and Ericka the slide.

PROCEDURE. The methodology used for chain diffusion experiments is more constrained than group diffusion because participants are added to the chain one by one, and hence the order of transmission is predetermined. Nevertheless, this methodology has several advantages because the identities of both the model and observer are known during each link in the chain, making it possible to determine when, and infer why, breakdowns in transmission might occur. In order to be representative of a naturalistic learning environment, it was important to run each chain in a manner that was sensitive to social tolerance between successive pairings. For that reason, the order of each chain was based on the order in which the chimpanzees had succeeded in the panpipe study, as a rough measure of social dominance (see study 1). A decreasing order of dominance was preferred, because participants would act as both observers and models as the chains progressed. For models to retain control of the apparatus long enough to perform at least one full demonstration they had to be dominant to, yet tolerant of, the observer with whom they were paired. Subordinate observers then had to perform as dominant models for the next individual in the chain. While we don't

claim to have run an actual multigenerational study (that would take many decades), the progression in each chain did loosely simulate generational relationships, with models acting as "adults" demonstrating behavior to "younger" individuals. Changes to the predetermined order were made only if there were known social incompatibilities between certain pairs based on our weekly observations of each group.

Each diffusion chain was conducted with the model and observer in the same room, interacting with the same apparatus. This allowed the observer to watch the demonstrations in close proximity, choose their own position for observation, and most importantly, interact socially with the model—factors thought to be important for a naturalistic learning environment (de Waal 2001; van Schaik 2003). Once the model and observer were in the same room, the model performed 10 demonstrations. In some cases the observer was able to displace the model before they had performed all 10 demonstrations, but the chain continued as long as the observer had seen at least one full demonstration. After the tenth demonstration or model displacement, whichever happened first, the model was prompted to leave the room by offering a further food reward so that the observer could interact with the apparatus alone. Each observer had the opportunity to retrieve 10 rewards from the doorian fruit to determine whether they used the same method of door opening used by the model. If successful, the observer then became the model for the next individual in the chain, irrespective of the method used. This procedure was repeated for each link in the chain. Chimpanzees that were unsuccessful became side branches to the main chain (see below).

Results

Each method of door opening (lift or slide) was faithfully transmitted down a chain of six chimpanzees in FS1 and five chimpanzees in FS2, such that the last individual in each chain performed the same behavior as the first (Horner et al. 2006; see also figure 9.4). Each chain included a number of "side branches" representing observers who successfully learned the same technique as the model but declined to participate as models themselves (in FS1, Reinette; in FS2, Amos). The predetermined order of each chain successfully elicited learning with only one excep-

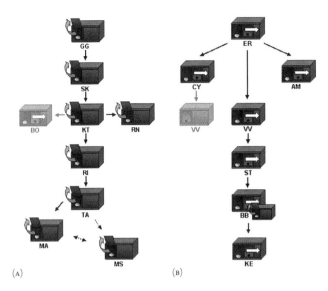

(A)   (B)

Figure 9.4 Schematic representation of the diffusion chains in (a) FS1 and (b) FS2. Each chain was initiated by a trained model (FS1, Georgia: lift; FS2, Ericka: slide). The method used by each chimpanzee is indicated by a gray arrow (lift) or a white arrow (slide). ID codes of each chimpanzee are in bold (see table 9.1). Side branches represent chimpanzees who successfully opened the doorian fruit, but declined to participate as models for the next individual in the chain. Borie (BO) and Vivienne (VV) are faded because they did not observe the behavior of the models with whom they were paired. Model Tara (TA) demonstrated for Mai (MA) and Missy (MS) at the same time because they are a highly affiliative mother-daughter pair and only volunteer to participate in studies if they are together. Reinette (RN) and Amos (AM) successfully learned from Georgia and Ericka respectively, but declined to participate as models themselves for the next chimpanzee in the chain.

tion in each chain. In FS1, Katie failed to retain control of the apparatus when paired with observer Borie, her grandmother. Borie displaced Katie without observing a single demonstration, and discovered the alternative slide method. While this does not represent social learning, it illustrates how easily the alternative method could be discovered and so highlights the strength of social learning in the FS1 chain for all the other cases in which observation took place. In FS2, model Cynthia was aggressive to observer Vivienne, who subsequently refused to observe Cynthia's behavior. However, when Vivienne was re-paired with Ericka, she successfully learned the slide method.

We recorded co-action during three links in each chain, including mother-daughter and non-kin pairs. It occurred when models were tolerant of observers touching either their hand or the door of the apparatus during demonstrations. It appears that, as in study 1, a high degree of social tolerance was important in the successful transmission of doorian fruit opening techniques.

The doorian fruit was also given to six control chimpanzees who did not have an opportunity to observe a knowledgeable conspecific. Three of the chimpanzees

failed to open the doorian altogether, one discovered slide, and two discovered lift, indicating that although the task was challenging, both methods could be discovered by exploration—further highlighting the significance of faithful transmission in each chain.

Discussion

Alternative doorian fruit opening techniques were faithfully transmitted along two diffusion chains of chimpanzees. In FS1 the lift method was passed along a chain of six chimpanzees with two side branches. In FS2 the slide technique was passed along a chain of five chimpanzees with two side branches, although each chain might have extended further had more chimpanzees volunteered to take part. In each linear chain, all participants performed the same techniques as the original model: the only variation occurred in FS2 when Barbie performed one lift (possibly accidentally), which was not transmitted further. Therefore, in order for a cultural behavior to persist over more than one generation, it seems that the capacity for linear transmission is an important component of chimpanzee culture.

As in study 1, there was a high degree of social tolerance between the models and the observers, including co-action. This was likely integral to the successful transmission of behavior. The side branches for each chain represented chimpanzees who (with the exception of Borie in FS1) copied the solution that they had seen, but were not motivated to participate further as models. In FS2, Vivienne failed to learn a solution when paired with Cynthia, but learned successfully when re-paired with Ericka, thus indicating a lack of social tolerance rather than a deficit in social learning. This observation lends support to the Bonding and Identification–based Observational Learning model (de Waal 2001), in which learning must be preceded by a desire to act like the model, as evidenced by Vivienne's apparent motivation to learn from Ericka but not Cynthia. This possibility should be taken into account when interpreting negative data in dyadic studies of chimpanzee cognition. This observation also supports theoretical models and field observations predicting that potential learners are choosy about whom they learn from and that social tolerance is integral to successful social learning (Boesch and Tomasello 1998; Coussi-Korbel and Fragaszy 1995; Lonsdorf et al. 2004).

Study 3: The Token Study—Arbitrary Conventions

Approach

This experiment employed a similar methodology to the panpipe group diffusion outlined in study 1. The crucial difference was that this study was designed to investigate the transmission of a previously arbitrary sequence of actions whose reproduction was rewarded by a human experimenter (for more details see Bonnie et al. 2007).

Once again, Georgia and Ericka initiated each new behavior. They were trained to collect cylindrical orange plastic "tokens" thrown into the outside enclosure of their respective groups and deposit them in either a rectangular blue bucket (Georgia, in FS1) or a tubular white pipe (Ericka, in FS2), both of which were attached side-by-side to the fence of the outside enclosure. Collecting and depositing tokens in either receptacle resulted in a food reward being thrown to the performer by an experimenter standing on an observation tower above the apparatus. The study was conducted during 30-minute sessions on 20 consecutive days during which the model could demonstrate her trained behavior and all group members could interact freely with the apparatus and tokens. The method used by each new practitioner (bucket or pipe) was noted. Since the bucket, pipe, and tokens were always available, preferential replication of the method used by the trained model would indicate social transmission.

Results and Discussion

New practitioners in each group (with one exception; see below) used the same method that had been introduced by the original trained model (Bonnie et al. 2007). A finding of particular relevance to this chapter is the behavior of the exceptional chimpanzee, Daisy, in the FS2 group. On day four, Daisy discovered and successfully implemented the alternative method to that used by the original model, Ericka. Up until that point, all chimpanzees in FS2 had put tokens into the pipe (figure 9.5). Daisy, on the other hand, used the bucket in 12 of the 20 sessions, and although several chimpanzees acquired the task after her discovery, all new performers used the original pipe method performed by Ericka.

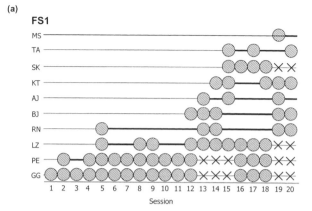

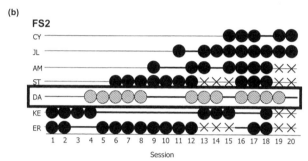

Figure 9.5 Differential spread of the Bucket and Pipe traditions in (a) the FS1 group, and (b) the FS2 group. Gray circles indicate deposits to Bucket; black circles indicate deposits to Pipe. Horizontal bars indicate that no deposit was made during that session; X indicates the individual was kept indoors during the session. Individuals are listed from bottom to top in order of their first successful deposit. Individuals GG and ER were the trained models (Bonnie et al. 2007).

Chimpanzees in FS2 who learned a solution after Daisy's discovery of the bucket on day 4 had the opportunity to observe different models performing alternative yet equally successful solutions. Why then did they preferentially copy the behavior introduced by Ericka? One possibility is that, as in study 1, new practitioners who acquired the behavior after day four were influenced by conformity because three chimpanzees (including Ericka) were using the pipe, compared to only Daisy using the bucket. Another possibility is that the identity of each model influenced the likelihood that their behavior would be copied by others. Learning theory predicts that an innovator's rank can have a significant effect on whether their behavior will be passed on, with the behavior of low-ranking individuals being least likely to spread (Boesch and Tomasello 1998; Henrich and Gil-White 2001). Reader and Laland (2001) also report that in chimpanzees, unlike other primate species, low-ranking individuals are more likely to invent new behaviors than are higher-ranking individuals, possibly driven by the need to develop new solutions to existing problems to circumvent competition with dominant group mates. Daisy may therefore have initially attempted to use the bucket to avoid competition with others around the pipe. Despite her success, however, her group mates may have preferred to copy Ericka and other successful individuals instead, due to Daisy's low rank and social status. It therefore seems that motivation on the part of observers to attend to and act like a social model is a third component of chimpanzee culture.

General Discussion

The three studies presented in this chapter provide evidence, using experimentally controlled methodologies, that chimpanzees can learn new behaviors accurately from one another in a manner that can lead to behavioral differentiation between groups. These results are supportive of the view that the geographic patterning of behavior found in wild chimpanzees represents cultural variation.

Cognitive Mechanisms of Cultural Transmission

Studies 1 and 2 used puzzle boxes that were designed to uncover the social learning mechanisms in operation during the transmission of behavior. For each piece of apparatus, alternative behavioral variants were introduced to each group by training the original model to manipulate part of the apparatus in one of two different ways. For the panpipe (study 1), each model moved the block inside the apparatus in a different way (poke or lift). Similarly, the door on the front of the doorian fruit (study 2) could be opened in two different ways (lift or slide). In each case, since the same part of the apparatus was manipulated with a different method, simple forms of social learning such as enhancement were not sufficient to maintain accurate transmission within each group. It is therefore likely that imitation and/or emulation learning were involved, which are thought to be at the cognitively complex end of the spectrum of possible learning strategies. In study 3, differential use of either the bucket or pipe can be explained by less cognitively complex learning strategies such as local and stimulus enhancement (Spence 1937; Thorpe 1956), whereby the actions of each trained model differentially drew observers' attention to either the bucket or the pipe. Nevertheless, all

three studies provide evidence for cultural transmission, drawing into question previous assumptions that culture must rely on complex cognitive strategies such as imitation (Tomasello 1999; see also chapter 13). The data provide further evidence that chimpanzees have a suite of social learning strategies at their disposal which they can employ selectively to acquire different skills in different social contexts (Horner and Whiten 2005). If culture can arise and be maintained by simple social learning mechanisms, it is likely that cultural behavior may be more widespread within the animal kingdom than previously thought. Future comparative investigation of culture across taxa may reveal the social and environmental circumstances that can lead to the convergent evolution of culture within the animal kingdom (see chapter 14).

Social Fads and Social Conformity

In study 1, chimpanzees from FS1 and FS2 differentially learned alternative lift and poke foraging techniques that were introduced by a trained conspecific. As well as being influenced by the original "inventor," the chimpanzees were also influenced by each other, as evidenced by the Poke subculture which sprang up in the FS2 group. Similar "fads" have been reported in wild primate populations, such as the rise and attenuation of social customs in capuchin monkeys (Perry et al. 2003), the gradual drift in potato washing techniques used by Japanese macaques (Watanabe 1994), and a change in the social context of leaf clipping by chimpanzees from the Taï forest (Boesch 1995). In FS2, the Poke subculture diminished over time because Pokers conformed to the original Lift behavior. In humans, conformers are assumed to have a selective advantage over nonconformers because they are more likely to adopt behaviors which are adaptive for their environment (Boyd and Richerson 1985). Given arguments referencing evolutionary continuity, it is possible that the same selective pressures would favor the development of a conformity bias in other species, and may therefore underlie the conformity observed in the panpipe study. Panpipe behavior spread within each group via a complex pattern of star and linear transmission that incorporated information from the original model and information from the ongoing actions of others. A similarly complex pattern of diffusion likely underlies the transmission of chimpanzee cultures in the wild.

The Importance of Social Tolerance and Social Interaction in Cultural Transmission

The successful transmission of behavior in each study required a high degree of social tolerance between models and observers. In studies 1 and 2, the observers closely watched the behavior of proficient models and in many cases engaged in co-action, touching the hand of the model or part of the apparatus as it was being manipulated. Social tolerance between individuals is believed to be one of the key factors in increasing the probability that a new behavior will spread, by allowing close observation and creating a relaxed learning environment which promotes attention to the task (van Schaik et al. 1999; van Schaik 2003). Correspondingly, tolerance of younger individuals by adults appears to play an important role in the acquisition of tool-use skills such as honey dipping (Hirata and Celli 2003), termite fishing (Lonsdorf 2006) and nut cracking (Matsuzawa 2001; see also chapter 12) as well as the development of cooperative foraging in bonobos and hunting in chimpanzees (Hare 2007). Social tolerance appears to be so important in the daily lives of chimpanzees that Wittig (see chapter 17) proposes that restoration of tolerance between former opponents is one of the primary motivations for consolation and reconciliation in chimpanzees.

Studies 1 and 2 also highlight the importance of physical contact between models and observers. Several of the chimpanzees who successfully learned to solve the panpipe and doorian fruit tasks engaged in co-action with a proficient model before their own success. In the wild, chimpanzees likely gain useful information about skills such as tool use by observation and through being involved in the ongoing actions of conspecifics. For example, nut cracking is learned by young chimpanzees over a period of many years through a process of apprenticeship (Matsuzawa 2001). During this time, infants have opportunities to interact with unused hammers and anvils and broken nutshells as well as with the ongoing actions of tolerant adult performers, possibly gaining valuable physical and visual information about successful execution of behavior. Indeed, when captive chimpanzees were shown the solution to the panpipe task in the absence of a demonstrator, they failed to learn a solution, thus emphasizing the importance of the social context of learning (Hopper et al. 2007).

How Social Tolerance Can Affect Transmission Pathways

Study 2 suggests that patterns of dissemination can be strongly influenced by the level of social tolerance and affinity between individuals. Chimpanzees were able to accurately pass alternative foraging behaviors down two chains, indicating that behavior patterns can be passed in a linear sequence between successive pairs of individuals. This pattern of linear diffusion could theoretically support the transmission of behavior over more than one generation. Breakdowns in the diffusion chains were also highly informative. In the FS2 chain, Vivienne failed to learn a solution when paired with Cynthia, but learned successfully when re-paired with Ericka, thus indicating a lack of social tolerance rather than a deficit in social learning. It is therefore important not to confuse *motivational* issues with *cognitive* capacity, particularly when interpreting the failure of individuals to learn in a dyadic social learning experiment (de Waal 2001). Individuals may completely fail to acquire novel behaviors, or fail to learn accurately because a lack of social tolerance inhibits their ability to gain the necessary visual and social information required to solve the task, or because they are simply not motivated to learn from the model. This possibility should be taken into account when interpreting negative data in dyadic studies of chimpanzee cognition. Because the order of each chain was predetermined, breakdowns in transmission resulted in dead-end "side branches" and "kinks" in the main chain. In a more naturalistic environment such as study 1, however, it is possible that these side branches might become side chains. Thus, social *intolerance* between individuals may act to fracture or split lines of transmission, resulting in complex transmission patterns which incorporate both star and linear diffusion.

Study 3 investigated the transmission of an arbitrary behavior in the FS1 and FS2 groups using a similar methodology to that of study 1. These results indicate an extremely high level of fidelity in the transmission of behavior within each group, demonstrating that chimpanzees are able to learn a previously arbitrary sequence of actions, perhaps analogous to social customs in the wild (McGrew and Tutin 1978; Nakamura and Nishida 2006). However, despite employing a methodology similar to that of study 1, the order in which chimpanzees learned to use the panpipe (study 1) and deposit tokens (study 3) differed in FS1 and FS2, suggesting that different transmission patterns occurred (Whiten et al. 2007). One explanation is that social tolerance is more likely to effect the transmission of behaviors which require close observation, such as tool use, while behaviors that can be effectively observed over greater distances, such as gestures and communication signals, require less close contact and are therefore less influenced by social tolerance (van Schaik 2003). Since the differences between the behavioral variants in study 3 were conspicuous and could potentially be observed from the other side of the enclosure ("put tokens in the bucket" or "put tokens in the pipe"), social tolerance is less likely to have constrained observation and execution of token behavior, resulting in a different pattern of transmission in study 3.

Innovator Identity and the Probability of Transmission

Study 3 highlights the importance of the identity of the original "innovator" in influencing the probability that their behavior will be learned by others. Chimpanzees from FS2 may not have learned from Daisy either because they were influenced to conform to the behavior being used by the majority of other chimpanzees in their group (as seen in study 1), they viewed Ericka's solution as more proficient, or they simply did not attend to her behavior since her low rank rendered her socially less salient than others (see chapters 4 and 5). It seems most likely that her lack of influence as a model resulted from a combination of all three possibilities, and that a similar bias may influence the transmission of inventions in the wild. Most importantly, study 3 highlights the critical role played by both Georgia and Ericka in the transmission of the Yerkes cultures. The high rank, competence, and level of social tolerance shown by both chimpanzees are likely key to the success of transmission. Without such good social models, it is possible that none of the studies might have provided the positive evidence they did for chimpanzee cultural minds. This further illustrates the power of social relationships in the transmission of chimpanzee culture, with factors such as tolerance and dominance acting to facilitate, fracture, or even prevent the transmission of novel behavior.

Implications and Future Directions

Social transmission can be simulated and studied under controlled experimental conditions to provide insights into the cultural minds of chimpanzees. However, such experimentation can only be informative if care is taken to respect the social structure of the group, acknowledge differing levels of social tolerance between participants, and understand the importance of social and physical interaction in a naturalistic learning environment. Studies that fail to respect these factors are unlikely to yield informative data, thus misrepresenting chimpanzee cognition. If these criteria are carefully considered, however, captive studies can provide an informative compliment to observational studies of wild chimpanzee behavior—and when combined together, they are likely to build the most representative picture of chimpanzee cultural minds.

Future studies of chimpanzee culture should focus on all levels of resolution; addressing how relationships between chimpanzees at the individual level (such as rank and tolerance) can have fundamental effects on the transmission of the behaviors we see at the population level. As new populations are studied, revealing new and complex behaviors (e.g., Pruetz and Bertolani 2007; see also chapter 11), this approach will be an important component in building a comprehensive understanding of chimpanzee culture and the evolution of our own cultural abilities.

Acknowledgments

This research was supported by the National Institutes of Health (RR-00165), the Living Links Center of Emory University, a project grant from BBSRC to Andrew Whiten, and the University of St. Andrews. I would like to thank my colleges and co-contributors, Frans de Waal, Andrew Whiten, and Kristin Bonnie, for their expertise and assistance in conducting and presenting each of the studies. I would also like to thank Devyn Carter and the animal care staff of Yerkes Field Station for their support and assistance with the chimpanzees. I am grateful to Andy Burnley for his expert construction of the panpipe and doorian fruit apparatuses, and to Yerkes Field Station Engineering for their logistical support. Yerkes is fully accredited by the American Association for Accreditation for Laboratory Animal Care.

Literature Cited

Bartlett, F. C. 1932. *Remembering*. Oxford: Macmillan.

Biro, D., N. Inokue-Nakamura, R. Tonooka, G. Yamakoshi, C. Sousa, and T. Matsuzawa. 2003. Cultural innovation and transmission of tool use in wild chimpanzees: Evidence from field experiments. *Animal Cognition* 6:213–23.

Boesch, C. 1995. Innovation in wild chimpanzees. *International Journal of Primatology* 16(1):1–16.

———. 2007. What makes us human (*Homo sapiens*)? The challenge of cognitive cross-species comparison. *Journal of Comparative Psychology* 121(3):227–40.

Boesch, C., and M. Tomasello. 1998. Chimpanzee and human cultures. *Current Anthropology* 39(5): 591–614.

Bonnie, K. E., and F. B. M. de Waal. 2006. Affiliation promotes the transmission of a social custom: Handclasp grooming among captive chimpanzees. *Primates* 47:27–34.

Bonnie, K. E., V. Horner, A. Whiten, and F. B. M. de Waal. 2007. Spread of arbitrary conventions among chimpanzees: A controlled experiment. *Proceedings of the Royal Society of London B* 274:367–72.

Boyd, R., and P. J. Richerson. 1985. *Culture and the Evolutionary Process*. Chicago: University of Chicago Press.

Coussi-Korbel, S., and D. M. Fragaszy. 1995. On the relation between social dynamics and social learning. *Animal Behaviour* 50:1441–53.

Curio, E., U. Ernst, and W. Veith. 1978. The adaptive significance of avian mobbing. *Zeitschrift fuer Tierpsychologie* 48:184–202.

De Waal, F. 2001. *The Ape and the Sushi Master: Cultural Reflections by a Primatologist*. New York: Basic Books.

De Waal, F. B. M., C. Boesch, V. Horner, and A. Whiten. 2008. Comparing Social Skills of Children and Apes. *Science* 319: 569.

Galef, B. G. Jr. 2003. "Traditional" foraging behaviors of brown and black rats (*Rattus norvegicus* and *Rattus rattus*). In *The Biology of Traditions: Models and Evidence*, edited by D. Fragaszy and S. Perry. Cambridge: Cambridge University Press.

Giraldeau, L., and L. Lefebvre. 1987. Scrounging prevents cultural transmission of food finding behaviour in pigeons. *Animal Behaviour* 35:387–394.

Goodall, J. 1963. Feeding behaviour of wild chimpanzees: A preliminary report. *Symposia of the Zoological Society of London* 10:39–47.

———. 1986. *The Chimpanzees of Gombe: Patterns of Behaviour*. Cambridge, MA: Harvard University Press.

Hare, B., A. P. Melis, V. Woods, S. Hastings, and R. Wrangham. 2007. Tolerance allows bonobos to outperform chimpanzees on a cooperative task. *Current Biology* 17(7):619–23.

Henrich, J., and R. Boyd. 1998. The evolution of conformist transmission and the emergence of between-group differences. *Evolution and Human Behavior* 19:215–41.

Henrich, J., and F. J. Gil-White. 2001. The evolution of prestige: Freely conferred deference as a mechanism for enhancing the benefits of cultural transmission. *Evolution and Human Behavior* 22:165–96.

Hirata, S., and M. L. Celli. The role of mothers in the acquisition of tool-use behaviors by captive infant chimpanzees. *Animal Cognition* 6 (2003): 235–44.

Hopper, L., A. Spiteri, S. P. Lambeth, S. J. Schapiro, V. Horner, and A. Whiten. 2007. Experimental studies of traditions and the underlying transmission processes in chimpanzees. *Animal Behaviour*, in press.

Horner, V., and A. Whiten. 2005. Causal knowledge and imitation/emulation switching in chimpanzees (*Pan troglodytes*) and children (*Homo sapiens*). *Animal Cognition* 8:164–81.

Horner, V., A. Whiten, E. Flynn, and F. B. M. de Waal. 2006. Faithful replication of foraging techniques along cultural transmission chains by chimpanzees and children. *Proceedings of the National Academy of Science USA* 103:13878–83.

Imanishi, K. 1952. The evolution of human nature. In *Man*, edited by K. Imanishi. Tokyo: Mainichi-Shinbunsha.

Laland, K. N. 2004. Social learning strategies. *Learning and Behavior* 32(1):4–14.

Laland, K. N., and V. M. Janik. 2006. The animal cultures debate. *Trends in Ecology and Evolution* 21(10): 542–47.

Laland, K. N., and H. C. Plotkin. 1993. Social transmission of food preferences among Norway rats by marking of food sites and by gustatory contact. *Animal Learning and Behavior* 21:35–41.

Laland, K. N., and K. Williams. 1998. Social transmission of maladaptive information in the guppy. *Behavioral Ecology* 9(5): 493–99.

Lonsdorf, E.V. What is the role of mothers in the acquisition of termite-fishing behaviors in wild chimpanzees (*Pan Troglodytes Scheinfurthii*)? *Animal Cognition* 9 (2006): 36–46.

Lonsdorf, E. V., L. E. Eberly, and A. E. Pusey. 2004. Sex differences in learning in chimpanzees. *Nature* 428:715–16.

Lycett, S. J., M. Collard, and W. C. McGrew. 2007. Phylogenetic analyses of behavior support existence of culture among wild chimpanzees. *Proceedings of the National Academy of Science USA* 104(45):17588–92.

Matsuzawa, T. 2001. Emergence of culture in wild chimpanzees: Education by master apprenticeship. In *Primate Origins of Human Cognition and Behavior*, edited by T. Matsuzawa. New York: US Publishing.

Matsuzawa, T, and G. Yamakoshi. 1998. Comparison of chimpanzee material culture between Bossou and Nimba, West Africa. In *Reaching into Thought: The Minds of the Great Apes*, edited by A. E. Russon, K. Bard, and S. Parker. Cambridge: Cambridge University Press.

McGrew, W.C. 1977. Socialization and object manipulation of wild chimpanzees. In *Primate Bio-social Development: Biological, Social, and Ecological Determinants*, edited by S. Chevalier-Skolnikoff and F. E. Poirier. New York: Garland Publishing, Inc.

———. 2004. *The Cultured Chimpanzee: Reflections of a Primatologist*. New York: Cambridge University Press.

McGrew, W. C., and C. E. G. Tutin. 1978. Evidence for a social custom in wild chimpanzees. *Man* 13:234–51.

Menzel, E. W. Jr., R. K. Devenport, and C. M. Rogers. 1972. Proto-cultural aspects of chimpanzees' responsiveness to novel objects. *Folia Primatologica* 17:161–70.

Nakamura, M., and S. Uehara. Proximate factors of different types of grooming hand-clasp in Mahale chimpanzees: Implications for chimpanzee social customs. *Current Anthropology* 45(2004): 108–14.

Nakamura, M., and T. Nishida. Subtle behavioral variation in wild chimpanzees, with special reference to Imanishi's concept of *kaluchua*. *Primates* 47 (2006): 35–42.

Nishida, T. 1968. The social group of wild chimpanzees in the Mahale mountains. *Primates* 9:167–224.

Perry, S., M. Baker, L. Fedigan, J. Gros-Luis, K. Jack, K.C. MacKinnon, J.H. Manson, M. Pagner, K. Pyle, and L. Rose. 2003. Social conventions in wild white-aced capuchin monkeys: Evidence for traditions in a neotropical primate. *Current Anthropology* 44(2): 241–68.

Pruetz, J. D., and P. Bertolani. 2007. Savanah chimpanzees, *Pan troglodytes*, hunt with tools. *Current Biology* doi: 10.1016/j.cub.2006.12.042.

Reader, S. M., and K. N. Laland. 2001. Primate innovation: Sex, age and social rank differences. *International Journal of Primatology* 22:787–805.

Richerson, P. J., and R. Boyd. 2005. *Not by Genes Alone: How Culture*

Transformed Human Evolution. Chicago: University of Chicago Press.

Russon, A. E. 2003. Developmental perspectives on great ape traditions. In *The Biology of Traditions: Models and Evidence*, edited by D. M. Fragaszy and S. Perry. Cambridge: Cambridge University Press.

Spence, K.W. 1937. Experimental studies of learning and higher mental processes in pre-human primates. *Psychological Bulletin* 34:806–50.

Thorpe, W. H. 1956. *Learning and Instinct in Animals*. London: Methuen.

Tomasello, M. 1999. *The Cultural Origins of Human Cognition*. Cambridge, MA: Harvard University Press.

Tomasello, M., E. S. Savage-Rumbaugh, and A. Kruger. 1993. Imitative learning of actions on objects by children, chimpanzees and enculturated chimpanzees. *Child Development* 64:1688–1705.

Van Lawick-Goodall, J. 1973. Cultural elements in a chimpanzee community. In *Precultural Primate Behaviour*, edited by E. W. Menzel, 144–84. Basel: Karger.

Van Schaik, C.P. 2003. Local traditions in orangutans and chimpanzees: Social learning and social tolerance. In *The Biology of Traditions: Models and Evidence*, edited by D. M. Fragaszy and S. Perry. Cambridge: Cambridge University Press.

Van Schaik, C. P., M. Ancrenaz, G. Borgen, B. Galdikas, C. D. Knott, I. Singleton, A. Sazuki, S. S. Utami, and M. Merrill. 2003. Orangutan cultures and the evolution of material culture. *Science* 299:102–5.

Van Schaik, C. P., R. O. Deaner, and M. Y. Merrill. 1999. The conditions for tool use in primates: Implications for the evolution of material culture. *Journal of Human Evolution* 36:719–41.

Visalberghi, E., and D. Fragaszy. 1990. Do monkeys ape? In *Language and Intelligence in Monkeys and Apes: Comparative Developmental Perspectives*, edited by S. T. Parker and K. Gibson. Cambridge: Cambridge University Press.

Watanabe, K. 1994. Precultural behavior of Japanese macaques: Longitudinal studies of the Koshima troops. In *The Ethnological Roots of Culture*, edited by R. A. Gardner, A. B. Chiatelli, B. T. Gardner, and F. X. Plooji. Dordrecht: Kluwer Academic.

Westergaard, G. C., and D. M. Fragaszy. 1987. The manufacture and use of tools by capuchin monkeys (*Cebus apella*). *Journal of Comparative Psychology* 101:159–68.

Whiten, A., J. Goodall, W. C. McGrew, T. Nishida, V. Reynolds, V. Sugiyama, C. E. G. Tutin, R.W. Wrangham, and C. Boesch. 1999. Culture in chimpanzees. *Nature* 399:682–85.

Whiten, A., V. Horner, and F. B. M. de Waal. 2005. Conformity to cultural norms of tool use in chimpanzees. *Nature* 437:737–40.

Whiten, A., V. Horner, C. A. Litchfield, and S. Marshall-Pescini. 2004. How do apes ape? *Learning and Behavior* 32(1):36–52.

Whiten, A., A. Spiteri, V. Horner, K. E. Bonnie, S. P. Lambeth, S. J. Schapiro, and F. B. M. de Waal. 2007. Transmission of multiple traditions within and between chimpanzee groups. *Current Biology* 17:1038–43.

IO

How Are Army Ants Shedding
New Light on Culture in Chimpanzees?

Tatyana Humle

A chimpanzee manufactures from surrounding vegetation a long wand, which he inserts into a nest of army ants. The ants attack the intruding object, either by biting the wand with their strong mandibles or running up the length of the tool. After a few seconds, the chimpanzee removes the wand, clasps the proximal end of the tool with his other hand, and swipes the length of the wand through his closed fingers, thus bringing the ants to his mouth (the pull-through technique). The chimpanzee then vigorously chews on the ants while rapidly brushing away the ones that have gone astray and are biting his face and hands. Another chimpanzee approaches a trail of ants that are moving swiftly on the ground on a hunting raid. This chimpanzee similarly modifies vegetation, generating a much shorter tool which she places among the traveling ants. Once again, the ants attack the tool. Upon removing the tool, however, she proceeds to consume the ants by wiping the wand directly through her lips and/or teeth (the direct mouthing technique). The aim is the same—to gather ants. The methods, however, are quite different. Why?

The above description of two different methods for gathering ants is a prime example of the behavioral diversity that is seen across communities of chimpanzees in Africa. Based on these two accounts of ant dipping, we can safely deduce that neither chimpanzee described above comes from the Taï forest in Côte d'Ivoire, although chimpan-

zees at this site customarily perform ant dipping. Although the first description certainly portrays ant dipping as it is customarily performed in the Gombe community in Tanzania, we can confidently infer that the second description does not (see figure 10.1). However, both accounts depict ant dipping as it is performed by some members of the Bossou community in Guinea. Because of these apparent differences in tool length and technique between communities and unique community profiles, ant dipping has often been cited as one of the best examples of culture in chimpanzees (Boesch & Boesch, 1990; McGrew, 1992). But is ant dipping really cultural, or can we provide other more parsimonious explanations for the remarkable variations in its practice between sites?

The Issue of Culture

Ant dipping is one of many behaviors in chimpanzees that have been described as potential cultural variants on the basis that they represent group-typical behavioral patterns, dependent on social learning processes for their maintenance and transmission (definition adapted from Laland and Hoppitt 2003). Four decades of field studies of wild chimpanzees in Africa have revealed substantial differences in behavioral repertoires at the subspecies, population, and community level (for reviews see Whiten et al. 1999, 2001; see also chapter 8).

(A) (B)

Figure 10.1 Bossou chimpanzees ant dipping at a trail of intermediate species of *Dorylus*. (a) Placing the tool among the moving ants. Photo © Gaku Ohashi. (b) Performing the direct-mouthing technique to consume the gathered ants from the tool.

However, due to the difficulty in identifying social learning processes involved in the transmission of behavior in natural settings, and in firmly excluding environmental differences as an explanation for some observed variations in behavior, the attribution of culture to chimpanzees has often been criticized and challenged, and is still the source of much heated debate (Galef 1992; Tomasello 1999; Laland and Janik 2006; see also chapters 8 and 14). In recent years this debate has fueled numerous studies both in captivity and in the wild. These have shed important insights into the cognitive abilities of chimpanzees, including their remarkable capacity for innovation and social learning and their ability to select raw materials and manufacture tools appropriate for the task at hand during their daily lives. Just as with the nature-versus-nurture debate used to describe human behavioral variants, I propose in this chapter that the premise of an "environment–versus-culture" debate is futile, since many cultural variants are the product of a mixed interplay between environmental influences and constraints on the one hand and social influences and constraints on the other.

Insights from Captive Studies

Studies with captive chimpanzees have shed important and useful insights into chimpanzees' capacity for social transmission of complex skills and tool-use behaviors, including arbitrary behavioral patterns (reviewed by Whiten et al. 2004; see also chapters 8 and 9).

Captive studies generally indicate that chimpanzees (1) rely on observation of others when learning a novel be-

havior or skill, (2) can imitate and emulate, that is, copy the end or the outcome of an action sequence—although it is equally apparent that both social and individual learning mechanisms are important in the learning process, depending on context and complexity of the novel behavior—and (3) are capable of transmitting behavior with a relatively high degree of fidelity along a sequential chain of individuals or even between groups. These studies have also revealed important constraints on the transmission of behavior posed by (1) the saliency and social relationship of the demonstrator to the naïve individual, (2) the possibility for co-action or joint interaction, (3) the type of actions or degree of complexity presented by the task, (4) the duration and frequency of exposure, and (5) the age and sex of the subjects used (see chapter 9). Finally, although studies in captivity have markedly contributed to our understanding of the capacity for social learning and transmission of behavior in chimpanzees, they have unfortunately revealed little about (1) sources of individual differences in behavioral performance and competence, (2) the interplay between ecological and social influences on behavioral transmission both within and between groups, and (3) the details of behavioral interactions between knowledgeable and naïve individuals on the learning trajectory of young.

Methods Employed in Field Settings

In the field, the identification of a cultural variant in nonhuman animals has typically relied on identifying geographical variations in behavior across communities that are independent of genetic or ecological differences between sites (Whiten et al. 1999, 2001; chapter 8). This approach is known as the ethnographic method (Wrangham et al. 1994), the group comparison method (Fragaszy and Perry 2003), or the elimination method (van Schaik 2003).

Some studies have also focused on comparing proximate quantifiable environmental variables that might explain observed behavioral differences between sites (e.g., McGrew et al. 1997; Humle and Matsuzawa 2004). If none of the variables compared can explain the absence of the behavioral pattern at a proximal level, the most parsimonious hypothesis retained is that the behavior is cultural. In some cases, however, this approach has its limitations since the list of alternative ecological hypotheses can sometimes be quite exhaustive and nearly impossible to test empirically.

In recent years, researchers have adopted supplemental approaches, aimed more specifically at investigating the intergenerational transmission and maintenance of behaviors among wild subjects. One approach, innovated and developed by Matsuzawa and colleagues, has been to stimulate the occurrence of tool-use behaviors in an outdoor laboratory setting (see Biro et al. 2003, and also chapter 1). In this setting, located within the home range of the chimpanzees, important variables such as tool and food availability, distribution, and type can readily be manipulated. When the chimpanzees visit the outdoor laboratory, any occurrence of tool use is then video-recorded. Consistent data on the same tool-use behaviors can thus be gathered longitudinally across several years. Manipulations can also stimulate the emergence of behavioral innovations, rarely observed firsthand in the wild. This approach can therefore provide invaluable information on the dynamics of diffusion of novel behaviors both between and within a community beyond what can be gathered by following the chimpanzees from dawn to dusk.

Nevertheless, careful and detailed observations of the chimpanzees during the course of their daily activities still remain quintessential and provide the most socially and ecologically valid context in which to study culture in chimpanzees (e.g., Nakamura and Uehara 2004; Lonsdorf 2005; Humle et al. 2009). This approach is more laborious, and is highly dependent on the natural frequency of occurrence of the focal behaviors and level of habituation of the community. Through video recordings of the focal behavior, this approach, like the natural experimental approach previously described, provides a powerful tool for detailed analysis of the role of social-mediation in behavioral transmission and individual behavioral performance and competence. This field approach is in fact the most appropriate for an accurate and comprehensive exploration of the environmental, social, and developmental influences and constraints on behavioral transmission and variation among individuals.

Finally, not only have field studies yielded regional, population, and community differences in behavioral repertoire, they have also highlighted in some cases important and interesting individual behavioral variations within communities. Some field studies have also shed light on

the learning trajectory of complex skills in young, highlighting as well the high propensity of young for innovation. In addition, although some studies have reported the diffusion of behavioral patterns between communities (see chapter 13), some have also indicated that transmission of complex tool-use behavioral patterns between neighboring communities do not necessarily correlate with geographical proximity (e.g., Humle and Matsuzawa 2001, 2004). These findings raise important questions regarding the respective roles of ecological and social influences and constraints on the process of transmission and maintenance of behaviors, both within and between communities, that no captive study has yet been able to address. I will discuss these influences and constraints below, based on our current understanding of army ant consumption in chimpanzees, as well as chimpanzees' social behavior and ecology.

Ecological Influences and Constraints

Ecological Influences: The Cost of Army Ant Consumption

Army ants are ubiquitous across all sites where chimpanzees have been studied. Their consumption by chimpanzees has so far been observed at 5 study sites in Africa and reported present at 11 others, while absent at 5 long-term study sites (Schöning et al. 2008; figure 10.2). One striking characteristic of army ant consumption by chimpanzees is that its performance exhibits a great deal of variation. Some chimpanzees may target the eggs and brood via manual extraction, which involves the insertion of the hand or arm directly into the ants' nest. Others may instead harvest the adults with the aid of a tool: ant dipping. McGrew (1974) suggested that the reliance on a tool allows chimpanzees to harvest these biting ants more effectively and less painfully. Ant dipping itself exhibits a great deal of variation in how and where it can be performed. Indeed, mean site tool length ranges between 23.9 cm and 84.6 cm (Schöning et al. 2008). There are also significant differences between sites in the relative frequency or in the presence or absence of (1) ant dipping at nests versus trails (foraging or migrating), and (2) the pull-through (McGrew 1974) versus the direct mouthing technique, which includes two variants: swiping sideways through the teeth and lip and frontal plucking (figure 10.1). Sites may exhibit a range of one to

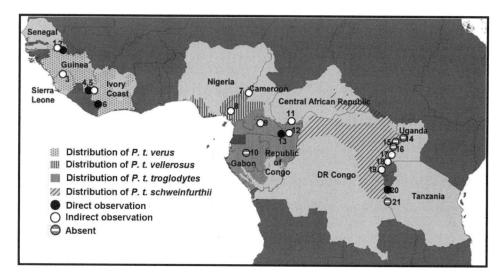

Figure 10.2   Sites across equatorial Africa where ant dipping by chimpanzees has been observed either directly or indirectly (based on the recovery of tool artifacts), and long-term sites where ant dipping has never been confirmed. Adapted from Schöning et al. 2008.

1. Assirik, Senegal; 2. Fongoli, Senegal; 3. Tenkere, Sierra Leone; 4. Bossou, Guinea; 5. Seringbara, Guinea; 6. Taï, Ivory Coast; 7. Gashaka, Nigeria; 8. Ntale, Cameroon; 9. Dja, Cameroon; 10. Lopé, Gabon; 11. Ngotto, Central African Republic; 12. Ndakan, Republic of Congo; 13. Goualougo, Republic of Congo; 14. Budongo, Uganda; 15. Semliki, Uganda; 16. Kibale, Uganda; 17. Kalinzu, Uganda; 18. Bwindi, Uganda; 19. Kahuzi-Biega, DR Congo; 20. Gombe, Tanzania; 21. Mahale, Tanzania.

six different species of army ants, and chimpanzees across Africa are known so far to consume a total of 12 species (Schöning et al. 2008). The species of army ants that are consumed by chimpanzees can be classed into two different lifestyles (Schöning et al. 2008). The species with an epigaeic lifestyle hunt for animal prey on the ground and in the vegetation, and produce conspicuous trails and earth nests. The other species have an intermediate lifestyle. These species hunt in the leaf litter, but never in the vegetation. While they also form conspicuous trails in open areas, their nests are not as readily detectable as those of epigaeic species.

Ant dipping at Bossou, a long-term study site of wild chimpanzees (*Pan troglodytes verus*) in southeastern Guinea, has been studied in detail since 1997 (Humle and Matsuzawa 2002; Yamakoshi and Myowa-Yamakoshi 2003; Humle 2006). Chimpanzees at this site employ both the pull-through and the direct mouthing techniques to gather ants off the tool. Chimpanzees at Bossou dip on five different species of army ants (three epigaeic species and two intermediate species) both at nests and trails (Humle, 2006). Since ants are present in high densities at nests with colonies containing up to nine million individuals, dipping at nests poses a greater risk to the chimpanzees than dipping at trails (Humle 2006). Based on the behavior

of the ants as assessed via a series of human ant-dipping experiments (Humle and Matsuzawa 2002), epigaeic species are also more gregarious and aggressive than intermediate species. Morphological data (cf. Schöning et al. 2008) additionally support the conclusion that epigaeic species can inflict more severe bites than intermediate species. When dipping at the nest, Boussou chimpanzees clearly adopt specific behavioral strategies to circumvent these risks by either (1) positioning themselves more above ground when dipping at nests than at trails or (2) using longer tools, particularly when dealing with ants at nests or with the more aggressive epigaeic species (Humle and Matsuzawa 2002; Humle 2006). However, no significant difference in tool length emerged between the two lifestyles at trails (Schöning et al. 2008; figure 10.3). In addition, the pull-through technique was almost exclusively associated with tools more than 50 cm long, whereas tools 50 cm long or shorter were solely associated with direct mouthing.

Schöning et al. (2008) explored the relationship between tool length and technique as a function of prey lifestyle and dipping condition—that is, nest versus trail—across 13 sites across eastern, central, and west Africa (four where ant dipping was observed directly and nine where it was only recorded indirectly). As found at

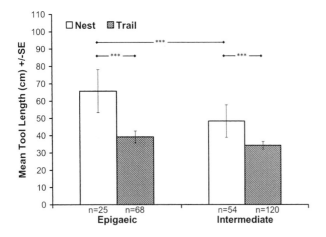

Figure 10.3   Mean length of tools used for ant dipping by chimpanzees more than 11 years old at Bossou between 2003 and 2006, on either epigaeic or intermediate army ant species (*Dorylus* spp.) at either nests or trails. ***: $P < 0.001$.

Bossou, epigaeic species at nests were dipped with longer tools, typically associated with the pull-through technique (e.g. Gombe, Tanzania), and intermediate species tended to be dipped with shorter tools coupled with the direct-mouthing technique (e.g. Taï, Côte d'Ivoire). Nevertheless, several important variations remained that could not be accounted for by microecological variables alone. The most remarkable differences lie between Bossou and Taï, two long-term study sites where the chimpanzees target the same five species of *Dorylus* ants (Schöning et al. 2008). Taï chimpanzees do not dip on epigaeic species at the nest (figure 10.4). In addition, although both Taï and Bossou chimpanzees dip on intermediate species, the Taï chimpanzees use significantly shorter tools for that purpose. They also do not dip on ants at trails, whereas Bossou chimpanzees do so customarily and preferentially. The only recorded instance of an individual dipping at a trail at Taï was performed by a young female chimpanzee (Boesch, personal communication). Finally, more than 70% of sessions at Taï were dedicated to brood and egg extraction on epigaeic species' nests, whereas at Bossou brood and egg extraction—although also predominantly focused on epigaeic species—was observed in less than 35% of sessions.

Möbius et al. (2008) assessed whether environmental differences in the availability, density, and behavior of the two types of species consumed at Bossou and Taï could account for the differences between the two study sites. We assessed differences in the speed and yield of the ants through a series of human ant-dipping experiments (*sensu*

Humle and Matsuzawa 2002), performed surveys to establish the availability and density of their nests and trails, and tested for differences in the accessibility of their eggs and brood at nests. We found no significant differences in the availability and density of nests and trails of either intermediate or epigaeic species between the two sites. Although insufficient data for analysis were gathered for intermediate species, the results for epigaeic species showed no significant differences in the ants' yield or speed, or in the accessibility of their brood and eggs, that could satisfactorily explain the observed variations in army ant consumption between Bossou and Taï. These results suggested to us that social learning might indeed play an important role in explaining these behavioral variations in army ant consumption between the two sites.

Finally, these studies taken together reveal that chimpanzees flexibly adjust their tool length and technique in response to microecological conditions, as reflected by differences in prey density and/or belligerence (i.e., biting risk), and between lifestyles and conditions (i.e., nest or trail; Humle 2006). Considering the remarkable cognitive abilities of chimpanzees in selecting suitable materials and/or manufacturing appropriate tools for various purposes (e.g., Boesch and Boesch 1990), the above therefore does not seem very surprising. However it is apparent that some variations cannot be explained on that basis, and thus we cannot rule out that the observed variations in army

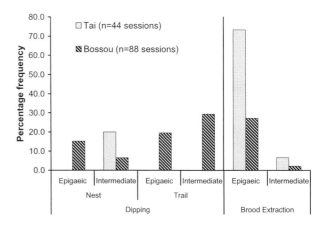

Figure 10.4   Comparison between Bossou and Taï in the percentage of frequency for which ant dipping and brood/egg extraction (nest only) were observed on either epigaeic or intermediate species at nest or trails. Taï data gathered by Yasmin Möbius from September 2003 to May 2004, and from February to May 2005; and by Tobias Deschner from December 2004 to January 2005. Bossou data gathered by Tatyana Humle from June 2003 to March 2004 and from July to September 2005 and 2006.

ant consumption between sites reflect cultural differences among chimpanzee communities.

Ecological Constraints

As first suggested by Sugiyama (1995) and shown above, local behavioral responses to prey characteristics and behavior do to some extent shape the observed variations in army ant consumption both within sites and between them. These findings logically raise the question of the relationship between ecology and cultural processes such as innovation, maintenance, and transmission of behavior within communities, and diffusion between communities. Seasonality, rarity, or unpredictability in availability (in time and space) of food resources are likely to affect the rate of innovation and demonstration as well as the opportunity for others to observe how these foods are processed. In addition, the availability of materials (if tool use is involved), the accessibility of target food, and the shareability of the resource may influence the acquisition of behavior, particularly among young individuals. Moreover, some foraging behaviors such as ant dipping and hunting for large mammalian prey (see chapter 18) can entail costs in their performance, thus creating a delicate balance between cost and benefit and possibly affecting the age and sex distribution of individuals who can model these behaviors, as well as affecting when these behaviors are acquired by young. If the nutritional and caloric gain is significant, such biases may in turn affect the age and sex-class distribution of individuals who perform alternative behaviors aimed at compensating for the nutritional and caloric loss. For example, at sites where hunting for mammalian prey is a group activity, hunting is typically a male-biased behavior (e.g., Gombe: McGrew 1979; see also chapters 15 and 18) and females tend to be more insectivorous (Gombe: McGrew 1979; Mahale: Uehara 1986). It has been argued that females at these sites focus more on the extraction of insects with the aid of tools (e.g., termite-fishing at Gombe: McGrew 1979; ant fishing at Mahale: Uehara 1986) to supplement the minimal amount of protein they gain through meat-eating. In contrast, at sites such as Bossou, where hunting for mammalian prey is a relatively rare, opportunistic, solitary, and non–sex-biased activity, adult males spent as much time as adult females performing ant dipping (Humle et al. 2009), thus providing further support for this "protein balance" hypothesis.

Yamakoshi (1998) also showed that when fruits are scarce, Bossou chimpanzees effectively increase their tool-use activities to gain access to otherwise inaccessible food resources and boost their energy intake. This pattern supports the idea that the likelihood of an innovation's diffusion may sometimes also depend on its adaptive value. Innovations are commonly observed, especially among young chimpanzees. Many of these novel behavioral variants are ephemeral, however, and are not transmitted and maintained—possibly because they have little adaptive value in a particular environment and time, either for the innovator or for other individuals who see them performed. Finally, ecological constraints—whether they be variability in resource availability, level of hazard, or the behavior's adaptive value—are likely to have an influence on when, how often, and by whom such behaviors may be innovated and/or acquired by a community's members.

Developmental and Social Influences and Constraints

Developmental and Social Influences

In most mammalian species, the ability to acquire foraging skills or knowledge about the environment is critical for young animals faced regularly with novel tasks or situations. For the young chimpanzee, as for humans and many other primate species, the primary socializing agent is the mother. For at least the first five years of life, the vast majority of the chimpanzee infant's interactions are with her (McGrew 1977). In addition, offspring and their mothers almost exclusively travel together for as many as 8 to 10 years (Pusey 1983). This prolonged period of association ensures that young are exposed to all of the mother's feeding and social activities at close range. Thus, the mother is likely to act as the prime model for her offspring, providing him or her with exposure to behaviors and opportunities to learn them (van Schaik et al. 2003).

Few studies have focused on the development of tool-use skills in wild chimpanzees. In a study exploring the role of mothers in the acquisition of ant dipping among the chimpanzees of Bossou, Humle et al. (2009) confirmed that for the first five years of a young chimpanzee's life, mothers were the prime model, although accessibility and exposure to other social models increased after weaning (after the age of five). Mothers clearly influenced the learning opportunity of their young at or before the age of five by dipping significantly more often at trails than at nests. Whether it was intentional or not, all mothers therefore provided their offspring with less hazardous conditions in which to observe and practice ant dipping, even though dipping at nests would supply a greater yield per unit of time than dipping at trails.

In addition, young with higher learning opportunity, as reflected by the mother's percentage of observed time spent ant dipping, started observing ant dipping sooner than young with less learning opportunity. High-opportunity young also first acquired ant dipping earlier than low-opportunity young. We cannot show with certainty that early observation influenced the onset of the behavior. But based on our understanding of the role of observation in the acquisition of other tool-use behaviors during ontogeny in chimpanzees (e.g., Lonsdorf 2005; Inoue-Nakamura and Matsuzawa 1997), it is likely that observation of ant dipping was vital in its acquisition, alongside the opportunity to practice the behavior under less risky conditions.

Learning opportunity also influenced dip success and proficiency in young. Young with greater learning opportunity committed fewer dipping errors (dips yielding no ants), especially during their formative years (between two and six years old). Dip duration was used as a measure of proficiency (i.e., the number of ants gathered per dip), since during a series of human ant-dipping experiments this measure correlated well with greater yield regardless of the ants' lifestyle or condition (Möbius et al. 2008). Young between 5 and 10 years old with greater learning opportunity demonstrated longer dip durations than low-opportunity young. Finally, the mother's proficiency and time spent ant dipping correlated positively with that of their offspring who were younger than five years old—indicating for the first time in chimpanzees a relationship between mothers and their progeny in time spent using tools and competence in doing so (Humle et al. 2009).

However, mothers and offspring did not match in tool length, although there was a trend for young with high learning opportunity to match their mother's tool length more closely than other young. Only a single mother more than 13 years old ever exhibited the pull-through technique when using tools more than 50 cm long, whereas all young more than 5 years old and all adult males exhibited this same technique. Young, therefore, did not acquire the pull-through technique by observing their mothers. Young between 5 and 10 years old experienced 61.8% (59/86) of their ant-dipping sessions in the presence of ant-dipping members of the community other than their mother, and thus they had ample opportunity to watch others employ the pull-through technique with tools more than 50 cm long (Humle et al. 2009). Therefore it is possible that young acquire this technique by observing others.

Lonsdorf (2006) found limited evidence for the relevance of learning opportunities (based on mothers' rate of tool use and level of sociality during tool-use activities) to the speed of acquisition of termite fishing in offspring at Gombe. Lonsdorf (2005) also reported sex differences in the development of termite fishing in chimpanzees at Gombe, while there was no such difference at Bossou for ant dipping. Although mothers at Gombe were equally tolerant of their male and female offspring when at the termite mound, young females spent significantly more time watching others than did young males. As a result, female offspring developed the skill earlier than males, and were more proficient at the skill once it was acquired. Also, they duplicated their mothers' depth of insertion of the tool (functional length) into the termite mound more closely than did young males. So apart from the underlying sex differences at Gombe, several patterns in the acquisition by young of termite fishing at Gombe were similar to what was observed for ant dipping at Bossou. Although neither study could precisely reveal the social learning mechanisms at work, both highlighted the importance of social learning opportunities in the acquisition of behavior (van Schaik et al. 2003). In addition, both studies revealed some behavioral matching related to behavioral competence between mother and offspring.

Developmental and Social Constraints

As mentioned above, the young chimpanzee's prime model is his or her mother. Vertical transmission can affect transmission of cultural variants both within and between communities. A young adult female immigrating into a new community may transfer her skills or knowledge to her offspring, who can then later serve as a model to other members of the community. In addition, as seen above, vertical transmission is essential for the intergenerational maintenance of cultural variants within a community.

Acquisition of a behavior by young may be influenced by factors that are both socially and behaviorally dependent, including (1) the opportunities for co-feeding, observation, and/or practice, whether simple manipulation of a tool or actual performance of skill elements, (2) opportunities for scrounging and passive food transfer, and (3) the behavior's hierarchical and manual complexity and the level of hazard it presents to the performer and learner. Although the role of all these variables has not yet been systematically explored across a wide range of behaviors, the few studies on the development of tool-use skill in chimpanzees do indicate that these variables are likely to dictate the age at which the behavior is fully acquired, as well as the individual differences among young in performance and competence levels.

The age at which some behaviors are acquired might also depend on a critical learning period, a window of cognitive receptivity to learning experiences. Through their longitudinal study of the development of nut-cracking behavior among Bossou chimpanzees, Inoue-Nakamura and Matsuzawa (1997) found that the critical learning period for the acquisition of nut cracking lies between the ages of three and five years. Beyond the age of five, acquisition of this complex tool-use task appears less probable, as is evidenced by the reported inability of some of that community's adolescents (between 8 and 11 years old) or adults (more than 11 years old) to demonstrate it. The evidence of a critical learning period for other complex behaviors has yet, however, to be established.

The study by Humle et al. (2009) of ant dipping at Bossou provides some preliminary support for the hypothesis of van Schaik et al. (2003) that individual differences in competence, and time spent in tool-use behaviors among adults, mirror the developmental experience of individual young. This hypothesis could also explain sex differences observed in adulthood. Lonsdorf's results (2005) reveal sex differences in the development of termite fishing that emulate sex differences in time spent performing that skill in adulthood (McGrew 1979). Similarly, the absence of sex differences in ant dipping at Bossou during development closely parallels the absence of sex differences in time spent in that activity by adults (Humle et al. 2009). However, the extent to which individual differences in performance and competence during ontogeny reflect individual differences in adulthood requires further investigation. Only long-term longitudinal data at other study sites, and

on other behaviors, will allow us to answer this question precisely.

Although Lonsdorf (2006) found no correlation between the proficiency of mothers and offspring, she found that by the age of 2.5 years there were extreme differences in offspring skill level that positively related to how much time mothers spent alone or with their families—which, in turn, predicted the amount of time mothers spent termite fishing. This result indicates that social constraints in some cases may also influence vertical transmission of behavior. For example, a low-ranking mother chimpanzee may only fish for termites when she is in a group with her maternal family, due to competition over mound holes when non-family members are present. Social constraints are also likely to have significant influence on transmission of behavior within and between communities both horizontally (between members of the same generation) and obliquely (from nonparental individuals of the parental generation to members of the filial generation, or vice versa) (Cavalli-Sforza et al. 1982). These modes of transmission are particularly relevant (1) when a community member invents a new behavior or (2) when an individual immigrates into a new community carrying with them a set of novel skills or behaviors absent from that community.

Coussi-Korbel and Fragaszy (1995) proposed that individuals in more egalitarian and tolerant social groups are more likely to learn socially and exhibit homogeneity in behavior, since they experience more opportunities for close behavioral coordination in space and time with other group members. Chimpanzees live in a fission-fusion social structure (Nishida 1968). This implies that at any given time, temporary and unstable parties are formed that represent only subsets of the whole community. Such a fluid and dynamic social structure is shaped by many different variables, including (1) the threat of predators, (2) the presence of neighboring communities, (3) the availability and distribution of resources—whether food, water, nesting sites, or reproductive females, (4) the patterns of affiliation and relatedness of individuals, and (5) the community's demographics (Wrangham et al. 1996).

These complex social dynamics, as well as differences in tolerance levels and affiliation among males and females in any one community, are likely to yield important differences in how behaviors are transmitted and maintained within and between communities. These differences may in turn dictate the degree of behavioral heterogeneity or homogeneity exhibited by a community. Conformity bias, an ability that chimpanzees demonstrate in captivity (Whiten et al. 2005; see also chapters 8 and 9), should result in homogeneity in behavior, and it is predicted to mirror social dynamics. Individuals who associate more with one another are also more likely to resemble one another in the behavioral variants they exhibit. Although the extent of such a conformity bias among wild chimpanzees still remains unclear, we do know that chimpanzees, especially young, generally have a high intrinsic motivation to observe and copy others (e.g., Biro et al. 2003; Matsuzawa et al. 2001). The homogeneity highlighted above in army ant consumption by Taï chimpanzees, and the rare observations of the direct-mouthing technique compared to the prime reliance on the use of long wands and the pull-through technique at Gombe, suggest that group norms likely exist in chimpanzees in the wild, especially when alternative behavioral variants are available. Why some communities exhibit greater homogeneity than others still remains to be investigated, but socioecological differences might allow us to elucidate some of these variations.

Other examples supporting cultural norms among wild chimpanzees lie in the domains of communication and social customs—for example, grooming behavior (Nakamura et al. 2000; see also chapter 13). The introduction of novel variants into a community might, however, encourage heterogeneity in behavior as observed in Mahale's M group with a pattern of hand-clasp (palm-to-palm) grooming having been introduced by an immigrating female from the K group (Nakamura and Uehara 2004). Apart from reflecting differential social dynamics in a community, homogeneity or heterogeneity in behavior is also likely to be influenced by (1) the number of variants available, either through innovation by a member of the community or the introduction of a novel variant by an immigrant individual; 2) the social learning mechanism(s) involved in the acquisition of the behavior, which will influence the degree of fidelity in the transmission of the behavior; and 3) the social learning opportunities during ontogeny.

Implications and Future Directions

Clearly ecological, developmental, and social influences and constraints may intermingle in shaping culture in chimpanzees (figure 10.5). However, this feature is not unique to chimpanzee culture. Many cultural anthropolo-

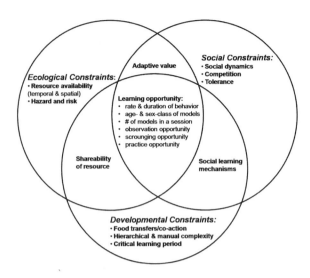

Figure 10.5 Variables influencing ecological, social, and developmental constraints on the transmission and maintenance of cultural variants both within and between communities.

gists and paleoanthropologists would argue that many aspects of human culture are shaped similarly (e.g., Best 2003). The results emanating from studies of army ant consumption within and between study sites illustrate the complexity of these interactions and how various cognitive abilities of chimpanzees may liaise in producing unique cultural community profiles.

Although these interactions render the study of animal culture in field settings quite complex, they also open up many interesting avenues for further exploration. These avenues, in turn, are important for understanding of the evolutionary origins of culture in humans. In this context, continued experimental studies in captivity have much to provide in helping us better understand the effect of the interplay between these different factors on cultural processes in chimpanzees (see chapters 8 and 9). In addition, the geographic technique complemented by video recordings of behaviors and inter-site exchanges of experienced field researchers (Nishida and Nakamura 2006) will continue to serve as a powerful tool for the study of culture and as an important guide for future research (see chapter 8).

To better understand the cultural mind of the chimpanzee, we also need more research at different field sites employing similar methodologies that rely on detailed and longitudinal behavioral recordings of the development of potential cultural variants among young, and their expression into adulthood. We must also supplement these studies with systematic data on the socioecology of these communities over time in order to fully understand the environmental and social influences and constraints that act within and between communities. Ideally such comparable data, complemented by knowledge of immigrants' behavioral repertoire before and after immigration, will provide the most pertinent insights. Nevertheless, witnessing the immigration of a known individual into a habituated community still remains a relatively rare phenomenon. Special focus on innovations and their pattern of diffusion and/or disappearance also has huge potential in teaching us about the cultural mind of the chimpanzee. Once again, however, such events are not frequently observed or identifiable, and cannot always readily be systematically monitored. As chimpanzees are studied at more and more sites across Africa, we can only hope that we will have the opportunity to witness and study such events and to evaluate their implications. In parallel, as more empirical data become available, modeling studies will also become increasingly valuable in helping us formulate or refine predictions and hypotheses (e.g., Laland and Kendal 2003), and understand more generally the interactions between ecology, cognition, and cultural processes.

Last, and not least, it is clear that if we wish to make progress in elucidating the cultural mind of chimpanzees, we as researchers must focus our energy and attention on their conservation. The extinction of chimpanzees would represent not only the extinction of our closest relative, but also the extinction of multiple cultures. Our responsibility to our subjects and our cousins is clear, and the clock is ticking.

Acknowledgments

I would like to thank the Ministère de l'Enseignement Supérieur et de la Recherche Scientifique, in particular the Direction Nationale de la Recherche Scientifique and l'Institut de Recherche Environnementale de Bossou (IREB), for granting me the permission to carry out research at Bossou. I am particularly grateful to Tetsuro Matsuzawa, Charles Snowdon, and William McGrew for their advice and support, and to Caspar Schöning, Kathelijne Koops, Gaku Ohashi, Gen Yamakoshi, Yasmin Möbius, Tobias Deschner, Christophe Boesch, and all the local assistants at Bossou for their invaluable contributions and collaboration. Finally I would like to acknowledge the financial support of the Ministry of Education, Science, and Culture, Japan (nos. 07102010, 12002009, and 10CE2005 to T. Matsuzawa), a Leakey Foundation Grant, and an NIH Kirschstein–NRSA Postdoctoral Fellowship (no. MH068906-01) to T. Humle.

Literature Cited

Best, A. 2003. *Regional Variation in the Material Culture of Hunter-Gatherers: Social and Ecological Approaches to Ethnographic Objects from Queensland, Australia.* Oxford: British Archaeological Reports International Series 1149.

Biro, D., N. Inoue-Nakamura, R. Tonooka, G. Yamakoshi, C. Sousa, and T. Matsuzawa. 2003. Cultural innovation and transmission of tool-use in wild chimpanzees: Evidence from field experiments. *Animal Cognition* 6:213–23.

Boesch, C., and H. Boesch. 1990. Tool use and tool making in wild chimpanzees. *Folia Primatologica* 54:86–99.

Cavalli-Sforza, L. L., M. W. Feldman, K. H. Chen, and S. M. Dornbusch. 1982. Theory and observation in cultural transmission. *Science* 218:19–27.

Coussi-Korbel, S., and D. M. Fragaszy. 1995. On the social relation between social dynamics and social learning. *Animal Behaviour* 50:1441–53.

Fragaszy, D., and S. E. Perry. 2003. *The Biology of Traditions: Models and Evidence.* Cambridge: Cambridge University Press.

Galef, B. G. Jr. 1992. The question of animal culture. *Human Nature* 3:157–78.

Humle, T. 2006. Ant-dipping in chimpanzees: An example of how microecological variables, tool-use and culture reflect the cognitive abilities of chimpanzees. In *Cognitive Development in Chimpanzees*, edited by T. Matsuzawa, M. Tomonaga, and M. Tanaka, 452–75. Tokyo: Springer-Verlag.

Humle, T. and T. Matsuzawa. 2001. Behavioural diversity among the wild chimpanzee populations of Bossou and neighbouring areas, Guinea and Côte d'Ivoire, West Africa. *Folia Primatologica* 72:57–68.

———. 2002. Ant dipping among the chimpanzees of Bossou, Guinea, and comparisons with other sites. *American Journal of Primatology* 58:133–48.

———. 2004. Oil palm use by adjacent communities of chimpanzees at Bossou and Nimba Mountains, west Africa. *International Journal of Primatology* 25:551–81.

Humle, T., C. T. Snowdon, and T. Matsuzawa. 2009. Social influences on the acquisition of ant-dipping among the wild chimpanzees of Bossou, Guinea, west Africa (*Pan troglodytes verus*). *Animal Cognition* 12:37–48.

Inoue-Nakamura, N., and T. Matsuzawa. 1997. Development of stone tool use by wild chimpanzees (*Pan troglodytes*). *Journal of Comparative Psychology* 111:159–73.

Laland, K. N., and W. Hoppitt. 2003. Do animals have culture? *Evolutionary Anthropology* 12:150–59.

Laland, K. N., and J. R. Kendal. 2003. What models say about social learning. In *The Biology of Traditions: Models and Evidence*, edited by D. M. Fragaszy and S. Perry, 33–55. Cambridge: Cambridge University Press.

Laland, K. N., and V. M. Janik. 2006. The animal cultures debate. *Trends in Ecology and Evolution* 21:542–547.

Lonsdorf, E. V. 2005. Sex differences in the development of termite-fishing skills in the wild chimpanzees, *Pan troglodytes schweinfurthii*, of Gombe National Park, Tanzania. *Animal Behaviour* 70:673–83.

———. 2006. What is the role of mothers in the acquisition of termite-fishing behaviors in wild chimpanzees (*Pan troglodytes schweinfurthii*)? *Animal Cognition* 9:36–46.

Matsuzawa, T., D. Biro, T. Humle, N. Inoue-Nakamura, R. Tonooka, G. Yamakoshi, and T. Matsuzawa. 2001. Emergence of culture in wild chimpanzees: Education by master-apprenticeship. In *Primate Origins of Human Cognition and Behavior*, edited by T. Matsuzawa, 557–74. Tokyo: Springer-Verlag.

McGrew, W. C. 1974. Tool use by wild chimpanzees in feeding upon driver ants. *Journal of Human Evolution* 3:501–8.

———. 1977. Socialization and object manipulation of wild chimpanzees. In *Primate Bio-Social Development*, edited by S. Chevalier-Skolnikoff and F. Poirier, 261–88. New York: Garland Press.

———. 1979. Evolutionary implications of sex differences in chimpanzee predation and tool use. In *The Great Apes*, edited by D. A. Hamburg and E. R. McCrown, 441–63. London: Benjamin Cummings.

———. 1992. *Chimpanzee Material Culture: Implications for Human Evolution*. Cambridge: Cambridge University Press.

McGrew, W. C., R. M. Ham, L. J. T. White, C. E. G. Tutin, and M. Fernandez. 1997. Why don't chimpanzees in Gabon crack nuts? *International Journal of Primatology* 18:353–74.

Möbius, Y., C. Boesch, T. Matsuzawa, K. Koops, C. Schöning, and T. Humle. 2008. Cultural and ecological factors intermingle to influence ant-dipping in West African chimpanzees. *Animal Behaviour* 76:37–45.

Nakamura, M., W. C. McGrew, L. F. Marchant, and T. Nishida. 2000. Social scratch: Another custom in wild chimpanzees? *Primates* 41:237–48.

Nakamura, M., and T. Nishida. 2006. Subtle behavioral variation in wild chimpanzees, with special reference to Imanishi's concept of *kaluchua*. *Primates* 47:35–42.

Nakamura, M., and S. Uehara. 2004. Proximate factors of different types of grooming hand-clasp in Mahale chimpanzees: Implications for chimpanzee social customs. *Current Anthropology* 45:108–14.

Nishida, T. 1968. The social group of wild chimpanzees in the Mahale Mountains. *Primates* 9:167–224.

Pusey, A. E. 1983. Mother-offspring relationships in chimpanzees after weaning. *Animal Behaviour* 31:363–77.

Schöning, C., T. Humle, Y. Möbius, and W. C. McGrew. 2008. The nature of culture: Technological variation in chimpanzee predation on army ants. *Journal of Human Evolution* 55:48–59.

Sugiyama, Y. 1995. Tool-use for catching ants by chimpanzees at Bossou and Monts Nimba. *Primates* 36:193–205.

Tomasello, M. 1999. *The Cultural Origins of Human Cognition*. Cambridge, MA: Harvard University Press.

Uehara, S. 1986. Sex and group differences in feeding on animals by wild chimpanzees in the Mahale Mountains National Park, Tanzania. *Primates* 27:1–13.

Van Schaik, C. P. 2003. Local traditions in orangutans and chimpanzees: Social learning and social tolerance. In *The Biology of Traditions: Models and Evidence*, edited by D. Fragaszy and S. E. Perry, 297–328. Cambridge: Cambridge University Press.

Van Schaik, C. P., E. A. Fox, and L. T. Fechtman. 2003. Individual variation in the rate of use of tree-hole tools among wild orang-utans: Implications for hominin evolution. *Journal of Human Evolution* 44:11–23.

Whiten, A., J. Goodall, W. C. McGrew, T. Nishida, V. Reynolds, Y. Sugiyama, C. E. G. Tutin, R. W. Wrangham, and C. Boesch. 1999. Cultures in chimpanzees. *Nature* 399:682–85.

———. 2001. Charting cultural variation in chimpanzees. *Behaviour* 138:1481–1516.

Whiten, A., I. Horner, C. A. Litchfield, and S. Marshall-Pescini. 2004. How do apes ape? *Learning & Behavior* 32:36–52.

Whiten, A., V. Horner, and F. B. M. de Waal. 2005. Conformity to cultural norms of tool use in chimpanzees. *Nature* 437:737–40.

Wrangham, R., C. Chapman, A. Clark-Arcadi, and G. Isabirye-Basuta. 1996. Social ecology of Kanyawara chimpanzees: Implications for understanding the costs of great ape groups. In *Great Ape Societies*, edited by W. C. McGrew, L. F. Marchant, and T. Nishida, 45–57. Cambridge: Cambridge University Press.

Wrangham, R. W., F. B. M. de Waal, and W. C. McGrew. 1994. The challenge of behavioral diversity. In *Chimpanzee Cultures*, edited by R. W. Wrangham, W. C. McGrew, F. B. M. de Waal, P. G. Heltne, and L. A. Marquardt, 1–18. Cambridge, MA: Harvard University Press.

Yamakoshi, G. 1998. Dietary responses to fruit scarcity of wild chimpanzees at Bossou, Guinea: Possible implications for ecological importance of tool use. *American Journal of Physical Anthropology* 106:283–95.

Yamakoshi, G., M. Myowa-Yamakoshi. 2003. New observations of ant-dipping techniques in wild chimpanzees at Bossou, Guinea. *Primates* 45:25–32.

11

The Complexity of Chimpanzee Tool-Use Behaviors

Crickette M. Sanz and David B. Morgan

There are constant sounds of chattering, whistling, and rustling as various creatures search for food, attempt to attract mates, or avoid predators in the dense forests of the Nouabalé-Ndoki National Park. However, the rhythmic sound of a chimpanzee using a large branch to pound open a beehive is distinct from all other forest noises. Such a reverberating echo of pounding sends us racing through the underbrush to catch a glimpse of a tool-using ape. We arrive at the origin of the sound to find a young adult female lounging near a beehive surrounded by a swarm of stingless bees. She picks up a large club lying on the branch beside her, holds it as if preparing to launch it as a javelin, but instead forcefully hits its end against the beehive. She repeatedly pounds with the club before inspecting the result of her efforts. After finding that the tool possibly is too large, she carefully places it in the canopy and manufactures a shorter tool that seems easier to maneuver. She then uses the two tools alternately to hammer and lever the hive. Suddenly appearing to abandon the task, the chimpanzee climbs up into the trees' leafy canopy and returns with a slender twig deftly fashioned into a dipping probe to extract the honey. She spends the rest of the afternoon enjoying the bounty of her technological skills, while we are left to reflect on the cognitive implications of such impressive "tool kits."

Tool use involves the relating of two objects to one another, and is therefore considered more cognitively com-

plex than other types of object use. However, some types of tool use are species-specific feeding adaptations that do not require any understanding of the task. The diversity and flexibility of tool use by primates once seemed to distinguish them from other taxonomic groups, but recent research on the tool behaviors of wild corvids has challenged such broad generalizations (Hunt 1996; Hunt et al. 2006; Hunt and Gray 2003). Parker and Gibson (1977) provide criteria for distinguishing complex foraging strategies that are "context-specific" from "intelligent" tool use that is characteristic of the more advanced stages of sensorimotor intelligence. Tomasello and Call (1997) state that understanding of the dynamic relationships between objects presumably enables flexibility in applying tool-use skills across contexts, which is indicative of intelligent tool use. The technological system of chimpanzees has been lauded as being the most sophisticated form of nonhuman material intelligence observed in natural environments. Various tool behaviors of wild chimpanzees, such as nut cracking and the use of tool sets in termite fishing, have been referred to as "complex," but there have been few systematic treatments of the composition or structure of chimpanzee tool use that allow us to assess the complexity of these tasks. In this chapter we examine several traditional notions of complexity, including the number of behavioral components (elements) needed to complete the tool task, structuring of actions, and hierarchical organization of object relations. We also provide some preliminary insights into the flexibility of different tool tasks shown by this species, which may provide some indication of their causal understanding of these tasks and the rules governing chimpanzee tool strategies.

Chimpanzee Tool Use in a Comparative Context

Among nonhuman primates, the material intelligence of chimpanzees and orangutans has been differentiated by the diversity of tools used in different contexts and relative regularity in which tools are used in comparison to their use by other taxa (van Schaik et al. 1999). Reports of tool use by New World monkeys have effectively challenged some long-standing assumptions about the phylogenetic distribution of particular features of tool-use skills. In particular, the behaviors of some wild capuchin monkey populations stand apart from the relatively rare and simple tool-use behaviors observed in other primates (Fragaszy et al. 2004; Phillips 1998; Waga 2006). Most recently, the tool-use behaviors of New Caledonian crows have further broadened our perspectives of complex tool use in other taxonomic groups. These corvids have a repertoire of several different tool behaviors (Hunt 1996), some of which differ in the complexity of their manufacturing sequences (Hunt et al. 2006; Hunt and Gray 2003). Future comparative examinations of the complexity of tool-use behaviors will provide insights into the cognitive capacities of these species. However, a necessary first step is to assess the tool technology within a species or population to determine the degree of variation in these systems and how this may inform us about ecological and social factors shaping different tool-use propensities.

Assessing Complexity of Tool Use

Sophisticated social networks and multistage food-processing techniques have been cited as complex behavior patterns that involve advanced cognitive functioning (Byrne and Byrne 1993; Byrne et al. 2001; Corp and Byrne 2002). However, there have been surprisingly few systematic treatments of complexity in wild chimpanzee tool use. Previous approaches to characterizing the complexity of particular tool tasks have included counting the number of behavioral components (elements), examining the sequential structuring of actions, and assessing the depth

and hierarchical organization of object relations. We proceed with a description of these methods before applying them to the tool-use behavior of wild chimpanzees in the Goualougo Triangle, Republic of Congo. This chimpanzee population has a large repertoire of tool-use behaviors, some of which are exhibited on a habitual or customary basis (Sanz and Morgan 2007). Through a systematic comparison of these tasks, we will assess whether the behaviors differ in their degree of complexity and what conclusions we can draw about the chimpanzees' understanding of these tasks.

Initial approaches to complexity in tool-use behaviors involved comparing the physical characteristics of tools and the modifications required in their manufacture. Oswalt (1976) proposed a method to systematically gauge the technological complexity of various hunter-gatherer populations that involved estimating the number of physically distinct structural configurations that contribute to the form of a tool. McGrew (1987) extended this analysis to wild chimpanzees, but the complexity of chimpanzee tool traditions was low on the human scale, and the variation between populations was too fine to be evaluated by Oswalt's technosystem. This is not to say that chimpanzee tools are not modified or manufactured toward a specific mental representation of a particular tool type, or even that they may not be comprised of multiple objects. Boesch and Boesch (1990) have shown that chimpanzees consistently fashion stick tools of specific lengths and diameters for particular tool-use tasks. Several stages of raw material modification may be necessary to produce a suitable ant-dipping rod at Taï (Boesch and Boesch 1990), a spear at Fongoli (Pruetz and Bertolani 2007), or a termite-fishing probe at Goualougo (Sanz and Morgan 2007; Sanz et al. 2009). Intriguingly, design complexity and number of nonrecapitulated modifications have been documented in the manufacture of tools from *Pandanus* tree leaves by New Caledonian crows to capture invertebrates (Hunt et al. 2006). The manufacture of tools with several "steps" or notches cut along the leaf edge is considered more complex than simple strip tools.

It has been suggested that estimations of behavioral complexity in natural systems can be accomplished by "repertoire counting," which involves an inventory of all the distinct components (elements) that comprise a behavior or task (Sambrook and Whiten 1997). The size of a behavioral repertoire is assumed to be positively related to the cognitive sophistication of the organism, in which the diversity of available choices indicates a propensity for innovation and an ability to select appropriate behaviors. In general, chimpanzees have a diverse behavioral repertoire that includes many different types of tool use. This can partially be attributed to the manual dexterity of primates, which aids in the formation and manipulation of external objects (van Schaik et al. 1999). Again, this has been challenged by the "tool kits" of New Caledonian crows (Hunt 1996). Rather than labeling behaviors as complex based on the number of their behavioral components, it is more important that particular skills are explicitly linked to cognitive capabilities, such as understanding of the task. The assemblage or organizational structure of behavioral components in different settings may assist in distinguishing rigidly fixed action patterns from more flexible manifestations of tool use which show that an organism has some understanding of the task.

Most sophisticated behavior patterns are interpreted as being sequentially or hierarchically organized such that later elements require the completion of one or more previous elements. However, context-specific or fixed patterns can also involve many elements that are linearly ordered with little or no evidence of strategic flexibility or understanding of causality between external objects. Boesch and Boesch-Achermann (2000) have suggested that this flexibility is a key component in inventing and developing tool repertoires such as those of humans and chimpanzees. Flow diagrams of decision processes or alternative pathways at natural junctions can provide important insights, such as propensities to reiterate strings of elements, and abilities to substitute elements and incorporate flexible responses for coping with new situations. It is also possible to superimpose quantitative data, such as frequencies or probabilities of element transitions, onto these flow diagrams of structural organization (Tonooka 2001). This provides insights into the statistical regularity of particular element combinations, and it may also highlight essential sequences of elements that are necessary to accomplish a task (Byrne 2003). The structured use of multiple tools to achieve a goal may necessitate a higher level of hierarchical organization than tasks involving only a single tool.

The use of multiple tools to achieve a common function is relatively rare in species other than humans, but it has been observed to be habitual in some chimpanzee populations. Sugiyama (1997) compiled reports of chimpanzees using more than one tool in sequence (serial tool use)

or a combination of tools together (composite tool use) to achieve a goal. The "tree structure analysis of hierarchical cognition" developed by Matsuzawa (1996) provides a way of describing the cognitive processes involved in a series of actions or behavioral patterns. The depth of nodes in the tree structure represents complexity of action, and the number of nesting clusters indicates hierarchical levels. In tool use, the depth of nodes increases with the number of objects used. For example, termite fishing consists of a single relationship between a fishing probe (tool) and a termite (target), whereas the use of metatools, observed in nut cracking at Bossou, consists of three nested object relationships (hammer, anvil, nut). Tree-structure analysis can be applied across cognitive domains to systematically document structure and hierarchical processes within or between taxa. This approach has also been used to depict cognitive modules of symbol use (Matsuzawa 1996) and social intelligence (de Waal 2003).

Comparisons of the catalogs of tool behaviors recorded at long-term study sites have shown that repertoires differ between populations and even between adjacent groups (Boesch and Boesch 1990; McGrew 1992; Sanz and Morgan 2007; Yamakoshi 2001). We have previously reported on the technological system of the chimpanzees in the Goualougo Triangle, which includes the habitual use of multiple tool techniques that have been described as being complex (Sanz and Morgan 2007; Sanz et al. 2004). In this chapter, we systematically compare the composition and structure of different tool-use behaviors within this wild chimpanzee population. Tool behaviors directed toward different targets are likely to differ in their element composition, organizational structure, and patterning of elements. However, different tool-use tasks are also compared to investigate whether these chimpanzees have rules that govern object relations or demonstrate flexibility in employing different tool strategies in the same or different contexts. Our aim is not only to better understand the technological sophistication of these chimpanzees, but also to elucidate some of the cognitive mechanisms which have led to the emergence of these fascinating behaviors.

Approach

Study Site and Population

The Goualougo Triangle is located within the Nouabalé-Ndoki National Park, Republic of Congo. The study area covers 380 km² of evergreen and semi-deciduous lowland forest with altitudes ranging between 330 and 600 m. The climate can be described as transitional between the Congo-equatorial and subequatorial climatic zones. The main habitats in the study area are mixed-species forest, monodominant *Gilbertiodendron* forest, and swamp. The main rainy season is from August through November, with a shorter rainy season in May.

Between February 1999 and December 2006, we spent a total of 88 months in the Goualougo Triangle habituating and studying wild chimpanzees. We conducted reconnaissance surveys in several community ranges, but the majority of our efforts were allocated to the Moto, Mopepe, and Mayele communities, which each consisted of 64 to 71 individuals (including immatures) during this period.

Data Collection

Tool behaviors were recorded ad libitum during direct observations with semi-habituated chimpanzees in the Goualougo study area. For all instances of tool behavior, we recorded the actor, behavior, type of object used, target of behavior, and outcome. We recorded digital video of tool-use behavior whenever possible. In addition, between 4 and 18 remote video-recording devices were used to conduct surveillance at termite nests of chimpanzee visitation between 2003 and 2006. (For precise details of the device used, see Sanz et al. 2004). The following tool-use behaviors were video-recorded during time spent with chimpanzees conducting direct observations or via remote video-recording devices that were installed in the forest:

HONEY GATHERING. Chimpanzees in the Goualougo Triangle have been observed to use dipping, levering, and pounding tools to gather honey from the hives of stingless bees and African honeybees (see Sanz and Morgan 2009 for a review). Inserting a probe into a bee nest to extract honey (dipping) is the most widespread tool-use strategy shown by chimpanzees in honey gathering and is seen in sites across Africa, from the Taï forest in Ivory Coast to Gombe in Tanzania (Bermejo and Illera 1999; Boesch and Boesch 1990; Boesch et al. 2009; Fay and Carroll 1994; Fowler and Sommer 2007; Hicks et al. 2005; Izawa and Itani 1966; Kajobe and Roubik 2006; Nishida and Hiraiwa 1982; Stanford et al. 2000; Tutin et al. 1995). Chiseling or lever-opening of arboreal bee nests to widen an access point to extract honey has been observed in Tanzania,

the Central African Republic, Gabon, and the Republic of Congo (Fay and Carroll 1994; Sanz and Morgan 2007; Wallauer, personal communication). Pounding or hammering of beehives with the end of a large club to break the hive structure has been observed rarely, but it seems to be exclusive to chimpanzee populations of the Congo Basin (Bermejo and Illera 1999; Boesch et al. 2009; Fay and Carroll 1994; Hicks et al. 2005; Sanz and Morgan 2007).

LEAF SPONGING. Leaf sponging involves using a mass of crushed or chewed leaves to sponge water from a tree basin. This behavior has been documented at several long-term study sites (Whiten et al. 1999, 2001), but is carefully distinguished from leaf folding by Tonooka (2001), who has compiled detailed data on the use of leaves for drinking water by wild chimpanzees in an outdoor laboratory in which the water was provisioned in a tree hollow. Reports of the natural use of leaves to drink water include the following: Gombe (Goodall 1964) and Mahale (Matsusaka et al. 2006) in Tanzania, Taï in Ivory Coast (Boesch and Boesch 1990), Bossou in Guinea (Sugiyama 1995), Lopé in Gabon (Tutin et al. 1995), Goualougo in the Republic of Congo (Sanz and Morgan 2007), Semliki in Uganda (McGrew et al. 2007), and Tongo in the Democratic Republic of Congo (Lanjouw 2002).

TERMITE FISHING AT ELEVATED NESTS. Several studies across equatorial Africa have reported that chimpanzees use fishing probes to extract termites from their earthen nests (see Sanz et al. 2004 for a review). Termite fishing typically involves inserting a flexible wand into a termite nest to extract termites that attack the invading object, but variations in this behavior have been documented between populations. For example, there are several populations of chimpanzees in central Africa that use fishing probes with a modified brush tip (Muroyama 1991; Fay and Carroll 1994; Suzuki et al. 1995; Bermejo and Illera 1999; Sanz et al. 2004, 2009). Another variation involves the use of a second tool; after unsuccessfully attempting to open termite exit holes manually, chimpanzees in the Goualougo Triangle have been observed to manufacture a perforating twig to open the exit holes on the surface of the nest (Sanz et al. 2004). The tip of the tool is pressed into the surface of the mound to clear soil from a closed exit hole. The chimpanzee then inserts a fishing probe into the cavities of the nest to extract termites. Perforating tools vary from small, straight twigs a few centimeters long to large, unwieldy branches with leafy twigs attached (average length = 32.9 ± 19.4 cm; range = 5; 91 cm, n = 54).

TERMITE FISHING AT SUBTERRANEAN NESTS. Yet another task involves extracting termites from subterranean nests, which necessitates a tool kit comprised of a puncturing stick and fishing probe. Although termite nest puncturing (also referred to as digging) and fishing tool assemblages have been recovered from sites in central Africa (Sabater Pi 1974; Sugiyama 1985; Muroyama 1991; Fay and Carroll 1994; Suzuki et al. 1995; Bermejo and Illera 1999), the first full descriptions of this tool behavior have only become available from the Goualougo Triangle (Sanz et al. 2004). The chimpanzee must first gain access to the subterranean chambers of the nest by inserting the length of a stout stick into the ground, holding the midsection of the tool with both hands, and often using a foot for additional leverage. The stick tool creates a long and narrow tunnel for insertion of the fishing probe. After removing the puncturing stick, the chimpanzee inserts a brush-tipped fishing probe to extract the termites. In contrast to perforating twigs, the tools used in puncturing are uniformly straight and smooth, and are manufactured from particular tree species.

Data Analysis

Individual chimpanzees were identified from their distinct physical characteristics and these data were compiled in a population history database. Video analysis was conducted using INTERACT Version 8.04 (Mangold 2006). Video recordings were scored as tool-use bouts, sessions, and episodes (definitions adopted from Yamakoshi and Myowa-Yamakoshi 2004). An episode began when the chimpanzee manufactured a tool (or at the first moment after which they were observed with the tool) and ended when the tool was discarded or the task was abandoned. Within an episode, any number of bouts or sessions could occur. A bout began when a chimpanzee used a tool to achieve a goal, and ended when they either succeeded or failed to achieve it. A session consisted of a series of bouts by an individual towards achieving a particular goal. The chimpanzee might make several attempts to achieve the goal (widen the entrance of a beehive, create a tunnel into a subterranean termite nest), but the session continued until they attained the goal, stopped using that tool to use

another, discarded the tool, or abandoned the endeavor. For all of the following analyses, we included only recordings of individuals who were capable of the task and complete sequences of tool behaviors.

We used Matsuzawa's (1996) tree structure analysis to compare the complexity of different tool tasks. Specifically, we examined the depth of object-relationship nodes as a measure of complexity and the number of nested clusters as indicative of hierarchical levels. In addition to the overall object-relationship structures, we also present both the elements and essential actions of each tool task in a traditional flow diagram. These diagrams are the traditional way of showing the structural organization of behavioral elements and essential actions (as in Tonooka 2001). Our flow diagrams focus on the target, tool, action, and goal of each tool task.

The behavioral elements in this study were an extension of the traditional ethogram approach. We defined elements as functionally distinct behavioral units, which we assumed to have biological meaning due to their seamless execution. Natural junctions were taken into consideration when defining behavioral elements. For example, we observed that the steps involved in removing a leaf sponge from the mouth and inserting/extracting it into a water basin were very rarely disjointed. Therefore, we did not split this behavior into distinct units as was done by Tonooka (2001), who divided this action into discrete steps. Behaviors associated with tool manufacture were not included in these measures of repertoire size, but have been previously published for these tool tasks (Sanz and Morgan 2007). We defined essential actions as those that were shared by all chimpanzees who successfully completed the tool task. Behavioral elements that involved the active use of the tool were defined as tool actions.

We defined a sequence as a continuous string of at least 30 behavioral elements employed toward accomplishing a task. This number of elements was more than sufficient to achieve each of the tasks, but longer sequence lengths were preferred for robust statistical analysis. We quantified nonrandomness in tool-action transitions by comparing observed first-order transition matrices with 1,000 randomly permuted matrices of the same data (custom software by R. Mundry). Matrix permutations preserve the patterning of elements within the matrix, whereas sequence permutations may alter the distribution of transitions in the matrix. However, we used sequence permutations when the number of columns was too small to allow matrix permutations. We also calculated the Shannon-Weaver entropy index for element transitions, which provided a standardized index ranging from 0 to 1, with lower values indicating more structured relations within the matrix.

Kruskal-Wallis tests were used to test differences in repertoire sizes of behavioral elements between tool tasks. Representation of each individual was limited to a particular tool task to meet assumptions of independent data points.

Results

Our data set was comprised of 27 video-recorded episodes of leaf sponging (adult/subadult females = 10; adult/subadult males = 2; juveniles = 2), 24 recordings of honey gathering (adult/subadult females = 6, adult/subadult males = 2; juveniles = 4), 25 recordings of termite fishing at elevated nests (adult/subadult females = 5, adult/subadult males = 4; juveniles = 4), and 21 recordings of termite fishing at subterranean nests (adult/subadult females = 2, adult/subadult males = 6). A subset of these data were composed of continuously recorded episodes that could be used for sequence analysis (11 segments of leaf sponging, 24 segments of honey gathering, 33 segments of termite fishing at elevated nests, and 21 segments of termite fishing at subterranean nests).

Tree-Structure Analysis

Figure 11.1 depicts the hierarchical tree-structure analysis of tool tasks analyzed in this study. The relationships between objects ranged from level 1 to level 3 as described by Matsuzawa (1996), and the hierarchical organization differed within and between tasks. Leaf sponging is consistently a level 1 tool use that involves a leaf sponge (tool) directed toward water in a tree basin (target). Several leaf sponges could be used in a bout, but we only observed one type of tool in this context. Termite fishing at elevated termite nests most often involved a relationship between two objects—the brush-tipped fishing probe (tool) and the termites (target)—which is a level 1 tool use. However, we also observed the use of a perforating twig (tool) to open the surface of a termite nest (target) prior to termite fishing, which is a second relationship between objects in this context. The serial order of two tools to open a subterranean nest (target) with a puncturing stick (tool), and then

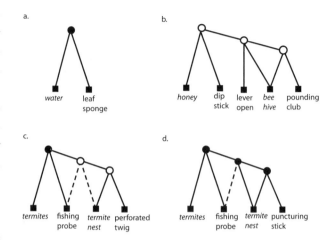

Figure 11.1   Tree-structure analysis of different tool tasks. We have adapted the notation analysis of Matsuzawa (1996) to accommodate serial tool use and multiple-function tool use. Objects are represented by solid squares, with targets italicized. A node that depicts an obligatory object relationship is represented by a solid circle. A node that represents an optional object relationship is shown as an open circle. A tool with a multiple function is connected to the target by a dotted line. Figure 11.1a depicts leaf sponging, which involves a single and consistent relationship between two objects. Figure 11.1b depicts honey gathering, which can involve multiple tools and targets but most often involves only a pounding club directed at a beehive. Figure 11.1c depicts termite fishing at an elevated termite nest, which requires the relationship between the fishing probe tool and the termites, but we have also frequently observed the use of a perforating twig to open the nest surface. A fishing probe can be used for multiple functions in this context when its orientation is reversed and the blunt end is used to clear the fishing hole. Figure 11.1d represents termite fishing at a subterranean termite nest, which requires two relationships between objects, a puncturing stick to create a tunnel into the nest, and a fishing probe to extract termites. We have also observed reversal of the fishing probe to clear the tunnel in this context.

use of a brush-tipped fishing probe (tool) to extract termites (target) was obligatory except in cases where termites were exiting their nests to forage. A third object relationship occurred in these termite fishing contexts when the reverse end of a fishing probe was used to clear debris from a termite tunnel. Three object relationships can also be detected in honey gathering, which may involve a pounding club (tool) and a lever stick (tool) to open the entrance of the beehive (target), and then also a dipstick (tool) to extract the honey (target). These episodes of multiple tool use within a task were serially ordered and temporally distinct, rather than occurring simultaneously as described in the metatool use of Bossou chimpanzees (Matsuzawa 1996, 2001).

Repertoire Size and Novelty of Elements

There were consistent and significant differences in the repertoire sizes of behavioral elements associated with each tool task (Kruskal-Wallis H-test, chi-square = 12.20, df = 3, p = 0.007). Figure 11.2 compares the average repertoire sizes of chimpanzees in different tool tasks with the relative

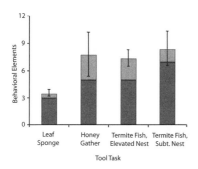

Figure 11.2 Comparison of average repertoire sizes associated with each of the tool tasks. Vertical bars indicate standard deviations. Proportions of the repertoires that consist of essential elements are highlighted in dark gray. Fewer elements and less variation are associated with leaf sponging for water than with other tool tasks. Extraction of termites from subterranean nests had the largest repertoire sizes and numbers of essential elements. The greatest variation in repertoire size was observed in honey gathering.

proportion of essential actions differentiated from additional variations in tool-use techniques. Leaf sponging was comprised of the fewest number of elements and a narrow range of associated variation (average = 3.5 ± 0.4), which indicates that this behavior may have few steps that are executed with high fidelity. Further, the majority of these behaviors were essential for achieving the task. More than twice as many elements were associated with all other tool behaviors (honey gathering = 7.8 ± 2.4; termite fishing at elevated nests = 7.4 ± 0.9; termite fishing at subterranean nests = 8.5 ± 1.9). Termite fishing at subterranean nests required two types of tool use (puncturing the nest, fishing to extract termites), and therefore was associated with the largest repertoire sizes. The number of essential elements necessary to accomplish each task was slightly lower, but the resulting depiction of task complexity was similar to that produced from estimating the entire repertoire size.

Figure 11.3 shows the overlap between the tool types and essential elements associated with each task. Leaf sponging and honey gathering did not share any tool types or elements with other tool tasks. Termite gathering at subterranean and elevated nests was differentiated by puncturing and perforating tool use. All of the fishing elements were shared between the different nest contexts—with the exception of the sweeping of termites, which was an essential behavior only at elevated nests.

Structural Organization

In addition to depicting the structural organization of behavioral elements and essential actions (figure 11.4), we

quantified transitions between tool actions and goal behaviors in observed tool sequences using the permutation method described above, which is similar to the analysis of stone tool use in nut cracking undertaken by Inoue-Nakamura and Matsuzawa (1997).

Leaf sponging required the fewest essential actions to achieve the goal. These were arranged in a simple linear structure that involved manufacture of a single tool (leaf sponge) and few associated tool actions (inserting the sponge into a basin, and then extracting it) to obtain drinking water. The consistent pattern observed in leaf-sponging sequences was seen in the fact that more than 75% of them differed from randomized data.

Termite fishing at elevated nests had a more complex structural organization of behavioral elements than leaf sponging. The chimpanzee must execute more steps by straightening the tool's fibers before inserting it into the termite nest. We also found that there were more options in this tool task than in the previous one, such as the choice of whether to gather termites directly from the tool or by sweeping a hand along its length. Despite variations observed in real sequences, all of these element transitions differed from randomly generated matrices. The transition between straightening the brush fibers to inserting them into the nest yielded a particularly strong signal, with more than 75% of observed sequences differing from randomized data.

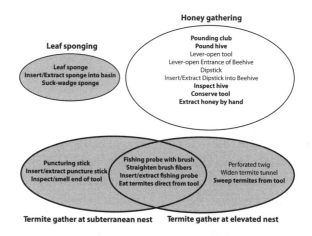

Figure 11.3 Overlap between the tool types and essential elements (in boldface) associated with each task. The region in which the oval diagrams overlap shows the elements that are shared between tool tasks. The graphical presentation is adapted from Takeshita's (2001) depiction of similarity of behavioral patterns between individuals. We found no overlap in elements of leaf sponging or honey gathering. Termite gathering at subterranean and elevated termite nests shared all the elements that were associated with fishing tool use, but diverged in their use of puncturing and perforating tools. Also, sweeping termites from the tool was an essential feeding technique only at elevated nests.

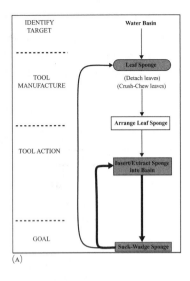

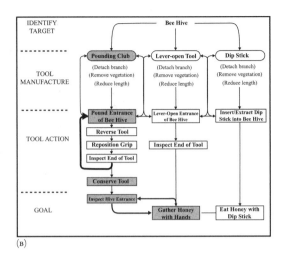

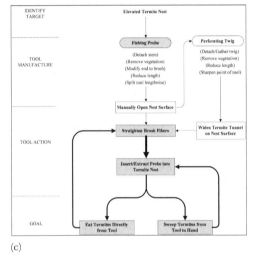

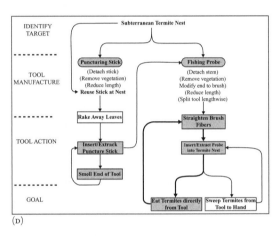

(A)

(B)

(C)

(D)

Figure 11.4   Structural configurations of the different tool-use behaviors: leaf sponging (figure 11.4a), honey gathering (figure 11.4b), termite fishing at elevated nests (figure 11.4c), and termite fishing at subterranean nests (figure 11.4d). The names of tools appear within ovals, with text in parentheses indicating steps in their manufacture or modification (Sanz and Morgan, 2007). Rectangles indicate actions toward achieving a target. Essential elements in each tool task are highlighted in gray. Transitions between tool actions and goals are quantified from first-order transition matrices. Thin lines between elements indicate observed transitions. Lines of medium thickness indicate that 50 to 75% of observed transitions differed from random permutations of the data. The thickest lines indicate that more than 75% of observed transitions differed from random permutations of the data.

Although the use of a puncturing stick is necessary to gain access to subterranean termite nests, the most salient transitions in this task were those associated with fishing. More than 50% of the observed transitions involving these elements (brush straightening, inserting the probe into a nest, gathering termites by eating them directly from the tool) differed from randomized data. In contrast to the feeding techniques shown at elevated nests, there was a stronger tendency for chimpanzees to eat termites directly from the tool rather than sweep them with their hands at the subterranean nests.

The structural complexity of honey gathering is a direct result of the increase in the number of different types of tools. Despite the various tools and action combinations that we observed, only three observed transitions differed consistently from random patterns. Repeated pounding of the beehive was the most consistent transition observed between two elements, with more than 75% of observed transitions differing from randomized data. The cycle of inspecting the hive and gathering honey was also consistent in our observations of honey gathering.

Sequence Analysis

Tool behaviors described here differed in Shannon Weaver values (H_s) of first-order transitions (Kruskal-Wallis Test, chi-square $= 12.47$; df $= 3$; p $= 0.006$). Leaf sponging was associated with the lowest H_s values (average $= 0.35 \pm 0.02$), which indicates that patterns were defined by few elements with stable first-order transitions, in contrast to the other tool tasks which showed more variation in behavioral elements and their organization within a sequence (honey gathering $= 0.45 \pm 0.10$; termite fishing, elevated $= 0.56 \pm 0.10$; termite fishing, subterranean $= 0.53 \pm 0.05$).

Discussion

We found significant differences in the composition and structural organization of different tool behaviors shown by the wild chimpanzees of the Goualougo Triangle. Not only do their tool behaviors vary in complexity, but the chimpanzees also employ different strategies to accomplish them. As expected, the tool-use behaviors were comprised of various elements necessary for tasks directed at different targets, but we also found that the sizes of element repertoires associated with each task were significantly different from each other. Viewed from the "repertoire counting" approach, this indicates that leaf sponging was a simpler task than puncturing at termite nests, which was associated with twice as many elements. However, we also found that a large number of elements could be shared between different tool-use tasks. Comparisons of the structure of these different tasks showed that the chimpanzees had the capacity to use tools in highly standardized sequential patterns (leaf sponging, termite fishing at elevated nests) and also in more flexible structural configurations (termite fishing at subterranean nests, honey gathering). Tree-structure analysis supported our finding that tool tasks differed in their complexity, and also that these chimpanzees showed flexibility in organizing them. Leaf sponging was depicted as the simplest tool task, consisting of a single relationship between tool and target. Multiple object relationships were shown in honey gathering and termite fishing at elevated and subterranean nests. Differences in the nesting of object relationships indicated that chimpanzees within this population had several strategies of multiple object use. In addition to showing that their tool behaviors varied in their degree of complexity, we found that they were capable of flexibly executing different tool-use strategies and generalizing object relationships across different contexts. Although further research is warranted, our conclusion is that the varying levels of task complexity

and flexibility in applying technological skills across different contexts demonstrates the intelligent use of tools by wild chimpanzees.

Examining Complexity in Chimpanzee Tool Behaviors

Systematic comparisons of the composition and structure of different chimpanzee tool tasks provided insights into differences that were not evident in our previous ethnographic descriptions of these behaviors (Sanz and Morgan 2007). Tool-use behaviors differed in the composition of element behaviors and sizes of specific tool-task repertoires. Within the context of unlimited choices, a selective and well-composed repertoire of behaviors associated with a particular task could indicate a chimpanzee's degree of insight into it. However, small repertoires have also been associated with task simplicity. We found that leaf sponging had the smallest number of elements with seemingly invariant transitions. It should also be taken into consideration that this type of analysis may be sensitive to differences in defining behavioral elements which can artificially inflate or reduce levels of complexity and invalidate comparisons between studies. The relative simplicity of the leaf-sponging task is supported not only by the agreement between different measures of complexity, but also by previously published reports. It has been claimed that leaf sponging may not be a difficult tool use to innovate because it has been observed in almost all well-studied chimpanzee populations (Whiten et al. 1999, 2001) and similar behaviors have been invented by captive chimpanzees (Kitahara-Frisch and Norikoshi 1982). As shown by tree-structure analysis, this behavior consists of a single object relationship without alternative structural elements. Tonooka (2001) showed that tool use in water drinking was acquired at a relatively young age, but that social facilitation played a role in its acquisition. Juvenile chimpanzees showed more inclination to watch others during this tool behavior than did adults. There is also indication that some type of social facilitation may be responsible for the sudden increase in frequency of leaf sponging in the M group of chimpanzees at Mahale in Tanzania (Matsusaka et al. 2006). Together with analysis of the other tool tasks, these findings have prompted our current research on the relationship between task complexity and social learning.

Although repertoire sizes were larger in termite and honey-gathering contexts, we found that many of elements could be shared between the different tasks. Generalization of elements across tool tasks could make learning of the tasks less cognitively demanding, but that depends on how the elements are executed within tool sequences. We found that the object relationships and element sequences in the fishing portions of the subterranean and elevated termite nest tasks were nearly identical, but that the necessity and timing of the incorporation of puncturing stick and perforating twig were extremely different. Flow diagrams of termite-gathering sequences at elevated and subterranean nests clearly illustrate the structural differences between these tool tasks (figure 11.4). Multiple tools are used in both tasks, but it is obligatory to use two tools in gathering termites only at subterranean nests. The entire length of a stout puncturing stick is inserted into the ground to create an access tunnel into the subterranean chambers of a termite nest, and then the fishing probe is used to extract the termites. Besides the obvious differences in form and function of the puncturing and perforating tools, the perforating twigs are used only occasionally to clear debris from an existing termite exit tunnel. Tree-structure analysis shows that these tasks have similar targets but involve different tools and relationships. We conclude that chimpanzees can apply termite fishing knowledge flexibly in different settings, and that they can implement these skills appropriately relative to other types of tools. Furthermore, the chimpanzees seemed to anticipate the different tasks by arriving at each nest with the appropriate tools. They often arrived at a subterranean nest with both the puncturing stick and the herbaceous fishing tools they would need. We have never observed them arriving at elevated nests with puncturing tools; rather they arrived only with fishing probes and then manufactured perforating twigs from nearby vegetation if needed.

Multiple tool use is relatively rare in nonhumans, which is another reason why the regular use of tool sets in the Goualougo is intriguing. We have observed chimpanzees using multiple tools to gather termites and open beehives to gather honey. Although tree-structure analysis shows that the depth of nodes may be similar to those in the termite-gathering tasks in this study and in the metatool use described by Matsuzawa (1996, 2001), the hierarchical structuring of object relationships in honey gathering differentiate it from the other tool-use tasks (see leaf sponging in figure 11.4a, and termite gathering in figures 11.4c and 11.4d). In contrast to the simultaneous use of multiple tools in metatool use by chimpanzees at Bossou, described by Matsuzawa (1996, 2001), the tool behaviors we observed

consisted of temporally distinct episodes of serial tool use. Although the types of object relationships varied between metatool use in nut cracking and multiple tool use in honey gathering, they had similar degrees of complexity, as evidenced by the depth of nodes produced in each case. The object relationships in the termite-gathering tasks involved the predictable use of certain tool sets in a particular context, whereas the use of tools in honey gathering seemed to be in response to the characteristics of a given bee hive. Honey gathering was also shown to be more flexible than the other tool tasks analyzed in this study, with a large element repertoire and few fixed transitions. A highly flexible tool-use strategy could be related to the highly variable physical characteristics of target beehives in the dense lowland forests of the Congo Basin. It is possible that ecological features have shaped the differences between the tool behaviors documented in this study, but direct comparisons of element repertoires, structural configurations of elements, and object relationships in different contexts may provide insight into the depth of chimpanzees' understanding in such situations.

Factors that Shape Complex Tool Use

There are several indications that technological skills similar to those described in this study exist in several chimpanzee populations in western equatorial Africa. Descriptions of tool sets used by chimpanzees to extract termites from their earthen nests have been reported from Gabon, the Republic of Congo, Cameroon, the Central African Republic, and Equatorial Guinea (Bermejo and Illera 1999; Deblauwe et al. 2006; Fay and Carroll 1994; McGrew and Rogers 1983; Muroyama 1991; Sabater Pi 1974; Suzuki et al. 1995). There is also evidence of multiple tool use in honey gathering from several sites in this region (Boesch et al. 2009; Bermejo and Illera 1999; Fay and Carroll 1994; Hicks et al. 2005). This raises the question of which specific factors could have shaped the broad and complex tool technology of these chimpanzees, and how those traditions are maintained.

Within this region of the Congo Basin we have found that there are ample ecological opportunities to use tools, and potential interspecific feeding competition that may force chimpanzees to adopt unique foraging niches. The Ndoki forests offer a relative abundance of ecological opportunities for tool use. Several species of *Macrotermes* build various types of nests, ranging from completely subterranean nests to conspicuous towers. With a total of seven species of army ants, the assemblage of *Dorylus* in the Goualougo Triangle is the most diverse ever recorded at a chimpanzee study site (Sanz et al. 2009). Ants are harvested by chimpanzees at their bivouacs or while traveling through the forest. We have also documented the presence of at least six different bee species in the Goualougo study area, with honey gathering observed at the hives of African honeybees and three species of melipones. The presence of particular targets does not mean, however, that chimpanzees will use tools to exploit these food resources. Motivation and technological knowledge are also prerequisites. Chimpanzees reside in sympatry with western lowland gorillas (*Gorilla gorilla gorilla*) throughout much of western equatorial Africa. The high degree of dietary overlap reported between these apes at this site and several others (Morgan and Sanz 2006; Tutin and Fernandez 1985, 1993) could prompt chimpanzees to adopt innovative foraging strategies, such as tool use, that give them access to embedded food items that are not exploited as efficiently by gorillas. Gorilla densities reported from the Goualougo Triangle study area are some of the highest in this region (Morgan et al. 2006) and it is conceivable that the diverse and complex tool strategies of the chimpanzees who also inhabit that area provides a means of coping with interspecific feeding competition. At Bossou, chimpanzees were shown to use tools least frequently during periods of high fruit availability (Yamakoshi 1998). The influence of feeding competition on the frequency of chimpanzees' tool use at sites in central Africa could be assessed by comparing tool use over periods with differing degrees of dietary overlap between chimpanzees and gorillas.

Van Schaik and Pradhan (2003) have proposed a theoretical model for tool-use traditions in primates which suggests that the likelihood of an individual showing a particular tool behavior is dependent upon the probability of asocial acquisition, the probability of social learning, and the individual's opportunities for observational learning. The probability that a task is socially learned is depicted as the inverse probability of innovation, meaning that the task is acquired without social input. Asocial learning mechanisms involve individual discovery of novel information, invention of new behaviors, or elaboration on existing themes (Reader and Laland 2001). Such innovations may occur in social contexts, but they are produced without input from other individuals. Intuitively, task complexity should be inversely related to the likeli-

hood of individual invention. Preliminary analysis of social interactions that occurred during our tool sequences showed that more complex tool tasks were associated with increased attendance and facilitation. We found that more individuals were in proximity and attending to the tool user at subterranean termite nests than in other tool settings. Further, facilitation of the task by sharing of tools or targets was more common in the termite tool-use context than in leaf sponging or honey gathering.

If one is willing to accept the premise that some aspect of complex tool-use behavior is facilitated by social transmission of information, then the social networks within and between groups are also an important factor in the maintenance of these behaviors over time and space. Undisturbed social networks across several intact chimpanzee communities are likely to exist in the Congo Basin, which harbors some of the largest remaining tracts of intact forest in the world, in contrast to the devastating fragmentation of chimpanzee habitats in west Africa (Kormos et al. 2004) and the isolation of small chimpanzee populations in some areas of east Africa (Pusey et al. 2007). However, the conservation context of apes in the Congo Basin is rapidly changing with the advance of mechanized logging, mining, conversion to farmland, and human settlement (Tutin et al. 2005). Expanding human influence on wild orangutans and their behaviors prompted van Schaik (2001) to propose the fragile cultures hypothesis, which suggests that local extinction, hunting pressure, selective logging, and habitat loss affect the transmission of traditional behaviors among wild apes. It is likely that chimpanzee tool traditions would be vulnerable to similar disturbances, which underscores the importance of considering the rich behavioral diversity and social inheritance systems of our closest living relations when developing long-term conservation strategies to ensure their survival.

Implications and Future Directions

Our comparison of composition, structure, and hierarchical organization of object relationships in the tool-use behaviors of chimpanzees in the Goualougo Triangle demonstrates that there is much to be learned from the intricate complexity of these skilled behavior patterns (Byrne 2007). Most previous studies of chimpanzee tool use have defined a tool by a general description of its use, rather than analyzing the actual sequence of actions that involve its use. Our data set included repeated observations of "multiple-function" tools, with single tools having relationships with multiple targets and different functions in each of many contexts. It is also likely that recent research to identify the putative cultural variants within particular species may have masked important differences within and between populations by compiling and comparing general catalogues of behaviors (Whiten et al. 1999, 2001; Hohmann and Fruth 2003; van Schaik et al. 2003). Fascinating differences have been found in more specific treatments of some of these behaviors, however, such as the careful differentiation of leaf-folding from leaf-sponging techniques to drink water (Tonooka 2001) and locale-specific grooming patterns of wild chimpanzees (Nishida et al. 2004). Future research should be devoted to refining analysis methods for comparisons of behavior between populations or species (see chapter 13).

Several other types of research emerge from the ongoing study of chimpanzee tool technology in this population and others. Recent research has reported a significant relationship between laterality in tool manufacture and the design complexity of tools used by New Caledonian crows (Hunt et al. 2006). This would be a reasonable extension of the current study, which would also contribute to the discussion of population-wide handedness in wild chimpanzees (Lonsdorf and Hopkins 2005). Although it is not discussed in this study, we have found that chimpanzees exhibit a high degree of material selectivity for some types of tools (Sanz and Morgan 2007). Detailed investigations should be conducted to examine the basis of these raw material choices in different contexts, which may reveal new insights into the cognitive abilities of chimpanzees in their natural habitats.

Acknowledgments

We are deeply appreciative of the opportunity to work in the Nouabalé-Ndoki National Park and especially the Goualougo Triangle. This work would not be possible without the continued support of the Ministère de l'Economie Forestière of the Government of the Republic of Congo and Wildlife Conservation Society–Congo. Special thanks are due to J. M. Fay, P. Elkan, S. Elkan, B. Curran, P. Telfer, M. Gately, E. Stokes, P. Ngouembe, and B. Dos Santos. We would also like to recognize the tireless dedication of J. R. Onononga, C. Eyana-Ayina, S. Ndolo, M. Mguessa, I. Singono, and the Goualougo tracking team. S. Gulick is to be credited for the remote-monitoring technology that was used in this study. R. Mundry provided essential statistical expertise to this project. Insightful conversations with C. Boesch and R. Byrne have also contributed to this work. Grateful acknowledgment of funding is due to the United States Fish and Wildlife Service, the National Geographic Society, the Wildlife Conservation Society, Columbus Zoological Park, Brevard Zoological Park, Lowry Zoological Park, and the Great Ape Trust of Iowa. During the preparation of the manuscript, C. Sanz was supported by a Richard Carley Hunt Fellowship from the Wenner-Gren Foundation for Anthropological Research.

Literature Cited

Bermejo, M., and G. Illera. 1999. Tool-set for termite-fishing and honey extraction by wild chimpanzees in the Lossi Forest, Congo. *Primates* 40:619–27.

Boesch, C., and H. Boesch. 1990. Tool use and tool making in wild chimpanzees. *Folia Primatologica* 54:86–99.

Boesch, C., and H. Boesch-Achermann. 2000. *The Chimpanzees of the Taï Forest: Behavioural Ecology and Evolution.* Oxford, Oxford University Press.

Boesch, C., J. Head, and M. Robbins, M. 2009. Complex tool sets for honey extraction among chimpanzees in Loango National Park, Gabon. *Journal of Human Evolution* 56:560–69.

Byrne, R. W. 2003. Imitation as behaviour parsing. *Philosophical Transactions of the Royal Society B* 358:529–36.

———. 2007. Culture in great apes: Using intricate complexity in feeding skills to trace the evolutionary origin of human technological prowess. *Philosophical Transactions of the Royal Society B* 362:577–85.

Byrne, R. W., and J. M. E. Byrne. 1993. Complex leaf-gathering skills of mountain gorillas *Gorilla g. beringei*: Variability and standardization. *American Journal of Primatology* 31:241–61.

Byrne, R. W., N. Corp, and J. M. E. Byrne. 2001. Estimating the complexity of animal behaviour: How mountain gorillas eat thistles. *Behaviour* 138:525–57.

Corp, N., and R. W. Byrne. 2002. Leaf processing by wild chimpanzees: Physically defended leaves reveal complex manual skills. *Ethology* 108:673–96.

De Waal, F. B. M. 2003. Social Syntax: The If-Then Structure of Social Problem Solving. In F. B. M. de Waal and P. L. Tyack, eds., *Animal Social Complexity: Intelligence, Culture, and Individualized Societies,* 230–48. Cambridge, MA: Harvard University Press.

Deblauwe, I., P. Guislain, J. Dupain, and L. van Elsacker. 2006. Use of a tool-set by *Pan troglodytes troglodytes* to obtain termites (*Macrotermes*) in the periphery of the Dja Biosphere Reserve, Southeast Cameroon. *American Journal of Primatology* 68:1191–96.

Fay, J. M., and R. W. Carroll. 1994. Chimpanzee tool use for honey and termite extraction in central Africa. *American Journal of Primatology* 34:309–17.

Fowler, A., and V. Sommer. 2007. Subsistence technology of Nigerian chimpanzees. *International Journal of Primatology* 28(5): 997–1023.

Fragaszy, D., P. Izar, E. Visalberghi, E. B. Ottoni, and M. Gomes de Oliveira. 2004. Wild capuchin monkeys (*Cebus libidinosus*) use anvils and stone pounding tools. *American Journal of Primatology* 64:359–66.

Goodall, J. 1964. Tool-using and aimed throwing in a community of free-living chimpanzees. *Nature* 201:1264–66.

Hicks, T. C., R. S. Fouts, and D. H. Fouts. 2005. Chimpanzee (*Pan troglodytes troglodytes*) tool use in the Ngotto Forest, Central African Republic. *American Journal of Primatology* 65:221–37.

Hohmann, G., and B. Fruth. 2003. Culture in bonobos? Between-species and within-species variation in behavior. *Current Anthropology* 44:563–71.

Hunt, G. R. 1996. Manufacture and use of hook-tools by New Caledonian crows. *Nature* 379:249–51.

Hunt, G. R., and R. D. Gray. 2003. Diversification and cumulative evolution in New Caledonian crow tool manufacture. *Proceedings of the Royal Society B* 270:867–74.

Hunt, G. R., M. C. Corballis, and R. D. Gray. 2006. Design complexity and strength of laterality are correlated in New Caledonian crows' pandanus tool manufacture. *Proceedings of the Royal Society B* 273:1127–33.

Inoue-Nakamura, N., and T. Matsuzawa. 1997. Development of stone tool use by wild chimpanzees (*Pan troglodytes*). *Journal of Comparative Psychology* 111:159–73.

Izawa, I., and J. Itani.1966. Chimpanzees in the Kasakati Basin, Tanganyika. 1. Ecological study of the rainy season. *Kyoto University African Studies* 1:73–156.

Kajobe, R., and D. W. Roubik. 2006. Honey-making bee colony abundance and predation by apes and humans in a Uganda forest reserve. *Biotropica* 38(2): 210–18.

Kitahara-Frisch, J., and K. Norikoshi. 1982. Spontaneous sponge-making in captive chimpanzees. *Journal of Human Evolution* 11:41–47.

Kormos, R., C. Boesch, M. I. Bakarr, and T. M. Butynski. 2003. *West African Chimpanzees: Status and Conservation Action Plan.* Washington, DC: Conservation International.

Lanjouw, A. 2002. Behavioural adaptations to water scarcity in Tongo chimpanzees. In C. Boesch, G. Hohmann and L. F. Marchant, eds., *Behavioural Diversity in Chimpanzees and Bonobos*, 52–60. Cambridge: Cambridge University Press.

Lonsdorf, E. V., and W. D. Hopkins. 2005. Wild chimpanzees show population-level handedness for tool use. *Proceedings of the National Academy of Sciences* 102(35): 12634–38.

Mangold 2006. INTERACT software, version 8.04. Arnstorf, Germany.

Matsusaka, T., H. Nishie, M. Shimada, N. Kutsukake, K. Zamma, M. Nakamura, and T. Nishida. 2006. Tool-use for drinking water by immature chimpanzees of Mahale: Prevalence of an unessential behavior. *Primates* 47:113–22.

Matsuzawa, T. 1996. Chimpanzee intelligence in nature and captivity: Isomorphism of symbol use and tool use. In W. C. McGrew, L. F. Marchant and T. Nishida, eds., *Great Ape Societies*. Cambridge: Cambridge University Press.

———. 2001. Primate foundations of human intelligence: A view of tool use in nonhuman primates and fossil hominids. In T. Matsuzawa, ed., *Primate Origins of Human Cognition and Behavior*, 3–25. Tokyo: Springer.

McGrew, W. C. 1987. Tools to get food: The subsistants of Tasmanian Aborigines and Tanzanian chimpanzees compared. *Journal of Anthropological Research* 43:247–58.

———. 1992. *Chimpanzee Material Culture: Implications for Human Evolution.* Cambridge: Cambridge University Press.

McGrew, W. C., L. F. Marchant, and K. D. Hunt. 2007. Etho-archaeology of manual laterality: Well digging by wild chimpanzees. *Folia Primatologica* 78:240–44.

McGrew, W. C., and M. E. Rogers. 1983. Chimpanzees, tools, and termites: New record from Gabon. *American Journal of Primatology* 5:171–74.

Morgan, D., and C. Sanz. 2006. Chimpanzee feeding ecology and comparisons with sympatric gorillas in the Goualougo Triangle, Republic of Congo. In G. Hohmann, M. Robbins, and C. Boesch, eds., *Primate Feeding Ecology in Apes and Other Primates: Ecological, Physiological, and Behavioural Aspects*, 97–122. Cambridge: Cambridge University Press.

Morgan, D., C. Sanz, J. R. Onononga, and S. Strindberg. 2006. Ape abundance and habitat use in the Goualougo Triangle, Republic of Congo. *International Journal of Primatology* 27:147–79.

Muroyama, Y. 1991. Chimpanzees' choice of prey between two sympatric species of *Macrotermes* in the Campo Animal Reserve. *Human Evolution* 6:143–51.

Nishida, T., and M. Hiraiwa. 1982. Natural history of a tool-using behavior by wild chimpanzees in feeding upon wood-boring ants. *Journal of Human Evolution* 11:73–99.

Nishida, T., J. C. Mitani, and D. P. Watts. 2004. Variable grooming behaviour in wild chimpanzees. *Folia Primatologica* 75:31–36.

Oswalt, W. H. 1976. *An Anthropological Analysis of Food-Getting Technology.* New York: John Wiley.

Parker, S. T., and K. R. Gibson, K. R. Object manipulation, tool use, and sensorimotor intelligence as feeding adaptations in Cebus monkeys and great apes. *Journal of Human Evolution* 6:623–41.

Phillips, K. A. 1998. Tool use in wild capuchin monkeys (*Cebus albifrons trinitatis*). *American Journal of Primatology* 46:259–61.

Pruetz, J. D., and P. Bertolani. 2007. Savanna chimpanzees, *Pan troglodytes verus*, hunt with tools. *Current Biology* 17:1–6.

Pusey, A.E., L. Pintea, M. Wilson, S. Kamenya, and J. Goodall. 2007. The contribution of long-term research at Gombe National Park to chimpanzee conservation. *Conservation Biology* 21:623 34.

Reader, S. M., and K. N. Laland. 2001. Primate innovation: Sex, age and social rank differences. *International Journal of Primatology* 22:787–805.

Sabater Pi, J. 1974. An elementary industry of the chimpanzees in the Okorobiko Mountains, Rio Muni (Republic of Equatorial Guinea), West Africa. *Primates* 15:351–64.

Sambrook, T., and A. Whiten. 1997. On the nature of complexity in cognitive and behavioural science. *Theory and Psychology* 7:191–213.

Sanz, C., J. Call, and D. Morgan. 2009. Design complexity in termite-fishing tools of chimpanzees (*Pan troglodytes*). *Biology Letters* 5:293–96.

Sanz, C., and D. Morgan. 2007. Chimpanzee tool technology in the Goualougo Triangle, Republic of Congo. *Journal of Human Evolution* 52(4): 420–33.

———. 2009. Flexible and persistent tool-using strategies in honey gathering by wild chimpanzees. *International Journal of Primatology* 30:411–27.

Sanz, C., D. Morgan, and S. Gulick. 2004. New insights into chimpanzees, tools, and termites from the Congo basin. *American Naturalist* 164:567–81.

Stanford, C. B., C. Gambaneza, J. B. Nkurunungi, and M. L. Goldsmith. 2000. Chimpanzees in Bwindi-impenetrable National Park, Uganda, use different tools to obtain different types of honey. *Primates* 4:337–41.

Sugiyama, Y. 1985. The brush-stick of chimpanzees found in south-west Cameroon and their cultural characteristics. *Primates* 26:361–74.

———. 1995. Drinking tools of wild chimpanzees at Bossou. *American Journal of Primatology* 37:263–69.

———. 1997. Social tradition and the use of tool-composites by wild chimpanzees. *Evolutionary Anthropology* 6:23–27.

Suzuki, S., S. Kuroda, and T. Nishihara. 1995. Tool-set for termite-fishing by chimpanzees in the Ndoki forest, Congo. *Behaviour* 132:219–34.

Takeshita, H. 2001. Development of combinatory manipulation in chimpanzee infants (*Pan troglodytes*). *Animal Cognition* 4:335–45.

Tomasello, M., and J. Call. 1997. *Primate Cognition*. Oxford, UK: Oxford University Press.

Tonooka, R. 2001. Leaf-folding behavior for drinking water by wild chimpanzees (*Pan troglodytes verus*) at Bossou, Guinea. *Animal Cognition* 4:325–34.

Tutin, C. E. G., and M. Fernandez. 1985. Foods consumed by sympatric populations of *Gorilla gorilla* and *Pan troglodytes* in Gabon: Some preliminary data. *International Journal of Primatology* 6:27–43.

———. 1993. Composition of the diet of chimpanzees and comparisons with that of sympatric lowland gorillas in the Lope Reserve, Gabon. *American Journal of Primatology* 30:195–211.

Tutin, C. E. G., R. Ham, and D. Wrogemann. 1995. Tool-use by chimpanzees (*Pan t. troglodytes*) in the Lopé Reserve, Gabon. *Primates* 36(2): 181–92.

Tutin, C. E. G., E. Stokes, C. Boesch, P. Walsh, D. Morgan, C. Sanz, S. Blake, and R. Kormos. 2005. *Regional Action Plan for the Conservation of Gorillas and Chimpanzees in Western Equatorial Africa*. Conservation International.

Van Schaik, C. P. 2001. Fragility of traditions: The disturbance hypothesis for the loss of local traditions in orangutans. *International Journal of Primatology* 23:527–38.

Van Schaik, C. P., M. Ancrenaz, G. Borgen, B. Galdikas, C. D. Knott, I. Singleton, A. Suzuki, S. S. Utami, and M. Merrill. 2003. Orangutan cultures and the evolution of material culture. *Science* 299:102–5.

Van Schaik, C. P., R. O. Deaner, and M. Merrill. 1999. The conditions for tool use in primates: Implications for the evolution of material culture. *Journal of Human Evolution* 36:719–41.

Van Schaik, C. P., and G. R. Pradhan. 2003. A model for tool-use traditions in primates: Implications for the coevolution of culture and cognition. *Journal of Human Evolution* 44: 645–64.

Waga, I. C. 2006. Spontaneous tool use by wild capuchin monkeys (*Cebus libidinosus*) in the Cerrado. *Folia Primatologica* 77:337–44.

Whiten, A., J. Goodall, W. C. McGrew, T. Nishida, V. Reynolds, Y. Sugiyama, C. Tutin, R. Wrangham, and C. Boesch. 1999. Cultures in chimpanzees. *Nature* 399:682–85.

———. 2001. Charting cultural variation in chimpanzees. *Behaviour* 138:1481–1516.

Yamakoshi, G. 1998. Dietary responses to fruit scarcity of wild chimpanzees at Bossou, Guinea: Possible implications for ecological importance of tool use. *American Journal of Physical Anthropology* 106:283–95.

———. 2001. Ecology of tool use in wild chimpanzees: Toward reconstruction of early hominid evolution. In T. Matsuzawa, ed., *Primate Origins of Human Cognition and Behavior*, 537–56. Tokyo: Springer.

Yamakoshi, G., and M. Myowa-Yamakoshi. 2004. New observations of ant-dipping techniques in wild chimpanzees at Bossou, Guinea. *Primates* 45:25–32.

12

Tools, Traditions, and Technologies: Interdisciplinary Approaches to Chimpanzee Nut Cracking

Dora Biro, Susana Carvalho, and Tetsuro Matsuzawa

Traveling east and then southeast from Conakry, Guinea, to a destination close to the Liberian border takes you on an 18-hour journey through a series of villages, looking as typically west African as they would in anyone's imagination. Women stand by hourglass-shaped wooden containers next to round huts with thatched roofs, and pound the contents with long wooden poles in rhythmic up-and-down motions. If you stop for food, fried plantain, fried dough, and fried meats are sold by the roadside from cooking pots over open fires. You smell hot palm oil; it is everywhere. The women you saw pounding were working to extract the oil from the fresh fruity outer layers of oil-palm nuts, an essential resource around here. You reach Bossou, your destination, and enter the forest. There is a rhythmic pounding here too. Chimpanzees use stones—anvils and hammers— to crack the very same nuts as the villagers. They know the kernels are precious; they are oily and full of high-energy nutrients. How long have they been doing this, hidden away in the forest? How do they know what to do?

Regional differences in the behavioral repertoires of wild chimpanzee communities are considered paradigmatic in our current burgeoning interest in nonhuman cultures (McGrew 1992; Whiten et al. 1999, 2001). While population- or community-specific differences in behavior have been documented in a variety of taxa, both primate and non-primate, the sheer unparalleled scale of the variability across tool-use, social, and self-maintenance domains singles out the chimpanzee as the most obvious first step in the search for the evolutionary origins of human culture (McGrew 2004). Nowhere else in the animal world does belonging to a specific community appear to have such diverse influences on individual behavior.

Each chimpanzee community possesses a unique set of skills that makes its members adept at dealing with a multitude of problems posed by the environment—sometimes presenting ingeniously different solutions for similar problems across different groups—but to what extent do these observed patterns in the regional distribution of behaviors really represent a form of culture? Suggestions that neither ecological nor genetic differences among groups can fully explain such inter-community differences point strongly towards social learning as being responsible for the emergence and maintenance of the variation—hence the "culture" label. However, ruling out ecological constraints as determinants of large-scale differences (for example, the absence or presence of a given tool-use behavior) is often difficult. Similarly, factors underlying differences in the precise characteristics of the tools used by members of a particular community compared to those preferred by another community for superficially similar purposes are often problematic to ascertain. In addition, while evidence for the involvement of social learning in within-community maintenance of behaviors can to some extent be explored by careful, intensive studies of individuals and their interactions in particular communities, little empirical evidence exists concerning both the initial emergence and any subsequent between-community spread of such behaviors in the wild.

One well-known chimpanzee behavior that lends itself well to addressing these issues is nut cracking. Technologically among the most advanced examples of wild chimpanzee tool use (although chimpanzees do not hold a monopoly on this behavior among nonhuman primates: see Fragaszy et al. 2004 for details of nut cracking by capuchin monkeys), it involves combining three or, occasionally, more objects in the environment in the appropriate spatiotemporal configuration: a hard-shelled nut is placed on an anvil, and a hammer is then used to pound the nut until the edible kernel is exposed. Large-scale patterns in the distribution of this behavior across Africa indicate clear regional clustering: only communities in west Africa have been observed to crack nuts (Boesch et al. 1994; though see also Morgan and Abwe 2006 for a recent revision of the geographical limits to the behavior) while those in central and east Africa have not—a surprising finding, given that many communities in those regions have similar easy access to the raw materials involved (McGrew et al. 1997). On a smaller scale, among those groups that do exhibit nut cracking, inter-community differences exist also in the precise nature of the targets, tools, and techniques characterizing the behavior. Explanations for these differences are also at the heart of the culture debate. Questions regarding the history of the initial invention (including whether nut cracking has single or multiple origins) and the behavior's subsequent spread and potential extinction in certain regions of Africa remain open to speculation (Wrangham 2006).

Our attempts at contributing to the debate surrounding wild chimpanzee cultures have used nut cracking as a case study, and have entailed the integration of evidence from various distinct yet complementary strands of research, including regional surveys to examine the contribution of ecological factors to inter-community differences, intensive longitudinal studies of developmental and individual-cognitive aspects of the skill, and field experiments exploring issues of innovation and within-community propagation. Taken together, these have been useful in allowing us to speculate about mechanisms involved in the emergence and maintenance of tool-use traditions in wild chimpanzee communities. A further, recent endeavor—taking an archaeological perspective on chimpanzee percussive technology—now adds another angle to our work.

In this chapter our aim is to summarize our findings so far, to provide an update on our recent progress in the field experiments, and to argue for a more rigorous treatment of regional survey data and its relevance to illuminating inter-community differences in tool-use traditions. We will outline our attempts to synthesize work from these distinct but interconnected strands of research, and discuss the contribution that archaeological methods can make to furthering our understanding of tool use in our closest living relatives.

Approach

Our study area is located in the southwestern corner of the Republic of Guinea, west Africa (see chapters 10 and 27). Our main study site for intensive observations, near the village of Bossou, is situated close to a three-way border with Côte d'Ivoire and Liberia. Here the long-term study of chimpanzees began in 1976 with the work of Yukimaru Sugiyama of Kyoto University. Since that time, the size of the community has remained in the range of 16–22 individuals, although as a result of deaths from a flu-like epidemic in 2003 (Matsuzawa et al. 2004), numbers recently dropped to their lowest ever—12 individuals—and currently the group consists of only 13 chimpanzees. The core area of the Bossou community comprises approximately 6 km² of primary and secondary forest and is surrounded entirely by savanna and cultivated fields.

A wide range of tool-use behaviors have been observed at Bossou, some of which appear so far to be unique to this community (see Ohashi 2006 for a review). Here, nut cracking was first recorded systematically by Sugiyama and Koman (1979), and was found to be targeted towards nuts of the oil palm (*Elaeis guineensis*), a resource also much utilized by the local human population. Nut-cracking behavior has been studied intensively by Matsuzawa and colleagues since 1988, with the help of a unique field facility: the so-called "outdoor laboratory" (Matsuzawa 1994; see figure 12.1). This comprises a natural clearing at the summit of one of the hills within the Bossou group's core area, which the chimpanzees visit spontaneously and reliably about once or twice a day during normal ranging activities.

Figure 12.1   The "outdoor laboratory" at Bossou, where intensive longitudinal observations of tool use and field experiments on nut cracking were conducted. Chimpanzees in the foreground (center and left) crack oil palm nuts, while those on the right select stones from among the set provided by the experimenters. Piles of oil palm nuts are scattered around the clearing, also provided by experimenters. At top left, chimpanzees perform a different kind of tool use: leaf folding (the use of a clump of leaves folded inside the mouth to extract water from a tree hole). The opportunity to simultaneously observe several different tool-use activities in individuals across consecutive bouts, days, or even years makes the outdoor laboratory a useful tool for tracing individuals' performance continuously over long periods. Photograph by T. Matsuzawa.

At this site, experimenters have been providing stones and oil-palm nuts during the dry season of each year in an attempt to encourage nut-cracking activity in chimpanzees visiting the clearing, thereby providing an opportunity to observe individuals' performance from close range and in the absence of thick occluding vegetation which often hampers detailed observations in a typical chimpanzee habitat. In-situ observations and video recordings of each group member have made it possible to trace the development of the skill in young individuals from one year to the next, and to elaborate on features characterizing the nut-cracking activity of skilled adult members of the group.

In parallel with the long-running observations at the outdoor laboratory, three additional lines of work were adopted. These aimed to explore (1) the contribution of ecological factors—in this case, nut availability—to the distribution of tool-use traditions, (2) processes of innovation and within-community spread of novel tool-use behaviors through field experiments, and (3) the potential contribution of archaeological techniques to the study of chimpanzee nut cracking.

Chimpanzees at west African sites are known to crack various species of nuts (Boesch et al. 1994). At Bossou, only oil-palm nuts are cracked—unsurprisingly, since the oil palm is the only nut-bearing species available naturally in the locality. However, is availability the only factor contributing to a community's profile in terms of nut species cracked? We set out to explore both the availability of different nut species and evidence of cracking by chimpanzees at sites near Bossou (for specific methodological details see Matsuzawa and Yamakoshi 1996; Matsuzawa et al. 1999; Humle and Matsuzawa 2001). Our aim was to explore whether patterns in the latter could be attributed purely to ecological factors, or whether any mismatch in the distribution of availability and cracking could alternatively be explained in cultural terms.

Besides studies of individual tool-use performance (described above), the outdoor laboratory at Bossou also provided a setting in which we could examine processes of behavioral innovation (and subsequent within-community spread thereof) in the context of nut cracking. While a field setting provides little opportunity for rigorously controlled experimental treatments (see chapters 8 and 9), we have tried to move beyond a purely observational approach. Our field experiments involved providing Bossou chimpanzees with nuts unfamiliar to them, with the aim of exploring how novel variations on an existing theme (the cracking of previously unknown nuts within an already nut-cracking community) can become incorporated into a group's tool-use repertoire (Matsuzawa 1994; Biro et al. 2003).

Finally, a recent extension of our ongoing work has involved further intensive observations of chimpanzee nut cracking, this time with an archaeological perspective. Exploiting archaeology's potential contribution to understanding the significance of chimpanzee material culture in the context of human evolution is not a novel concept (e.g., Wynn and McGrew 1989; Sept 1992; Joulian 1996), but it has in the last few years seen a surge in interest (e.g., Mercader et al. 2002, 2007; Heaton and Pickering 2006; Haslam et al. 2009). Not being able to make direct observations, archaeologists must seek clues to aid their understanding of the ways in which lithic tools were used from the analysis of tool use in living communities of nonhuman primates. Primatologists can, in turn, look to whether a specific raw material is selected in a systematic and patterned way; whether a landscape is exploited to obtain the desired resource; whether that resource has systematically identifiable similar characteristics; whether the raw material is always used in the same way; and whether there is a real technological pattern or a functional standardization reflected in the tool. With such questions in mind, newly initiated research at Bossou and at a nearby site (Diecké) by Carvalho and colleagues has been examining technological variability across different chimpanzee communities, while at the same time attempting to identify variables influencing regional diversity in tool typology and technology (Carvalho 2007; Carvalho et al. 2008). As part of this study, and with a view to illuminating the phasal structure of the nut-cracking process, the Chaîne Opératoire concept (Mauss 1967), originally borrowed from ethnography and appropriated later by archaeology to understand the technological process of tool making, was applied for the first time to chimpanzee lithic technology. Our main aim was to detect an operational sequence during wild chimpanzee nut cracking, recreating the technical history of the object (Tixier 1980; Boëda et al. 1990; Lucas 2000). The potential nut-cracking behavioural sequence was recorded directly at the outdoor laboratory and analyzed (in terms of raw material selection, stone selection, transport, use, reuse, tool fracturing, unintentional flake extraction, and discard), as well as indirectly at several "natural" nut-cracking sites around Bossou. Individual tool analysis (with respect to macro use traces, typology, refitting, mobility) was also carried out for each of the sites sampled, in addition to daily monitoring of chimpanzee visits and tool movements. Simultaneously, the chimpanzee group was followed in an attempt to directly record nut-cracking activity at the sampled sites.

Results

Longitudinal Records: Learning to Solve a Problem

Our longitudinal records gathered over almost two decades of intensive observation at the outdoor laboratory has been helpful in illuminating both the developmental stages that young chimpanzees go through on their way to acquiring the skill, and the features that characterize skilled adult performance of the behavior. These in turn inform us of the ways in which the chimpanzee mind deals with problem solving in a natural setting. Our results have been summarized recently in detail (see Biro et al. 2006), so here we restrict ourselves to a condensed version of our findings.

Successful nut cracking is first observed in young chimpanzees between the ages of about three and seven years. Prior to that, individuals go through various stages of learning: different forms of manipulation of the objects involved in nut cracking that gradually approximate the final form of the behavior. Inoue-Nakamura and Matsuzawa (1997) summarize this as a progression starting with the manipulation of single objects (such as the rolling of a nut on the ground), followed by two objects in combination (placing a nut on a stone, or a stone on top of another stone), and finally all three objects (anvil, hammer, and nut). However, even three-object combinations do not necessarily yield a reward. Initially they may still involve incorrect spatial configurations, or the hammer blows may not have enough force to crack the hard shell. Once successful, efficiency at cracking continues to increase gradually: youngsters need, on average, between three to six blows (or occasionally many more) to crack open a single nut, while adults require on average only one or two (see Boesch and Boesch 1984, who also report a gradual improvement in efficiency in the cracking of two different types of nuts by chimpanzees of the Taï forest). Those individuals that do not succeed by the age of about seven years (in our case, 4 of the 26 individuals who remained at Bossou beyond this age over the course of the study, constituting around 15%), apparently never acquire the skill later in life, suggesting that there may exist a sensitive period for learning.

Nut cracking in young chimpanzees is also characterized by various behaviors much reduced or absent in adult performers. For example, while adults very rarely change their hammers or anvils during a bout of cracking, young chimpanzees do so often. In particular, when adult group members move on after a bout of cracking, their abandoned tools become attractive targets for youngsters who then claim them and favor them over their own stones. Also, infants and juveniles are attracted by adults' nut-cracking activities, which they often observe intensively and from close range—something that is tolerated by all members of the community (although somewhat less so as the juveniles get older). This is sometimes accompanied by attempts to scrounge freshly cracked nuts from adults; this again is tolerated, although only in the case of infants.

Young chimpanzees change not only their stones mid-bout, but occasionally also the hand they use to hold the hammer. In contrast, all skilled adult performers exhibit perfect laterality during nut cracking: the same hand is used to hold the hammer from nut to nut, bout to bout, and year to year. Across all individuals whose hand preferences we have been able to record over the years, right-handers outnumber left-handers 15 to 10. This is not a statistically significant community-level bias; however, it has been suggested that given larger sample sizes, population-level biases can emerge (Lonsdorf and Hopkins 2005; see also chapter 6). Comparing the distribution of right- and left-handers among related individuals reveals two interesting patterns. Mothers and their offspring show no obvious correspondence in laterality: in only about half the cases do they share the identity of the hammering hand. Among

siblings, however, there is a pattern of concordance that is beyond what would be expected by chance (although in comparison to previous reports, we now have two juveniles who developed hand preferences different from their siblings). This echoes findings from captive settings, where full siblings and maternal half-siblings were found also to share hand preferences while no such patterns were found between mothers and their offspring (Hopkins 1999). We believe it is likely that inter-sibling correspondence arises as a result of some aspect of maternal behavior directed towards the offspring—consistent across successive offspring from the same mother, but varied across different mothers—early in life. At Bossou we have yet to identify these aspects influencing laterality, although cradling position has already been shown to be a reliable correlate in captivity (Hopkins 2004).

In terms of the cognitive demands of the task, Bossou chimpanzees have taken an already complex form of tool use (indeed, the only example of a level 2 tool according to Matsuzawa's (1996) definition, meaning one that necessitates the correct spatiotemporal combination of three separate objects) one step further: they have been observed to use "metatools" during nut cracking. In other words, they employ a tool-for-a-tool. An additional stone—a wedge—may be inserted under the anvil, stabilizing an otherwise slanted upper surface and allowing it to serve as an effective anvil. Recently we have even observed a single instance of a four-stone composite tool, in which two smaller stones acted to steady an anvil (figure 12.2). It is important to note, however, that the use of wedge stones is extremely rare: we estimate it to be in the region of once every 100 visits by chimpanzee parties to the outdoor laboratory (this rate has remained consistent across different study periods). In addition, it is difficult for the observer to ascertain the degree of purpose involved in constructing the composite: it is likely that they are often accidental, coming about after a nut-cracking individual attempts to adjust an anvil stone by rolling it over on the ground. While rarity itself may suggest a lack of deliberateness, it is also interesting that in a recent survey of a "natural" nut-cracking site at Bossou (i.e., one where the availability of stones was not controlled by human experimenters and the terrain was less even), Carvalho (2007) reports a much higher incidence of wedge use, suggesting that environmental circumstances may readily affect the frequency of the behavior. From a cognitive point of view, both wedge

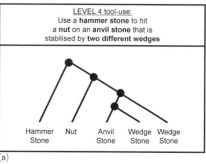

(A)   (B)

Figure 12.2   (a) Four-stone composite tool used in nut cracking. Note that the anvil (large stone on the right) is supported by two smaller stones, making its upper surface level. The stone on the left is the hammer; the white notation along its side is part of our numbering system for naturally occurring stones found to have been used by chimpanzees. Broken nutshells are visible on the ground, scattered around the anvil. Photograph by S. Carvalho. (b) Tree-structure analysis (after Matsuzawa 1996) of the four-stone composite tool shown in (a). The nodes represent associations between the objects: "level 4" refers to the total number of nodes. To our knowledge, this is the only example of a level 4 tool in the wild.

use and adjustments to anvils and hammers during the execution of the cracking task can themselves be highly informative regarding chimpanzees' flexibility in using tools and their understanding of the physical properties of the objects involved.

Regional Survey: Ecology and Tradition

Looking beyond Bossou, our local regional survey examined nut species availability and evidence of cracking at three additional sites located within 50 km from our main study site. Initial efforts focused on recording the presence of (a) nut-bearing species within the flora and (b) any evidence of cracking by chimpanzees, such as broken nutshells together with stones showing signs of wear in areas not visited by humans.

The results show that the distribution of cracking does not correlate perfectly with availability (table 12.1). We focused on three nut species: oil palm, coula (*Coula edulis*), and panda (*Panda oleosa*). Of the 12 possible combinations of site and species used in cracking, four are ecologically impossible (coula and panda at Bossou and Seringbara, since neither nut is naturally available at those loca-

tions), five are taken advantage of (oil palm is cracked at Bossou and Yealé, coula at Yealé and Diecké, and panda at Diecké), while three appear to show non-use despite the nuts' presence (oil palm is not cracked at Seringbara and Diecké, nor is panda at Yealé). However, our most recent follow-up work also highlights the fact that nut availability itself must be considered with care. For example, the

Table 12.1   Species of nuts cracked by wild chimpanzees at Bossou, Seringbara, Yealé, and Diecké.

Site	Distance and direction from Bossou	Species of nut		
		oil palm (*Elaeis guineensis*)	coula (*Coula edulis*)	panda (*Panda oleosa*)
Bossou		yes	–	–
Seringbara	6 km, ESE	no	–	–
Yealé	12 km, SE	yes	yes	no
Diecké	50 km, SW	no[a]	yes	yes

"Yes": the target nut species is available at the site and evidence of cracking by chimpanzees has been found. "No": no evidence of cracking by the chimpanzees has so far been found even though the nuts are available. Superscript ([a]): oil palm nuts at Diecké are available only in the periphery of the chimpanzees' range. Dash indicates that the target nut species is not available at the site.

oil palms present at Diecké are found overwhelmingly in locations peripheral to the chimpanzees' core area (Carvalho et al. 2008). It is as yet unclear how often chimpanzees actually visit these areas; access to the nuts may in fact be very limited. Nevertheless, a detailed study taking into account oil palm densities and distributions at Seringbara and Bossou failed to reveal any differences, suggesting that the chimpanzees' access to oil palm nuts is similar at these two sites—yet the authors also identified very different utilization patterns for the two communities (Humle and Matsuzawa 2004). The conclusion that at least in some cases inter-community differences can be attributed to cultural rather than ecological factors seems therefore likely. Further work in this area aims to continue assessing the extent to which lack of evidence of cracking indeed reflects non-use despite availability, and whether more subtle ecological factors are involved in determining patterns of use.

Field Experiment: Who Innovates and Who Cares?

Beginning in 1993, the outdoor laboratory has also been providing a setting for an experiment aimed at examining flexibility in nut cracking, the appearance of behavioral innovations, and channels for the within-community diffusion of novel behaviors. For this purpose, nuts unfamiliar to members of the Bossou group—coula and panda nuts, both unavailable at Bossou but cracked at other sites nearby—were presented alongside familiar oil-palm nuts (figure 12.3a). For a few weeks of the dry season in the years 1993, 1996, 2000, 2002, 2005, and 2006 we provided small piles of coula nuts at the outdoor laboratory. Panda were presented only once, in 2000. We examined group members' responses to these novel items, classifying them into three categories: "ignore" (no interest shown towards nuts), "explore" (look closely at, sniff, mouth, or handle nuts, but without any attempts to crack them), and "hit" (place nut on anvil stone and hit with hammer, irrespective of whether the nut is cracked or not). The distribution of responses among three different age classes—infants, juveniles, and adults—revealed some interesting tendencies.

When considering the initial presentation of the two nuts, juveniles (5–8 years) were the age group most likely to investigate the unfamiliar objects: all juveniles present

in 1993 explored or attempted to crack coula nuts, and two-thirds did so with panda in 2000 (see Biro et al. 2003, 2006 for a detailed breakdown of the results). Adults (9 years and above) were more likely to ignore, with only a half cracking or exploring coula on the first encounter, and a fifth doing so in the case of panda. Infants (0–4 years), in most cases not yet able to crack, generally ignored the unfamiliar objects, with the exception of a single infant who showed exploratory attempts towards panda in 2000.

Over the years that followed the initial presentation of coula nuts, the proportion of crackers increased among both juveniles and adults (figure 12.3b). Thus, although the established form of the behavior—the cracking of oil palm nuts—did not immediately generalize to novel target items, over repeated presentations the behavior gradually became incorporated into the group's repertoire. With one exception, an adult female, all the individuals who crack oil palm nuts also currently crack coula. In addition, this is now done by almost all individuals without any form of exploratory behavior prior to cracking, even though such behaviors had been a very prominent part of early attempts. In fact, only one individual, an adult female named Yo, showed no sign of sniffing or mouthing on her first encounter with a novel nut: in 1993 she proceeded immediately to crack the newly introduced coula nuts with all the signs of familiarity. This has prompted the suggestion that Yo may be an immigrant female who arrived at Bossou sometime prior to 1976 from a community where coula cracking was habitual (Matsuzawa 1994). The observation that in the case of panda, she, like all others who attempted to crack, did so after extensive sniffing and mouthing of the objects suggests that her natal community was not familiar with panda nuts. It is also noteworthy that the chimpanzees' attempts to crack panda were much more short-lived than their attempts to crack coula: in 2000, of the four individuals (two juveniles and two adult females) who attempted to crack, adults abandoned their efforts after the very first encounter, and juveniles after just four or five, despite continued presentation of the nuts over the following week.

Chimpanzees engaged in nut cracking often attract the attention of other group members (figure. 12.3c). Given that for any socially acquired behavior, observation of another individual performing the behavior is essential, we examined patterns in the occurrence of such observa-

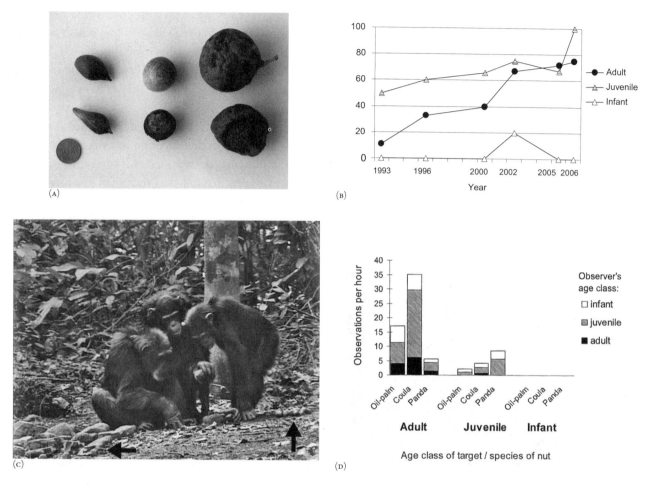

Figure 12.3   Field experiments with unfamiliar nuts. (a) Three species of nut presented in the outdoor laboratory. From left to right: oil palm, coula, and panda. Top row consists of fresh fruits, bottom row nuts ripe for cracking. The shells of the teardrop-shaped oil palm nut and the spherical coula are comparable in hardness (Boesch and Boesch, 1983, report that slightly more force is needed to crack open the oil-palm), while the larger, round panda are by far the toughest to crack. Only oil palm nuts are available naturally at Bossou; the other two species are found at nearby sites and are cracked by different chimpanzee communities. Photograph by T. Matsuzawa. (b) Percentage of individuals in the three age classes who attempted to crack coula nuts on the six occasions when these nuts were presented at the outdoor laboratory. (c) Conspecific observation during nut cracking. An adult female, Yo, is cracking coula nuts (a pile of these is visible on the right; up arrow) while two juveniles observe her actions closely. Also visible near the bottom left is a row of three panda nuts (left arrow), which all three individuals ignored on this occasion. Photograph by D. Biro. (d) Patterns in conspecific observation according to age class of observer and target. Bars of different colors correspond to each observer's age class. Rates were calculated as number of episodes of observation divided by the amount of time individuals in the three different age classes spent in the outdoor laboratory handling or cracking nuts. Data from 2000.

tions, with a view to illuminating possible channels for the within-community diffusion of behavioral traditions. Figure 12.3d reveals various striking patterns. Adults were by far the most likely to be observed by fellow group members during their nut-cracking activities, while most of the observing itself was done by juveniles and infants. Overall, the general tendency was for individuals to pay attention to the nut cracking of conspecifics in the same age class or older, but not younger than themselves. Given these general rules of thumb (the same patterns are evident in the case of other tool-use activities: Biro et al. 2006), we may be able to speculate about the direction in which behaviors

acquired at least in part through socially mediated learning may spread within a community.

Contributions from Archaeology

Our latest study, a combination of primatology and archaeology, can be considered an interdisciplinary tool, which strives for a new insight into human evolution through a form of etho-archaeology: *the study of the behavior or behavioral products of living nonhuman species as proxies for extinct ones* (McGrew 2004, p. 37). With this in mind, we propose that detailed knowledge of nut-cracking technologies and typologies in different Western chimpanzee groups can contribute important data for discussing, among others, the following questions: (1) Does natural selection exert an influence on the evolution of material culture? (2) Did ecological variables (e.g., raw material availability) create typological and technological constraints during the emergence of the first lithic industries? (3) Are chimpanzee nut-cracking sites technologically comparable to ancestral human sites? (4) What are the differences between *Pan* and hominin sites in the archaeological record? (Carvalho 2007).

We examined tool and usage characteristics under three different conditions: at the outdoor laboratory in Bossou, at naturally occurring nut-cracking sites in other areas of Bossou, and at natural nut-cracking sites in Diecké (Carvalho et al. 2008). Applying the concept of "chaîne opératoire" to chimpanzee nut cracking in two distinct communities (Bossou and Diecké) has allowed us to argue that the behavior is characterized by a sequence of several acts, performed repeatedly to achieve a goal. These are: raw material selection, tool construction, transport, utilization, reutilization, and discard. In addition, echoing our previous work at the outdoor laboratory, we found that chimpanzees clearly were making discriminative use of the available tools at both the experimental and the more "natural" nut-cracking sites. The discrimination seemed to be based on certain tool features, including the width and weight of the objects as well as their material (chimpanzees preferred diorite and granite over the softer quartz, for example). Such selectivity can be an indicator that chimpanzees are capable of distinguishing between some morphological features characterizing the objects, thus attributing them a specific function. Furthermore, the collected data provide evidence that chimpanzees discard or change the tool function after fracture according to stone morphology.

A comparison between tools analyzed in Bossou and Diecké provided evidence of regional diversity in stone tool typology and technology. In Bossou, to process the local soft oil-palm nuts, chimpanzees used smaller stone hammers, while in Diecké, to process the hard panda nuts, chimpanzees selected larger hammers (table 12.2). These results agree with those of Boesch and Boesch (1983), who found, when comparing characteristics of coula-nut and panda-nut cracking among chimpanzees of the Taï forest, that individuals selected heavier stone hammers for the harder nut (panda). In addition, they found that more for panda than for coula, the selection of anvils was biased towards outcrop rock rather than surface root. In a related comparison, Bossou chimpanzees overwhelmingly relied on loose anvils (detached stones), while at Diecké their anvils consisted exclusively of embedded rocks. Such differentiation in anvil use may again reflect differences in the hardness of the nuts cracked (see Boesch and Boesch 1983), but it may also be a constraint imposed by the environment: at Diecké loose stones suitable as anvils are rare in the habitat, while at Bossou outcropping rocks tend to be of weathered granite that easily disintegrates. The latter constraint may therefore have been a driving force behind a form of technological innovation at Bossou: the use of mobile, transportable anvils. Taken together, these marked differences between stone tool characteristics at the two sites may indicate some "optimal" combination of raw material, size, and weight with typological and technological

Table 12.2   Measurements of chimpanzee hammers and anvils at three of the sites sampled by Carvalho (2007): the outdoor laboratory at Bossou, a naturally occurring cracking site in Bossou forest, and a naturally occurring cracking site in Diecké forest.

Site	Length (cm)		Width (cm)		Height (cm)		Weight (kg)	
	hammer	anvil	hammer	anvil	hammer	anvil	hammer	anvil
Bossou, outdoor lab	11.9±1.9	17.2±3.2	7.7±1.5	13.9±1.1	5.5±1.5	9.0±2.5	0.68±0.21	3.20±1.40
Bossou forest, SA13 site	8.4±1.5	11.8±2.9	13.9±5.4	16.2±3.9	6.6±2.0	6.3±2.5	1.38±0.79	2.10±0.78
Diecké forest, SB1 site	16.2±6.3	44.4±14.7	11.5±3.9	26.4±4.5	n.a.	n.a.	n.a.	n.a.

"n.a.": height and weight measurements were not taken at Diecké (neither was possible in the case of anvils, as they were comprised exclusively of embedded rocks; length and width measurements for these anvils relate to the visible outcropping).

adaptations, with differences in tool morphologies that seem related to the nature of the food to be consumed, and with the type of raw material available.

On a related note, one way in which the archaeological perspective can add new insights and enrich the primatological view of tool use is in the definition of tool itself. In archaeological functional analysis, the essential characteristics that define a tool are the use and function of an object (Semenov 1981)—hence the implication that use wear and functional analysis can tell us much about human and nonhuman tools, and also about what in fact defines the tool. Hammers and anvils both at Bossou and Diecké showed evident macro-use wear traces (such as pitting or depressions), suggesting that these objects had been used for extended periods of time. It may therefore be safe to conclude that they must have been functioning efficiently: the goal was achieved, and the tools were often reused and not abandoned. Therefore, use wear should be considered an important variable in defining tools and successful technology, more so than typology or morphology. This is in accord with Karlin and Pelegrin's (1988, p. 823) definition of tool: *intentionally (or purposefully) made objects, or any natural object or knapping debris which show use wear at the macro or micro scale.*

Furthermore, of three tool categories put forward by Karlin and Pelegrin—*outil*, *outil composit* and *instrument*—the *outil composit* shows close affinities with what we might define as a "tool construction": the purposeful association of two or more objects (whether transformed or not) that must be used in combination in order to function and achieve a specific goal. This is a notion also adopted by primatology (Sugiyama 1997). Consider grinders, for instance, as prehistoric equivalents: a tool made up of two stone components which together have the function of grinding cereals but cannot work separately. Incidentally, the exception to the rule that hammers and anvils do not work separately seems to be afforded by chimpanzees smashing baobab fruits directly against an anvil (Marchant and McGrew 2005). Accordingly, this could be considered tool use but not tool construction. Thus, the ecological constraint (the fruit species consumed) may in this case also be considered a significant driving force behind technological diversity. When the desired food is embedded inside a large shell that allows efficient percussion with no need for the use of a hammer, the degree of technological complexity involved naturally need not be as high as when

the sequential use of up to five items (four stones and a nut; see figure 12.2) is necessary for extraction.

In a recent study (Carvalho et al. 2009), we attempted to address the issue of the definition of "tool" with data gathered at Bossou's outdoor laboratory over five separate field seasons. Given that we know chimpanzees discriminate among available raw materials when selecting stones for nut cracking along a number of parameters (such as size, weight, and material; Sakura and Matsuzawa 1991; Biro et al. 2003; Carvalho et al. 2008), we asked if they also discriminated among *combinations* of stones. Did chimpanzees preferentially reuse specific tool-composites (hammer-anvil combinations)? In other words, did they choose the "nutcracker" as a whole? Our analysis of the frequencies with which specific hammer-anvil pairings were repeatedly combined by chimpanzees showed that most such reuse of tool composites arose incidentally out of individuals' preferences for specific stones, but that at least in a few cases, chimpanzees did seem to select particular hammer-anvil combinations more often than would be expected on the basis of preferences for their individual elements. Chimpanzees may thus have recognized that "*this* hammer works particularly well with *this* anvil," allowing us to argue that they evaluated the nutcracker as a single tool, composed of two elements but functional only as a whole.

Discussion

Our work at Bossou and neighboring areas reported here is concerned mainly with exploring chimpanzee tool use from a number of different angles: acquisition processes of complex behaviors during individual development, what such behaviors can tell us about chimpanzees' perception of their natural environment and of the physical properties of the objects around them, processes of cultural innovation and transmission within and between wild chimpanzee communities, and finally, the potential contributions of chimpanzee studies to our understanding of human evolution.

Ontogeny, Physical Understanding, and Social Learning Processes in Tool Use

Longitudinal records from the outdoor laboratory have allowed us to formulate a model describing possible mechanisms behind young chimpanzees' acquisition of the skill, taking into account both individual learning processes and contributions from the social environment within which such learning takes place. Matsuzawa et al. (2001) coined the term "education by master-apprenticeship" as a way of characterizing key features of the learning process. On the assumption that individual acquisition is in part mediated by the social environment, we suggest that young chimpanzees' learning is aided by extended exposure to the tool-use activities of skilled individuals (particularly their own mothers), which they are allowed to observe from close range (for systematic evaluations of the role of mothers on young chimpanzees' acquisition of tool use, see also Lonsdorf 2006 and Hirata and Celli 2003, both emphasizing high levels of tolerance towards infant observers during the acquisition of termite fishing at Gombe and honey fishing in a captive environment, respectively). Given the relatively long period during which young chimpanzees gradually approximate the final form of a behavior, it seems that the acquisition process has more in common with emulation learning than imitation. "Masters" perform no active teaching (cf. Boesch 1991), but their actions do draw observers' attention to the tools and targets involved, as well as to the possible outcomes of the behavior. Even in the face of a lack of any direct reward (for the first few years, young chimpanzees' trial-and-error attempts yield not a single kernel), the "apprentice's" efforts continue. This may be due to an intrinsic motivation driving youngsters to copy other group members' behavior. Such copying may also be thought of as a form of social canalization (Boesch 2003), whereby the social environment (more specifically, the techniques observed within it) acts as a limitation on the techniques eventually adopted by learners. That the observation of conspecifics' techniques can indeed have a potent effect on the acquisition of tool use (including the stage of tool construction) has been shown convincingly even in the case of demonstrations observed by learners via video footage (Price et al. 2009).

At the individual level, the detection of a chaîne opératoire in skilled performers' nut cracking signifies that this behavior presents a complex pattern of systematic and sequential actions, which implies a high degree of mental organization including the capacity for anticipating tasks (Stokes and Byrne 2001) and for combining different elements to make the tool work (Carvalho et al. 2008). Furthermore, underlying the process of an operational se-

quence are complex systems of subsistence economy that defined the first activities of hominins and can be now rethought in regards to chimpanzee material culture. The processes of tool selection and use reveal information about the optimal use of raw materials in a process that provides better nutrition and may therefore improve individuals' chances of survival—with obvious implications for the evolutionary success of the individual adopting such behavior.

At the level of inter-community comparisons, data from our regional survey confirmed that chimpanzee groups can differ not only in *whether* they perform nut cracking but also in the precise details of that behavior. At the sites we studied, communities varied in their tool typologies and in the targets of their behavior: different raw materials were used and different species of nuts were cracked at different sites. While to some degree this was explained by ecological constraints, such constraints could not explain the full extent of the variation. For example, why do Seringbara chimpanzees completely neglect oil-palm nuts while their closest neighbors at Bossou devote a considerable amount of their feeding budget to cracking them? What would it take for nut cracking to appear and become established at Seringbara?

Our experiments with novel nuts demonstrated that there is considerable flexibility in tool use within the context of nut cracking; individuals can apply existing cracking skills to objects in the environment that are similarly hard-shelled, even in the absence of any obvious evidence that their contents are edible. However, how can a novel behavior—or a novel variation on an existing behavior, such as the cracking of a previously neglected species of nut—become integrated into a community's repertoire? Such innovations can appear in two main ways: existing community members can invent a new behavior, or immigrants to the community can bring their natal group's knowledge with them. In both cases, however, only subsequent spread within the community can ensure the behavior's continued persistence.

Based on our observations during the presentation of unfamiliar nuts at Bossou, we propose that innovations are most likely to originate among juveniles, rather than among the more conservative adults, and that if the innovations spread at all, they will do so among individuals in the same age group or younger (i.e., among other juveniles and infants). The innovations are unlikely to spread "upwards" to adults, since adults seem to pay little attention to younger group members' activities. This echoes, in part, what we know of patterns of transmission in the washing of sweet potatoes by Japanese monkeys on Koshima Island (see Hirata et al. 2001 for a recent review). There too the innovation—the cleaning of sand-covered sweet potatoes by dipping them in water—appeared and initially spread among young members of the group, and the behavior was transmitted to a limited extent only to adults (primarily to mothers of sweet-potato–washing young). Nevertheless, even in the absence of transmission to older group members this horizontal/downward spread, if maintained, would be sufficient to establish the behavior over time. The same applies to the spread of an immigrant's knowledge, potentially with faster rates of diffusion if they are adults more readily observed by fellow group members. Hence Yo—assuming she was indeed an immigrant, already familiar with coula nuts—may have contributed to coula cracking becoming established at Bossou: from the first time she proceeded to crack the nuts, she remained a reliable model for other group members to observe. In the case of panda nuts, attempts to crack them died out within a single field season, potentially due to the absence of a comparably salient performer. Together these results emphasize how the rates of innovation, migration, and within-community spread can affect how quickly novel behaviors are assimilated by wild chimpanzee communities. Further work on each of these aspects should help us illuminate why, for example, neighboring communities that share migrants and similar ecological backdrops do not necessarily possess identical cultures.

Novelty of the Archaeological Approach at Bossou

Prior to the work of Carvalho and colleagues, no study focused on the existence of operational sequences *during* the use of stone tools by chimpanzees in their natural habitat. Existing archaeological studies of chimpanzee stone tools were not designed to monitor, both directly and indirectly, nut-cracking activity on a daily basis (Joulian 1996; Mercader et al. 2002, 2007). We must also stress that Bossou is the only long-term study site where chimpanzees systematically use *movable stone* hammers and anvils to crack nuts open; no archaeological study has been undertaken before in this area. In addition, the outdoor laboratory provides the archaeologist with a unique opportunity to record tool-use details both at the individual and at the tool level.

Carvalho et al. (2008) was the first study to focus specifically on the question of reduction strategies in the use of stone tools by wild chimpanzees, and also pioneered daily archaeological drawing and mapping of spatial tool distribution at currently functioning nut-cracking sites, thus allowing us to detect a pattern of three different resource exploitation systems, and revealing similarities with known Oldowan strategies (Kimura 1999).

A good example illustrating the advantage of complementing archaeological methods with real-time observation regards the transport of tools: this was confirmed both within and outside the outdoor laboratory, through combining direct and indirect records. From the perspective of etho-archaeological study, the frequencies of tool transportation can be properly evaluated only when simultaneously combining the two types of observation. Indirect data only allow us to record the last place where a tool was left, while saying nothing about the number and type of movements that occurred during the nut-cracking process (Carvalho 2007; Carvalho et al. 2008). In a similar vein, the finding that chimpanzees may be selecting tool composites on the basis of the quality-efficacy of particular combinations of hammer and anvil stones (Carvalho et al. 2009) would suggest the kind of frequent reuse that may lead to the amplification of use-wear traces and an increased likelihood of fracturing and detaching objects by battering.

What the Archaeological Approach Can Tell Us about Early Hominins

Observations of synchronic differences in present-day cultures show that for *Homo sapiens*, modern technology can coexist with rudimentary technology. While doing fieldwork in Guinea in 2006, Carvalho (2007) observed that local people crack oil-palm nuts using the same raw materials, the same tool types, and the same techniques as chimpanzees use in the vicinity, confirming previous anecdotal reports (Kortlandt and Holzhaus 1987). Thus, while *Homo sapiens* possesses a much larger brain capacity than the chimpanzee, this in itself does not constitute a major influential variable: these so-called rudimentary tools must be considered successful and efficient, given that they still function in the 21st century as part of a local tradition.

Archaeology draws more and more on evolutionary theory (Jones 2006; Lycett et al. 2007), and typological and technological studies should profit if material cultures are analyzed more synchronically, through a model that can be described metaphorically as a "tree with many branches," rather than as a linear, progressive view of successive technologies. Through our combined approach we are able to evaluate chimpanzee behavior directly, to understand *how* and *why* our closest living relatives are using and producing specific types of stone tools, and what the variables are that influence the existence of cultural diversity—thus doing more than simply examining the final morphological product.

The forested regions of the tropics have long been regarded as extreme environments where substantial cognitive and technological abilities are necessary to survive (Bailey et al. 1989). Hence, it was assumed that such costly and risky places were likely to be avoided by early hominins. It is now known, however, that prehistoric hunter-gatherers were able to occupy such forests from at least the Last Glacial Maximum, approximately 20,000 BP (Mercader 2003). For early hominins, a multidiversity and multiregional focus of cultural emergence may be detectable—for example, the first lithic industries were probably not a monopoly of *Homo habilis* (Semaw et al. 1997; Delagnes and Roche 2005). Furthermore, by applying archaeology to modern chimpanzee stone-tool use to help reconstruct early hominin tool-use behavior, we are advancing the possibility of the first extensive techno-morphologic comparison on the subject of percussive technology between chimpanzees and early hominins.

Future Directions

Ongoing work at Bossou currently comprises further observations and field experiments on individual and social aspects of skill acquisition, both in nut cracking and in other forms of tool use (Biro et al. 2006; see also chapter 10). Our interdisciplinary approach will also focus on the description of typological and technological variation between different chimpanzee communities, each having diverse ecological constraints (Bossou and Diecké: Carvalho et al. 2008). This will allow us to consider, with regard to early hominins, the presence of diversity and multiregionality, and the emergence of culture. In addition, tool characteristics and technological features will continue to be studied, focusing on raw-material type availability. As an extension of our experiments with novel nuts, we also aim to introduce novel raw materials in an attempt to explore whether the quality of available raw material is a constraint

on, for example, the occurrence of knapping, as well as to examine the degree of physical understanding that chimpanzees possess about the objects involved in tool use.

In addition to work at Bossou itself, neighboring communities continue to be explored in order to build a comprehensive picture of the cultural life of chimpanzees in this corner of Africa (e.g., Humle and Matsuzawa 2004, Koops et al. 2006). Genetic analyses of these populations are also helping to illuminate local migration patterns (Shimada et al. 2004) and thus possible channels for the flow of knowledge between communities. More concretely, the ongoing habituation of the Seringbara group, for example, will hopefully lead to reliable direct daily observation of individuals, potentially allowing us to rediscover chimpanzees who have recently disappeared (and presumably emigrated) from Bossou, and examine tentatively the impact that their knowledge may have had on the community that received them. Evidence from the field that the knowledge of a chimpanzee from one community can spread to those in another is currently lagging behind experimental demonstrations in captivity (Whiten et al. 2007). Hence, this line of work, although ambitious, would represent a significant step forward in verifying a crucial element of models of culture in wild chimpanzees.

Acknowledgments

We thank the Direction National de la Recherche Scientifique et Technique, République de Guinée, for permission to conduct fieldwork. The research was supported by grants-in-aid for scientific research from the Ministry of Education, Science, Sports, and Culture of Japan (Grants 07102010, 12002009, 10CE2005, and the 21COE program). D.B. is grateful to the Royal Society for financial support; S. C. to FCT-Portugal, the Wenner-Gren Foundation, and Cias-Portugal. We also thank the following people who have been involved in research at Bossou over the years: Yukimaru Sugiyama, Gen Yamakoshi, Rikako Tonooka, Noriko Inoue-Nakamura, Tatyana Humle, Hiroyuki Takemoto, Satoshi Hirata, Gaku Ohashi, Makoto Shimada, Takao Fushimi, Osamu Sakura, Masako Myowa-Yamakoshi, Kat Koops, Claudia Sousa, and Misato Hayashi. We are extremely grateful to our guides Guanou Goumy, Tino Camara, Paquilé Cherif, Pascal Goumy, Marcel Doré, Bonifas Camara, Gilles Doré, Henri Zougbila, Jnakoi Malamu, Jean Marie Kolié, Cé Koti, and Albert Kbokmo for assistance in the field.

Literature Cited

Bailey, R.C., G. Head, M. Jenike, B. Owen, R. Retchman, and E. Zechenter. 1989. Hunting and gathering in tropical rain forest: is it possible? *American Anthropologist* 91:59–82.

Biro, D., N. Inoue-Nakamura, R. Tonooka, G. Yamakoshi, C. Sousa, and T. Matsuzawa. 2003. Cultural innovation and transmission of tool use in wild chimpanzees: Evidence from field experiments. *Animal Cognition* 6:213–23.

Biro, D., C. Sousa, and T. Matsuzawa. 2006. Ontogeny and cultural propagation of tool use by wild chimpanzees at Bossou, Guinea: Case studies in nut-cracking and leaf-folding. In T. Matsuzawa, M. Tomonaga, and M. Tanaka, eds., *Cognitive Development in Chimpanzees*, 476–508. Tokyo: Springer.

Boëda, E., J.-M. Geneste, and L. Meignen. 1990. Identification de chaînes opératoires lithiques du paléolithique ancient et moyen. *Paleo* 2:43–80.

Boesch, C. 1991. Teaching among wild chimpanzees. *International Journal of Primatology* 41:530–32.

———. 2003. Is culture a golden barrier between human and chimpanzee? *Evolutionary Anthropology* 12:82–91.

Boesch, C., and H. Boesch. 1983. Optimisation of nut-cracking with natural hammers by wild chimpanzees. *Behavior* 83:256–86.

———. 1984. Possible causes of sex differences in the use of natural hammers by wild chimpanzees. *Journal of Human Evolution* 13:415–40.

Boesch, C., P. Marchesi, N. Marchesi, B. Fruth, and F. Joulian. 1994. Is nut-cracking in wild chimpanzees a cultural behavior? *Journal of Human Evolution* 26:325–38.

Carvalho, S. 2007. *Applying the Concept of Chaîne Opératoire to Nut-Cracking: An Approach Based on Studying Communities of Chimpanzees (Pan Troglodytes Verus) in Bossou and Diecké (Guinea)*. Master's thesis. University of Coimbra, Portugal.

Carvalho, S., D. Biro, W. C. McGrew, and T. Matsuzawa. 2009. Tool-composite reuse in wild chimpanzees (*Pan troglodytes*): Archaeologically invisible steps in the technological evolution of early hominins? *Animal Cognition* 12:103–14.

Carvalho, S., E. Cunha, S. Sousa, and T. Matsuzawa. 2008. Chaînes opératoires and resource exploitation strategies in chimpanzee nut-cracking (*Pan troglodytes*). *Journal of Human Evolution* 55:148–63.

Delagnes, A., and H. Roche. 2005. Late Pliocene hominid knapping skills: The case of Lokalalei 2C, West Turkana, Kenya. *Journal of Human Evolution* 48:435–72.

Fragaszy, D., P. Izar, E. Visalberghi, E. B. Ottoni, and M. G. de Oliveira. 2004. Wild capuchin monkeys (*Cebus libidinosus*) use anvils and stone pounding tools. *American Journal of Primatology* 64:359–66.

Haslam, M., A. Hernandez-Aguilar, V. Ling, S. Carvalho, I. de la Torre, A. DeStefano, A. Du, B. Hardy, J. Harris, L. Marchant, T. Matsuzawa, W. McGrew, J. Mercader, R. Mora, M. Petraglia, H. Roche, E. Visalberghi, and R. Warren. 2009. Primate archaeology. *Nature* 460:339–344.

Heaton, J. L., and T. R. Pickering. 2006. Archaeological analysis does not support intentionality in the production of brushed ends on chimpanzee termiting tools. *International Journal of Primatology* 27:1619–33.

Hirata, S., and M. L. Celli. 2003. Role of mothers in the acquisition of tool-use behaviors by captive infant chimpanzees. *Animal Cognition* 6:235–44.

Hirata, S., K. Watanabe, and M. Kawai. 2001. "Sweet-potato washing" revisited. In T. Matsuzawa, ed., *Primate Origins of Human Cognition and Behaviour*, 487–508. Heidelberg: Springer Verlag.

Hopkins, W. D. 1999. Heritability of hand preference in chimpanzees: Evidence from a partial interspecies cross-fostering study. *Journal of Comparative Psychology* 109:291–97.

———. 2004. Laterality in maternal cradling and infant positional biases: Implications for the development and evolution of hand preferences innon-human primates. *International Journal of Primatology* 25:1243–65.

Humle, T. 2010. How Are Army Ants Shedding New Light on Culture in Chimpanzees? In E. V. Lonsdorf, S. R. Ross, and T. Matsuzawa, eds., *The Mind of the Chimpanzee: Ecological and Experimental Perspectives*. Chicago: University of Chicago Press.

Humle, T., and T. Matsuzawa. 2001. Behavioral diversity among the wild chimpanzee populations of Bossou and neighbouring areas, Guinea and Cote d'Ivoire, West Africa. *Folia Primatologica* 72:57–68.

———. 2004. Oil-palm use by adjacent communities of chimpanzees at Bossou and Nimba Mountains, West Africa. *International Journal of Primatology* 25:551–81.

Inoue-Nakamura, N., and T. Matsuzawa. 1997. Development of stone tool use by wild chimpanzees (*Pan troglodytes*). *Journal of Comparative Psychology* 111:159–73.

Jones, M. 2006. Archaeology and the genetic revolution. In J. Bintliff, ed., *A Companion to Archaeology*, 39–45. London: Blackwell.

Joulian, F. 1996. Comparing chimpanzee and early hominid techniques: Some contributions to cultural and cognitive questions. In P. Mellars and K. Gibson, eds., *Modelling the Early Human Mind*, 173–89. Cambridge, UK: McDonald Institute Monographs.

Karlin, C., and J. Pelegrin. 1988. Outil. In A. Leroi-Gourhan, ed., *Dictionnaire de la préhistoire*. Paris: Quadrige, Presses Universitaires de France.

Kimura, Y. 1999. Tool-using strategies by early hominids at Bed II, Olduvai Gorge, Tanzania. *Journal of Human Evolution* 37:807–31.

Koops, K., T. Humle, E. H. M. Sterck, and T. Matsuzawa. 2006. Ground-nesting by the chimpanzees of the Nimba Mountains, Guinea: Environmentally or socially determined? *American Journal of Primatology* 69:407–19.

Kortlandt, A., and E. Holzhaus. 1987. New data on the use of stone tools by chimpanzees in Guinea and Liberia. *Primates* 28:473–96.

Lonsdorf, E. V. 2006. What is the role of mothers in the acquisition of termite-fishing behaviors in wild chimpanzees (*Pan troglodytes schweinfurthii*)? *Animal Cognition* 9:36–46.

Lonsdorf, E., and W. D. Hopkins. 2005. Wild chimpanzees show population-level handedness for tool use. *Proceedings of the National Academy of Sciences, USA* 102:12634–38.

Lucas, G. 2000. *Les industries lithiques du Flageolet I (Dordogne): Approche économique, technologique, fonctionelle et analyse spatiale*. Thèse présentée a L`Université Bordeaux I pour obtenir le grade de Docteur. Volume I. Bordeaux.

Lycett, S. J., M. Collard, and W. C. McGrew. 2007. Phylogenetic analyses of behavior support existence of culture among wild chimpanzees. *Proceedings of the National Academy of Sciences, USA* 104:17588–92.

Marchant, L.F., and W. C. McGrew. 2005. Percussive technology: Chimpanzee baobab smashing and the evolutionary modelling of hominid knapping. In V. Roux and B. Bril, eds., *Stone Knapping: The Necessary Conditions for a Uniquely Hominid Behaviour*, 341–50. McDonald Institute for Archaeological Research, University of Cambridge, Cambridge.

Matsuzawa, T. 1994. Field experiments on use of stone tools by chimpanzees in the wild. In R. Wrangham, W. McGrew, F. de Waal, and

P. Heltne, eds., *Chimpanzee Cultures*, 351–70. Cambridge, MA: Harvard University Press.

———. 1996. Chimpanzee intelligence in nature and captivity: Isomorphism of symbol use and tool use. In W. C. McGrew, L. F. Marchant, and T. Nishida, eds., *Great Ape Societies*, 196–209. Cambridge: Cambridge University Press.

Matsuzawa, T., D. Biro, T. Humle, N. Inoue-Nakamura, R. Tonooka, and G. Yamakoshi. 2001. Emergence of culture in wild chimpanzees: Education by master-apprenticeship. In: T. Matsuzawa, ed., *Primate Origins of Human Cognition and Behavior*, 557–74. Tokyo: Springer.

Matsuzawa, T., T. Humle, K. Koops, D. Biro, M. Hayashi, C. Sousa, Y. Mizuno, A. Kato, G. Yamakoshi, G. Ohashi, Y. Sugiyama, and M. Kourouma. 2004. Wild chimpanzees at Bossou-Nimba: Deaths through a flu-like epidemic in 2003 and the Green Corridor Project. *Primate Research* 20:45–55 (in Japanese with English summary).

Matsuzawa, T., H. Takemoto, S. Hayakawa, and M. Shimada. 1999. Diecke forest in Guinea. *Pan Africa News* 6:10–11.

Matsuzawa, T., and G. Yamakoshi. 1996. Comparison of chimpanzee material culture between Bossou and Nimba, West Africa. In A. E. Russon, K. Bard, and S. Taylor Parker, eds. *Reaching into Thought: The Minds of the Great Apes*, 211–32. Cambridge: Cambridge University Press.

Mauss, M. 1967. *Manuel d'etnographie*. Paris: Éditions Payot.

McGrew, W. C. 1992. *Chimpanzee Material Culture: Implications for Human Evolution*. Cambridge: Cambridge University Press.

———. 2004. *The Cultured Chimpanzee: Reflections in Cultural Primatology*. Cambridge: Cambridge University Press.

McGrew, W. C., R. M. Ham, L. J. T. White, C. E. G. Tutin, and M. Fernandez. 1997. Why don't chimpanzees in Gabon crack nuts? *International Journal of Primatology* 18:353–74.

Mercader, J., ed. 2003. *Under the Canopy: The Archaeology of Tropical Rain Forests*. New Brunswick, NJ: Rutgers University Press.

Mercader, J., M. Panger, and C. Boesh. 2002. Excavation of a chimpanzee stone tool site in the African rainforest. *Science* 296:1452–55.

Mercader, J., H. Barton, J. Gillespie, J. Harris, S. Kuhn, R. Tyler, and C. Boesch. 2007. 4,300-year-old chimpanzee sites and the origins of percussive stone technology. *Proceedings of the National Academy of Sciences of the USA* 104:1–7.

Morgan, B., and E. Abwe. 2006. Chimpanzees use stone hammers in Cameroon. *Current Biology* 16:632–33.

Ohashi, G. 2006. Behavioral repertoire of tool use in the wild chimpanzees at Bossou. In T. Matsuzawa, M. Tomonaga, and M. Tanaka, eds., *Cognitive Development in Chimpanzees*, 439–51. Tokyo: Springer.

Pelegrin, J., C. Karlin, and P. Bodu. 1988. Chaînes opératoires: Un outil pour le préhistorien. *Technologie Préhistorique, Notes et Monographies* 25:63–70.

Price, E. E., S. P. Lambeth, S. J. Shapiro, and A. Whiten. 2009. A potent effect of observational learning on chimpanzee tool construction. *Proceedings of the Royal Society B*, published online.

Sakura, O., and T. Matsuzawa. 1991. Flexibility of wild chimpanzee nut-cracking behaviour using stone hammers and anvils: An experimental analysis. *Ethology* 87:237–248.

Semaw, S., P. Renne, J. W. K. Harris, C. S. Feibel, R. L. Bernor, N. Fesseha, and K. Mowbray. 1997. 2.5-million-year-old stone tools from Gona, Ethiopia. *Nature* 385:333–36.

Semenov, S.A. 1981. *Tecnología prehistórica (Estudio de las herramientas y objetos antiguos através de las huellas de uso)*. Madrid: Akal Universitária.

Sept, J. M. 1992. Was there no place like home? A new perspective on early hominid archeological sites from the mapping of chimpanzee nests. *Current Anthropology* 33:187–207.

Shimada, M. K., S. Hayakawa, T. Humle, S. Fujita, S. Hirata, Y. Sugiyama, and N. Saitou. 2004. Mitochondrial DNA genealogy of chimpanzees in the Nimba Mountains and Bossou, West Africa. *American Journal of Primatology* 64:261–75.

Stokes, E. J., and R. W. Byrne. 2001. Cognitive capacities for behavioral flexibility in wild chimpanzees (*Pan troglodytes*): The effect of snare injury on complex manual food processing. *Animal Cognition* 4:11–28.

Sugiyama, Y. 1997. Social tradition and the use of tool composites by wild chimpanzees. *Evolutionary Anthropology* 6:23–27.

Sugiyama, Y., and J. Koman. 1979. Tool-using and making behavior in wild chimpanzees at Bossou, Guinea. *Primates* 20:513–24.

Tixier, J., M.-L. Inizan, and H. Roche. 1980. *Préhistoire de la pierre taillée, terminologie et technologie*. Paris: Éditions du Cercle de Recherches et d'Études Préhistoriques.

Whiten, A., J. Goodall, W. C. McGrew, T. Nishida, V. Reynolds, Y. Sugiyama, C. E. G. Tutin, R. W. Wrangham, and C. Boesch. 1999. Cultures in chimpanzees. *Nature* 399:682–85.

———. 2001. Charting cultural variation in chimpanzees. *Behavior* 138:1481–1516.

Whiten, A., A. Spiteri, V. Horner, K. E. Bonnie, S. P. Lambeth, S. J. Schapiro, and F. B. M. de Waal. 2007. Transmission of multiple traditions within and between chimpanzee groups. *Current Biology* 17:1038–43.

Wrangham, R. W. 2006. Chimpanzees: The culture-zone concept becomes untidy. *Current Biology* 16:R634–35.

Wynn, T., and W. C. McGrew. 1989. An ape's view of the Oldowan. *Man* 24:383–98.

13

Ubiquity of Culture and Possible Social Inheritance of Sociality among Wild Chimpanzees

Michio Nakamura

An adult female chimpanzee, Miya, grooms the back of another female, Fatuma, on the observation trail while Fatuma grooms her infant daughter sleeping in her lap. Their juvenile daughters, Mitsue and Flavia, play-wrestle nearby. Miya softly pushes on the back of Fatuma, who stops grooming her baby and leans forward. While Miya scratches Fatuma's back with her right hand, Fatuma shows no apparent reaction but just continues to sit still. After closely inspecting the area she has just scratched, Miya continues grooming Fatuma's back.

The above peaceful scene is not an uncommon occurrence at Mahale, my study site in Tanzania. Many readers of this book may accept that cognitively demanding skills such as different types of tool use (e.g., McGrew 1992; see also chapters 10, 11 and 12) are the representation of chimpanzee culture. However, the above scene also depicts one of the many cultural variations found among chimpanzees, albeit without the same cognitive implications. As I am a field primatologist, I will not address chimpanzee cognition directly in this chapter. Instead, I will provide an outsider's view of the current focus on complex skills (implying complex cognitive abilities) in discussions of chimpanzee cultural capability. As part of this viewpoint, I question what I call the "individual agent-based information transfer model" of culture. That is, investigations of animal culture or tradition are most often viewed as an individual's ability to learn from others (Fragaszy and Perry 2003). Even though no one has expressed such a model explicitly, the

basic idea permeates studies of animal culture, perhaps tacitly. When we think of two animals in an experiment of social learning, one as a learner and the other as a model, we are testing whether the information is transferred from the latter individual to the former. Anthropologists, who have been struggling with culture in the sister species of chimpanzees, often criticize viewing nonhuman culture in this way as "decidedly anachronistic" (Ingold 2001). My concern is that although culture is collectively achieved, and we thus consider this phenomenon as being beyond the individual level, it is often converted to the question of individual ability when it comes to mechanisms. In this chapter I will outline a viewpoint that suggests taking a more holistic approach to the study of behaviors that broadens our perspective beyond the current emphasis on subjectivity and intelligence of individual agents.

Individual Behaviors and Collective Behaviors

Whether studying chimpanzees in experimental settings or in the wild, we can only approach their minds or culture indirectly by observing their behaviors. Thus, in this section I briefly introduce the matter of understanding behaviors in general. By doing this, I confirm that we cannot attribute all behaviors to an individual agent and in fact can explain some behaviors only collectively.

Itani (1981) made an important point about how we can understand the behaviors of nonhuman primates. He emphasized that "any behaviors are expressed with our language. In this sense, all the behaviors are included within our language system." Then he typified eight strata of verbs, or words to describe specific behaviors. Rather than focus on the relative importance of the strata, I wish to focus on how some behaviors can be understood at the individual level while others can only be understood at the collective level. Take the term "groom," for example. If individual A is grooming B, we can say "A grooms B," or in the passive voice, "B is groomed by A." We can also use the verb to denote that "A and B are grooming," expressing that these two individuals are simultaneously participating in and sharing the same social situation regardless of the role of each participant. This latter viewpoint cannot be described in the passive voice. That is, in the former two statements the subject of the verb is individual A, who takes on a positive role as performer while B assumes a passive role as recipient. In the latter statement, however, "grooming" cannot be attributed to any one individual, and this interaction is only understood when we see A and B as a set of interactants (e.g., Nakamura 2003). In animal behavioral sciences, we usually analyze a particular behavior at the individual level and assume that it is the outcome of an individual's independent strategy. This view is consistent with sociobiology, where we basically interpret behavior as something that maximizes individual fitness.

Chimpanzee Culture as Collective Behavior

Collective aspects of behavior are beginning to be addressed in cultural primatology, especially by studies on chimpanzee culture (e.g., Nishida 1987; Whiten et al. 1999; McGrew 2004; see also chapters 8 and 14). Without doubt culture has something to do with collectivity, as in McGrew's (2004) simple definition, "the way we do things," where "we" refers to collectivity. In this section I raise two questions from this standpoint regarding the current trends in studies of culture in nonhuman animals, especially chimpanzees.

The first point is that the question of culture is often converted to the question of intelligence or a complex cognitive ability. It is true that cultural behaviors in chimpanzees include complex and sophisticated behaviors that strongly imply an underlying intelligence. We are often fascinated by marvelous and varied examples of tool technologies developed by wild chimpanzees (e.g., McGrew 1992; see also chapters 10, 11, and 12). However, human culture

encompasses not only our intellectual behavior but also our daily mundane behavior such as what and how to eat, what to wear, and how and when to greet others—actions that we may even sometimes perform subconsciously. For example, as a Japanese male I feel slightly uneasy when an American female friend hugs me when we greet, because in Japanese culture women do not typically hug upon greeting. I was not aware of this uneasiness until I was hugged, even though I knew that the behavior was a normal part of American culture. This means that having knowledge about a behavior in a particular culture is not always the same as actually performing or experiencing that behavior. That is, culture may be associated with intelligence, but there are also other aspects to culture, such as shared social experiences and the society's history.

The second point is that we have thus far mostly dealt with distinctive behavioral patterns that can easily be labeled and listed (e.g., Whiten et al. 1999). Again, tool use nicely fits this type of analysis. We can easily distinguish a particular tool usage from other behavioral patterns by defining a tool as a detached object (Beck 1980). We can also differentiate between tools of different materials (e.g., wooden hammers versus stone hammers) and sizes, or by target food (e.g., termite fishing versus ant fishing), because we can easily measure or recognize such differences. When we think of human cultural differences, however, they do not have to be restricted to such distinctive patterns. For example, Hall (1966) argued that people maintain differing degrees of personal distance depending on their cultural background. The difference in social distance is continuous, but there is no doubt that culture affects such aspects of sociality. If we accept the ubiquity of cultural variation in the lives of chimpanzees, as I describe in a later section, it may be time to take our investigations beyond just distinctive behavioral patterns. This certainly comes back to what Imanishi (1952) proposed for culture in nonhuman animals. He questioned whether group living, or sociality itself, could be explained in terms of culture, and whether social organization could be inherited through culture (see Nakamura and Nishida 2006 for English review).

Approach

To investigate elements of chimpanzee culture that do not necessarily rely on complex cognitive abilities or overt individual action patterns, I collected behavioral data in Mahale Mountains National Park, Tanzania, intermittently

Table 13.1 Densities of within-sex grooming and numbers of grooming partners.

Male-male

group	group size	# males[a]	density[b]	# male partners[c] median	range	mean ± SD	# overall partners[e] median	range	mean ± SD	obs. hrs	obs. periods	obs. methods[d]	Source[e]
Mahale	ca 100	10 (9)	77.8 +	7.0	3–8	6.2 ± 1.5	–	–	–	–	1978–1979	Ad	1
Mahale	95	12 (10)	73.3 +	7.0	3–8	6.6 ± 1.6	–	–	–	205	8–10/1981	Fm	2
Mahale	82	9	91.7	8.0	6–8	7.3 ± 0.9	–	–	–	390	9–11/1992	Fm	3
Mahale	53	7	81.0	5	4–6	4.9 ± 0.9	24.0	17–36	24.7 ± 6.5	476	7/1996–4/1997	Fm, Ff	4
Mahale	53	7 (6)	73.3 +	4.5	4–6	4.7 ± 0.8	19.5	0–22	25.5 ± 6.7	476	7/1996–4/1997	Fm, Ff	4
Mahale	60	8	57.1	4.5	0–5	4.0 ± 1.7	–	–	16.9 ± 7.2	272	8–10/2004	Ff	4
Ngogo	>130	21	51.0	–		11.6 ± 4.3	–			342	6–12/1995–6–8/1996	Fm	5
Ngogo	>130	24	36.0	–		8.4 ± 4.1	–			212	6–8/1997	Fm	5
Budongo	50–51	9–11	77.8	7.5	2–9	–	13.0	6–16	–	ca 783	9/1998–10/1999	Sc, Fm	6
Gombe	44	13 (11)	85.5 +	9.0	7–10	8.5 ± 1.2	–			304	5/1969–4/1970	Ca, Fm	7
Gombe (groom)	57	7? (6)	51.3*	5.5	4–6	5.2 ± 1.0	–				1978	Ca	8
Gombe (be groomed)	57	7? (6)	58.9*	5.5	5–6	5.5 ± 0.5	–				1978	Ca	8
Bossou	17–22	3	100.0	2.0	2–2	2.0 ± 0.0	–			500	12/1976–5/1977	Ad	9
Bossou	17–22	2	100.0	1.0	1–1	1.0 ± 0.0	–			245	12/1982–3/1983	Ad	9

Female-Female

group	group size	# females[a]	density[b]	# female partners[c] median	range	mean ± SD	# overall partners[e] median	range	mean ± SD	obs. hrs	obs. periods	obs. methods[d]	Source[e]
Mahale	53	18	44.6	12.5	0–7	11.4 ± 3.3	27.5	6–35	24.9 ± 7.8	476	7/1996–4/1997	Fm, Ff	4
Mahale	53	18 (7)	90.5 +	14.0	11–16	13.6 ± 1.5	30.0	24–35	29.7 ± 3.5	476	7/1996–4/1997	Fm, Ff	4
Mahale	60	22	44.2	10.5	0–19	9.3 ± 5.0	21.0	1–40	20.3 ± 11.0	272	8–10/2004	Ff	4
Mahale	60	22 (10)	71.1 +	12.0	6–19	12.3 ± 3.9	28.5	15–40	28.1 ± 8.8	272	8–10/2004	Ff	4
Budongo	50–51	11–12	3.0	0.17	0–1	–	2.0	2–6	–	ca 783	9/1998–10/1999	Sc, Fm	6
Gombe (groom)	57	19 (7)	6.2*	5.0	1–9	5.1 ± 2.8	–				1978	Ca	8
Gombe (be groomed)	57	19 (7)	5.1*	5.0	1–9	4.7 ± 2.5	–				1978	Ca	8
Bossou	17–22	7	81.0	5.0	4–6	4.9 ± 0.7	–			500	12/1976–5/1977	Ad	9
Bossou	17–22	7	85.7	5.0	3–6	5.1 ± 1.1	–			245	12/1982–3/1983	Ad	9
Kanyawara	>=37	13? (13)	0.0	–		–	–			85	2–12/1989	Ff	10

a. Numbers in parentheses indicate numbers of focal or target individuals.

b. Numbers with + show percentage of grooming dyads per possible dyads of focal individuals, numbers with * are not observed but expected percentage of grooming dyads from mean of individuals, and other numbers show actual percentage of grooming dyads per all the possible dyads of the sex.

c. Undefined numbers include non-target individuals as partners.

d. Ad: *Ad libitum*; Fm: Focal on males; Ff: Focal on females; Sc: Scan, Ca: Camp data

e. 1 Kawanaka 1990; 2 Takahata 1990; 3 Nishida & Hosata 1996; 4 This study; 5 Watts 2000; 6 Arnold & Whiten 2003; 7 Simpson 1973; 8 Goodall 1986; Sugiyama 1998; 10 Wrangham et.al. 1992.

from 1994 to 2007 on a well-habituated population of chimpanzees known as the M group (see Nishida 1990 for details of the research site). For comparison, I also conducted field observations at Bossou, Guinea, for two months in 2003 (see Nakamura and Nishida 2006 for details). For the purposes of this chapter I will focus on two main areas of investigation: (1) summarizing our recent findings of behavioral variation of wild chimpanzees outside of tool use to show that cultural behaviors are widespread in various domains including social behaviors, and (2) comparing the grooming densities between chimpanzee study sites and age classes to investigate the possible culture transmission of sociality. Since all chimpanzees groom, grooming is considered a chimpanzee universal—that is, something ubiquitous across populations. In this sense, grooming behavior itself is never considered a cultural variant according to the classification system used by Whiten et al. (1999; see also chapter 8). However there is some variation in grooming types that are widely accepted as cultural (see below). I argue here that differences in grooming networks, both in size and in type, can be acquired socially. Such variations, if detected, must at least be relevant to cultural learning, although it may be difficult to prove that in the field study.

In addition to data from two study periods in Mahale based on my own fieldwork, I used the grooming data extracted from the literature for different chimpanzee study sites. Whenever possible, I extracted data on group size and the numbers of males and/or females, same-sex grooming partners, and overall grooming partners (table 13.1). A grooming partner was defined as any individual observed either grooming or being groomed at least once during the study period. In some cases I calculated figures directly from the raw data. Table 13.1 summarizes the observational hours, periods, and methods. I calculated a metric I term "grooming density" as the number of observed grooming dyads divided by the possible dyads of the sex cohort. When using focal data, density was calculated only among focal targets of the same sex. Note that this density is usually underestimated but never overestimated. I made this rough comparison because the observability and methods of observation employed were quite different among various study sites.

Finally, I recorded the number of grooming partners in each age class as substitute for documenting the developmental process of individual grooming patterns. These data were collected from focal follows on females with dependent offspring for one study period at Mahale (272 hours of observation, August through October 2004). Based on these data, I counted the number of overall partners for all the individuals of the M group.

Ubiquity of Cultural Behaviors

Below I describe four broad behavioral categories that may or may not be determined in part or in whole by cultural transmission. Note that none of them are the sort of cognitively demanding behaviors that are often found in the list of cultural behaviors. Instead, I argue here that local variation in more ubiquitous behaviors, sometimes unrelated to quantifiable function, may be as fertile a ground for these types of investigations.

Grooming

As with many other primates, social grooming is one of the most frequent social interactions among chimpanzees. Chimpanzees' grooming is unique in that they often groom face to face simultaneously and often groom in large gatherings (Nakamura 2003). Because chimpanzee grooming is highly social and collectively performed, it is expected that some local variations will be found. The grooming handclasp (McGrew and Tutin 1978) has been the classic example of cultural behavior outside of feeding skills. In the grooming handclasp, two chimpanzees sit face to face, raise their corresponding hands in the air to form an A-frame posture, and then groom each other's underarm areas. This behavior is performed by most group members in Mahale (K and M groups), Kibale (Uganda: Kanyawara and Ngogo groups), Kalinzu (Uganda), and Lopé (Gabon). Only some individuals exhibit this behavior in Taï (Côte d'Ivore), and none do in Gombe (Tanzania), Budongo (Uganda: Sonso group), or Bossou (Guinea) (Nakamura 2002). Handclasp grooming has also been documented in at least one captive chimpanzee colony at Yerkes Primate Research Center (de Waal and Seres 1997).

McGrew et al. (2001) reported two different types of grooming handclasp: palm-to-palm (figure 13.1a) and non–palm-to-palm (figure 13.1b). In the former, two chimpanzees clasp each others' hands with mutual palmar contact, while in the latter, one or neither hand clasps the other, and usually the hands are instead flexed with one limb resting on the other. Although the authors argued

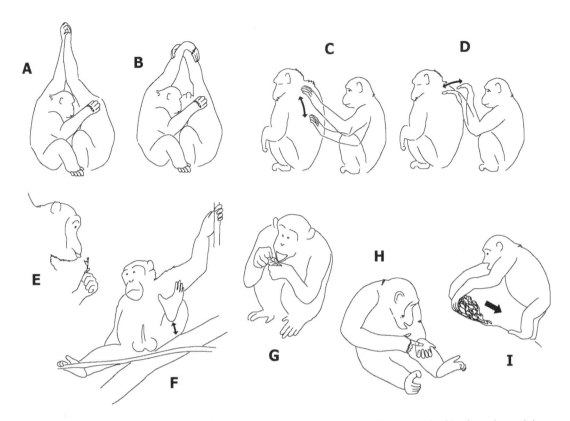

Figure 13.1 Examples of local variation outside of feeding skills. A: grooming handclasp (palm-to-palm); B: grooming handclasp (non-palm-to-palm); C: social scratch (stroke type); D: social scratch (poke type); E: leaf clipping; F: heel tap; G: leaf grooming; H: index to palm; I: leaf-pile pulling.

that palm-to-palm dominated in the Mahale K group but was not observed in the M group, a follow-up study revealed that it was also performed in the M group, although all observed cases involved one old female who had immigrated from the K group (Nakamura and Uehara 2004).

Social scratch, described at the beginning of this chapter, is another grooming pattern observed consistently in Mahale (Nakamura et al. 2000). A different type of social scratching is customarily observed at Ngogo. Mahale chimpanzees employ stroke scratching (figure 13.1c), whereas Ngogo chimpanzees employ poke scratching (figure 13.1d) (Nishida et al. 2004). Although Shimada (2002) reported that three individual Gombe chimpanzees performed social scratching, there are no reports of this behavior from other study sites.

The sounds produced during grooming also differ among populations. Ngogo chimpanzees "sputter," which sounds as though they were forcing air through their lips, while Mahale chimpanzees do not (Nishida et al. 2004). This sound has also been documented in a few individuals in Bossou, albeit mostly in a single juvenile male (Naka-mura and Nishida 2006), and thus the behavior may not have reached the group level.

Courtship Displays

Courtship displays solicit copulation by a male or an estrous female from an individual of the opposite sex (Nishida 1997). Because sexual context is often obvious from penile erection of the male and the tumescence of the sexual skin of the female, any attention-getting behaviors can be used as courtship displays. However, the fact that courtship displays often converge within a group but vary among groups suggests underlying cultural processes. Leaf clipping (figure 13.1e)—that is, clipping a leaf or leaves to produce an audible sound—at Mahale is a famous example of such a courtship display made to attract the attention of the prospective mate (Nishida 1980). Bossou chimpanzees also exhibit this behavioral pattern, but according to Sugiyama (1981) it is used to express frustration. Taï chimpanzees also clip leaves, but only before they perform buttress drumming (Boesch 1996).

At Bossou, mature males often perform a heel tap (figure 13.1f) in the context of courtship (Nakamura and Nishida 2006), rhythmically tapping a tree bough, rock, or the ground with the heel. This is not the usual stamping, because the sole makes no contact with the substrate but instead is held upright facing forward, and only the heel makes contact. This behavior produces a conspicuous sound. At Mahale, stamping is common but the heel tap has never been observed. According to earlier observations by Sugiyama (1989), immature individuals also exhibit this behavior in the context of inviting play.

Ectoparasite Squashing

As mentioned above, grooming serves a social function in chimpanzees, but at the same time it also serves the hygenic function of removing ectoparasites and debris from the body. After removing an ectoparasite, chimpanzees often inspect and squash it. This behavior also shows variations across the different groups in which it has been observed. At Mahale, chimpanzees perform leaf grooming (figure 13.1g), in which they pick up ectoparasites with their lips during grooming, place them on leaves, and then squash them with their thumbs (Zamma 2002). Chimpanzees in Gombe (Goodall 1965) and Budongo (Assersohn et al. 2004) exhibit a similar pattern. At Bossou, chimpanzees place the parasites in the palm of one hand and then smash them with the index finger of the other hand (figure 13.1h) (Nakamura and Nishida 2006). Similarly, chimpanzees in Taï smash the parasites against a forearm, using the index finger of the free hand (Boesch 1995).

These behaviors may have some social functions. Both leaf grooming at Mahale and index-to-palm squashing at Bossou sometimes attract the interest of other individuals who gather around and watch. Moreover, leaf grooming at Mahale sometimes occurs outside of the grooming context (Nakamura, personal observation), and thus it is possible that this behavior could also be performed without actually smashing any ectoparasites.

Play

Play is another social behavior that is ubiquitous in chimpanzees. Cultural variation in play is not well documented at most study sites, perhaps because the function of play is often difficult to define and young individuals are rarely

the focus of study. However, leaf-pile pulling (figure 13.1i) (Nishida and Wallauer 2003) at Mahale and Gombe may be an example of cultural variation in play behavior. In leaf-pile pulling, young chimpanzees turn around and walk backwards while raking a pile of dry leaves with both hands down a slope and producing a lot of noise. Although this could be interpreted as solitary play, the performer often faces another individual who is immediately following him or her and attracts social attention.

Another local variant involved in play is the use of tools for drinking (e.g., Tonooka 2001), whereby a chimpanzee uses leaves or sticks to obtain water. This has been observed only recently at Mahale, and only among immature chimpanzees (Matsusaka et al. 2006). They most often exhibit this behavior during the rainy season, when running water is plentiful and thus it does not seem necessary to use tools for obtaining water. Therefore, it seems that tool use in this context may be not out of necessity but for play.

Example of Quantitative Local Differences: Grooming Densities of Females and Development of Grooming Networks

Above I have described potential cultural variation in four different behavioral domains that are far from complex skills. These examples, however, are still distinctive patterns we can easily separate from other behaviors. For example, the grooming handclasp can potentially be a cultural behavior only because it is qualitatively different from any other forms of grooming. In reality, however, it is not isolated but rather is embedded in the grooming sequence. Therefore, I turn now to the possibility of quantitative rather than qualitative local differences through a case study of grooming networks across chimpanzee study sites. Using the data sets shown in table 13.1, the densities of male-male and female-female grooming are plotted against the numbers of individuals of each sex in figure 13.2. Male grooming networks in general are very dense. Densities of Ngogo males seem lower than others, but this is likely because the large number of males at Ngogo prevents each individual from distributing his grooming efforts to all others.

The networks among females, on the other hand, are relatively sparse at Gombe and in the Kanyawara (Kibale) and Sonso (Budongo) groups. The only dense network of females found in literature is from Bossou. However,

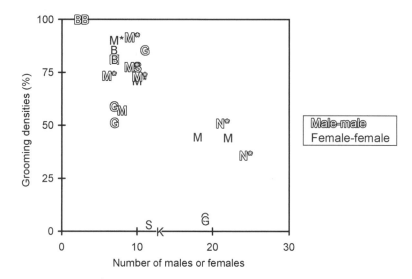

Figure 13.2 Densities of within-sex grooming plotted against the number of males or females. Data on male-male grooming are shown with outlined letters; data on female-female grooming are shown with solid letters. Names of the groups are as follows. B: Bossou; G: Gombe; K: Kanyawara (Kibale); M: Mahale; N: Ngogo (Kibale); S: Sonso (Budongo). Densities calculated only within focal individuals are marked with asterisks. There is an outlined *M* just behind the solid *B*.

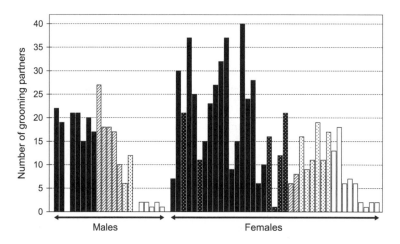

Figure 13.3 Number of grooming partners of all 60 individuals at Mahale in August through October 2004. One adult male had no grooming partners because he had been ranging alone for several years and seldom met other individuals. Males and females are arranged from left to right in descending order of age. Black bars: adult males and adult females with offspring (mothers). Black bars with white dots: adult females without offspring (nulliparous and probably some sterile females). Oblique-lined bars: adolescents. Dotted bars: juveniles. White bars: infants.

the grooming densities of Mahale females are also as high as those of Ngogo males. In fact, when the data are limited to focal females only, the densities become much higher (i.e., solid M* at upper left in figure 13.2).

To investigate possible social influences on the development of grooming networks, I calculated the overall number of grooming partners of all members of the Mahale

M group during a two-month study period (figure 13.3). Mothers have up to 40 grooming partners, averaging about 20.3 partners each. Although I gathered these data while following mothers and thus the result is biased toward females, grooming by adult males was also constantly observed, and they had an average of about 19.3 partners (excluding the former alpha male who has been ranging alone for several years). This is consistent with my results (Nakamura, unpublished data) in a different study period, during which I followed both males and females and found that they had about the same number of grooming partners. The number of partners of some adult females is more than two-thirds of the total group membership, perhaps because these females also groom with immature individuals while adult males often do not. Juvenile and older infant females had 10 to 20 grooming partners (mean = 14.3) during this short study period.

Discussion

With the accumulation of empirical evidence, we now accept that several animal species exhibit local behavioral variation that cannot be simply explained by environmental or genetic factors (Japanese monkeys: Itani and Nishimura 1973; chimpanzees: Whiten et al. 1999; cetaceans: Rendell and Whitehead 2001; capuchin monkeys: Perry et al. 2003; orangutans: van Schaik et al. 2003). When we define culture as collectively performed behaviors, previous standpoints appear rather biased toward cognitively demanding and distinctive behaviors to which a researcher can simply answer yes or no regarding their existence.

Beyond Individual Intelligence and Information Transfer

It is true that animal species that exhibit behavioral variation tend to be "smart" animals with large brains (van Schaik 2006). This does not mean, however, that cultural behaviors are always intellectual. Rather, locality-specific behaviors are ubiquitous in every domain of behavior, including very mundane ones.

This connection of culture with intelligence can often be linked to the individual agent-based information-transfer model of skills and knowledge. The idea that a cultural packet, or something similar to a duplicable "meme" (Dawkins 1976; Blackmore 1999), is stored in an individual's mind and then transferred to another individual's mind is very common in the Darwinian viewpoint of culture. It is often assumed that such duplication is only possible when an individual of the species possesses some complex cognitive mechanism. For example, Blackmore (1999) argued that true imitation plays a role in the transferring of memes among humans. However, behavior is not always performed individually, as discussed earlier in regards to Itani's eight strata of behaviors (Itani 1981). For example, in the case of the grooming handclasp, the individual agent-based transfer model does not fit well because it is not very clear what useful information is transferred from individual to individual. Moreover, the handclasp behavior cannot be performed by one individual alone, but only with the presence and acceptance of another. Thus, even when an individual knows how to perform this behavior, she or he cannot always do so. In the case of social scratch, if we break down the behavior into performers and receivers, there seems to be no benefit for the performers; if there is benefit at all, only the receiver seems to be rewarded (Nakamura et al. 2000). Then why and how would a performer acquire this behavior?

It is also interesting that some cultural behaviors seem to be determined arbitrarily rather than rationally. Take ectoparasite squashing, for example. Removing ectoparasites from their bodies may be important to chimpanzees, but whether they squash the parasites on leaves or on their own arms does not seem to be critical to survival. Yet each population exhibits only one pattern of this behavior. Thus, even though these behaviors are performed individually, the choice of which pattern is arbitrarily selected is likely determined at the group level.

Note that I am not arguing that all of the behaviors described above are determined only by culture. For example, we still do not know which ectoparasite species are squashed in different populations (or even whether they are in fact ectoparasites). Differences in ectoparasite species could be associated with different types of squashing. Although recent cultural studies have attempted to exclude ecological explanations of behavioral diversity for the sake of strictness of interpretation (e.g., Whiten et al. 1999), cultural explanations do not necessarily exclude other possible explanations. There are likely social and ecological influences and constraints on all cultural behaviors (see chapter 10).

It is becoming clear that local variation exists in various behavioral domains, and that it is neither always directly related to particular functions nor well explained by the information-transfer model between individual agents. Some behaviors must be performed jointly, as in the case of the grooming handclasp, while some seemingly arbitrary behaviors performed individually are somehow shared among members of the society.

Beyond Distinct Behavioral Patterns

If we were to assess "female grooming" as a possible cultural behavior and check its existence or absence in the different study sites, we would see that all groups exhibit it. Female sociality has never been questioned in terms of being a cultural trait. This does not mean, however, that there is no variation in the behavior between populations.

Some authors have emphasized the solitary nature of females, stating that "many females meet each other rarely" (Wrangham 1997) or that "females spend most of their time with their offspring, rarely joining large foraging parties" (Stanford 2006). It is often generalized that female chimpanzees, or at least east African females, are less gregarious than males, and therefore they are sometimes depicted as being less social than males or even asocial. However, females of at least some groups have grooming networks as dense as those of males. On the other hand, females at Gombe and in the Kanyawara (Kibale) and Sonso (Budongo) groups seem to have sparser networks than males do irrespective of the potential number of female partners, which concords with the notion that females interact with each other less than males do. Before my report on the grooming densities of Mahale females, only Bossou females had been an exception to this tendency. We were tempted to explain the exception as the result of the Bossou group's relatively small number of members, especially males, and the absence of neighboring groups. This condition resembles that of many captive colonies, in which females are socially as active as males (de Waal 1982). However, the addition of the Mahale data complicates the group-size explanation because the group at Mahale is three times larger than the one at Bossou.

Another possible explanation for the apparent exception of Bossou females was that it resulted from a difference between two subspecies: eastern and western chimpanzees (e.g. Yamakoshi 2004, Furuichi 2006). Indeed, Boesch (1991) describes a social group pattern in which females tend to move together (Boesch 1991), which suggests that Taï females (of the western chimpanzee subspecies) also have dense grooming networks like those at Bossou but distinct from eastern chimpanzees. However, this subspecies differentiation is weakened in light of data showing that females from the Mahale M group have high grooming densities similar to those at Bossou. According to the authors who emphasized the subspecies differences, Ma-

hale females are typical eastern females with low grooming densities similar to those at Gombe and in the Kanyawara group at Kibale. These authors usually cited Nishida's early works (e.g., Nishida 1979), in which observations were conducted mostly at feeding stations. Females, especially those nursing young, are typically less bold than males, and so it is understandable that they may be observed less often grooming each other at artificial feeding stations. Thus, my tentative conclusion is that Bossou, Mahale, and probably Taï females have denser social networks than females at Gombe, Kanyawara, and Sonso, and that this difference may not be due simply to differences in group sizes or subspecies.

Of course, there are other potential explanations to consider. First, females usually take longer than males to become habituated to observation. In my experience at Mahale, some older mothers never fully tolerated human observers to the degree that males did. The degree of habituation might affect the observation of grooming, for which individuals must be relatively relaxed. Thus it may be possible that low grooming densities of females in the Kanyawara and Sonso groups represent less female habituation. It appears, however, that Gombe females had been fully habituated at the time of study (Goodall 1986), so this argument does not apply to all populations. Secondly, the issue of variable sampling methods among the studies must be considered (see table 13.1). However, the Kanyawara data were taken from female follow observations in which no focal females groomed with other females. In contrast, when I followed females at Mahale I frequently observed them grooming with other females even within a day. Thus it is unlikely that the differences in grooming densities can be attributed only to observation methods. It would be premature to conclude whether the observed differences in female grooming densities represent local cultural variation, different degrees of habituation, or differences in observation methods. However, I propose that further studies on female sociality and other social behaviors that may show cultural variation are a fruitful avenue for research.

Possible Social Inheritance of Sociality

To my knowledge, there are no similar data from other populations on the development of grooming partners as I have shown for Mahale (figure 13.3). I emphasize that even juvenile and older infant females had 10 to 20 grooming partners within a two-month period. Certainly, these numbers could not have been achieved if the chimpanzees had limited their grooming to kin only. It is possible that the wide range of grooming networks of mothers led to their offspring also having a wide range of grooming partners. Indeed, when mothers groomed with adults of either sex, their daughters often joined in and groomed, or were groomed by the other adults. The juveniles also often groomed each other while their mothers did the same among themselves. If mothers tended to be solitary and their social networks were small, as has been argued for other east African populations, the Mahale offspring would not exhibit such interaction with so many individuals either. As such, it may be possible that the social breadth of females is socially inherited. Of course this is speculation at the moment, but I believe this hypothesis warrants further investigation.

Implications and Future Directions

Few scholars accept the nature-versus-nurture distinction; it is generally accepted that some intellectual behaviors (e.g., complex feeding skills) are the result of a combination of the innate propensity to learn and the social environment, and that these two factors may not be mutually exclusive. However, it seems that more basic behaviors, such as female association patterns, are often explained as direct responses to natural variables by individuals who independently choose what is best in the current situation. Considering the ubiquity of culture in every domain of behavior in chimpanzees and perhaps other nonhuman animals, we must entertain the possibility that sociality may also be socially inherited, and that the propensity toward such behavior in a particular social situation may be, in part, socially shared among the members of the society (see also Matsuzawa 2001; de Waal 2001). In that sense, we can never fully understand chimpanzee society and chimpanzees themselves without understanding the history and culture of their society.

Studies on the collective aspects of behavior have just begun. Although we definitely need more data, it is not unreasonable at this point to consider the possibility that females of some groups have a wider grooming network than females in other groups. If we accept that there are differences among populations in such a basic social behavior, then what is causing these differences? Some ecological factors, such as food availability and distribution or reproductive and/or demographic parameters, may or may not explain the difference. However, if we begin to accept that cultural differences may play a significant role in every domain of chimpanzee behavior, we must also explore the possibility that even very basic sociality is affected by culture.

In the famous words of Köhler (1925), "It is hardly an exaggeration to say that a chimpanzee kept in solitude is not a real chimpanzee at all." Thus, even the chimpanzee mind is not isolated in its brain or only expressed with its individual behavior. We would accept that the human mind is deeply embedded in our society; why should that not also hold true for the chimpanzee mind?

Acknowledgments

I would like to thank the Tanzania Commission for Science and Technology, the Tanzania Wildlife Research Institute, Tanzania National Parks, Mahale Mountains National Park, and the Mahale Mountains Wildlife Research Centre, of Tanzania, and Direction Nationale de la Recherche Scientifique et Technique, of Guinea, for permission to conduct the field research at Mahale and Bossou respectively. I thank Dr. T. Nishida for continuous support and advice on my research. The study was financially supported by grants from the Japanese Ministry of Education, Culture, Sports, Science, and Technology (#16255007, #19255008 to T. Nishida and #16770186 to MN), the Biodiversity Research of the 21st Century Centers of Excellence (A14), and the Global Environment Research Fund (#F-061).

Literature Cited

Arnold, K., and A. Whiten. 2003. Grooming interactions among the chimpanzees of the Budongo Forest, Uganda: Tests of five explanatory models. *Behaviour* 140:519–52.

Assersohn, C., A. Whiten, Z. T. Kiwede, J. Tinka, and J. Karamagi. 2004. Use of leaves to inspect ectoparasites in wild chimpanzees: A third cultural variant? *Primates* 45:255–58.

Beck, B. B. 1980. *Animal Tool Behavior: The Use and Manufacture of Tools by Animals*. New York: Garland STPM Press.

Blackmore, S. 1999. *The Meme Machine*. Oxford: Oxford University Press.

Boesch, C. 1991. The effects of leopard predation on grouping patterns in forest chimpanzees. *Behaviour* 117:220–42.

———. 1995. Innovation in wild chimpanzees (*Pan troglodytes*). *International Journal of Primatology* 16:1–16.

———. 1996. Three approaches for assessing chimpanzee culture. In A. E. Russon, K.A. Bard, and S. T. Parker, eds., *Reaching into Thought: The Minds of the Great Apes*. Cambridge: Cambridge University Press, 404–29.

Dawkins, R. 1976. *The Selfish Gene*. Oxford: Oxford University Press.

De Waal, F. B. M. 1982. *Chimpanzee Politics: Power and Sex among Apes*. Baltimore: Johns Hopkins University Press.

———. 2001. *The Ape and the Sushi Master: Cultural Reflections of a Primatologist*. New York: Basic Books.

De Waal, F. B. M., and M. Seres. 1997. Propagation of handclasp grooming among captive chimpanzees. *American Journal of Primatology* 43:339–46.

Fragaszy, D.M., and S. Perry. 2003. Towards a biology of traditions. In D. M. Fragaszy and S. Perry, eds., *The Biology of Traditions: Models and Evidence*. New York: Cambridge University Press, 1–32.

Furuichi, T. 2006. Evolution of the social structure of hominoids: reconsideration of food distribution and estrus sex ratio. In H. Ishida, R. Tuttle, M. Pickford, N. Ogihara, and M. Nakatsukasa, eds., *Human Origins and Environmental Backgrounds*. New York: Springer, 235–48.

Goodall, J. 1965. Chimpanzees of the Gombe Stream Reserve. In I. De Vore, ed., *Primate Behavior*. New York: Holt, Reinhart, and Winston, 425–73.

———. 1986. *The Chimpanzees of Gombe: Patterns of Behavior*. Cambridge, MA: Harvard University Press.

Hall, E.T. 1966. *Hidden Dimension*. New York: Doubleday.

Imanishi, K. 1952. [Evolution of humanity.] In K. Imanishi, ed., [*Man.*] Tokyo: Mainichi Shinbunsha, 36–94. Published in Japanese.

Ingold, T. 2001. The use and abuse of ethnography. *Behav Brain Sci* 24:337.

Itani, J. 1981/1987. [Behaviors that make social structure.] Reprinted in [*Evolution of Primate Societies.*] Tokyo: Heibonsha, 223–45. Published in Japanese.

Itani, J., and A. Nishimura. 1973. The study of infrahuman culture in Japan. In E. W. Menzel, ed., *Precultural Primate Behavior*. Basel: Karger, 26–50.

Kawanaka, K. 1990. Alpha male's interactions and social skills. In T. Nishida, ed., *The Chimpanzees of the Mahale Mountains: Sexual and Life History Strategies*. Tokyo: University of Tokyo Press, 171–87.

Köhler, W. 1925. *The Mentality of Apes*. London: Paul, Trench, Trubner & Company.

Matsusaka, T., H. Nishie, M. Shimada, N. Kutsukake, K. Zamma, M. Nakamura, and T. Nishida. 2006. Tool-use for drinking water by immature chimpanzees of Mahale: Prevalence of an unessential behavior. *Primates* 47:113–22.

Matsuzawa, T., D. Biro, T. Humle, N. Inoue-Nakamura, R. Tonooka, and G. Yamakoshi. 2001. Emergence of culture in wild chimpanzees: Education by master-apprenticeship. In T. Matsuzawa, ed., *Primate Origins of Human Cognition and Behavior*. Tokyo: Springer, 557–74.

McGrew, W.C. 1992. *Chimpanzee Material Culture: Implications for Human Evolution*. Cambridge: Cambridge University Press.

———. 2004. *The Cultured Chimpanzee: Reflections on Cultural Primatology*. Cambridge: Cambridge University Press.

McGrew, W. C., L. F. Marchant, S. E. Scott, and C. E. G., Tutin. 2001. Intergroup differences in a social custom of wild chimpanzees: The grooming hand-clasp of the Mahale Mountains. *Current Anthropology* 42:148–53.

McGrew, W. C., and C. E. G. Tutin. 1978. Evidence for a social custom in wild chimpanzees? *Man* 13:234–51.

Nakamura, M. 2002. Grooming-hand-clasp in Mahale M group chimpanzees: Implication for culture in social behaviors. In C. Boesch, G. Hohmann, and L. F. Marchant, eds., *Behavioural Diversity in Chimpanzees and Bonobos*. Cambridge: Cambridge University Press, 71–83.

———. 2003. "Gatherings" of social grooming among wild chimpanzees: Implications for evolution of sociality. *Journal of Human Evolution* 44:59–71.

Nakamura, M., W. C. McGrew, L. F., Marchant, and T. Nishida. 2000. Social scratch: Another custom in wild chimpanzees? *Primates* 41:237–48.

Nakamura, M., and T. Nishida. 2006. Subtle behavioral variation in wild chimpanzees, with special reference to Imanishi's concept of *kaluchua*. *Primates* 47:35–42.

Nakamura, M., and S. Uehara. 2004. Proximate factors of different types of grooming hand-clasp in Mahale chimpanzees: Implications for chimpanzee social customs. *Current Anthropology* 45:108–14.

Nishida, T. 1979. The social structure of chimpanzees of the Mahale Mountains. In Hamburg, D.A., McCown, E.R., eds., *The Great Apes*. Menlo Park: Benjamin/Cummings, 73–121.

———. 1980. The leaf-clipping display: A newly-discovered expressive gesture in wild chimpanzees. *Journal of Human Evolution* 9:117–28.

———. 1987. Local traditions and cultural transmission. In B. B. Smuts, D. L. Cheney, R. M. Seyfarth, R. W. Wrangham, and T. T., Struhsaker, eds., *Primate Societies*. Chicago: University of Chicago Press, 462–74.

———, ed. 1990. *The Chimpanzees of the Mahale Mountains: Sexual and Life History Strategies*. Tokyo: University of Tokyo Press.

Nishida, T., and K. Hosaka. 1996. Coalition strategies among adult male chimpanzees of the Mahale Mountains, Tanzania. In W. C. McGrew, L. F. Marchant, and T. Nishida, eds., *Great Ape Societies*. Cambridge: Cambridge University Press, 114–34.

Nishida, T., J. C. Mitani, and D. P. Watts. 2004. Variable grooming behaviours in wild chimpanzees. *Folia Primatologica* 75:31–36.

Nishida, T., and W. Wallauer. 2003. Leaf-pile pulling: An unusual play pattern in wild chimpanzees. *American Journal of Primatology* 60:167–73.

Perry, S., M. Baker, L. Fedigan, J. Gros-Louis, K. Jack, K. C. MacKinnon, J. H. Manson, M. Panger, K. Pyle, and L. Rose. 2003. Social conventions in wild white-faced capuchin monkeys. *Current Anthropology* 44:241–68.

Rendell, L., and H. Whitehead. 2001. Culture in whales and dolphins. *Behavioral and Brain Sciences* 24:309–82.

Shimada, M. 2002. Social scratch among the chimpanzees of Gombe. *Pan Africa News* 9:21–23.

Simpson, M. J. A. 1973. The social grooming of male chimpanzees. A study of eleven free-living males in the Gombe Stream National Park, Tanzania. In R. P. Michael and J. H. Cook, eds., *Comparative Ecology and Behaviour of Primates*. London: Academic Press, 411–505.

Stanford, C. B. 2006. The behavioral ecology of sympatric African apes: implications for understanding fossil hominoid ecology. *Primates* 47:91–101.

Sugiyama, Y. 1981. Observations on the population dynamics and behavior of wild chimpanzees at Bossou, Guinea, in 1979–1980. *Primates* 22:435–44.

———. 1988. Grooming interactions among adult chimpanzees at Bossou, Guinea with special reference to social structure. *International Journal of Primatology* 9:393–407.

———. 1989. Description of some characteristic behaviors and discussion on their propagation process among chimpanzees of Bossou, Guinea. In Y. Sugiyama, ed., *Behavioral Studies of Wild Chimpanzees at Bossou, Guinea.* Inuyama: KUPRI, 43–47.

Takahata, Y. 1990. Social relationships among adult males. In T. Nishida, ed., *The Chimpanzees of the Mahale Mountains: Sexual and Life History Strategies.* Tokyo: University of Tokyo Press, 149–70.

Van Schaik, C. P. 2006. Why are some animals so smart? *Scientfic American* 294(4):64–71.

Van Schaik, C. P., M. Ancrenaz, G.. Borgen, B. Galdikas, C. D. Knott, I. Singleton, A. Suzuki, S. S. Utami, and M. Merrill. 2003. Orangutan cultures and the evolution of material culture. *Science* 299:102–5.

Watts, D. P. 2000. Grooming between male chimpanzees at Ngogo, Kibale National Park. I. Partner number and diversity and grooming reciprocity. *International Journal of Primatology* 21:189–210.

Whiten, A., J. Goodall, W. C. McGrew, T. Nishida, V. Reynolds, Y. Sugiyama, C. E. G.. Tutin, R. W. Wrangham, and C. Boesch. 1999. Cultures in chimpanzees. *Nature* 399:682–85.

Wrangham, R. W. 1997. Subtle, secret female chimpanzees. *Science* 277:774–75.

Wrangham, R. W., A. P. Clark, and G. Isabirye-Basuta. 1992. Female social relationships and social organization of Kibale Forest chimpanzees. In T. Nishida, W. C. McGrew, P. Marler, M. Pickford, and F. B. M., de Waal, eds., *Topics in Primatology, Vol. 1: Human Origins.* Tokyo: University of Tokyo Press, 81–98.

Yamakoshi, G.. 2004. Food seasonality and socioecology in *Pan*: Are West African chimpanzees another bonobo? *African Study Monographs* 25:45–60.

Zamma, K. 2002. Leaf-grooming by a wild chimpanzee in Mahale. *Primates* 43:87–90.

14

New Theaters of Conflict in the Animal Culture Wars: Recent Findings from Chimpanzees

W. C. McGrew

At Gombe, in Tanzania, a chimpanzee en route between food-patches comes to a stream. It is clear and shallow, less than 20 cm deep and 2 m wide, but rather than wade across, the ape leaps from rock to rock, avoiding the water. His companion adopts a more circuitous detour: She climbs up a tree on the near side, clambers overhead via an intervening vine, then climbs down a tree on the far side. Meanwhile, more than half a continent away, at Fongoli in Senegal, a group of chimpanzees piles into a pool created by recent rainfall (Bertolani, personal communication). They sit immersed up to their chests, grooming and playing in a sort of "pool party." How are we to explain this extreme variation in response to surface water, from apparent hydrophobia to hedonism?

Cultural primatology has made remarkable advances since the last gathering of chimpanzees researchers in 2000 for the Animal Social Complexity conference (de Waal and Tyack 2003), but many findings have been steeped in continuing controversy (e.g., whether or not only humans depend on culture; see below), hence the derivative title used here. The aim of this essay is to scrutinize and to update new results from research on our nearest living relations, *Pan*, in the light of ongoing debate, especially as they reflect the cognitive abilities and limitations of the apes. Nuanced cultural diversity is arguably the single most impressive manifestation of the cognitive capacities of large-brained social organisms. The somewhat adversarial framework used here will be similar to the one used then (McGrew 2003), but with the acknowledgement that there have been intervening updates (McGrew 2004; Byrne et al. 2004). My viewpoint is admittedly opinionated and biased toward observational field studies; other contributors to this volume are far more qualified than I to discuss experimental studies from zoos and laboratories.

Who Is Cultural?

The humanist view that culture is uniquely human (without specifying exactly when humanity emerged: e.g., pre-*Homo*, earliest *Homo*, later *Homo*, *Homo sapiens*, or anatomically and behaviorally modern *Homo sapiens sapiens*) probably predominates, at least in the social sciences. One way to probe this is the "textbook test." That is, what are the latest editions of the mainstream textbooks that dominate the North American market and shape the minds of the current generation of undergraduates saying about cultural primatology, and especially about chimpanzees?

An unsystematic sample of textbooks written for general or cultural anthropology courses yields mixed results, with most still resisting the idea of culture in chimpanzees (much less in other species; see below). Kottak (2008) still defines culture as "distinctly human" but includes a text box from the *New York Times* that dares to label orangutan tool use as cultural. He also devotes several pages to description of wild chimpanzee behavior (ethnography by any other name; see below), but these are riddled with errors—for example, the claim that many Gombe chimpanzees never master termite fishing. Nanda and Warms (2007) are more cautious, giving over a few paragraphs to chimpanzee tool use, but making only a passing mention of the possibility of its expression being cultural. On the other hand, Bodley (2005) meets the challenge head-on with a section entitled "Chimpanzee Culture?," which gives both sides of the argument—sometimes confusedly—but avoids drawing conclusions. Similarly, Harris and Johnson (2007) grant culture to nonhumans but rely on outdated and misunderstood results on tool use only, and conclude that there is a qualitative—not quantitative—difference between human and nonhuman culture. Recurring issues in most of these treatments are whether humans rely on culture while for nonhumans it is optional, and whether human culture is cumulative while nonhuman culture is not.

Not surprisingly, introductory textbooks in biological anthropology give cultural primatology much more extensive and positive treatment and more accurate accounts (e.g., Boyd and Silk 2006, Relethford 2005, Stanford et. al. 2006).

On the other hand, a new and potentially confusing subdiscipline within anthropology has emerged in recent years: ethnoprimatology. Whereas cultural primatology is best seen as analogous to any other area of inquiry applied to primates—for example, behavioral primatology, medical primatology, and so on—ethnoprimatology is a fusion of sociocultural anthropology and primatology. It focuses on the interaction between (usually) traditional peoples and the nonhuman primates with whom they coexist in daily life, in contexts that are positive (e.g., ceremonial), negative (e.g., crop raiding), or both (e.g., ecotourism). Cormier's (2003) *Kinship with Monkeys* is a stellar example: The same spider monkeys that are hunted mercilessly yield the monkey orphans that are then reared as if they were human children.

What is clear is that the ethnography of nonhuman primates (here restricted to *Pan*) is piling up impressively, as old study-sites like Gombe are reworked and new ones like Fongoli are established. For example, after more than four decades of study, termite-fishing techniques at Gombe have

been shown to entail social learning (Lonsdorf 2005, 2006). Study of the fourth subspecies of chimpanzee, *Pan troglodytes vellerosus*, is now underway (Schoening et al. 2006), yielding the most insectivorous of all ape populations. Sanz and Morgan (2007) at Goualougo have reported that rainforest chimpanzees have novel twists on old patterns, such as using tool sets to tap reservoirs of underground termites. Nut cracking with hammer and anvil, long thought to be confined to far west Africa, has been found in Cameroon (Morgan and Abwe 2006). Finally with the first-time habituation of savanna-dwelling chimpanzees at Fongoli, a host of new patterns have come to light: spearing in bush baby hunting, picnicking in caves, etc. (e.g., Pruetz and Bertolani 2007). Bonobos are now known to hunt regularly, both for ungulates and monkeys (Hohmann and Fruth 2008). Anyone who thinks that we have finalized even the basic database for *Pan troglodytes*, much less *Pan paniscus*, is not reading *Current Biology* these days!

The universalist view is that culture is commonplace throughout the animal kingdom, being found apparently in all vertebrate classes (e.g. Laland and Hoppitt 2003, Laland and Janik 2006). This view reaches beyond the extensions of membership in the "culture club" in cultural primatology, such as *Pongo* (van Schaik et al. 2003), *Cebus* (Perry et al. 2003), *Pan paniscus* (Hohmann and Fruth 2003). It also occurs in other mammals, such as cetaceans (Rendell and Whitehead 2001), and in birds, such as the New Caledonian crow (Hunt and Gray 2003).

Universalist arguments are based on granting cultural capacity to a species once it is shown to be capable of social learning. Thus, a guppy learning the correct route through a maze by following another guppy is accorded cultural status. Therefore, for the universalists, social learning is both a necessary and a sufficient condition for culture. This makes a mockery of culture as the complex phenomenon it was originally defined to be by anthropologists, in which mechanisms such as arbitrariness, conformity, diffusion (not just dissemination), and transmission fidelity (or lack of it) are needed to try to explain institutions or conventions such as ostracism, xenophobia, ritual, and taboos (McGrew 2004). Such a dumbed-down version of culture may be needed for some organisms, but for humans and (arguably) great apes and cetaceans, social learning is merely the starting point—that is, a necessary condition but not a sufficient one.

Continuing Issues in Culturology

In recent years the term "tradition" has increasingly been used interchangeably with, or as an alternative to, the term "culture" (e.g., Perry et al. 2003). The former usage is confusing, and the latter requires careful definition. If tradition is thought of as cumulative cultural change over generations through vertical or oblique transmission (a strong definition; see McGrew 2004), then clearly it cannot be synonymous with culture, as culture can also be intragenerational and horizontally transmitted (Richerson and Boyd 2005). If tradition is an alternative to culture, and is defined weakly as "socially learned behavioral diversity," then important distinctions regarding long-term temporal continuity are lost in vagueness. If tradition is preferred as a label for what nonhumans do while culture is reserved for what humans do, then transparent working definitions are needed, or the dichotomy smacks of speciesism.

Teaching has been offered more than once as a uniquely human mechanism of cultural transmission, but like all complex cognitive phenomena it must be handled with care (McGrew 2004; Csibra 2007). Caro and Hauser's (1992) heuristic criteria are hard to test empirically, and in the intervening years alternative definitions have been broached. The upshot of this proliferation is that teaching is not an either/or phenomenon, but instead a constellation of mechanisms (facilitate, mould, scaffold, train, etc.) that have in common the (apparently) purposive transmission of knowledge from informed tutor to naïve pupil, the success of which is judged by the pupil's improved performance. Given this, claims for teaching have been advanced for a range of species from meerkats (Thornton and McAuliffe 2006) to ants (Leadbetter et al. 2007). Teaching may be overrated as a means of cultural transmission, but its employment (or lack of it) is still to be documented in apes (McGrew 2004).

It is received wisdom in anthropological textbooks that at least for humans, language and culture are inextricable. For non-linguistic nonhumans, this obligatory link could be a damning exclusion from the culture club, if language were both a necessary and sufficient condition for culture. Recent findings from field and laboratory studies of chimpanzees suggest, however, that the picture is far richer than we realized. No longer do we need to rely on the contrived data from "pongo-linguistics"—that is, the teaching of artificial languages to apes in laboratories. Instead, there is a new generation of elegant, often experimental studies of natural communication (Slocombe and Zuberbuehler 2005, 2007; see also chapter 16) suggesting that the gap between human and nonhuman communication is not so wide. If these phenomena turn out to be enabled or even modified by cultural transmission, then a rethinking may be in order.

Reliance on culture is repeatedly asserted to be uniquely human (see above), apparently on the basis of its obviously crucial role in human diversity and extreme adaptiveness. Humans, with their widespread global distribution, could not survive (it is claimed) in polar climes or barren deserts without ingenious technological innovations (although of course other mammals do so without such technology). However, the question is not whether all human societies rely on culture, which is hardly disputable, but whether other species might also be similarly dependent. Logically, there is an obvious test of the idea: Find a thriving population of wild chimpanzees that lacks culture, and the hypothesis of reliance is disproven. No such population has yet been found, from the hot, dry, and open savannas to the cool, wet, and closed rain forests, as the chimpanzee ethnographic record continues to accumulate. The most parsimonious summary of current knowledge is that both humans and chimpanzees—and perhaps other species— are adapted to be cultural.

Almost a decade ago, Tomasello (1999) proposed the ratchet effect as a distinguishing feature of human culture: unlike nonhumans, whose social learning is limited to the here and now, humans accumulate their culture with incremental improvements built on prior accomplishments. (This notion of relentless progress as being inherent in the cultural process rings some alarm bells for anthropologists, who have argued strongly that it is simplistic and potentially prejudicial.) In fact, several primate species, such as the Japanese macaque (Hirata et al. 2001; Nahallage and Huffman 2007), as well as nonprimate species such as the New Caledonian crow (Hunt and Gray 2003), show evidence of cumulative culture. Much of the human evidence of the ratchet effect comes from prehistory, and now that chimpanzees have been shown to have an archaeological record (Mercader et al. 2007), another avenue of investigation is opened up.

None of these issues shows consensus, and for none is the evidence yet conclusive. But all of them are live wires, subject to empirical, hypothesis-driven study. No longer is it possible merely to dismiss them a priori; instead tighter, evidence-driven arguments are the mainstream of cultural primatology.

Vexed Definition

It is now commonplace to say that culture has been defined many ways and that there is no consensus, but this does not stop researchers and scholars from wrangling over definitions. Alarmingly, instead of convergence upon essential elements, there seems to be picayunish proliferation, reminiscent of the hair splitting that plagues hominin taxonomy, when each scientist champions his or her chosen taxon. I will not try to disentangle this spaghetti here, but some comments may be helpful. One approach is oversimplification, as in my epigrammatic five-word phrase, "the way we do things" (McGrew 2004, p. 24); this distillation sought to focus efforts on essentials. Another approach is to list elements and features (each of which demands its own definition); but if only one of these is problematic, then the whole constellation suffers accordingly. Consider Laland and Janik's (2006, p. 542) definition of culture: ". . . all group-typical behaviour patterns, shared by members of animal communities, that are to some degree reliant on socially learned and transmitted information." At first glance this seems operationally useful, but the final element is a problem: What exactly is *information* here, and how is one to recognize and record it, especially in nature?

It is likely that as with other complicated phenomena (e.g. language, intelligence, etc.), seeking a single, all-encompassing definition is like questing for a holy grail. Instead, it must be recognized that different definitions are needed for different exercises, so that operational definitions are crucial for empirical studies, just as conceptual definitions are needed for theoretical explorations.

But how to choose among the array of elements? I suggest that the crux of culture, in the richest sense of the term—which critics will doubtless dismiss as being anthropocentric (e.g., Laland and Janik 2006, Byrne 2007)—is collectivity, and ultimately therefore identity. That is, it seems likely that natural (or, in this case, group; see Richerson and Boyd 2005) selection has advantaged collectives of cognitively advanced creatures who ultimately showed more reproductive success than less cohesive aggregations of individuals. Further, I suggest that the key mechanism

in the functioning of this collectivity is self-perceived common identity—that is, a way to conceive of being "us versus them." To reach this collective status presumably involved an evolutionary process starting from group-typical statistical norms but progressing to institutional rituals such as rites of passage and linguistic labels. Such collectivity-based identity demands a certain level of social cognition that includes mutual self-awareness (for an evolutionarily based treatment of identity, see McElreath et al. 2003).

Note that this is more than shared dyadic idiosyncrasy (e.g., my godfather taught me to fish in this particular way, however unusual it may be), lineage bias (e.g., in our family we always use this recipe, not any other), or even polyadic bonding (e.g., we prefer sticking together as a community, clan, or family, no matter what). Chimpanzee examples of these might be daughters learning maternal caregiving techniques from their mothers, adolescent males copying the display characteristics of the alpha male, termite-fishing techniques being passed down matrilineally (Lonsdorf 2005, 2006), or individuals coming together to nest in the same tree or grove. All of these phenomena can be accounted for in individual selectionist terms, so none is conclusively cultural.

So how should one seek to test collectivity and identity? This calls for the same sort of imaginative ingenuity that led Byrne and Whiten (1988) to operationalize tactical deception. For chimpanzees, one might start by focusing on the interactive details of social hunts and boundary patrols as well as on the details of emigration and immigration, by longitudinally following transferring females and their behavioral modification, or lack of it, as they move from natal to breeding community.

Discerning Culture

Even if we could agree that chimpanzees in principle might be culture bearers, there remains the task of recognizing chimpanzee culture in practice. Disagreement abounds. Experimentalists working on simpler taxa in the controlled circumstances of captivity have always queried the observational data of field workers (Galef 1992; Laland and Hoppitt 2003; Laland and Janik 2006). Essentially they have posed an inescapably catch-22–like syllogism: Alternative (to cultural) explanations to behavioral complexity and flexibility can be tested only in controlled experimental settings, and conclusive experiments in nature are logically impossible, so field research therefore can never demonstrate the existence of culture.

Consider the issue of behavioral diversity and what it means for culture. Whiten et al. (2001 and chapter 8) showed that behavior varies across populations of chimpanzees, in comparing up to nine groups of chimpanzees from across Africa. In seeking to explain diversity, they sought to exclude the relatively less interesting possibilities that it resulted from either genetic or environmental determinism. What remained were variants that were candidates for cultural traits. This approach has a long history in primatology: It seems to have been first put forward in typically elegant style by Kummer (1971, pp. 9–16) as a field worker, but it was echoed by Galef (1976, 1977) as a laboratory worker. Both focused on the trinity of genes, environment, and sociality; and the application of these three alternatives by Whiten et al. (2001) was an attempt to put them to the test using ethnographic data.

It was always clear, however, that this trio of elements was not so easily distinguishable either in principle or in practice. Logically, genetic influences can be elucidated through population genetic approaches such as experimental cross-fostering of breeding and rearing, although this is ethically repugnant and logistically impractical for apes, especially in nature. Similarly, genomic approaches might isolate a set of genes responsible for cognitive culture processes, and this could be pursued through genetic engineering, especially now that the human and chimpanzee genomes are known. However, the exclusion of environmental influences can never be complete, and even if one sought to control for every relevant independent environmental variable, one could never be sure of not having missed the crucial one that explained the variation. By definition, this control could never be done in nature, but any results achieved in captivity are suspect on the grounds of lack of ecological validity. So the ground is prepared for endless debate.

Consider the example of ant dipping, the form of elementary technology by which chimpanzees harvest army ants (McGrew 1974). This form of extractive foraging is present in some wild populations of apes and absent in others, while all co-occur with the *Dorylus* ants. On this basis, ant dipping was proposed to be cultural (McGrew 1992). Continuing ethnography from other populations yielded further variation in both technique and tools, which was interpreted as further evidence for cultural diversity. Then,

Humle and Matsuzawa (2002) reported that ant dipping varied as a function of the ants (part of the environment), in that some types of ants reacted differently to chimpanzee predation. This raised the possibility that variation in the predator's behavior was being "driven" by that of the prey, and so it need not have been socially learned or transmitted. Critics immediately seized on these new data to dismiss ant dipping as no longer being cultural (Laland and Hoppitt 2003; Byrne 2007). To an anthropologist this was a strange conclusion, as if anyone had ever claimed that human culture was purely social and independent of the environment—for example, that dietary taboos were totally unrelated to foodstuffs (cf. Aunger 1994). As it happens, Humle and Matsuzawa's results spurred a new, wider, and more detailed study of sympatric *Pan* and *Dorylus*, encompassing 13 populations of apes who exploited the ants (Schoening et al. 2007). Their results confirmed Humle and Matsuzawa's findings, but also showed that even when the suite of ant species present is identical in two populations of chimpanzees, there are still differences in the tools and techniques used to acquire the prey. Similarly, they found that the same tools and techniques are used to harvest different sorts of ants at different sites (see chapter 10).

Some confusion arises from the lack of an accepted term for the analytical approach advocated by Kummer and Galef (see above). Kruetzen et al. (2007) called it the "method of exclusion," which is accurate but not explicit enough. Byrne (2007) termed it "patterns of local ignorance." Others (Laland and Janik 2006; Byrne 2007) have termed it the "ethnographic method," but this sows discord in several ways. One is that ethnographic methods are well established as the core of sociocultural anthropology, where they are taught in every undergraduate curriculum as applying only to humans. Ethnography in this sense refers to the process of description whereby the ethnographic record of human behavioral diversity is compiled. This is typically done by interviewing informants from other cultures. The resulting database is referential, not analytical, and its investigatory use is termed ethnology. Thus the phrase has a prior use, established by precedence.

To suggest that cultural primatologists use the ethnographic method is true in the sense that they, too, are systematically compiling a compendium of databases on different populations of primates, species by species. Equally, cultural primatologists are beginning to use this compendium to do ethnological studies, as in van Schaik et al.'s (2003) preliminary test of geographical proximity as an alternative explanation for cultural similarity. However, the particular form of analysis-by-exclusion outlined above is foreign to sociocultural anthropologists, who typically discount any form of genetic determinism in their explanations of behavioral diversity. To characterize cultural primatology's use of the method of exclusion as being "ethnographic" is to misrepresent it, and to stir up already troubled waters with regard to cultural primatology's relationship to its cognate discipline of cultural anthropology.

Byrne (2007) has recently advanced an alternative view for tackling at least the material culture of nonhuman species, which emphasizes the technical demands of foraging in primate life. He advocates focusing on behavior that shows intricate complexity and local near-ubiquity, at least as a way of modeling the evolutionary origins of human technology. His examples are from food processing, however, and not from tool use.

So how are we to advance cultural primatology? One way is to apply cladistic methods that have proven successful in explaining diversity in human material culture—although ultimately drawn from evolutionary biology—to chimpanzee databases. Such systematic analyses can help to disentangle the nature and nurture of culture—for example, by comparing intra- and interspecific variation in trait expression (Lycett et al. 2007).

Expanding Cultural Primatology

Until recently, it could be said that cultural primatology was trapped in the present, so that any application of archaeological methods or principles was limited to ethoarchaeology (McGrew et al. 2003). That is, one could use indirect approaches developed to deal with ephemeral phenomena from the past as a way to deal with elusive phenomena in the present. Extinct creatures no longer think or behave, but some of their artifacts (as the products of thought and behavior) persist and so can act as proxies. Similarly, absent creatures such as unhabituated apes yield little or no behavioral data, but their artifacts can be revealing. The methods of analysis may be identical, especially if applied to the same sort of objects, such as stone tools.

Now we have an archaeological record for the chimpanzee. That is, we have recovered artifacts by excavation from sites that can be dated by standard archaeometric techniques, such as radiocarbon dating, that range from hundreds (Mercader et al. 2002) to thousands (Mercader et al. 2007) of years old. Such archaeology yields modified lithic materials and organic residues that give clues to their function. Finally, we may be able to tackle key questions about possible interactions between ancestral apes and humans: Did apes learn techniques from humans? Or humans from apes? Or neither, with both converging on similar solutions to the same tasks? Resolving these alternatives requires being able to distinguish between the two types of archaeological record, which is no small challenge (e.g., Harris 2006).

At the last gathering of chimpanzee researchers in Chicago (the Animal Social Complexity conference), it could be said that solid data on the mechanisms involved in social learning were better known for rodents or fish than for primates. "Solid" here means well-designed, replicable experiments that meet criteria for reliability and validity, and that enable the testing of predictions drawn from well-grounded hypotheses. Now such results are available for chimpanzees, from a series of studies of mechanisms of transmission done on laboratory populations of group-living apes (see chapter 8). Whiten et al. (2005) showed the spontaneous emergence of conformity in within-group acquisition of a novel foraging task that offered two alternative solutions. Horner et al. (2006) found multiple faithful replications along transmission chains of the same sort of task. Bonnie et al. (2007) showed that similar processes applied to arbitrary acts and not just problem-solving solutions. Hopper et al. (2007) found that by using a "ghost control" condition, these transmissions were based on imitation and not emulation. Whiten et al. (2007) demonstrated that similar transmission could occur not only within groups but across chains of groups. Thus the experimental data on the processes underlying cultural transmission are now better known for chimpanzees than for any other species, including *Homo sapiens*.

Implications and Future Directions

So where is cultural primatology, especially as applied to chimpanzees, to go from here? It is risky to try to be prescriptive, but useful signposts may be derived from the evolution of cultural anthropology, which in less than a century made the transition from natural history to ethnography to ethnology (McGrew 2004). Something comparable has happened in cultural primatology, albeit sometimes unwittingly. Contrast, for example, the treatments of ape tool use in Goodall (1964) and in van Schaik and Pradhan (2003). However, not all precedents need to be followed: Current sociocultural anthropology largely has become theoretical and has rejected empirical research, even to the extent of becoming antiscientific (Mesoudi et al. 2006).

Here are some speculative thoughts, following the order of presentation of issues in this essay:

Armchair cultural primatologists might consider monitoring the continuing presentation of cultural primatology in textbooks, whether these be in general, sociocultural, archaeological, or biological anthropology. These could be contrasted with corresponding treatments of the discipline in biology (e.g., ethology) and psychology (e.g., comparative). Similarly, one might compare such reporting in North America with that in the rest of the world, where big, comprehensive textbooks play a far less important role in higher education. Why concentrate on textbooks? Because like it or not, and for whatever reason, they are the first and only point of contact for most students seeking academic knowledge about primates. My sporadic inquiries into textbook presentation of cultural primatology are discouraging in their findings of inaccuracy, perhaps because most textbook authors (with important exceptions, e.g., Stanford et al. 2006, Boyd and Silk 2006) are not primatologists.

The emerging field of ethnoprimatology could articulate usefully with cultural primatology. Cultural change by one party probably influences the cultural status of the other, and vice versa, in what is likely to be a perpetual feedback loop (Hockings et al. 2006, 2007, chapter 27). That is, if apes develop new tactics for crop raiding in response to humans clearing forests and planting new types of crops, then farmers will develop new responses to those tactics, and so on. This might create a cognitive as well as behavioral "arms race"—for example, as expressed in perceptions and attitudes directed toward the other party. Such research demands training that bridges disciplines that normally go their separate ways. How many primatologists are well informed about sociocultural anthropology, and vice versa? Cormier (2003) has shown that it can be done, in her study of Guaja foragers as hunters and howler monkeys as prey in the Amazon, but her work remains exceptional. Such research needs ethological methods, as it is not enough to interview informants about their impressions. Why is all this important? Because such arms races may eventually result in one or both parties losing out. Crop raiders may be exterminated, and villagers reliant on forest products may find that their resources disappear—for example, in the absence of seed-dispersing primates who are necessary for their regeneration.

To study traditions, there is no substitute for longitudinal study that follows individuals and groups through successive generations. There is no shortcut to this, but archival data available on the web would greatly facilitate cross-cultural research. Ultimately what will be needed is some chimpanzee equivalent to the Human Relations Area File and Standard Cross-Cultural Sample (McGrew 2004). With more than 50 field sites at which chimpanzees have been studied, who otherwise can keep straight the various data sets that have been laboriously collected at no small cost in blood, sweat, and tears? The Collaborative Chimpanzee Cultures Project (CCCP) is leading the way (see chapter 8), but it covers only a fraction of groups of a fraction of populations of chimpanzees. In principle, a comparable range of behavioral diversity in captivity is available from the massive accumulation of data in the ChimpanZoo database maintained by the Jane Goodall Institute, but analyses of these results in mainstream publications have yet to materialize.

The extent to which teaching is important in cultural transmission in primates remains to be seen, as it is deucedly difficult to isolate in field conditions. That is, depending on definition, to rule out other means of social learning without being able to control the knowledge states of individuals, and without being sure of their prior experience, is a stiff challenge for field workers. Studies in captive environments, especially of apes living in demographically valid groups, could give answers. What remains to be done is an experiment in which an informed individual is put into a context where it is advantageous to pass on knowledge (not objects or behaviour patterns) to another, less informed individual. If this can be induced, and if it then generalizes from the training context to a novel testing one, then teaching will have been shown.

Studies of vocal communication in chimpanzees have gone beyond long, loud calls (i.e., pant-hoots), and the benefits of such studies are clear, so one hopes that this trend will continue—to include, for example, the soft grunts that occur so often in peaceful social contexts. Diversity in vocal learning, as with grooming sounds (Nishida et al. 2004), needs to be pursued, especially as CCCP is not yet tackling this area of chimpanzee behavior. However, logistical and ethical constraints apply to some aspects of this research, such as broadcast and playback paradigms. Nonvocal acoustic (e.g., drumming) and spontaneous gestural (e.g., begging) communication require more detailed investigation. The extent to which any or all of these types of research are relevant to questions of language remains to be seen. For example, seeking recursiveness in the natural utterances of nonhumans is a challenge (Hauser et al. 2002).

Knowledge of the extent to which primates rely on cultural knowledge and practice clearly needs to go beyond presence/absence data on individual features or overall repertoires. When culture is crucial may be revealed in extreme or marginal environments, such as near drinking water in savanna sites, or in extreme perturbations in climatic conditions, such as drought in normally well-watered sites. However, these data will always be correlational and cannot be causal, given the lack of opportunity to exercise random assignment to treatment in the field. Studies in captivity may show the way, perhaps by "seeding" different groups of primates with cultural practices of varying degrees of efficacy and then following the differential payoffs and their impact.

Such research could well be combined with experimental investigation of the ratchet effect. (None of the field studies cited above, whether of ape or crow, has actually demonstrated cultural change over time. Instead they have inferred past change on the basis of current diversity that varies in complexity, which is sometimes graded geographically over a species' range.) Groups seeded with the same elementary extractive foraging techniques can be followed to see whether innovation occurs spontaneously or can be manipulated, and to what effect (Reader and Laland 2003). There have been notable suggestive studies of innovation, but the key is to extend these to cumulative cultural change.

Many of the questions of the origins of behavioral diversity—that is, what combination of genetic, environmental, or social influences yields cultural processes—cannot be answered conclusively with currently used meth-ods. For example, we surely need to follow emigrating female chimpanzees as they disperse and then immigrate elsewhere, but this requires habituated neighboring groups and decades of patient ontogenetic research. Even then, the determining variables cannot be totally disentangled. Experimental translocation of such females would sharpen the picture, but we resist this on ethical and economical grounds, at least in nature. These constraints do not apply to human-initiated transfers in captivity, although cultural factors seem not to be taken into account in species survival plans (SSPs) for management of captive populations. Such cultural variants exist, as in the case of bonobos from San Diego Zoo spreading hand clapping, but these seem not to have been studied systematically as cultural diffusion.

Interdisciplinary cooperation in cultural primatology remains largely untapped, but pioneering studies (Mercader et al. 2002, 2007) show the way. The average primatologist is clueless about how to document a work site in terms that would allow comparison with archaeological sites, past or present. The average paleoanthropologist has never seen a wild ape, much less collected systematic data on obviously relevant features of chimpanzee behavior, such as the use of anvils to process hard-shelled fruits (as opposed to the far more famous use of hammer and anvil to process nuts) (McGrew et al. 2003). The solution is simple. Send primatologists on archaeological digs and send palaeoanthropologists to chimpanzee field sites, and have them work side by side. Instead of talking retrospectively, why not plan truly collaborative projects prospectively? The same sorts of arguments apply within cultural primatology, of course, with regard to exchanges between researchers in nature and in captivity.

All in all, the future of cultural primatology is bright, being increasingly well grounded in accumulating findings on the chimpanzee mind as studied both in captivity and in nature.

Acknowledgments

The author thanks Linda Marchant for critical comments; Caroline Phillips and Tim Webster for help with manuscript preparation; and the National Science Foundation (HOMINID Program) for a Revealing Hominid Origins Initiative grant to the late F. Clark Howell and Tim D. White for funding.

Literature Cited

Aunger, R. 1994. Are food avoidances maladaptive in the Ituri Forest of Zaire? *Journal of Anthropological Research* 50:277–310.

Bodley, J. H. 2005. *Cultural Anthropology: Tribes, States, and the Global System.* 4th edition. New York: McGraw-Hill.

Bonnie, K. E., V. Horner, A. Whiten, and F. B. M. de Waal. 2007. Spread of arbitrary conventions among chimpanzees: A controlled experiment. *Proceedings of the Royal Society B* 274:367–72.

Boyd, R., and J. B. Silk. 2006. *How Humans Evolved.* 4th edition. New York: W. W. Norton.

Byrne, R. W. 2007. Culture in great apes: Using intricate complexity in feeding skills to trace the evolutionary origin of human technical prowess. *Philosophical Transactions of the Royal Society B* 362:577–85.

Byrne, R. W., P. J. Barnard, I. Davidson, V. M. Janik, W. C. McGrew, A. Miklosi, and P. Wiessner. 2004. Understanding culture across species. *TRENDS in Cognitive Sciences* 8:341–46.

Byrne, R. W., and A. Whiten. 1988. *Machiavellian Intelligence: Social Expertise and the Evolution of Intellect in Monkeys, Apes, and Humans.* Oxford: Oxford University Press.

Caro, T. M., and M. D. Hauser. 1992. Is there teaching in nonhuman animals? *Quarterly Review of Biology* 67:151–74.

Cormier, L. 2003. *Kinship with Monkeys: The Guaja Foragers of Eastern Amazonia.* New York: Columbia University Press.

Csibra, G. 2007. Teachers in the wild. *TRENDS in Cognitive Sciences* 11:95–96.

De Waal, F. B. M., and P. J. Tyack, eds. 2003. *Animal Social Complexity: Intelligence, Culture, and Individualized Societies.* Cambridge, MA: Harvard University Press.

Galef, B. G. 1976. Social transmission of acquired behaviour: A discussion of tradition and social learning in vertebrates. *Advances in the Study of Behavior* 6:77–100.

———. 1992. The question of animal culture. *Human Nature* 3:157–78.

Goodall, J. 1964. Tool-using and aimed throwing in a community of free-living chimpanzees. *Nature* 201:1264–66.

Harris, D. R. 2006. The interplay of ethnographic and archaeological knowledge in the study of past human subsistence in the tropics. *Journal of the Royal Anthropological Institute* 12:S63–78.

Harris, M., and O. Johnson. 2007. *Cultural Anthropology.* 7th edition. Boston: Pearson.

Hauser, M. D., N. Chomsky, and W. T. Fitch. 2002. The faculty of language: What is it, who has it, and how did it evolve? *Science* 298:1569–79.

Haviland, W. A., H. E. L. Prins, D. Walrath, and B. McBride. 2007. *The Essence of Anthropology.* Belmont, CA: Thomson Wadsworth.

Hirata, S., K. Watanabe, and M. Kawai. 2001. "Sweet-potato washing" revisited. In *Primate Origins of Human Cognition and Behavior,* 487–508. Tokyo: Springer.

Hockings, K. J., J. R. Anderson, and T. Matsuzawa. 2006. Road crossing in chimpanzees: A risky business. *Current Biology* 16: R668–70.

Hockings, K. J., T. Humle, J. R. Anderson, D. Biro, C. Sousa, G. Ohaski, and T. Matsuzawa. 2007. Chimpanzees share forbidden fruit. *PLoS One* 2:e886.

Hohmann, G., and B. Fruth. 2003. Culture in bonobos? Between-species and within-species variation in behaviour. *Current Anthropology* 44:563–71.

———. 2008. New records on prey capture and meat eating by bonobos at Lui Kotal, Salonga National Park. *Folia Primatologica* 79:103–10.

Hopper, L. J., A. Spiteri, S. P. Lambeth, S. J. Schapiro, V. Horner, and A. Whiten. 2007. Experimental studies of traditions and underlying transmission processes in chimpanzees. *Animal Behaviour* 73:1021–32.

Horner, V., W. Whiten, E. Flynn, and F. B. M. de Waal. 2006. Faithful replication chains of foraging techniques along cultural transmission chains by chimpanzees and children. *Proceedings of the National Academy of Sciences of the United States of America* 103:13878–83.

Humle, T., and T. Matsuzawa. 2002. Ant-dipping among the chimpanzees of Bossou, Guinea, and comparisons with other sites. *American Journal of Primatology* 58:133–48.

Hunt, G. R., and R. D. Gray. 2003. Diversification and cumulative evolution in tool manufacture by New Caledonian crows. *Proceedings of the Royal Society B* 270:867–74.

Kottak, C. P. 2008. *Cultural Anthropology.* 12th edition. Boston: McGraw-Hill.

Kruetzen, M., C. van Schaik, and A. Whiten. 2007. The animal cultures debate: Response to Laland and Janik. *TRENDS in Ecology and Evolution* 22:6.

Kummer, H. 1971. *Primate societies: Group Techniques of Ecological Adaptation.* Chicago: Aldine.

Laland, K. N., and W. Hoppitt. 2003. Do animals have culture? *Evolutionary Anthropology* 12:150–59.

Laland, K. N., and V. M. Janik. 2006. The animal cultures debate. *TRENDS in Ecology and Evolution* 10:542–47.

Leadbetter, E., N. E. Raine, and L. Chittka. 2006. Social learning: Ants and the meaning of teaching. *Current Biology* 16:R323–25.

Lonsdorf, E. V. 2005. Sex differences in the development of termite-fishing skills in the wild chimpanzees, *Pan troglodytes schweinfurthii,* of Gombe National Park, Tanzania. *Animal Behaviour* 70:673–83.

———. 2006. What is the role of mothers in the acquisition of termite-fishing behaviors in wild chimpanzees (*Pan troglodytes schweinfurthii*)? *Animal Cognition* 9:36–46.

Lycett, S. J., M. Collard, and W. C. McGrew. 2007. Phylogenetic analyses of behavior support existence of culture among wild chimpanzees. *Proceedings of the National Academy of Sciences of the United States of America* 104:17588–92.

McElreath, R., R. Boyd, and P. J. Richerson. 2003. Shared norms and the evolution of ethnic markers. *Current Anthropology* 44:122–29.

McGrew, W. C. 1974. Tool use by wild chimpanzees in feeding upon driver ants. *Journal of Human Evolution* 3:501–8.

———. 1992. *Chimpanzee Material Culture: Implications for Human Evolution.* Cambridge: Cambridge University Press.

———. 2003. Ten dispatches from the chimpanzee culture wars. In *Animal Social Complexity: Intelligence, Culture, and Individualized Societies,* ed. F. B. M. de Waal and P. L. Tyack, 419–39. Cambridge, MA: Harvard University Press.

———. 2004. *The Cultured Chimpanzee: Reflections on Cultural Primatology.* Cambridge: Cambridge University Press.

McGrew, W. C., P. J. Baldwin, L. F. Marchant, J. D. Pruetz, S. E. Scott, and C. E. G. Tutin. 2003. Ethoarchaeology and elementary technology of unhabituated wild chimpanzees at Assirik, Senegal, West Africa. *PaleoAnthropology* 1:1–20.

Mercader, J., H. Barton, J. Gillespie, J. Harris, S. Kuhn, R. Tyler, and C. Boesch. 2007. 4,300-year-old chimpanzee sites and the origins of percussive stone technology. *Proceedings of the National Academy of Sciences of the United States of America* 104:3043–48.

Mercader, J., M. A. Panger, and C. Boesch. 2002. Excavation of a chimpanzee stone tool use site in the African rainforest. *Science* 296:1452–55.

Mesoudi, A., A. Whiten, and K. N. Laland. 2006. Towards a unified science of cultural evolution. *Behavioral and Brain Sciences* 29:329–83.

Morgan, B. J., and E. E. Abwe. 2006. Chimpanzees use stone hammers in Cameroon. *Current Biology* 16:632–33.

Nahallage, C. A. D., and M. A. Huffman. 2007. Acquisition and development of stone handling behaviour in infant Japanese macaques. *Behaviour* 144:1193–1215.

Nanda, S., and R. L. Warms. 2007. *Cultural Anthropology*. Belmont, CA: Thomson Wadsworth.

Nishida, T., J. C. Mitani, and D. P. Watts. 2004. Variable grooming behaviours in wild chimpanzees. *Folia Primatologica* 75:31–36.

Perry, S. E. 2006. What cultural primatology can tell anthropologists about the evolution of culture. *Annual Review of Anthropology* 35: 171–90.

Perry, S. E., M. Baker, L. Fedigan, J. Gros-Louis, K. Jack, K. C. MacKinnon, J. H. Manson, M. A. Panger, K. Pyle, and L. Rose. 2003. Social conventions in wild white-faced capuchin monkeys: Evidence for behavioral traditions in a neotropical primate. *Current Anthropology* 44:241–68.

Pruetz, J. D., and P. Bertolani. 2007. Savanna chimpanzees, *Pan troglodytes verus*, hunt with tools. *Current Biology* 17:412–17.

Reader, S. M., and K. N. Laland, eds. 2003. *Animal Innovation*. Oxford: Oxford University Press.

Relethford, J. H. 2005. *The Human Species: An Introduction to Biological Anthropology*, 6th edition. Boston: McGraw-Hill.

Rendell, L., and H. Whitehead. 2001. Culture in whales and dolphins. *Behavioral and Brain Sciences* 24:309–82.

Richerson, P. J., and R. Boyd. 2005. *Not by Genes Alone: How Culture Transformed Human Evolution*. Chicago: University of Chicago Press.

Sanz, C. M., and D. B. Morgan. 2007. Chimpanzee tool technology in the Goualougo Triangle, Republic of Congo. *Journal of Human Evolution* 52:420–33.

Schoening, C., D. Ellis, A. Fowler, and V. Sommer. 2006. Army ant availability and consumption by chimpanzees (*Pan troglodytes vellerosus*) at Gashaka (Nigeria). *Journal of Zoology* 271:125–33.

Schoening, C., T. Humle, Y. Moebius, and W. C. McGrew. 2007. The nature of culture: Technological variation in chimpanzee predation on army ants. *Journal of Human Evolution* 55:48–59.

Slocombe, K. E., and K. Zuberbuehler. 2005. Agonistic screams in wild chimpanzees (*Pan troglodytes schweinfurthii*) vary as a function of social role. *Journal of Comparative Psychology* 119:67–77.

———. 2007. Chimpanzees modify recruitment screams as a function of audience composition. *Proceedings of the National Academy of Sciences of the USA* 104:17228–33.

Stanford, C. B., J. S. Allen, and S. C. Anton. 2006. *Biological anthropology. The Natural History of Humankind*. Upper Saddle River, NJ: Pearson Prentice Hall.

Thornton, A., and K. McAuliffe. 2006. Teaching in wild meerkats. *Science* 313:227–29.

Tomasello, M. 1999. The human adaptation for culture. *Annual Review of Anthropology* 28:509–29.

Van Schaik, C. P., M. Ancrenaz, G. Borgen, G. Galdikas, C. D. Knott, I. Singleton, A. Suzuki, S. S. Utami, and M. Merrill. 2003. Orangutan cultures and the evolution of material culture. *Science* 299:102–5.

Van Schaik, C. P., and G. R. Pradhan. 2003. A model for tool-use traditions in primates: Implications for the coevolution of culture and cognition. *Journal of Human Evolution* 44:645–64.

Whiten, A., J. Goodall, W. C. McGrew, T. Nishida, V. Reynolds, Y. Sugiyama, C. E. G. Tutin, R. W. Wrangham, and C. Boesch. 2001. Charting cultural variation in chimpanzees. *Behaviour* 138:1481–1516.

Whiten, A., V. Horner, and F. B. M. de Waal. 2005. Conformity to cultural norms of tool use in chimpanzees. *Nature* 437:737–40.

Whiten, A., A. Spiteri, V. Horner, K. E. Bonnie, S. P. Lambeth, S. J. Schapiro, and F. B. M. de Waal. 2007. Transmission of multiple traditions within and between groups of chimpanzees. *Current Biology* 17:1038–43.

PART III

Social Minds:
Ecological Perspectives

15

Chimpanzee Minds in Nature

John C. Mitani, Sylvia J. Amsler, and Marissa E. Sobolewski

A small party of chimpanzees gathers during the middle of the day. Bartok, the alpha male, grooms quietly with his long-term friend and ally, Hare. The two rest comfortably beside each other, reaffirming their social bond while a third chimpanzee, the beta male Hodge, sits in the distance surveying the group. Just hours earlier, these three joined several other males on a patrol deep into their neighbors' territory, where together they encountered, attacked, and killed a rival male.

Suddenly another party of chimpanzees calls, having found a nearby tree laden with ripe fruit. Startled, Bartok and Hare cease grooming and quickly move off in the direction of the calls to join the others. When they arrive at the fruit tree, they form a coalition and chase off Hodge, with whom they had earlier fought side-by-side against the neighboring male. In this series of constantly changing social and ecological contexts, how are chimpanzee minds adapted to meet the challenges of their natural world?

Chimpanzees fascinate the scientific and lay communities alike, in part because of their evolutionary relationship with humans. They are our closest living relatives, and as a result they resemble us in several ways. Anatomical similarities between chimpanzees and humans were described more than 150 years ago (Huxley 1863), while striking genetic similarities between them and us

have been documented more recently (King and Wilson 1975; Chimpanzee Sequencing and Analysis Consortium 2005). Similarities between chimpanzees and humans are also known to extend to behavior. Jane Goodall's pioneering observations (1963) of wild chimpanzees making and using tools, and hunting and eating meat, fundamentally changed our definition of what it is to be uniquely human. As other chapters in this volume attest, behavioral parallels exist with respect to cooperation (see chapters 20 and 21), communication (see chapter 16), and even culture (see chapters 8, 9, 14, and 23).

Behavior is an observable aspect of the phenotype, and it is a relatively simple exercise to document behavioral similarities between chimpanzees and humans. Probing the minds of chimpanzees and seeking clues as to whether similarities exist therein is much more difficult. Fieldwork alone is unlikely to improve our understanding of the chimpanzee mind. As shown elsewhere in this volume, findings along these lines will emerge from experimentation conducted in captivity, rather than from behavioral observations made in the wild. Observations of the behavior of wild chimpanzees are nevertheless important for studies of the chimpanzee mind for two conceptual and empirical reasons. Conceptually, field observations provide an ecological framework to guide the design of experiments. Empirically, they reveal the kinds of problems chimpanzee minds must solve in the real world. In this paper, we review some social and ecological challenges chimpanzees face in the wild. A consideration of these problems leads directly to a set of questions about how chimpanzees use their minds in nature. We highlight these questions and propose that pursuing answers to them will yield insights into the cognitive processes of our closest living relatives. Other animals deal with some of the same social and ecological problems that confront chimpanzees, and in our discussion we review how they respond to these challenges. This comparison suggests that several parallels are likely to exist between the cognitive skills of chimpanzees and other animals, and also highlights how chimpanzees and other organisms differ. Our goal in this chapter is to pose questions that reveal gaps in our understanding of the chimpanzee mind. We do not intend to provide answers to these questions per se, but instead use them to highlight potential avenues for further investigation. Other contributors in this section furnish additional insights into some of these questions.

The examples that we will use are drawn largely from fieldwork that we have conducted at Ngogo, Kibale National Park, Uganda. During the past 15 years, we have carried out observations on an unusually large community of chimpanzees there (Mitani et al. 2002a; Mitani 2006). Struhsaker (1997) provides a detailed description of the Ngogo study site. The Ngogo chimpanzee community has consisted of approximately 140 to 160 individuals during the past 15 years of our study. The extremely large size of the community makes it relatively easy to find and follow chimpanzees there, thus reducing some of the logistical problems typically associated with fieldwork.

Like many other organisms, chimpanzees are confronted with a series of ecological and social problems. These challenges have frequently been hypothesized to act as primary selective factors that account for the evolution of intelligence in primates and other organisms (Whiten and Byrne 1997). In order to grow, maintain themselves, and survive, chimpanzees must acquire enough food to satisfy their nutritional requirements, take effective steps to avoid becoming someone else's food, and engage in a variety of social interactions and relationships. Getting along with others in the context of social interactions is essential for chimpanzees to acquire mates and reproduce. We outline below seven questions related to how chimpanzees use their minds in nature to deal with these challenges (table 15.1).

Table 15.1   Seven questions about social and ecological problems chimpanzees face in nature.

1. How do chimpanzees navigate their ever-changing social environment?

2. To what extent do chimpanzees possess information about conspecifics who are not present?

3. What cognitive mechanisms do chimpanzees use to track their social relationships over the short term and long term?

4. How do chimpanzees monitor third-party relationships?

5. How and for how long do chimpanzees keep records of past interactions, and how do they evaluate exchanges made in different currencies?

6. What knowledge do chimpanzees possess about the territories of others?

7. What knowledge do chimpanzees have about other ecologically relevant sympatric species?

Fission Fusion Societies: How Do Chimpanzees Navigate Their Ever-changing Social Environment?

Given the ecological and social problems that wild chimpanzees face, an immediate challenge for chimpanzees arises due to the nature of their society. Unlike many other social primates, chimpanzees live in open, fluid "unit groups" or communities whose members split and fuse to form temporary subgroups or "parties" that vary in size and composition (Nishida 1968; Goodall 1986; Wrangham 2000; Boesch and Boesch-Achermann 2000; Mitani et al. 2002b). At Ngogo, party sizes average around 10 individuals (Mitani et al. 2002b). There is considerable temporal variation in the sizes of parties that form over different months (Kruskal-Wallis test, $\chi^2 = 154$, $df = 15$, $P < 0.0001$; figure 15.1). In addition, parties are extremely labile over the short term, displaying considerably more variation within months than between months (variance components: within months $= 72\%$, between months $= 28\%$; figure 15.1). As chimpanzees develop and maintain social relationships with others, they must constantly assess and reassess their social worlds in relation to those around them. The cognitive mechanisms required to cope with this ever-changing social context remain a fruitful area for future research.

Vocal Mediation of Social Interactions: To What Extent Do Chimpanzees Possess Information about Other Conspecifics Who Are Not Present?

Several years ago one of us (JCM), working together with Toshisada Nishida, suggested that chimpanzees use their

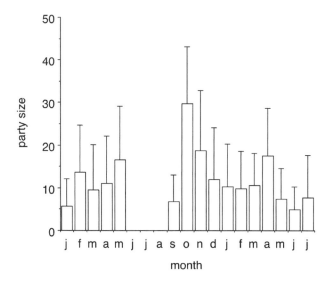

Figure 15.1 Temporal variation in chimpanzee party sizes at Ngogo between January 1998 and July 1999. Means + one SD are shown. Adapted from Mitani et al. 2002b.

vocal behavior to monitor and track changes in their social worlds. The long-distance pant-hoot is the call uttered most frequently by chimpanzees (Goodall 1986). In observations carried out at the Mahale Mountains National Park, Tanzania, Mitani and Nishida (1993) showed that male chimpanzees called much more frequently when their friends were nearby and within earshot than when those same individuals were absent from the party that day (figure 15.2). In contrast, calling rates of males did not differ when randomly selected control males were nearby or absent. While these data suggest that chimpanzees who are out of sight may not always be out of mind, they still leave open an important question about how chimpanzees monitor others who are not present. The selective advantage of possessing such knowledge is immediately obvious,

and this represents a second question that clearly deserves further scrutiny.

Temporal Variation in Social Relationships: What Cognitive Mechanisms Do Chimpanzees Use to Track Their Social Relationships over the Short and Long Term?

While the fission-fusion societies of chimpanzees create a set of unusual social challenges, it is important to note that chimpanzees operate within a web of differentiated social relationships that are manifest in several different behaviors, including association, proximity maintenance, coalition formation, and grooming interactions (Muller and Mitani 2005). The strength of these social relationships can be quantified in different ways. In the past, we have done so utilizing a pairwise affinity index (Pepper et al. 1999). Numerically this index is

$$\frac{I_{ab}^{*}\Sigma s_i(s_i-1)}{\Sigma a_i(s_i-1)^{*}\Sigma b_i(s_i-1)}$$

where I_{ab} = the number of appearances of individuals a and b together, a_i = the number of appearances of individual a, b_i = the number of appearances of individual b, and s_i = the size of group i.

We can employ this index along with a randomization procedure to evaluate whether pairs of male chimpanzees prefer to engage in social behaviors with each other more than one would expect on the basis of chance. We applied this technique to assess grooming relationships between adult male chimpanzees at Ngogo during two different time periods, 2000–01 and 2005–06. Results of these analyses, depicted in figure 15.3, show that males at Ngogo groomed as few as 3 and as many as 15 other males more often than chance expectation.

Figure 15.3 also clearly indicates that the number of preferred grooming partners changed over time for each male. This number decreased for several males, while other individuals expanded their grooming networks to include more partners. The identities of preferred partners also changed over time. We implemented a simple index, $1 - n / t$, where n = the number of individuals who were preferred grooming partners in both sampling periods and t = the total number of individuals who were preferred grooming partners in both periods, to assay the percentage change in grooming partners for each male. Using this index, we found that each male experienced considerable change in

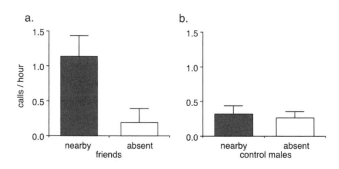

Figure 15.2 Effects of social context on the production of pant-hoots: (a) calling rates of male chimpanzees when friends are nearby and absent; (b) calling rates of male chimpanzees when randomly selected control males are nearby and absent. Means ± one SE are shown. Adapted from Mitani and Nishida 1993.

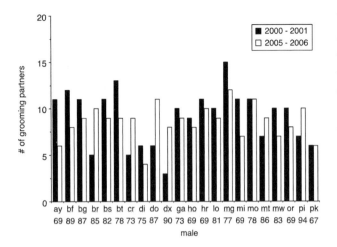

Figure 15.3 Temporal variation in grooming relationships. Shown are the numbers of preferred grooming partners of 22 male chimpanzees at Ngogo during two periods, 2000–01 and 2005–06. Preferred grooming partners are individuals whom males groomed more often than was expected by chance as determined via a randomization procedure. Numbers below two-letter designations of each male indicate the percentage change in the number and identity of preferred grooming partners of male chimpanzees across two periods, 2000–01 and 2005–06.

the number and identity of preferred grooming partners over time (figure 15.3), ranging from 67 to 94% (X = 78%). These results indicate that social relationships are unstable, and suggest that chimpanzees possess cognitive abilities to track their social relationships over time.

Third-Party Relationships: How Do Chimpanzees Monitor Them?

Chimpanzee social relationships change over time. As is the case in humans, today's friend can be tomorrow's foe (de Waal 1982; Nishida 1983). Viewed within this context, chimpanzee social relationships are also likely to be affected by other factors known to influence human behavior. Consider how our behavior as humans is affected by third parties. Our actions in the company of friends and relatives often differ substantially from what we do in the presence of individuals with whom we are less familiar.

To what extent does the presence of third parties affect the behavior of chimpanzees? To examine this question, we investigated the behavior of two adult male chimpanzees at Ngogo: Bartok, the current alpha male, and Hare (figure 15.4). Over the years, Bartok and Hare have maintained a strong friendship and alliance. They frequently associate, form coalitions, share meat, and groom each other. We can assess the effect that the presence of Hare has on Bartok's grooming relationships by implementing

the pairwise affinity index. Using observations made during five months in 2005, we found that Bartok groomed eight males, including Hare, at levels that exceeded chance expectation (figure 15.4a). His grooming interactions changed quite dramatically in Hare's absence. While four males from the original grooming cluster remained, an additional seven other males filled the gap left by Hare (figure 15.4b).

A very different scenario unfolded when we asked what effect Bartok had on Hare's grooming interactions. With Bartok present, Hare groomed with 10 males more than expected by chance (figure 15.4c). In contrast, Bartok's absence resulted in a situation where Hare's grooming cluster shrank (figure 15.4d). Three males from the original group remained, and three others stepped in to fill the void left by Bartok. Interestingly, one of these three new males turned out to be Hare's younger maternal brother, Morton.

These examples reveal that third parties have tangible and important effects on the behavior of chimpanzees (cf. chapter 17). In the case illustrated here, Morton's behavior with his brother was obviously affected by the presence of the alpha male, Bartok. These observations result in our fourth question regarding whether and how chimpanzees recognize third-party relationships.

How and for How Long Do Chimpanzees Keep Records of Past Interactions, and How Do They Evaluate Exchanges Made in Different Currencies?

Thus far, we have shown that chimpanzee social relationships change over time and that third parties can have a significant impact on them. Some of the most complex social interactions that take place in chimpanzee society occur in the context of coalitions. Additional observations suggest that coalitionary behavior is likely to create heavy demands on the mental scorekeeping abilities of chimpanzees.

As is the case in other primates, coalitions in chimpanzees involve situations where two or more individuals jointly direct aggression toward others (de Waal and Harcourt 1992). Coalitions are especially important for male chimpanzees, because they have significant fitness consequences. For example, the receipt of coalitionary support by male chimpanzees at Ngogo improves their mating success with periovulatory females (Spearman $r = 0.49$, $N = 17$ males, $P < 0.05$; figure 15.5). Both of these variables are

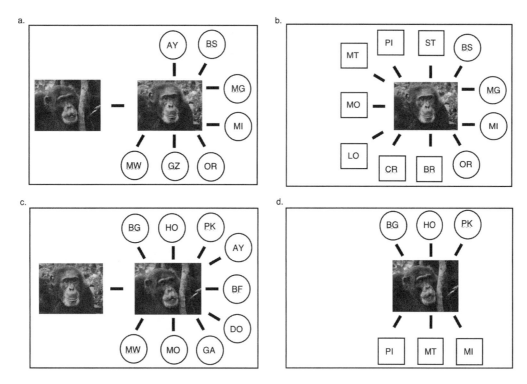

Figure 15.4   Effects of third parties on grooming relationships: (a) alpha male Bartok's preferred grooming partners in parties where his friend and ally, Hare, is present; (b) Bartok's preferred grooming partners in parties where Hare is absent (circles denote preferred grooming partners in the presence and absence of Hare, while squares denote preferential grooming relationships formed only in Hare's absence); (c) Hare's preferred grooming partners in parties where Bartok is present; (d) Hare's preferred grooming partners in parties where Bartok is absent (circles denote preferred grooming partners in the presence and absence of Bartok, while squares denote preferential grooming relationships formed only in Bartok's absence). Preferred grooming partnerships were determined via randomization using a pairwise affinity index.

positively correlated with male dominance rank, but the relationship between male mating success and coalitionary support persists after controlling for the effect of rank.

Because they derive important fitness benefits via coalitions, male chimpanzees compete for coalitionary partners. They do so by engaging in a series of reciprocal exchanges of behaviors that are both similar and different in kind. Our observations at Ngogo indicate that males there reciprocated coalitionary support at the group level (Mitani 2006). In addition, male chimpanzees at Ngogo reciprocally exchanged behaviors that were entirely different in kind, trading grooming and meat for coalitionary support (Mitani 2006).

These data suggest that chimpanzees track interactions with others using goods and services in similar and different currencies. What cognitive abilities do chimpanzees require and use to keep records of prior interactions? If trading is done in different currencies, how do chimpanzees track and evaluate the relative costs and benefits of these exchanges?

Territoriality: What Knowledge Do Chimpanzees Possess about the Territories of Others?

Our previous discussion has focused on challenges that wild chimpanzees face in the contexts of their social interactions and relationships. That chimpanzees possess remarkable knowledge about their environment is suggested by observations of their territorial behavior. Territoriality is a conspicuous aspect of chimpanzee behavior; interactions between members of different communities are frequently aggressive (reviewed in Wilson and Wrangham 2003). Occasionally these interactions escalate, leading to fatal consequences when male chimpanzees make coalitionary attacks on their neighbors.

Boundary patrol behavior is an integral part of chimpanzee territoriality (Watts and Mitani 2001; Mitani and Watts 2005). During patrols, chimpanzees typically move in a single file toward the periphery of their territory. When they arrive there, their behavior changes dramatically. They fall completely silent. They scan the environ-

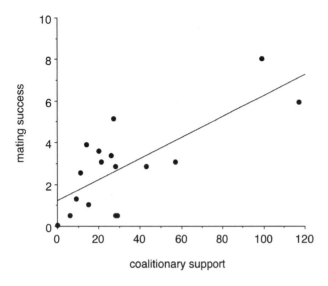

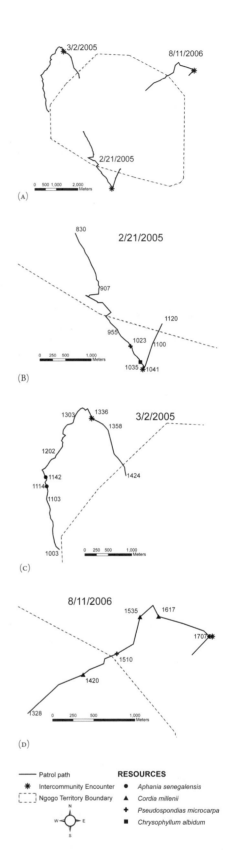

Figure 15.5 Relationship between male mating success and receipt of coalitionary support. Male mating success was assayed as the arcsin transformed percentage of all matings with periovulatory females. Coalitionary support was assayed by the number of times males received support.

ment and startle at any movement in the treetops. They also sniff the ground and vegetation, and inspect signs left behind by conspecifics such as nests, food wadges, urine, and feces. Chimpanzees rarely feed during patrols. Instead they alter their normal foraging movements, occasionally making deep incursions into the territories of their neighbors. Patrolling chimpanzees sometimes encounter their neighbors, and will attack them if they outnumber them (review in Wilson and Wrangham 2003).

Patrolling chimpanzees display keen awareness of the territories of their neighbors. Specifically, they show non-random movements and frequently visit sites well outside of their own territory where neighboring chimpanzees may congregate, such as fruit trees. A few examples will serve to illustrate (figure 15.6a). On February 21, 2005, a large party of Ngogo chimpanzees assembled in the southern part of their territory (figure 15.6b). There they began a patrol at 8:30 a.m. Moving steadily southward, they crossed the boundary of their territory, arriving at a fruit tree that lay well inside their neighbors' territory. This tree, *Pseudospondias microcarpa*, was a favored food source of chimpanzees at Kibale (Mitani and Watts 2001). Continuing to move southward, they encountered another important food tree in fruit, *Chrysophyllum albidum*. Six minutes later, they encountered a small party of chimpanzees from the neighboring community. During this encounter they attacked a female, grabbed her infant, and killed it (Mitani and Amsler, unpublished data).

Figure 15.6 Territorial boundary patrols by chimpanzees at Ngogo. Maps indicate the paths taken by patrollers and the locations of fruit trees that were inspected. (a) Map of three patrols in relation to the Ngogo territory boundary. (b) The path of a patrol on February, 21 2005. (c) The path of a patrol on March 2, 2005. (d) The path of a patrol on August 11, 2006.

A few days later, on March 2, 2005, the Ngogo chimpanzees gathered in the northwest portion of their territory (figure 15.6c), and at midmorning began another patrol. Moving progressively northward, they inspected a food tree of yet another species (*Aphania senegalensis*) in fruit, situated deep within the territory of their neighbors. Continuing on, they passed another tree of the same species about 30 minutes later. They resumed moving northward for another two hours before meeting chimpanzees from an adjacent community. During this encounter they chased a female with an infant and beat another female before returning to their territory (Amsler, unpublished data).

Events of August 11, 2006, furnish a third example (figure 15.6d). While ranging in the northeast portion of their territory, a large group of Ngogo chimpanzees initiated another patrol, this time in the early afternoon. They started by inspecting two fruit trees (*Cordia millenii* and *Pseudospondias*) in the area of overlap between their own territory and that of an adjacent community. Continuing to move northeast and making a deep incursion into the neighboring territory, they visited three other *Cordia* and *Pseudospondias* trees in the next two hours. At the last of these trees, they encountered chimpanzees from another community. The ensuing encounter resulted in the Ngogo chimpanzees attacking and killing an adult male from the rival group (Mitani, Pav, and Sarringhaus, unpublished data).

Previous researchers have noted that primates have an exceptional understanding of their own home ranges and territories (e.g., Milton 1988, Menzel 1997). The preceding observations suggest that chimpanzees have very good knowledge of the territories of their neighbors as well. Targeted visits to specific food trees that lie well inside the territories of others are a striking aspect of their patrols. These anecdotal observations provide only hints of the extent to which chimpanzees are aware of the ecological conditions inside the territories of others. What precise knowledge they do possess constitutes a sixth question that clearly warrants further investigation.

Hunting Behavior and Patrols: What Knowledge Do Chimpanzees Have about Other Ecologically Relevant, Sympatric Species?

While chimpanzees appear to have a good understanding of conspecifics with whom they live (see above), additional observations suggest that they also possess some knowledge about the movements and activities of other species. Hints that this knowledge exists come from studies of chimpanzee hunting behavior.

Chimpanzees are well known for their predatory behavior (reviews in Uehara 1997; Stanford 1998; see chapter 18). Their favored prey is red colobus monkeys. Wherever chimpanzees live sympatrically with red colobus, they hunt them avidly and with high success. Hunting success rates average over 50% across study sites (review in Mitani and Watts 1999). This high degree of success depends in part on encountering prey. Such encounters, however, occur under different circumstances across study sites. Chimpanzees at Gombe and Mahale hunt red colobus opportunistically after meeting them during their regular foraging movements and activities (Stanford 1998; Hosaka et al. 2001). In contrast, chimpanzees in the Tai National Park and at Ngogo actively search for red colobus during "hunting patrols" (Boesch and Boesch 1989; Mitani and Watts 1999). At Ngogo these patrols last two hours on average, and can sometimes take up to 5 hours of a 12-hour waking day (S. Amsler and J. Mitani unpublished data).

Observations of hunting patrols suggest that chimpanzees may possess a mental map of the activity and locations of red colobus prey. Here again a few examples will serve to illustrate this point (figure 15.7). On November 3, 2004, the Ngogo chimpanzees captured one red colobus monkey in the northeastern portion of their territory (figure 15.7a). Two days later they initiated a patrol in the same area. Passing the previous hunting site, they moved a few meters to the north before encountering a group of red colobus. A hunt ensued, with four monkeys falling victim.

Three days later, on November 8, 2004, the Ngogo chimpanzees hunted again, this time in the far southeastern edge of their territory (figure 15.7b). They returned to the general vicinity three days later while on another hunting patrol. After meeting a group of red colobus, the Ngogo chimpanzees hunted and killed one more monkey. Two years later, on June 21, 2006 (figure 15.7c), the Ngogo chimpanzees preyed upon five red colobus monkeys in the same area. Returning again to the exact spot about three weeks later while on a hunting patrol, the Ngogo chimpanzees encountered red colobus monkeys again. Two monkeys subsequently succumbed to predation.

The non-random pattern of these patrols suggests that chimpanzees may possess a mental map of where they have encountered and successfully hunted red colobus prey in the past. These examples generate our seventh and final

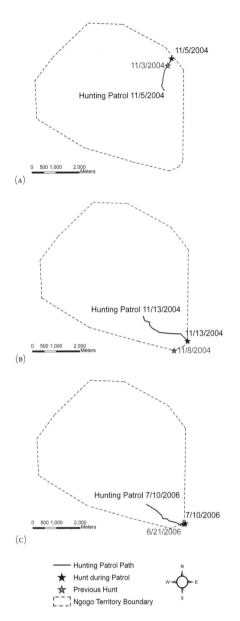

Figure 15.7 Hunting patrols by chimpanzee at Ngogo. (a) The path of a hunting patrol on November 5, 2004, which passed by the location of a hunt that had taken place two days earlier. (b) The path of a hunting patrol on November 13, 2004, which returned to the general vicinity of another hunt that had occurred three days earlier. (c) The path of a hunting patrol on July 10, 2006, which returned to the exact spot of a hunt that had taken place three weeks earlier.

question about the knowledge chimpanzees possess about other species with whom they live sympatrically.

Discussion

The preceding discussion highlights the kinds of social and ecological problems that chimpanzees typically encounter in nature. Many of them are not unique to chimpanzees, and prior research reveals how other animals respond to

similar challenges. Several studies show that primates and nonprimates use calls to negotiate social interactions, even in cases where individuals are physically separated and not within visual range of each other. For example, recent research indicates that baboons engage in "vocal alliances" to support close relatives in conflicts, and grunt to reconcile with individuals who have engaged in aggression with kin (Wittig et al. 2007a, 2007b; see also chapter 7). When attacked, rhesus macaques emit acoustically distinct screams that encode information about the genealogical and dominance rank relationships of opponents engaged in the dispute; other monkeys use the information to decide whether to intervene (Gouzoules et al. 1984). Similarly, birds "eavesdrop" on the singing contests of males to evaluate their competitive ability, dominance status, and mate quality (Peake et al. 2002; Mennill et al. 2002). Relatively few data exist regarding chimpanzee vocal behavior, though recent studies by Slocombe (see chapter 16) are beginning to show, perhaps not surprisingly, that chimpanzees use calls to monitor social interactions in ways similar to those of other animals.

Chimpanzees are also not unusual with respect to third-party relationships. These relationships play an important role not only in chimpanzee society, but in the lives of many other animals as well (review in Cheney and Seyfarth 2007). Several studies indicate that primates recognize the rank, kinship, coalitionary, and mating relationships between other individuals (Cheney et al. 1995; Cheney and Seyfarth 1999; Silk 1999; Crockford et al. 2007). Additional research indicates that baboons are able to simultaneously integrate information about third-party dominance and kin relationships between other group members (Bergman et al. 2003). Birds and non-primate mammals also recognize relationships in which they are not directly involved (Paz-y-Miño et al. 2004; Engh et al. 2005). Given these findings, it is quite likely that chimpanzees also recognize third-party relationships (see chapter 17). Although considerable research indicates that primates and other animals recognize third-party relationships, the cognitive mechanisms individuals employ to accomplish this are unknown and currently debated. While some researchers suggest that relatively simple contingency-based rules may be at work (e.g., Schusterman and Kastak 1998), others argue that the large number of individuals with whom social animals typically associate dictates the use of more complex cognitive mechanisms (e.g., Cheney and Seyfarth 2007). Further research is needed to clarify this issue.

While the preceding discussion indicates that primates and other animals possess good knowledge about their social worlds, which they monitor closely, some studies suggest that primates may not be particularly attentive to aspects of their physical environments. For example, vervet monkeys fail to respond to irrelevant aspects of the behavior of other species, such as that of hippopotamuses when out of water, or to secondary cues of danger furnished by predators (Cheney and Seyfarth 1985). Despite these findings, monkeys respond adaptively to the alarm calls of other primate species (Zuberbühler 2000), an ability displayed by several other non-primate mammals and birds (review in Caro 2005). The ecological knowledge of primates is best exemplified in terms of their food-finding skills. Under laboratory and field conditions, monkeys select more productive food sources over ones containing less food, with animals moving more directly and faster to the former than to the latter (review in Janson and Byrne 2007). Classic experiments conducted by Emil Menzel (1973) reveal that immature chimpanzees possess similar skills for finding hidden foods in a large outdoor enclosure. The knowledge wild chimpanzees possess about their environments remains largely unexplored. Anecdotes like those we have discussed above furnish hints that chimpanzees are well aware and cognizant of many aspects of their physical environment, but systematic research is required to validate this claim.

Implications and Future Directions

Wild chimpanzees face a set of social and ecological problems that are similar to those dealt with by many other animals. Because of this, they are likely to be unexceptional in some of their cognitive skills, such as in their use of calls to monitor conspecifics they cannot see, and in their ability to recognize third parties. Our results and discussion, however, highlight some unique social and ecological challenges that confront chimpanzees in the wild. These challenges place heavy demands on their mental capabilities and point to a significant gap in our knowledge about chimpanzee cognition.

The fission-fusion nature of chimpanzee society has few parallels in the animal kingdom, and is shared only with a few primates, carnivores, cetaceans, and ungulates (Mac-Donald 2001). Within chimpanzee groups, the labile social relationships between males that we have shown here differ from the extremely stable relationships that form between individuals in other primate species (e.g., Samuels et al. 1987). Chimpanzees are extremely long-lived, and social interactions and relationships unfold over extended periods. This adds another layer of temporal complexity to the challenges posed by chimpanzee social life. As we have noted in this paper, some of these challenges have important fitness consequences and are neither straightforward nor simple, as they involve the trade of goods and services that differ in kind. Taken as a whole, fission-fusion sociality and unstable social relationships create an ever-changing social milieu that places a selective premium on the ability to recall past social interactions, and perhaps plan future ones, with individuals who may be encountered only rarely over time.

Recent experiments indicate that chimpanzees process visual scenes extraordinarily quickly, and may possess a better short-term working memory than do humans (Inoue and Matsuzawa 2007). Additional studies suggest that they recall events and aspects of the environment over longer periods of time. For example, chimpanzees help others with whom they have cooperated earlier in the day (de Waal 1997), and individuals who learn a novel foraging technique retain the skill two months after acquisition (Whiten et al. 2005; see also chapter 9). Experiments with an 11-year old language-trained female chimpanzee reveal that she accurately recalled the identities and locations of objects that had been hidden up to 16 hours before, and spontaneously transmitted this information to caretakers who were unaware of the objects (Menzel 1999). These latter findings furnish the best evidence for long-term memory in chimpanzees, but they fall short of experiments conducted with food-caching jays who displayed "episodic-like" memory by remembering what, where, and when food was stored for up to five days (Clayton and Dickinson 1999).

Episodic memory refers to the ability to recall events based on personal experience (Tulving 1983). In humans this involves a kind of "mental time travel," with individuals consciously reexperiencing past events. Our discussion of the types of social and ecological problems encountered by wild chimpanzees makes the selective advantages of episodic memory immediately obvious. The ability to recall specific past events with regard to what happened, where, when, and with whom would permit chimpanzees to track complex social interactions and relationships involving a large number of individuals over both short and extended periods of time (see above and questions 1, 2, and 5 in table 15.1). In addition, episodic memories would furnish chimpanzees knowledge about the behavior of other sympatric species and the locations of food in their own territories as well as in areas visited infrequently, such as the territories of their neighbors (questions 6 and 7 in table 15.1). Future studies investigating the ability of chimpanzees to recall the what, where, when, and with whom of past events promise to provide insights into how chimpanzees use their minds to deal with the challenges of life.

In sum, we have drawn attention to several social and ecological problems that chimpanzees confront in the wild. We make no claim that this list is exhaustive. We do, however, believe that experiments designed to explore the chimpanzee mind will benefit by considering such problems, which constitute a set of challenges that the minds of chimpanzees have been designed—through the process of natural selection—to solve during the course of their evolution. The ecological validity of experiments conducted in captive settings depends critically on taking such problems into account, and doing so will ultimately improve our ability to generalize findings derived from such experiments.

Acknowledgments

Our fieldwork in Uganda has been sponsored by the Uganda Wildlife Authority and the Ugandan National Council for Science and Technology. We thank J. Kasenene, G. Isabirye-Basuta, J. Lwanga, and the Makerere University Biological Field Station for providing logistical support at Ngogo. A. Magoba, G. Mbabazi, L. Ndagizi, and A. Tumusiime have provided expert field assistance over the years. We are grateful to T. Bergman for discussion, and to the editors and four reviewers for their comments on the manuscript. Our fieldwork has been supported by grants from the Detroit Zoological Institute, the Little Rock Zoo, the Little Rock chapter of the American Association of Zookeepers, the L.S.B. Leakey Foundation, the National Geographic Society, the National Science Foundation (SBR-9253590, BCS-0215622, and IOB-0516644), the University of Michigan, and the Wenner-Gren Foundation. We gratefully acknowledge additional support from the NSF Graduate Research Fellowship Program to SJA and MES.

Literature Cited

Bergman, T., J. Beehner, D. Cheney, and R. Seyfarth. 2003. Hierarchical classification by rank and kinship in baboons. *Science* 302:1234–36.

Boesch, C., and H. Boesch-Achermann. 2000. *The Chimpanzees of the Taï Forest*. Oxford: Oxford University Press.

Boesch, C., and H. Boesch. 1989. Hunting behavior of wild chimpanzees in the Taï National Park. *American Journal of Physical Anthropology* 78:547–73.

Caro, T. 2005. *Antipredator Defenses in Birds and Mammals*. Chicago: University of Chicago Press.

Cheney, D., and R. Seyfarth, 1985. Social and non-social knowledge in vervet monkeys. *Philosophical Transactions of the Royal Society of London B* 308:187–201.

———. 1999. Recognition of other individuals' social relationships by female baboons. *Animal Behaviour* 58:67–75.

———. 2007. *Baboon Metaphysics*. Chicago: University of Chicago Press.

Cheney, D., R. Seyfarth, and J. Silk. 1995. The responses of female baboons to anomalous social interactions: Evidence of causal reasoning? *Journal of Comparative Psychology* 109:134–41.

Chimpanzee Sequencing and Analysis Consortium. 2005. Initial sequence of the chimpanzee genome and comparison with the human genome. *Nature* 437:69–87.

Clayton, N., and A. Dickinson. 1998. Episodic-like memory during cache recovery by scrub jays. *Nature* 395:272–78.

Crockford, C., R. Wittig, R. Seyfarth, and D. Cheney. Baboons eavesdrop to deduce mating opportunities. *Animal Behaviour* 73:885–90.

De Waal, F. 1982. *Chimpanzee Politics*. New York: Harper and Row.

De Waal, F., and A. Harcourt. 1992. Coalitions and alliances: A history of ethological research. In *Coalitions and Alliances in Humans and Other Animals*, ed. A. Harcourt and F. de Waal, 1–27. Oxford: Oxford University Press.

De Waal, F. 1997. The chimpanzee's service economy: Food for grooming. *Evolution and Human Behavior* 18:375–85.

Engh, A., E. Siebert, D. Greenberg, and K. Holekamp. 2005. Patterns of alliance formation and post-conflict aggression indicate spotted hyenas recognize third party relationships. *Animal Behaviour* 69:209–17.

Goodall, J. 1963. Feeding behaviour of wild chimpanzees: A preliminary report. *Symposium of the Zoological Society of London* 10:39–48.

———. 1986. *The Chimpanzees of Gombe*. Cambridge: Belknap Press.

Gouzoules, S., H. Gouzoules, and P. Marler. 1984. Rhesus monkey (*Macaca mulatta*) screams: Representational signaling in the recruitment of agonistic aid. *Animal Behaviour* 32:182–93.

Hosaka, K., T. Nishida, M. Hamai, A. Matsumoto-Oda and S. Uehara. 2001. Predation of mammals by the chimpanzees of the Mahale Mountains, Tanzania. In *All Apes Great and Small. Volume 1. African Apes*, ed. B. Galdikas, N. Briggs, L. Sheeran, G. Shapiro, and J. Goodall, 107–30. New York: Kluwer Academic Publishers.

Huxley, T. 1863. *Evidence as to Man's Place in Nature*. London: Williams and Norgate.

Inoue, S., and T. Matsuzawa. 2007. Working memory of numerals in chimpanzees. *Current Biology* 17:R1004–5.

Janson, C., and R. Byrne. 2007. What wild primates know about resources: Opening up the black box. *Animal Cognition* 10:357–67.

King, M. C., and A. Wilson. 1975. Evolution at two levels in humans and chimpanzees. *Science* 188:107–16.

MacDonald, D. 2001. *The Encyclopedia of Mammals.* New York: Facts on File.

Mennill, D., L. Ratcliffe, and P. Boag. 2002. Female eavesdropping on male song contests in songbirds. *Science* 296:873.

Menzel, C. 1997. Primates' knowledge of the natural habitat: As indicated by foraging. In *Machiavellian Intelligence II: Extensions and Evaluations,* ed. A. Whiten and R. Byrne, 207–39. Cambridge: Cambridge University Press.

Menzel, E. 1973. Chimpanzee spatial memory organization. *Science* 182:943–45.

Milton, K. 1988. Foraging behaviour and the evolution of primate intelligence. In *Machiavellian Intelligence,* ed. R. Byrne and A. Whitten, 285–305. Oxford: Oxford University Press.

Mitani, J. 2006. Demographic influences on the behavior of chimpanzees. *Primates* 47: 6–13.

———. 2006. Reciprocal exchange in chimpanzees and other primates. In *Cooperation in Primates: Mechanisms and Evolution,* ed. P. Kappeler and C. van Schaik, 101–13. Heidelberg: Springer-Verlag.

Mitani, J., and T. Nishida. 1993. Contexts and social correlates of long distance calling by male chimpanzees. *Animal Behaviour* 45:735–46.

Mitani, J., and D. Watts. 1999. Demographic influences on the hunting behavior of chimpanzees. *American Journal of Physical Anthropology* 109:439–54.

———. 2001. Why do chimpanzees hunt and share meat? *Animal Behaviour* 61:915–24.

———. 2005. Correlates of territorial boundary patrol behaviour in wild chimpanzees. *Animal Behaviour* 70:1079–86.

Mitani, J., D. Watts, and M. Muller. 2002a. Recent developments in the study of wild chimpanzee behavior. *Evolutionary Anthropology* 11:9–25.

Mitani, J., D. Watts, and J. Lwanga. 2002b. Ecological and social correlates of chimpanzee party size and composition. In *Behavioural Diversity in Chimpanzees and Bonobos,* ed. C. Boesch, G. Hohmann and L. Marchant, 102–11. Cambridge: Cambridge University Press.

Muller, M., and J. Mitani. 2005. Conflict and cooperation in wild chimpanzees. *Advances in the Study of Behavior* 35:275–331.

Nishida, T. 1968. The social group of wild chimpanzees in the Mahale Mountains. *Primates* 9:167–224.

———. 1983. Alpha status and agonistic alliance in wild chimpanzees (*Pan troglodytes schweinfurthii*). *Primates* 24:318–36.

Paz-y-Miño, G., A. Bond, A. Kamil, and R. Balda. 2004. Pinyon jays use transitive inference to predict social dominance. *Nature* 430:778–81.

Peake, T., A. Terry, P. McGregor, and T. Dabelsteen. 2002. Do great tits assess rivals by combining direct experience with information gathered by eavesdropping? *Proceedings of the Royal Society of London B: Biological Sciences* 269:1925–29.

Pepper, J., J. Mitani, and D. Watts. 1999. General gregariousness and specific social preferences among wild chimpanzees. *International Journal of Primatology* 20:613–32.

Samuels, A., J. Silk, and J. Altmann. 1987. Continuity and change in dominance relations among female baboons. *Animal Behaviour* 35:785–93.

Schusterman, R., and D. Kastak. 1998. Functional equivalence in a California sea lion: Relevance to animal social and communicative interactions. *Animal Behaviour* 55:1087–95.

Silk, J. 1999. Male bonnet macaques use information about third-party rank relationships to recruit allies. *Animal Behaviour* 58:45–51.

Stanford, C. 1998. *Chimpanzee and Red Colobus: The Ecology of Predator and Prey.* Cambridge: Harvard University Press.

Struhsaker, T. 1997. *Ecology of an African Rain Forest.* Gainesville: University Press of Florida.

Tulving, E. 1983. *Elements of Episodic Memory.* New York: Oxford University Press.

Uehara, S. 1997. Predation on mammals by the chimpanzee (*Pan troglodytes*). *Primates* 38:193–214.

Watts, D., and J. Mitani. 2001. Boundary patrols and intergroup encounters in wild chimpanzees. *Behaviour* 138:299–327.

Whiten, A., and R. Byrne. 1997. Machiavellian intelligence II: Extensions and evaluations. Cambridge: Cambridge University Press.

Whiten, A., V. Horner, and F. de Waal. 2005. Conformity to cultural norms of tool use in chimpanzees. *Nature* 437:737–40.

Wilson, M., and R. Wrangham. 2003. Intergroup relations in chimpanzees. *Annual Review of Anthropology* 32:363–92.

Wittig, R., C. Crockford, R. Seyfarth, and D. Cheney. 2007a. Vocal alliances in chacma baboons, *Papio hamadryas ursinus. Behavioral Ecology and Sociobiology* 61:899–909.

Wittig, R., C. Crockford, E. Wikberg, R. Seyfarth, and D. Cheney. 2007b. Kin-mediated reconciliation substitutes for direct reconciliation in female baboons. *Proceedings of he Royal Society of London B: Biological Sciences* 274:1109–15.

Wrangham, R. 2000. Why are male chimpanzees more gregarious than mothers? A scramble competition hypothesis. In *Primate Males: Causes and Consequences of Variation in Group Composition,* ed. P. Kappeler, 248–58. Cambridge: Cambridge University Press.

Zuberbühler, K. 2000. Interspecies semantic communication in two forest primates. *Proceedings of the Royal Society of London, B: Biological Sciences* 267:713–18.

16

Vocal Communication in Chimpanzees

Katie Slocombe and Klaus Zuberbühler

Wilma, a fully estrous female of the Sonso community in the Budongo Forest, Uganda, had been followed by three dominant males most of the morning. They had all climbed to feed together, but Wilma then left, leaving the males behind. She traveled approximately 100 m on the ground, where she encountered Bob, a low-ranking male. Bob started soliciting for copulation, using the conventional gesture of branch-shaking and the more idiosyncratic gesture of pelvic thrusting. Wilma did not approach and present for a copulation, but instead started producing extremely high-pitched screams. Our research had already shown that these types of screams were normally given only by victims of severe aggression, usually while being physically assaulted, yet Wilma produced them in response to a young male who silently gestured to her at a distance of more than 3 m. In response to Wilma's screams, two of the dominant males charged into view, displaying. Bob rapidly left the area, and Wilma joined the dominant males and rested close to them.

Although it is just an anecdote, we think that the above observation is rather revealing about the potentially complex nature of chimpanzee vocal communication and underlying cognition. Wilma appears to have produced her calls strategically; her calling seemed to be based on detailed knowledge of how other group members would behave towards her, and how she could affect the behavior of others by producing vocal signals in certain ways.

Our current understanding of chimpanzee vocal communication is highly limited. This is in contrast to our understanding of much of the rest of their behavior and cognitive abilities, which have been documented in detail both in the wild and in captivity. A startlingly small amount of systematic research effort has been dedicated to what is arguably a key aspect of their biology: communication in the acoustic domain. A number of reasons may be able to account for this. First, chimpanzee vocalizations are difficult to analyze. Most of their calls are of a graded nature, requiring sophisticated acoustic and statistical analysis techniques. Second, there is a widespread belief that primate vocal communication is the result of hardwired, inflexible, and cognitively uninteresting behavior. Third, much of the existing research on chimpanzee vocal behavior has focused on one call type, the conspicuous pant-hoots, while other call types have not been investigated systematically despite their importance during social interactions. In an influential review, Mitani (1996) concluded a decade ago that wild chimpanzees had given no indication that they possessed special vocal skills, and that their communication system did not appear to be extraordinary.

The paucity of our knowledge about chimpanzee vocal communication and its cognitive underpinnings is especially striking in comparison with what we know from some monkeys and non-primate species. The seminal work of Cheney and Seyfarth first established that vervet monkeys produce acoustically distinct alarm calls to different predators, and recipients respond to these calls as if they have seen the corresponding predator themselves (Seyfarth et al. 1980). Similar findings have come from other species, such as Diana monkeys (Zuberbühler et al. 1997) and meerkats (Manser et al. 2002). Chickens and several species of primates have been found to produce calls in manner that reference the nature of the food source encountered (e.g., Evans and Evans 2007; tamarins, Di Bitetti 2003; macaques, Hauser 1998). Rhesus macaques also provide listeners with complex information about the nature of agonistic interactions with their scream calls (Gouzoules et al. 1984). Putty-nosed monkeys have two loud call types—"pyows" and "hacks"—that they combine into a distinctive sequence to convey specific meanings (Arnold and Zuberbühler 2006a). Experiments have shown that this "pyow-hack" sequence functions to initiate group travel (Arnold and Zuberbühler 2006b), the first documented experimental evidence of a nonhuman primate *naturally* combining signals in meaningful ways to refer to external events and social goals.

Monkeys also demonstrate considerable cognitive versatility when responding to other individuals' calls. For example, Diana monkeys often encounter signals with vague or ambiguous referents, requiring integration of additional information before an adaptive response can be selected. In the Taï forest, Côte d'Ivoire, these monkeys share their habitat with other mammalian and avian species, many of which produce their own alarm calls to predators. For example, both Diana monkeys and crested guinea fowls are hunted by leopards and humans. Unlike the monkeys, however, these birds respond with only one general alarm call to both predators. Experiments have shown that the monkeys distinguish between cases where the birds' alarm calls are caused by a leopard and those caused by a human hunter (Zuberbühler 2000). The monkeys, in other words, infer the cause of an individual's alarm calls rather than responding directly to the calls themselves.

Importance of Vocalizations in the Natural Habitat

Chimpanzees communicate in a number of modalities, with vocal and gestural communication being the most significant. In captivity, where they are usually kept in small groups with almost constant visual contact, individuals fre- quently use gestures as an important mode of communication, and researchers have noted the flexibility inherent in the production of these signals (Tomasello and Zuberbühler 2002; Call and Tomasello 2007). From these captive studies, it is tempting to conclude that vocal communication is simply not important to this species and that the relevant selective pressures have not been present to shape the vocal behavior into a complex system. However, when observing chimpanzees in their natural environment, the functional significance of the vocal channel of communication becomes very apparent. The visually dense nature of their natural rain-forest habitat, combined with their fission-fusion social structure, leads to a situation in which chimpanzees are out of visual contact with the majority of their fellow community members, often including individuals only 20 to 30 meters away. In such an environment a complex vocal communication system, which enables individuals to communicate about important social events, the discovery of food, or an encounter with a predator, is of considerable evolutionary advantage.

The Relation between Communication and Cognition

One core problem in animal communication research, especially with regard to the evolution of human language, concerns the relationship between animal cognitive capacities and communication skills. Do more intelligent species possess greater communication skills, or is communication a product of social evolution? Great apes are highly intelligent animals, and they live in highly complex social systems. One might reasonably predict, therefore, that their high intelligence and social complexity should be mirrored by equally complex communicative behavior. However, great apes so far have not been reported to perform remarkably in the vocal domain by any standard, despite several long-term field studies. As previously mentioned, this fact is even more striking in light of research conducted with several species of monkeys whose vocal skills have outperformed all great apes by any comparison (Cheney and Seyfarth 1990, 2007).

Two basic explanations could account for this chasm. First, as suggested before, it is possible that great apes rely on communication modes other than the vocal channel to accomplish their communicative needs. Researchers working with captive apes have long argued that gestural behavior in apes is complex and is in many ways a clear demonstration of their flexible and insightful minds (Call and Tomasello 2007). These studies have shown, for example, that apes carefully adjust their production of gestural signals depending on the attentional state of the targeted receiver. Before using visual gestures, chimpanzees position themselves such that they have visual contact with the receiver, or use tactile or acoustic gestures if the receiver is socially engaged elsewhere. Apes thus assess a targeted receiver not only in terms of biological and social categories, but also take into account psychological variables such as attention, capacity to help, or ability to comprehend. Vocal signals, in contrast, may not be linked to such higher cognitive processes, but may instead be the product of hardwired evolutionary predispositions, with a limited set of calls linked to a small number of contexts.

Another approach, which we will take in this chapter, is that with regard to vocal communication many interesting patterns have simply been overlooked, and that there is much greater sophistication in the vocal domain than is currently described in the literature. This prediction is based on the fact that natural selection will always favor signalers who are able to adjust signal production to maximize their own benefit. Vocal signals are highly suitable for this purpose in visually dense rainforests and in species that live in large and complex societies. Chimpanzees and bonobos are the obvious species in which to test these ideas, since they are most likely to realize any evolutionary potential due to their presence in forest habitats and complex social structures.

The Graded Vocal Repertoire of Chimpanzees

Vocal systems may be placed on a continuum, ranging from discretely organized to extensively graded. Rowell and Hinde (1962) first described the highly graded nature of the rhesus macaque vocal system and since then many species of primate, including Japanese macaques (Green 1975), red colobus monkeys, *Procolobus badius* (Marler 1970), and chimpanzees (Marler 1976), have also been classified as having graded vocal repertoires. In chimpanzees, the degree of grading is variable depending on call types (see table 16.1), with "waa-barks" being the most highly graded and laughter being the most discrete call type (Marler 1976). Even within discrete call types, there is still acoustic variation. Chimpanzee vocalizations, in other

words, show extensive grading both between and within call types. Although they are more difficult to analyze, graded repertoires offer the potential for encoding large amounts of information, provided that the relationship between signal grading and circumstances of production is patterned (Marler 1976).

The amount of information a graded system can convey is largely determined by how recipients perceive the continuum. Human speech contains significant grading between different phonemes, which are still perceived categorically (Abramson and Lisker 1970). It is very likely, therefore, that chimpanzees are equally capable of perceiving graded call types categorically, and that such call variants can function as carriers of discrete information. Secondly, in human speech the grading of sounds is tied not only to emotion or motivation but also to cognitive judgments. Again, it is certainly possible that different call variants in chimpanzees reflect differences in cognitive judgments rather than differences in arousal, a popular and widespread hypothesis of chimpanzee vocal behavior that is not based on any solid empirical grounds.

Very little research has been conducted to examine whether chimpanzees perceive their vocalizations in a categorical manner. While this constrains our understanding of their vocal communication, research on other primate species strongly suggests that chimpanzees should be able to perceive their calls categorically. Barbary macaques, *Macaca sylvanus*, for example, are able to categorize an acoustically continuous variation within one call type, the shrill bark (Fischer 1998) and the same has been described for baboons' bark variants (Fischer et al. 2001).

Describing the graded vocal repertoire of a species is a significant challenge. In the absence of naturally occurring discontinuities in acoustic structure, investigators often rely on their own categorization propensities or select exemplars of vocalizations that are far removed from each other in physical form (Green 1975). Many researchers, from Reynolds (1965) to Goodall (1986), have provided descriptions of the chimpanzee repertoire that rely on subjective human categorization and call context. Marler and Tenaza (1977) went significantly further by providing the first quantitative acoustic description of the chimpanzee repertoire, which they categorized into 13 broad call types. In the meantime, technology has advanced and modern computer-based sound-analysis techniques now allow rapid and large-scale analyses of much greater detail.

Some progress has already been made, particularly with regards to acoustic variation within the broad call types and their perceptual relevance for the receivers. This work is beginning to illustrate that the chimpanzee vocal system is capable of conveying a considerable amount of information to listeners.

In this chapter we will first describe and illustrate the vocal repertoire of the chimpanzee, in the tradition of Marler and Tenaza (1977), before reviewing recent research that has managed to identify meaningful variations within the different call types. We then discuss the function of different calls, as well as the significance of vocal behavior for understanding other aspects of cognition and behavior. Finally, we will assess the contribution such research has made toward an understanding of how human language may have evolved.

Approach

Study Sites and Data Collection

Data presented in this chapter were collected at two study sites, the Budongo Forest Reserve, in Uganda, and Edinburgh Zoo, in the United Kingdom. The first author collected data on a habituated study group in the Sonso region of the reserve (Reynolds 2005) for a total of 13 months between February 2003 and March 2006. Field assistant Raimond Ogen collected vocal rate data for an additional 4 months in 2003. Details of the location and ecology of the Budongo Forest (Eggeling 1947) and this study site (Newton-Fisher 1999) have been described elsewhere. Habituation of these chimpanzees to humans began in 1990, and provisioning has never been used. The community size increased from 53 individuals in 2003 to 72 individuals at the end of the study period in 2006.

We collected focal animal samples from adult and subadult males and from adult females (Altmann 1974), each sample lasting a full day. We recorded all vocalizations from the focal individual along with details of the accompanying context, behavior of the focal and behavior of others in the party. In addition, all vocalizations from known individuals whose behavior could be clearly observed were recorded in an ad libitum manner.

At Edinburgh Zoo we collected data on a group of captive chimpanzees for four months between May and September 2004. There were eleven individuals in the group: three adult males, one subadult male, two juvenile males, four adult females, and one subadult female. Data were collected from the outdoor enclosure to which the chimpanzees had free access at all times except during cleaning. In this portion of the study we recorded feeding vocalizations specifically. For analyses we only considered vocalizations produced by individuals feeding on a single type of food, and who were in physical contact with the food.

Recordings of Vocalizations

Vocalizations were recorded and transferred to computer for acoustic analysis using standard methodology and equipment (for precise details see Slocombe and Zuberbühler 2005a, 2005b, 2007). For the Budongo data we also determined the calling rate of the focal animal by monitoring its calling activity (bouts of calling) over observation time. We distinguished between four basic contexts in which calls were produced: while feeding, while not feeding, while in a group (two or more adults or subadults), and while alone. Vocalizations of the same type had to be separated by at least one minute to be counted as different bouts of calling. We only considered continuous observation periods of at least 100 minutes in a given context for each individual to ensure the sampling was representative. This data was collected by two observers, but an interobserver reliability test indicated that they were recording data in the same way (Crombach's α test = 0.86).

Acoustic Analysis

To examine the acoustic structure of vocalizations given in different contexts, it is necessary to obtain an objective description of the calls. There are no fixed rules as to which acoustic parameters need to be included in such an analysis, but typically any analysis consists of temporal measurements, such as call duration and number of calls per bout, as well as a number of frequency measures. Most chimpanzee calls are generated by vibrations of vocal chords and are then filtered by the vocal tract, which attenuates or accentuates certain components of the signal. The position and shape of lips, tongue, mandible, and larynx all affect the acoustic structure of the signal, in addition to the basic vibration rate of the vocal folds (Fitch and Hauser 2002). As a result, any acoustic analysis should consider two types

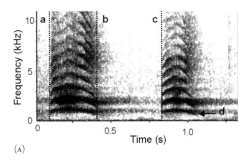

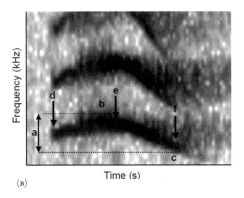

Figure 16.1 (a) Spectrogram illustrating temporal measurements and fundamental frequency. Call duration = b – a. Intercall duration = c – b. Fundamental frequency = d. (b) Spectrogram illustrating frequency measures taken on the fundamental frequency. Frequency modulation = (a); high frequency (b) – low frequency (c). Peak frequency = frequency of maximum acoustic energy at call middle (e). Transition onset = Frequency of maximum energy at call onset (d) – frequency of maximum energy at call middle (e). Transition offset = Frequency of maximum energy at call middle (e) – frequency at of max energy call offset (f).

of frequency measurements: the ones concerned with the fundamental frequency are determined by the rate at which the vocal chords vibrate, while the ones concerned with the filtering of the sound are determined by the shape and length of the vocal tract. The frequencies that the vocal tract enhances are called formants (Fitch and Hauser 2002). They are not always directly visible in the spectrogram but they can be determined using a specialized procedure, a so-called LPC (linear predictive coding) analysis, which is a standard feature of many acoustic software packages such as PRAAT (www.fon.hum.uva.nl/praat). Figures 16.1a and 16.1b illustrate some of the acoustic measurements commonly taken.

Statistical Analysis

To determine whether individuals reliably vary the acoustic structure of their calls according to context, it is necessary to show that the acoustic structures of calls given in different contexts are statistically different from each other. A standard approach is to first conduct one-way analyses of variance (ANOVA) on all acoustic measures to determine the key acoustic measures by which the calls differ. It is often then useful to conduct a discriminant function analysis to establish whether all the acoustic measures together, which provide a quantitative description of the call, predict to which group a call belongs—that is, the context in which it was produced.

Results and Discussion

How Much Do Chimpanzees Vocalize?

In order to determine the frequency of vocal behavior, and to provide an estimate of the importance of this mode of communication to this species, we examined the rate of call production over an entire year in the Budongo Forest, Uganda. We followed six different adult males, three subadult males, six adult noncycling females, and four adult cycling females for a total of 755 hours. We did not observe all individuals in all contexts, but the number of individuals contributing data to each condition is specified in figure 16.2.

While in a group, adult and subadult males and noncycling females were equally vocal (figure 16.2a). The cycling females were the most vocal in the group context, due to the frequent pant-grunts, squeaks, and screams that were given in response to male copulation solicitation and aggression. Both males and females were less vocal when feeding alone as opposed to when feeding in a group (figure 16.2b). In sharp contrast to females, males were most vocal when alone and not feeding, in accordance with the high number of pant-hoots they usually produce while traveling alone through the forest.

Although there was little difference in the overall number of calls male and female chimpanzees produced, the frequency with which they produced certain call types was heavily mediated by the sex and rank of the individual. This finding is consistent with what has been observed at other long-term field sites. In the Gombe population, although most call types were given by all age and sex classes, there were sex-biased trends in vocal production with males giving the majority of pant-hoots, rough grunts, grunts, and wraas and females giving the majority of pant-grunts, squeaks, screams, and barks (Marler and Tenaza 1977). In addition to these basic sex biases in production of different

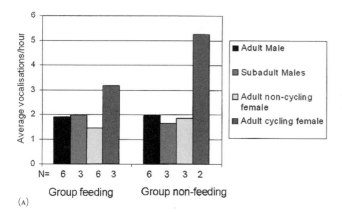

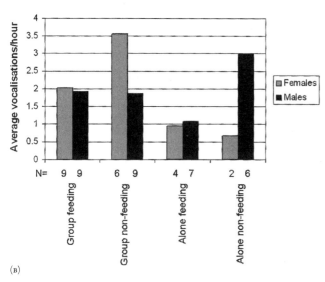

Figure 16.2 (a) Mean number of vocalizations per hour given by adult males, subadult males, cycling females, and noncycling females in groups during feeding and non-feeding activities (b) Mean number of vocalizations per hour by males and females in feeding and non-feeding contexts whilst in groups and alone

and three subadult males and nine adult females from the same community who produced an average of 1.97 vocalizations per hour (std dev = 1.00) in a group-feeding context and 2.54 vocalizations per hour (std dev = 2.25) in a group non-feeding context. A more detailed study of gestures in this community is ongoing, and those results are needed before strong conclusions can be drawn from this comparison.

Vocal Repertoire of Budongo Chimpanzees

Sonso chimpanzees produced all the call types identified by Marler and Tenaza (1977), except the "wraa." Wraas are therefore not illustrated or discussed further and instead the "huu" vocalization is included, which was not included in their original repertoire description. Acoustically, huu calls are similar to "hoo" calls, which are produced during whimper bouts. However, the amplitude and duration of huu calls observed in the Sonso community was much greater than that of hoo calls. In Budongo they served as alarm calls, and were produced in situations where chimpanzees in other communities (e.g., Gombe) might have produced wraas. This may represent a vocal "cultural" variant in this population, and as such it should be investigated further across sites. Table 16.1 presents acoustic descriptions, circumstances of production, and putative function of 13 call types. Each of these call types is spectrographically illustrated in Figure 16.3.

Meanings of Call Types That Have Been Systematically Investigated

PANT-HOOTS. For a considerable time, the pant-hoot vocalizations were the only chimpanzee vocal signals that had been analyzed in depth. In a comprehensive review, Mitani (1994) summarized much of the progress that had been made, and his analyses and interpretations of the evidence indicated that this call type functioned primarily to attract the caller's allies and close associates to the site. One important point was that chimpanzees did not produce pant-hoots to maintain contact indiscriminately with distant conspecifics, but instead targeted specific individuals who were valuable to them. This kind of behavior is evidence against a basic hypothesis in animal communication—namely, that animal signals are automatic responses to evolutionarily relevant external stimuli. More likely, these

call types, a study of the chimpanzees of the Kibale forest, Uganda, revealed that rank and social context also mediates individual rates of calling (Clark 1993). High-ranking males were highly vocal in all social contexts, but adult females and low ranking adult and subadult males were predominantly silent, unless they were in large mixed parties.

A second finding was that in this community, vocal communication was more frequent in comparison to visual or tactile gestures. A preliminary study, based on 29 hours of focal sampling across four adult and three subadult males and four adult females (mean/focal = 2.6 hrs) indicated that these chimpanzees produced an average of 0.95 gestures/hour (std dev = 1.16) that were visual or tactile, across contexts (Anna Roberts, personal communication). This is in contrast to the vocal rate of six adult

Table 16.1 Acoustic description, production context, and function of wild chimpanzee vocalizations from the Budongo Forest, Uganda.

Call type	Acoustic description	Circumstances of production	Function
Pant-hoot	Species-typical long-distance vocalisation with four distinct phases: introductory phase (low frequency hoo calls), buildup phase (increasingly loud panted hoo calls with energy in both the inhalation and exhalation), climax phase (screams or roars), and letdown phase (resembles the buildup phase but with progressively decreasing energy). Calls often do not include all four phases; introduction and letdown are commonly omitted.	*Who?* Mainly adult males. *When?* During display, travel, arrival at feeding sites, spontaneously during feeding and resting (including at night), in response to hearing other chimpanzees, during periods of social excitement, and when meeting other parties of chimpanzees.	Long-distance spacing function (Marler 1968); announces the location and general activity of individuals over a large area.
Whimper	A series of soft low-frequency hoo calls that can become higher in both frequency and amplitude as a bout progresses. The frequency and amplitude often rises and falls throughout a bout. Individual hoo calls are tonal signals with a variable number of harmonics. Whimpers are often produced with a pout expression, and they commonly grade into screams.	*Who?* Distressed individuals, typically juveniles. *When?* Juveniles when lost, separated from the mother, or facing the mother's refusal to carry or feed them; adults most often in competitive feeding or begging contexts.	Distress signal, to which other individuals will sometimes respond in a manner that alleviates the caller's distress (retrieval of lost juvenile, or sharing of food).
Scream	Loud, high-pitched, harmonic vocalization with varying degrees of tonality. They are almost always given in bouts, but the length of a bout can vary from a few seconds to several minutes. In most contexts acoustic energy is present only during exhalation, but during intense tantrums it is often present during inhalation as well. Screams grade into squeaks, whimpers and barks.	*Who?* All individuals, but mainly subordinates *When?* During agonistic encounters, by victims and by aggressors of equal or lower rank to the victim; during tantrums; during periods of intense social excitement or feeding; or in extreme alarm due to a very severe attack or encounter with a predator, such as a leopard	Agonistic screams function to recruit aid. Tantrum screams may work as noxious, aversive stimuli to change the recipient's behaviour to the caller's advantage. Screams given in situations of extreme alarm can attract individuals to the caller from considerable distances.
Squeak	High-pitched, short calls often given in fast succession to form short bouts. The calls are tonal signals, often with clear harmonic structures. Squeaks and screams often grade into one another.	*Who?* Mainly subordinates and females. *When?* Usually when elicited by mild threats of aggression. Calls given during copulations by females are acoustically most similar to squeaks, but have also been described as "squeals" (Tutin and McGrew 1973) and screams (Goodall 1986).	Signal of the caller's distress, which can elicit reassurance behaviors from others. The function of copulation squeaks is currently unclear, although they are likely to play a role in inciting sexual competition.
Bark	Sharp, loud calls with abrupt onsets. They are often noisy and are generally low-pitched vocalisations. Barks grade into screams, being lower in frequency and generally shorter in duration. They also grade into grunts at the lower end of the frequency spectrum.	*Who?* Females more than males *When?* In situations of social excitement and agonistic encounters. Males do not commonly produce barks, but they have been recorded to produce them in response to agonistic encounters, snakes, hunting, traveling, and auditory contact both with other communities and with distant parties from their own communities (Crockford and Boesch 2003)	Barks rarely elicit specific responses in listeners, so their function is unclear. It has been proposed that barking from onlookers in a fight signals support for the victim (Goodall 1986), but we currently lack empirical data to support this hypothesis.
Waa-bark	A distinct loud, intimidating bark variant in which the sound "waa" is clear. Acoustically distinct from the shorter barks and screams into which they often grade, waa-barks typically have a low frequency "w" introduction at call onset, and then a clear rise and short fall in pitch during the "aa" element of the call. Waa-barks are produced in isolation or in short series.	*Who?* Adults and subadults. *When?* During observation of an agonistic interaction, as threats to other species including baboons and bush pigs and as threats to distant opponents, most often by victims of aggression after the aggressor has retreated. Sometimes accompanied by threat gestures such as arm raising.	Aggressive threat, although a lack of clear responses to these calls by recipients makes defining their function difficult.
Cough	Low-frequency vocalization, similar to a grunt but rarely voiced. A short noisy signal with no harmonic structure.	*Who?* Dominant individuals. *When?* Given as a mild threat by an annoyed individual to a subordinate. It is often accompanied by a threat gesture such as a ground slap or an arm-raise.	The cough serves as a warning and is usually sufficient to secure the termination of the undesirable behavior in the subordinate.
Grunt	Short, soft low-frequency calls given singularly or in short bouts. Grunts grade into rough grunts, barks, and pant-grunts.	*Who?* Adults and subadults. *When?* During resting and the making of nests. Short grunt exchanges can also often be heard between affiliative individuals as they initiate travel, foraging, or rest, or they hear an approaching chimpanzee.	Grunts are poorly understood, but they may function to promote behavioural synchrony in affiliative individuals.
Rough grunt	An umbrella term that describes the vocalisations produced by individuals when approaching, collecting, or consuming food. Acoustically, rough grunts grade from low-frequency, unvoiced, noisy grunts to high-frequency tonal squeaks. They grade into screams in feeding situations accompanied by great social excitement.	*Who?* Predominantly adult males, but also subadults and females. *When?* Produced as individuals approach and climb feeding trees, and then typically in the first few minutes of feeding on a preferred food source.	"Food call" to attract other chimpanzees to a feeding site.

Table 16.1   (*continued*)

Call type	Acoustic description	Circumstances of production	Function
Pant-grunt	Noisy, low-frequency grunts or barks, panted in a rapid rhythm with audible energy in both inhalation and exhalation. Pant-grunts grade into pant-barks, which in turn grade into pant-screams.	*Who?* Given by subordinate individuals to dominant individuals. *When?* When greeting a dominant or signaling subordination to an approaching aggressive chimpanzee.	Sign of submission, vital for maintaining dominance relationships between group members.
Pant	Unvoiced, soft, low-frequency sounds. Temporal patterning is regular and rapid. Panting may grade into pant-grunting.	*Who?* Mainly adults. *When?* As part of a greeting or during grooming. In this context the caller habitually presses their open mouth against the other individual whilst panting. Panting also occurs in males during copulations and tends to be loudest in that context.	Seems to be an affiliatory signal in the grooming or greeting context.
Huu / alarm call	Tonal calls with most energy at onset and a rise and fall in frequency over the call. Huu calls can be loud, sometimes carrying over a hundred meters.	*Who?* Surprised or frightened individuals. *When?* In response to unusual events, such as a dead monkey in a snare, earth tremors, and seeing the waterproof cloak of a new researcher. The Budongo chimpanzees, who have never been recorded to produce "wraa" calls, also gave these as alarm calls in response to buffalo, pythons, and dying members of the community.	Alarm calls, which generally elicit approach and alert scanning behaviour from recipients. They attract other individuals and alert them to dangers in the environment.
Laughter	Noisy, low-frequency grunts and moans delivered in an irregular rhythm, reminiscent of hoarse, wheezing human laughter. Acoustic energy is audibly present in both inhalation and exhalation, with most voicing occurring during inhalation. The most discrete call type in the chimpanzee repertoire, as it rarely grades into any other call type.	*Who?* Predominantly infants and juveniles. *When?* In play contexts and particularly during physical-contact play, which may involve wrestling, tickling, and play-biting.	Primarily a reaction to physical contact in a play context, but it may also function to extend and maintain play bouts (Matsusaka 2004).

chimpanzee calls are the product of a caller's assessment of his or her social environment (Mitani 1996).

Another important point that arose from the research on pant-hoot vocalizations was that subtle but consistent structural differences emerged between the pant-hoots produced at two different study sites in east Africa, Gombe and Mahale (Mitani et al. 1992). These results were initially interpreted as evidence for different dialects. In other species, including songbirds and humans, dialects are usually the result of a vocal convergence process, and most likely the outcome of an underlying drive toward social conformity. Dialects require at least some degree of vocal learning abilities, something that is rarely described in nonhuman primates. Chimpanzees are extremely hostile towards neighboring communities, and dialectic variants in long-distance calls could serve as a behavioral mechanism to maintain group identity. The topic of dialects in nonhuman primates continues to be controversial and differences in genetics, habitat acoustics, and body size often provide more parsimonious explanations for such observations (Mitani et al. 1999). More recent work, however, has provided further support for the dialect hypothesis. Several groups of researchers have found evidence for communities converging on group-specific pant-hoot vocalizations both in captivity (Marshall et al. 1999) and in the wild (Crockford et al. 2004). Further evidence for vocal learning has come from studies showing flexible modification of pant-hoot structure in wild chimpanzees during male chorusing (Mitani and Brandt 1994; Mitani and Gros-Louis 1998).

Pant-hoots are also interesting because there is some evidence that they are given in context-specific ways. An early study suggested that pant-hoots produced during the arrival at fruiting trees (arrival pant-hoots) provide information regarding the status or rank of the party (Clark and Wrangham 1994). However, more detailed research has indicated that the context-specific usage of pant-hoots may provide listeners with relatively specific information regarding the caller's environment and behavioral context (Uhlenbroek 1996; Notman and Rendall 2004). Uhlenbroek (1996) established the existence of four acoustically distinct pant-hoot variants in the Gombe population and found that three of the four variants were most commonly given in distinct behavioral contexts: "wail" pant-hoots whilst feeding, "roar" pant-hoots during travel or display, and "slow roar" pant-hoots upon arrival at a plentiful food

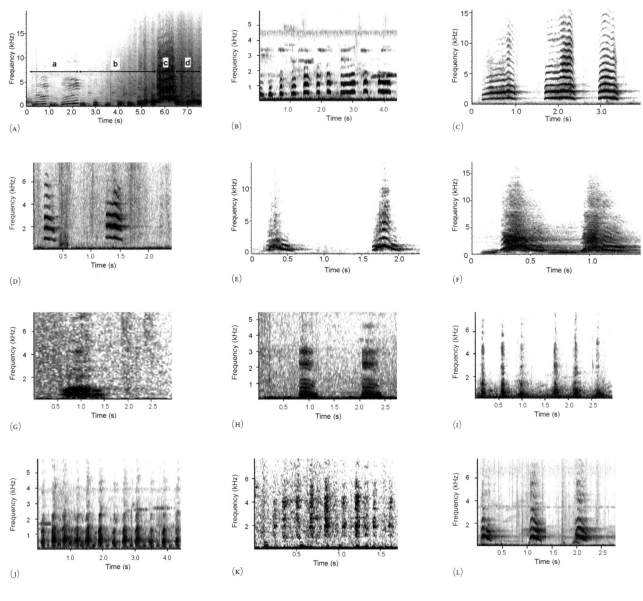

Figure 16.3   Spectrograms illustrating 13 main call types recorded from the Sonso community of chimpanzees, Uganda: (a) pant-hoot given by adult male Bwoba in a display, where a = introduction, b = buildup, c = climax, d = letdown; (b) whimper given by juvenile female Nora when lost; (c) scream given by adult female Flora in tantrum; (d) squeak given by adult female Janie during copulation; (e) bark given by adult female Kwera while watching display by adult male Nick; (f) waa-bark given by adult male Nick to baboons; (g) cough given by adult female Kwera to subadult playing too roughly with her juvenile son; (h) grunt given by subadult male Gashom before initiating travel; (i) rough grunts given by adult male Nick whilst feeding on *Chrysophyllum milicia excelsa*; (j) pant-grunt given by adult female Kalema to alpha male Duane; (k) pant given by Zefa during copulation with Janie; (l) huu given by Kalema in response to earth tremor; (m) laughter given by juvenile female Janet whilst wrestling on the ground.

source. A replication of this study with the Sonso community of the Budongo forest yielded similar results, although the data were interpreted somewhat differently (Notman and Rendall 2005). Context-specific use of different call variants generally carries the potential to provide listeners with information about the environment experienced by the caller, a fundamental prerequisite for referential signals. Whether or not receivers are able to discriminate the acoustic differences described by Uhlenbroek (1996) and Notman and Rendall (2005), and whether they can infer the corresponding context experienced by the caller, has not yet been investigated.

BARKS. Barks have been identified as context-specific calls, with the potential to function as referential signals because different types of acoustic variants appear to relate to discrete events and objects in the external world. A study with the wild chimpanzees of the Taï Forest, Côte D'Ivoire, examined the circumstances of bark production in adult males and identified different acoustic variants in response to snakes and while hunting (Crockford and Boesch 2003). In addition, when bark variants were combined with other calls or with drumming, then context specificity was even higher, between 93% and 100%. Barks, especially in combination with other call types, are produced in a highly context-specific manner and so have the potential to function as carriers of distinct meanings. As with the pant-hoots, it is currently not known whether receivers are able to discriminate between these context-specific call variants and extract meaningful information from them. Playback experiments will ultimately be needed to assess whether that is so.

ROUGH GRUNTS. "Rough grunts" is a term for a group of vocalizations that are produced in the feeding context. Receivers commonly respond to these signals by approaching the food source. Captive chimpanzees are more likely to give rough grunts when discovering a large and divisible food source than when discovering a small or indivisible one (Hauser et al. 1993), and callers appear to adjust call production depending on the audience present (chimpanzees: Hauser and Wrangham 1987; bonobos: Van Krunkelsven et al. 1996). Because these calls are produced exclusively in feeding contexts and because there is substantial acoustic variation within this call type, they may function as referential signals. Indeed, previous work with macaque monkeys has shown that this species produces acoustically different signals to food of different quality, and that these calls are meaningful to listeners (Hauser 1998).

In a set of recent studies we aimed to investigate whether chimpanzee rough grunts functioned to indicate more than the mere presence of food but also contained information about the nature of the discovered food source. We conducted these studies primarily with captive chimpanzees housed at Edinburgh Zoo, due to the difficulties in objectively measuring food quality, quantity, and individual

preferences in the wild. First, we constructed a food preference hierarchy. We did this by providing the chimpanzees with feeds consisting of two types of food, and then recorded which of the two foods each individual chose. This way we were able to objectively rank nine common foods in order of preference. We then focused our analysis on four individuals in the group, whose preferences closely converged. These four individuals were an adult male, an adult female, a subadult male, and a subadult female. We split the nine food types into three different value groups (high, medium, and low) and then recorded rough grunts produced by the four different individuals upon discovering the different foods. We had a total of 76 calling bouts of suitable quality to use for analysis. We first compared the acoustic structure of the rough grunts given to high, medium and low value foods (N = 108 calls). This analysis demonstrated that chimpanzee rough grunts were subtly but consistently different depending on the value of food encountered (see figure 16.4).

High-value food elicited high-frequency, tonal, long grunts whilst low-value food elicited low-frequency, noisy, short grunts. Medium-value food elicited grunts that were intermediate in structure. Chimpanzees thus reliably produced grunts that reflected the value of the food they were eating. We then examined whether the rough grunts could have an even greater degree of referential specificity and could act as unique labels for specific food types. In order to control for the effect preference has on the rough grunt structure, we examined whether the grunts given to the three food types that made up each value class differed in acoustic structure. We found no evidence for this in low- and medium-value foods, but within the high-value category we found, somewhat to our surprise, that the grunts given to bread, mango, and banana were acoustically distinct. We also found that the grunts given to bread and bananas remained stable across different feeding events, and thus that they could function as labels for food type (Slocombe and Zuberbühler 2006).

We then examined a number of rough grunts from the wild chimpanzees of the Budongo Forest, given in response to three medium-high preference foods (i.e. three foods with the highest proportion of total feeding time dedicated to their consumption). The acoustic structure of these grunts (n = 36) was comparable to the grunts given by captive chimpanzees to high- or medium-preference foods, giving the captive results a degree of ecological

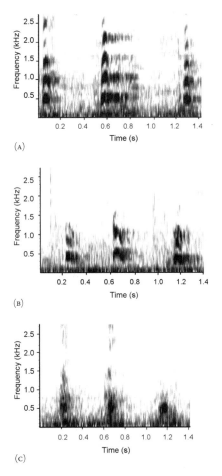

Figure 16.4   Example time-frequency spectrograms illustrating rough grunts given by captive adult male Louis in response to (a) bread, (b) chow, (c) carrots. Bread is a high-preference food, chow is a medium-preference food, and carrots are a low-preference food. The spectrograms depict a representative 1.4-second section of a 15-second calling bout given to bread, chow, and carrots respectively. Acoustic energy is depicted by the darkness of each image.

validity. We were, however, unable to replicate the finding that different food types elicited acoustically distinct grunts. Although we cannot conclude that this does not occur in the wild until we have better measures of food value, we hypothesize that the labeling of specific food types is likely to be a by-product of a preference-driven system, within the special conditions of captivity. In captivity the quality, quantity, and novelty of the different food types were relatively consistent, and food type may thus become the main determinant of food value in captive chimpanzees (Slocombe and Zuberbühler 2006). In summary, the most parsimonious interpretation we can offer is that chimpanzees produce rough grunts that reference the *value* of a discovered food source, although in certain circumstances the system has the capacity to label specific food items as well.

As highlighted earlier, a crucial piece of evidence con-

cerning the meaning and function of a communication system relates to the problem of how receivers process different call types produced by individuals in response to different events. In order to establish that a call references an event in the environment, it is necessary to show that the listeners understand its meaning and its environmental referent. We therefore conducted a playback study to examine a listener's response to rough grunt calls, to see if they meaningfully refer to food value (Slocombe and Zuberbühler 2005a). This study was also conducted at Edinburgh Zoo, where we first established two artificial trees in the chimpanzees' enclosure. These "trees" regularly dropped tubes, filled either with (low-value) apples or with (high-value) bread, into a gully in the study group's outdoor enclosure. We introduced the contingency that only one tree ever fruited successfully, with the other tree producing empty tubes. This ensured that the chimpanzees did not simply monopolize the preferred bread tree. Once the chimpanzees became accustomed to this rule we started the playback trials. Before any playback trial we waited until all individuals were in the indoor enclosure, usually to receive a small feed. While they were absent, we dropped empty tubes from both trees to simulate a successful fruiting event by one of them. As soon as the first individual emerged from the inside enclosure we played back rough grunts, originally given by a familiar group member to either bread or apples, from a loudspeaker that was positioned exactly halfway between the two trees. We then filmed the focal animal's response to monitor his or her foraging strategy. As it was impossible to separate individuals, we were only able to collect systematic data from one individual, a six-year-old male named Liberius, who was always the first individual to come out into the outside enclosure.

First we found that when Liberius heard examples of rough grunts from the speaker, he spent significantly more time sitting on the ladder leading into the outdoor enclosure, as if trying to figure out the event's significance, than he did in control trials where he heard no grunts. At the very least this gave us reassurance that Liberius was processing the stimuli we were using. In the first few trials he consistently visited the "correct" tree first—that is, the tree indicated by the rough grunts. Soon thereafter, however, he adopted an idiosyncratic foraging strategy, during which he consistently searched his favorite bread tree first, regardless of the types of rough grunts used as playback stimuli. We therefore analyzed subtler behavioral measures collected during all trials, to examine whether he was processing the rough grunts in a meaningful manner. We hypothesized that if he had an expectation about where to find food on the basis of the grunts he had heard, he should put different amounts of effort into searching the two locations. We found that when Liberius heard "apple grunts" (n = 9), he spent more time searching and tended to search more tubes under the correct (apple) tree than under the incorrect (bread) tree. In contrast, when he heard "bread grunts" (n = 8) he searched more tubes and tended to search longer under the correct (bread) tree than the incorrect (apple) tree. In control trials when Liberius heard no grunts (n = 10), he showed no such biases in search effort. We concluded that he understood that the grunts referred to foods of different value, and that he used this information to aid any search for food. This was the first experimental evidence that any member of the great ape species used their vocalizations in a functionally referential manner (Slocombe and Zuberbühler 2005a).

SCREAMS. Agonistic screams were first identified as functioning as referential signals in macaque monkeys. Macaques produce acoustically distinct screams depending on the severity of aggression and the rank of their opponent, and listeners seem to be able to extract this information from the calls (Gouzoules et al. 1984). We have recently investigated the screams produced by wild chimpanzees in the Budongo Forest, Uganda, during agonistic interactions. One study examined whether the social role of the caller in a fight was encoded in the acoustic structure of the screams they produced. Victims commonly screamed, and aggressors attacking individuals of equal or higher rank also screamed regularly. We compared the acoustic structure of screams produced by 14 individuals, in the roles of both victim and aggressor, and found subtle but consistent acoustic differences (Slocombe and Zuberbühler 2005b). The key differences in acoustic structure were related to the shape of the call, not the absolute frequency or temporal parameters; victim screams were very flat, with very little change in frequency over the call (figure 16.5). In contrast, aggressor screams had a distinct dip in frequency in the second half of the call, and to the ear these screams have a bark-like quality. Some anecdotal observations indicated that listening individuals were able to extract meaningful social information about the caller's role from the screams, because it seemed to influence the listeners' decisions about whether to intervene in a fight. Recent playback experiments conducted in captiv-

ity have shown that these calls are processed meaningfully by listeners (Slocombe, Kaller, Call, and Zuberbühler, in review).

A second study considered the more common victim screams in greater detail. We examined victim screams produced by 21 different individuals, and found that the calls varied in their acoustic structure as a function of the severity of the aggression they were receiving (Slocombe and Zuberbühler 2007). Victims receiving severe aggression (chasing or beating) gave longer bouts of screams, in which each call was longer in duration and higher in frequency than screams produced by victims of mild aggression (charges or postural threats). Recent playback experiments conducted in the wild have shown that listening individuals meaningfully distinguish between these two graded scream variants (Slocombe et al. 2009).

Inspired by the anecdote presented at the start of the chapter, we also examined whether individuals modified their screams according to the composition of the audience. We examined the behavior of 21 individuals, and found that victims receiving severe aggression were sensitive to the composition of the listening audience. If there was an individual of equal or higher rank than the aggressor (i.e.,

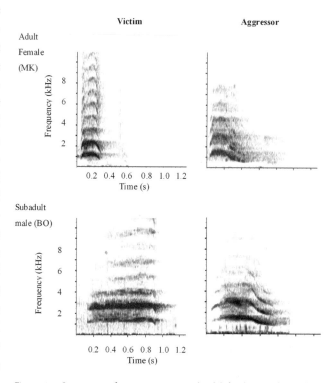

Figure 16.5 Spectrograms of agonistic screams given by adult female MK and subadult male BO in the roles of victim and aggressor. Victim and aggressor screams differ crucially in the shape of the call, with aggressor screams having a distinctive downturn in frequency during the second half of the call.

an individual who could effectively challenge the aggressor) present in the party, then the victim produced screams that were acoustically consistent with extremely severe aggression. This vocal exaggeration occurred only when the chimpanzees most needed aid—that is, when they encountered severe aggression rather than mild aggression. It seems that victim screams thus function to recruit aid. The mere presence of a high-ranking individual did not explain this audience effect, however—which indicated that chimpanzees have intricate knowledge of third-party relationships (see chapters 15 and 17), and that this knowledge of who can effectively challenge whom effects their vocal production (Slocombe and Zuberbühler 2007). From a cognitive perspective, this finding is very exciting. It adds to the body of evidence that chimpanzees are capable of tactical deception (Byrne and Whiten 1992), and provides the first systematic evidence of the exaggeration of distress to manipulate the behavior of group members. Currently we cannot be sure of the cognitive mechanism underlying this behavior; the results are consistent with a number of explanations, including complex context-specific vocal production and intentional deception. Ongoing work aims to tease apart these explanations to elucidate the mechanisms underlying this fascinating behavior. This study does, however, show how the study of vocal behavior in chimpanzees can be used to examine other elements of cognition, such as their awareness and understanding of third-party relationships and their capacity for deception.

Summary

Wild chimpanzees are avid producers of vocal signals, their primary mode of communication in their visually dense natural forest habitat. Technical advances in sound analysis and a renewed interest in chimpanzee vocal behavior have resulted in relatively detailed investigations of some of their calls. This work has been able to identify acoustic variants within the call types, and in some cases it has been possible to assign particular meanings and functions to the different call variants. Barks and pant-hoots are given in context-specific ways that may provide listeners with information about the behavior of the caller. Rough grunts, given in response to food, vary in acoustic structure as a function of food value, and in a captive setting this can even result in grunt variants effectively functioning as labels for different food types. A playback experiment has shown that a listening individual can infer the kind of food that has been discovered by the caller, and use that information to guide his or her own search for food. Screams given during agonistic interactions vary reliably with the caller's social role, with victims and aggressors giving distinct screams. The acoustic structure of victim screams also varies as a function of the severity of the aggression the caller is experiencing. Recent playback experiments have shown that listeners meaningfully process both the social role of the scream provider and the severity of aggression experienced by the caller. Victims of severe aggression also modify the structure of their screams if there is an individual in the audience who can effectively challenge the aggressor. Thus, the emerging picture of these vocalizations is one of a highly complex behavior—a reflection of chimpanzees' sophisticated intelligence that is much in line with their already well-documented behavioral and cognitive capacities.

Implications and Future Directions

As research begins to expose the complexities of the chimpanzee vocal system, it may become necessary to reevaluate the role and importance of vocal communication in many aspects of chimpanzee behavior. Some of the studies presented here suggest that chimpanzee listeners may be able to draw inferences about their social and ecological world that they cannot see. We may have underestimated the role that information about distant events, gained from auditory cues, has on the behavior of these animals. In the social realm, chimpanzees may be able to effectively monitor third-party relationships simply by listening to call exchanges made by others. Knowledge of the presence and quality of food patches could be obtained in similar ways, and that information may affect future travel patterns and foraging behavior. Previously, it has been accepted that inter-party vocalizations aid coordination of travel between distant individuals (Goodall 1986), but it is plausible that vocal behavior also provides individuals with much more complex information about social interactions and other important events in the environment.

Enhanced Understanding of Human Language Evolution

Research into chimpanzee vocal communication has an impact on our understanding of human language evolution.

The comparative approach to language evolution searches for precursors to human language in animals, commonly primates, in order to model the capacities of our common ancestors. This approach is built on the understanding that human language is incredibly complex, the product of a large set of sophisticated cognitive abilities that could not have evolved from scratch in the relatively short period of evolutionary time during which speech-related articulatory skills have been available to humans (Enard et al. 2002). Thus, we assume that many of the cognitive capacities involved in language processing are much older than language itself, with their phylogenetic roots deep in the primate lineage. Until now, comparative work on primate vocal communication has focused on various monkey species (Zuberbühler 2003). This work has been critical in suggesting that primates possess some of the precursors to human linguistic abilities, such as reference. However, it has always been a problem that comparable abilities have not been reported in any of the great apes. Without such findings in apes, the abilities found in monkeys are better interpreted as a case of convergent evolution rather than the expression of a general primate-specific ability, and therefore as not being particularly informative for theories of language evolution. The current evidence indicates that chimpanzees can produce referential calls, and that their calls are aimed at specific audiences and modified in response to them. As chimpanzees are the closest living approximation to the common human-chimpanzee ancestor (McGrew 1991), we are confident that future research of this species will identify other communication and cognitive skills that are relevant in this context, and which may further shape our understanding of how our own language evolved.

Cognitive Mechanisms Underlying Vocal Production in Chimpanzees

Although considerable advances have been made in understanding the function and meaning of many chimpanzee calls, the cognitive mechanisms underpinning their vocal behavior remain more elusive. Acts of communication involve at least a signaler and a receiver, and more typically a network of individuals, but many aspects of call processing and comprehension in chimpanzees are not well understood. For example, do individuals perceive graded calls in a categorical manner, and what are the relevant distinctions? How well can listening chimpanzees recognize familiar individuals from the different call types? What sorts of inferences about environmental events are they capable of making when hearing others call? Our playback experiment with rough grunts has shown that chimpanzees are able to engage in such inferential activities, but the level of sophistication is not well explored (Slocombe and Zuberbühler 2005a). Equally little is known about the ontogenetic processes—that is, how receivers learn to associate meanings with the different call types and variants. For example, we do not know whether chimpanzees engage in any kind of "fast mapping" between sounds and events, in a manner analogous to the way young children acquire language or the way some dogs acquire the meaning of human words (Kaminski et al. 2004).

As signalers, primates are remarkably unskilled at controlling the acoustic structure of their vocalizations (Riede et al. 2005). Although their vocal repertoire seems to be innately determined, chimpanzees must still learn to produce the correct call in the appropriate context. More research is needed to elucidate the range of social and ecological information contained in the acoustic variation of the various call types, and the sorts of psychological variables that callers are able to take into account before producing a call. Our recent work indicates that chimpanzees have a degree of flexibility and voluntary control over their vocal production. For example, they are capable of exaggerating the nature of ongoing events (Slocombe and Zuberbühler 2007). However, the intentions and motivations underlying their vocal production remain largely unclear. Future research is needed to elucidate the degree of intentionality that underlies chimpanzees' call production, as well as the degree to which they take each others' knowledge and likely behavioral responses into account before producing a call.

Acknowledgments

We thank Simon Townsend for useful discussion of this chapter. We are grateful to the staff of the Budongo Conservation Field Station, Vernon Reynolds, the Ugandan Wildlife Authority, the Ugandan National Council for Science and Technology, and the President's Office for permission to live and work in Uganda. We also thank the Royal Zoological Society of Scotland for providing key funding for the Budongo Conservation Field Station, and for permission to work at Edinburgh Zoo.

Literature Cited

Abramson, A. S., and L. Lisker. 1970. Discriminability along with voicing continuum: Cross language tests. In *Sixth International Congress of Phonetic Sciences*. Prague: Academia.

Altmann, J. 1974. Observational study of behaviour: Sampling methods. *Behaviour* 49:227–67.

Arnold, K., and K. Zuberbühler. 2006a. The alarm-calling system of adult male putty-nosed monkeys, *Cercopithecus nictitans martini*. *Animal Behaviour* 72:643–653.

———. 2006b. Language evolution: Semantic combinations in primate calls. *Nature* 441:303.

Byrne, R. W., and A. Whiten. 1992. Cognitive evolution in primates: Evidence from tactical deception. *Man* 27:609–27.

Call, J., and M. Tomasello. 2007. *The Gestural Communication of Apes and Monkeys*. Oxon, UK: Routledge.

Cheney, D. L., and R. M. Seyfarth. 1990. *How Monkeys See the World: Inside the Mind of Another Species*. Chicago: University of Chicago Press.

———. 2007. *Baboon Metaphysics: The Evolution of a Social Mind*. Chicago: University of Chicago Press.

Clark, A. P. 1993. Rank differences in the production of vocalizations by wild chimpanzees as a function of social-context. *American Journal of Primatology* 31(3): 159–79.

Clark, A. P., and R. W. Wrangham. 1993. Acoustic analysis of wild chimpanzee hoots: Do Kibale forest chimpanzees have an acoustically distinct food arrival pant hoot? *American Journal of Primatology* 31:99–110.

Crockford, C., and C. Boesch. 2003. Context-specific calls in wild chimpanzees, *Pan troglodytes verus*: Analysis of barks. *Animal Behaviour* 66:115–25.

Crockford, C., I. Herbinger, et al. 2004. Wild chimpanzees produce group-specific calls: A case for vocal learning? *Ethology* 110(3): 221–43.

Di Bitetti, M. S. 2003. Food-associated calls of tufted capuchin monkeys (*Cebus apella nigritus*) are functionally referential signals. *Behaviour* 140:565–92.

Eggeling, W. J. 1947. Observations on the ecology of Budongo rain forest, Uganda. *Journal of Ecology* 34:20–87.

Enard, W., M. Przeworski, S. E. Fisher, C. S. L. Lai, V. Wiebe, T. Kitano, A. P. Monaco, and S. Paabo. 2002. Molecular evolution of FOXP2, a gene involved in speech and language. *Nature* 418:869–72.

Evans, C. S., and L. Evans. 2007. Representational signalling in birds. *Biology Letters* 3:8–11.

Fischer, J. 1998. Barbary macaques categorize shrill barks into two different call types. *Animal Behaviour* 55:799–807.

Fischer, J., M. Metz, et al. 2001. Baboon responses to graded bark variants. *Animal Behaviour* 61:925–31.

Fitch, W. T. and M. D. Hauser. 2002. Unpacking "honesty": Vertebrate vocal production and the evolution of acoustic signals. In *Acoustic Communication*, ed. A. M. Simmons, R. R. Fay, and A. N. Popper. New York: Springer, 65–137.

Goodall, J. 1986. *The Chimpanzees of Gombe: Patterns of Behavior*. Cambridge, MA: Harvard University Press.

Gouzoules, S., H. Gouzoules, and P. Marler. 1984. Rhesus monkey (*Macaca mulatta*) screams: Representational signalling in the recruitment of agonistic aid. *Animal Behaviour* 32:182–93.

Green, S. 1975. Communication by a graded vocal system in Japanese monkeys. In *Primate Behavior*, ed. L. A. Rosenblum. New York: Academic Press.

Hauser, M. D. 1998. Functional referents and acoustic similarity: Field playback experiments with rhesus monkeys, *Animal Behaviour* 56:1309–10.

Hauser, M. D., P. Teixidor, et al. 1993. Food-elicited calls in chimpanzees: Effects of food quantity and divisibility? *Animal Behaviour* 45:817–19.

Hauser, M. D. and R. W. Wrangham. 1987. Manipulation of food calls in captive chimpanzees: A preliminary report. *Folia Primatologica* 48:24–35.

Kaminski, J., J. Call, and J. Fischer. 2004. Word learning in a domestic dog: Evidence for "fast mapping." *Science* 304:1682–83.

Manser, M. B., R. M. Seyfarth, and D. L. Cheney. 2002. Suricate alarm calls signal predator class and urgency. *Trends in Cognitive Sciences* 6:55–57.

Marler, P. 1968. Aggregation and dispersal: Two functions in primate communication. In *Primates: Studies in Adaptation and Variability*, ed. P. C. Jay. New York: Holt, Rinehart & Winston, 420–38.

———. 1970. Vocalizations of East African monkeys. I. Red colobus. *Folia Primatologica*, 13: 81–91.

———. 1976. Social organization, communication and graded signals: The chimpanzee and the gorilla. In *Growing Points in Ethology*, ed. P. P. Bateson and R. A. Hinde. Cambridge: Cambridge University Press.

Marler, P. and R. Tenaza. 1977. Signaling behavior of apes with special reference to vocalizations. In *How Animals Communicate*, ed. T. A. Sebeok. Bloomington: Indiana University Press, 965–1033.

Marshall, A., R. Wrangham, et al. 1999. Does learning affect the structure of vocalizations in chimpanzees? *Animal Behaviour* 58:825–30.

Matsusaka, T. 2004. When does play panting occur during social play in wild chimpanzees? *Primates* 45:221–29.

McGrew, W. 1991. Chimpanzee material culture: What are its limits and why? In *The Origins of Human Behavior*, ed. R. Foley. London: Unwin Hyman Ltd.

Mitani, J. 1994. Ethological studies of chimpanzee vocal behavior. In *Chimpanzee Cultures*, ed. R. W. Wrangham, W.C. McGrew, F. B. M. deWaal, and P.G. Heltne. Cambridge, MA: Harvard University Press.

Mitani, J. C. 1996. Comparative studies of African ape vocal behavior. In *Great Ape Societies*, ed. W. C. McGrew, L. F. Marchant, and T. Nishida,. Cambridge: Cambridge University Press, 241–54.

Mitani, J. C., and K. L. Brandt. 1994. Social factors influence the acoustic variability in the long-distance calls of male chimpanzees. *Ethology*, 96(3): 233–52.

Mitani, J. C., and J. Gros-Louis 1998. Chorusing and call convergence in chimpanzees: Tests of three hypotheses. *Behaviour* 135(8–9): 1041–64.

Mitani, J. C., T. Hasegawa, et al. 1992. Dialects in wild chimpanzees? *American Journal of Primatology* 27:233–43.

Mitani, J. C., K. L. Hunley, et al. 1999. Geographic variation in the calls of wild chimpanzees: A reassessment. *American Journal of Primatology* 47(2): 133–51.

Newton-Fisher, N. E. 1999. The diet of chimpanzees in the Budongo Forest Reserve, Uganda. *African Journal of Ecology* 37:344–54.

Notman, H., and D. Rendall. 2005. Contextual variation in chimpanzee pant hoots and its implications for referential communication. *Animal Behaviour* 70(1): 177–90.

Reynolds, V. 1965. *Budongo: A Forest and its Chimpanzees*. London: Methuen.

———. 2005. *The Chimpanzees of Budongo Forest: Ecology, Behaviour and Conservation*. Oxford: Oxford University Press.

Riede, T., E. Bronson, H. Hatzikirou, and K. Zuberbühler. 2005. Vocal production mechanisms in a non-human primate: Morphological data and a model. *Journal of Human Evolution* 48:85–96.

Rowell, T. E., and R. A. Hinde. 1962. Vocal communication by the rhesus monkey (Macaca mulatta). *Proceedings of the Zoological Society of London* 8:279–94.

Seyfarth, R. M., D. L. Cheney, and P. Marler. 1980. Monkey responses to three different alarm calls: Evidence of predator classification and semantic communication. *Science* 210:801–3.

Slocombe, K. E., T. Kaller, J. Call, and K. Zuberbühler, in review. Chimpanzees extract social information from agonistic screams.

Slocombe, K. E., S. W. Townsend, and K. Zuberbühler. 2009. Wild chimpanzees (*Pan troglodytes schweinfruthii*) distinguish between different scream types: Evidence from a playback study. *Animal Cognition* 12(3): 441–49.

Slocombe, K. E., and K. Zuberbühler. 2005a. Functionally referential communication in a chimpanzee. *Current Biology* 15(19): 1779–84.

———. 2005b. Agonistic screams in wild chimpanzees (*Pan troglodytes schweinfurthii*) vary as a function of social role. *Journal of Comparative Psychology* 119(1): 67–77.

———. 2006. Food-associated calls in chimpanzees: Responses to food types or food preferences? *Animal Behaviour* 72:989–99.

———. 2007. Chimpanzees modify recruitment screams as a function of audience composition. *Proceedings of the National Academy of Science* 104(43): 17228–33.

Tomasello, M., and K. Zuberbühler. 2002. Primate vocal and gestural communication. In *The Cognitive Animal*, ed. M. Bekoff, C. Allen, and G. M. Burghardt. Cambridge, MA: MIT Press, 293–99.

Tutin, C. E. G., and W. C. McGrew. 1973. Chimpanzee Copulatory Behavior, *Folia Primatologica* 19(4): 237–56.

Uhlenbroek, C. 1996. *The Structure and Function of the Long-Distance Calls Given by Male Chimpanzees in Gombe National Park*. PhD thesis, University of Bristol.

Van Krunkelsven, E., J. Dupain, et al. 1996. Food calling in captive bonobos (*Pan paniscus*): an experiment. *International Journal of Primatology* 17:207–17.

Zuberbühler, K. 2000. Causal cognition in a non-human primate: Field playback experiments with Diana monkeys. *Cognition* 76:195–207.

———. 2003. Referential signalling in non-human primates: Cognitive precursors and limitations for the evolution of language. *Advances in the Study of Behavior* 33:265–307.

Zuberbühler, K., R. Noë, and R. M. Seyfarth. 1997. Diana monkey long-distance calls: Messages for conspecifics and predators. *Animal Behaviour* 53:589–604.

17

The Function and Cognitive Underpinnings of Post-Conflict Affiliation in Wild Chimpanzees

Roman M. Wittig

It is dry season in the Taï National Park, Côte d'Ivoire. Faint hammering noises swirl through the air of the rain forest—pound, pound, crack. A group of female chimpanzees lounges under a huge tree while two more of them sit on the tree roots and crack open panda nuts using heavy stones as hammers. Pound, pound, crack—Castor, a low-ranking female, opens another nut and holds the open shell to her mouth to scoop out the nut's white flesh while holding the stone firmly in her other hand. A stone of that size is a rare and essential tool for opening the hard shell of a panda nut. Suddenly, food grunts emerge from the forest and another party of females arrives at the nut tree. One of them is Venus, a high-ranking female who approaches Castor and stares at her. Castor avoids Venus's gaze and continues cracking. About five meters behind Castor the alpha female, Mystère, is resting on the ground. She grunts, and Venus leaves Castor to sit down beside Mystère. After about 10 minutes Mystère gets up and leaves the nut-cracking site. As soon as Mystère is out of sight, Venus approaches Castor again and, with a leap over the root, supplants her and takes over her stone hammer.

Why did Venus delay taking over the stone hammer and the nut-cracking site? Did she need Mystère to leave before she would feel able to supplant the subordinate Castor? Did she delay her attack because she had a hostile relationship with Mystère, or because Mystère and Castor shared a supportive bond? Either way, it must have been advantageous for Venus to know exactly about the status of her own relationship to Mystère, and about the relationship between Mystère and Castor.

Monitoring and recognizing others' relationships seems to be one of the crucial characteristics of social intelligence. Group-living monkeys are known to recognize their own affiliative bonds and those of others; they track each other's short-term relationships and understand indirect causal inferences by combining several pieces of social knowledge. They even seem to recognize the social relationships between individuals of a different species (reviewed in Cheney and Seyfarth 2007). While studies have tried to compare the social intelligence of monkeys with that of carnivores (Holekamp 2007), cetaceans (Connor 2007), and birds (Emery 2006), our knowledge about the social intelligence of apes is rather patchy. Cognitive research in apes has focused mainly on instrumental cognition (see chapters 10, 11, and 12), culture (see chapters 8, 9 and 14), "theories of mind" (see chapters 7, 19, 22, and 23), and communication (see chapter 16). In this chapter I begin to investigate what chimpanzees know about their social environment within the framework of conflict management.

The Framework of Conflict Management

Living in groups not only provides benefits but entails costs. Social animals have to compete with their own group members over food, mating partners, and shelter

(Huntingford and Turner 1987). They face a dilemma: to compete over resources with the same individuals with whom they must cooperate to gain the benefits associated with group living (e.g., cooperation: see chapters 15, 21, and 28; social learning: see chapters 9, 12, and 14).

Facing a competitive situation, as when two group members both want to eat the same fruit, group-living animals have to "decide" which kind of action to take (de Waal 1996; Wittig and Boesch 2003a). Should one group member avoid the other and refrain from eating the fruit, tolerate the other and share the fruit, or act aggressively to gain possession of the fruit? Aggression may jeopardize a group's cohesion and entail costs for both combatants. Aside from its physical costs in injuries and energy consumption (Neat et al. 1998), aggression increases the levels of stress hormones (Sapolsky 1995; Wallner et al. 1999; Crockford et al. 2008) and entails social costs (de Waal 1996; Wittig and Boesch 2003a). After an aggressive interaction, opponents' levels of tolerance for mutual proximity are disturbed (Cords 1992; Wittig and Boesch 2005). It is unclear whether the victim of aggression avoids the aggressor for fear of further aggression, or whether the aggressor avoids the victim for fear of retaliation. In any case, beneficial cooperation between former opponents is rarely possible. Post-conflict affiliation, however, appears to reduce the negative repercussions of aggression and help maintain the benefits of group living (reviewed in Aureli and de Waal 2000).

Post-Conflict Affiliation

In 1979 de Waal and van Roosmalen published the first study on post-conflict affiliation, conducted on a colony of chimpanzees at Burgers' Zoo in Arnhem, the Netherlands. They found that despite the prediction that opponents would separate after aggression, they were actually more likely to affiliate with each other soon after the conflict. De Waal and van Roosmalen found also that victims of aggression were often targets of affiliative behavior by third parties soon after the conflict. Intuitively, they gave the behaviors functional instead of operational labels, and calling the post-conflict affiliation among former opponents "reconciliation" and the bystander-initiated post-conflict affiliation with the victim "consolation".

Over the next few decades, "reconciliation" was found in many primates and some nonprimate social mammal species (reviewed in Aureli and de Waal 2000). It became

increasingly apparent that the function implied by the label "reconciliation" was reflected in the post-conflict behavior of primates (the Oxford Dictionary's definition of "reconcile" is "to restore friendly relations between"). Several studies have since proved that tolerance among former opponents is disturbed after aggressive interactions, and that reconciliation is able to restore the tolerance between opponents to baseline levels (Cords 1992; Cheney and Seyfarth 1997; Wittig and Boesch 2005). These findings are best interpreted by the *valuable relationship hypothesis* (de Waal and Aureli 1997), which is based on the idea that social relationships are valuable tools to increase each partner's reproductive success. According to the valuable relationship hypothesis, the restoring of tolerance to normal levels between opponents should occur more often when the opponents are mutually valuable partners. Indeed, the relatedness of dyads and rates of affiliation are commonly correlated with the dyad's rate of reconciliation (reviewed in Aureli and de Waal 2000).

In contrast to reconciliation, the function of "consolation" has not yet been rigorously tested. The label implies stress reduction (the Oxford Dictionary's definition of *console* is "to comfort in a time of grief and disappointment"), and several studies have suggested that "consolation" alleviates the stress of the victim of aggression (de Waal and Aureli 1996; Aureli 1997; Wittig and Boesch 2003b; Kutsukake and Castles 2004; Palagi et al. 2004). When tested, the occurrence of consolation did decrease the rate of self-scratching in one colony of captive chimpanzees (Fraser et al. 2008), but not in another (Koski and Sterck 2007). Its function, therefore, remains under debate (reviewed in Fraser et al. 2009).

Two hypotheses, not mutually exclusive, shape our understanding of "consolation." First, the *consolation hypothesis* suggests that the bystander behaves in an empathetic manner to "console" the victim (de Waal and Aureli 1996; Aureli 1997) and alleviate his or her stress. This hypothesis predicts that empathetic "consolation" behavior is more likely to occur when the bystander and the victim share a close relationship. Since empathy seems to depend on complex cognitive abilities, such as that of attributing emotions to others that are different from one's own, "consolation" was once thought exclusive to apes (Preston and de Waal 2002). This assumption is reflected in many positive findings of "consolation" events in apes (e.g., *Pan troglodytes*: de Waal and van Roosmalen 1979, Kutsukake and Castles 2004, Koski and Sterck 2007; *Pan paniscus*: Palagi et al. 2004; *Gorilla gorilla*: Watts et al. 2000). However, "consolation" events have recently also been observed in monkeys (Call et al. 2002; Wittig et al. 2007) and probably in corvid birds (Seed et al. 2007) and dogs (Cools et al. 2008).

Second, the *relationship-repair hypothesis* proposes that "consolation" repairs the opponents' relationships as it substitutes for reconciliation by restoring the opponents' relationship to baseline levels (Watts et al. 2000; Wittig et al. 2007). A bystander who has a close bond with the aggressor may act as a "proxy," reconciling with the victim on behalf of the close associate (Wittig et al. 2007). The *relationship-repair hypothesis* predicts that "consolation" behavior is more likely to occur when the bystander and the aggressor share a close relationship. It also predicts that the victim recognizes the bond between the aggressor and the consoler, and therefore treats the consoler's action as a "proxy" for reconciliation. Evidence for the relationship-repair hypothesis is provided by experiments on kin-mediated reconciliation among savannah baboons (*Papio hamadryas ursinus*). Female baboons reconcile conflicts by emitting conciliatory grunts toward the victims of their aggression (Cheney and Seyfarth 1997). These grunts function as signals of benign intent and increase the likelihood that the aggressor and her victim will subsequently tolerate each other's close proximity (Silk et al. 1996). Recent playback experiments have demonstrated that reconciliatory grunts by the aggressor's close kin towards the victim (operationally equivalent to "consolation") have a similar effect (Wittig et al. 2007). Victims tolerated a former opponent more quickly after hearing the grunts of their opponents' close relatives than after hearing the grunts of unrelated females, which did not change their tolerance towards the aggressors.

Reconciliation in Wild Chimpanzees

Twenty years after the first study on reconciliation in a captive chimpanzee colony (de Waal and van Roosmalen 1979) the first such studies on wild chimpanzees emerged. By now, research has offered some insight into their reconciliation behavior from four different study sites across Africa (table 17.1; Budongo: Arnold and Whiten 2001; Taï: Wittig and Boesch 2003b, 2005; Mahale: Kutsukake and Castles 2004; Ngogo: Watts 2006). Although there is considerable variation between study groups, some features are the same across groups. First, the rate of reconciliation is rather stable across study sites. Between 15% and 19% of all aggressive interactions in the wild are reconciled, as measured by the corrected conciliatory tendency (CCT). Second, the mean latency to reconcile is generally around two minutes after the conflict. Third, the baseline frequency of affiliative behaviors between opponents seems to determine whether they reconcile their conflicts, except in the Mahale M-group. However, reconciliation rates in different communities appear to relate to different affiliative behaviors such as association, grooming rate, relationship benefit index, and support rate (table 17.1). This allows for different conclusions to be drawn as to why chimpanzees appear to reconcile.

Opponents who often associate may be more likely to reconcile simply because they are more likely to meet each other. Their other behaviors (grooming, support, and food sharing) may include social exchange of either reciprocity (exchange of same acts) or interchange (exchange of different acts). Although their benefits are immediate, these behaviors allow for interchange within a biological market (Noë et al. 2001). Correlations between reconciliation and these behaviors may reflect the need of former opponents to reconcile conflicts so that they can access the benefits of an undisturbed exchange of beneficial behaviors (Wittig and Boesch 2003a, 2005; Watts 2006).

In wild populations, however, grooming seems to have a reciprocal nature (Boesch and Boesch-Achermann 2000; Watts 2000; Arnold and Whiten 2003; but see Watts 2002). Taking into account that grooming has an ameliorating effect on stress hormone levels (Shutt et al. 2007; Crockford et al. 2008) and on heart rate (Aureli et al. 1999), a positive correlation between grooming and rates of reconciliation may indicate that an important function of reconciliation is stress reduction.

A problem in investigating the effect of affiliative behaviors is that they usually intercorrelate. A simple relation between a beneficial behavior and the rate of reconciliation does not necessarily mean that the behavior predicts reconciliation. Another beneficial behavior, positively correlated to the tested behavior, may actually be the true predictor for reconciliation. Here I will investigate the effects of all four affiliative behaviors on reconciliation rates, using multivariate analysis methods to control for intercorrelation. In this way overlapping effects will be minimized, thus allowing independent effects to emerge. The valuable relationship hypothesis would predict that the behavior that is most beneficial to a relationship—or, in

Table 17.1 Nature and patterns of reconciliation in four different study sites of wild chimpanzees across Africa.

	Location (country)	Budongo Forest (Uganda)[1]	Mahale Mountains (Tanzania)[2]	Kibale Forest (Uganda)[3]	Taï National Park (Côte d'Ivoire)[4-7]
Community		Sonso group	M group	Ngogo group	North group
Composition of community	adult	11♂, 12♀	8♂, 20♀	25♂	4♂, 11♀
	subadult	28	26	ca. 120 ♀ + sa.	16
Subadults included?		Y	Y (>4 years)	N (only ♂♂)	N
Data set	conflicts N	120	206	—	876
	reconcil N	31	44		178
CCT (corrected conciliatory tendency)	♂♂	18%	14%	—	15%
	♂♀	20%	22%		27%
	♀♀	0%	21%		6%
	all	19%	15%		16%
Latency of reconciliation	average (50%)	1 min.	2 min.	—	3 min.
Initiator of reconciliation	aggressor	11%	37%	—	60%
	victim	89%	61%		40%
Correlates of reconciliation		grooming ↑	physical aggression ↑	alliance ↑	benefit of relationship ↑
		association ↑	non-physical aggression ↓	grooming ↑	(association) ↑
				association ↑	

(1) Arnold and Whiten 2001; (2) Kutsukake and Castles 2004; (3) Watts 2006; (4) Wittig and Boesch 2003a; (5) Wittig and Boesch 2003b; (6) Wittig and Boesch 2005; (7) this chapter.

other words, the behavior that best reflects each individual's perception of a valuable relationship—determines the rate of reconciliation.

Compared with data from wild chimpanzees, captive studies have generally reported higher rates of reconciliation (between CCT = 41% and 22%; Preuschoft et al. 2002 and Koski et al. 2007) but similar periods of latency (3 min.: Koski et al. 2007). The higher rates of reconciliation may be a result of the captive situation. Chimpanzees living together in a small space must get along with each other independently of hostile relationships and disturbed tolerance levels. Wild chimpanzees, in contrast, can take advantage of the fission-fusion character of their group and separate from their opponents. In Taï, opponents separated after about 10% of all conflicts, one opponent leaving the party of the other before any post-conflict interaction occurred (Wittig, unpublished data). As with reconciliation and possibly "consolation," separation may have an effect that returns tolerance to baseline levels, though probably over a longer period of time. Unpublished data from Taï suggest that opponents are less likely to separate after conflicts over a temporary resource (e.g., food) than after fights over dominance. Reconciling a food

conflict may allow a second attempt to access a food source, whilst dominance-related fights may more likely result in renewed aggression. Higher rates of reconciliation in captive studies may be explained by the opponents' need to reconcile conflicts that would normally lead to separation in the wild (see Wittig and Boesch 2003b).

"Consolation" in Wild Chimpanzees

In contrast to reconciliation, "consolation" in wild chimpanzees appears still to be mostly a *terra incognita*. Only three studies allow an (incomplete) view of the nature and function of "consolation" (table 17.2). None focused entirely on the task of describing and analyzing "consolation" behavior; instead, they contrasted it with other post-conflict interactions. Even though the data are limited, wild chimpanzees show some general patterns. First, while "offered consolation" (initiated by the bystander) is a true post-conflict interaction since it is dependent on the preceding conflict, "solicited consolation" (initiated by the victim) does not seem to be a reaction to the aggressive interaction (Wittig and Boesch 2003b; Kutsukake and Castles 2004). Second, occurring after an average 7% of conflicts

Table 17.2 Nature and patterns of "consolation" in three different study sites of wild chimpanzees across Africa.

	Location (country)	Budongo Forest (Uganda)[1]	Mahale Mountains (Tanzania)[2]	Taï National Park (Côte d'Ivoire)[3-4]
Community		Sonso group	M group	North group
Composition of community	adult	11♂, 12♀	8♂, 20♀	4♂, 11♀
	subadult	28	26	16
Subadults included		yes	yes (>4 years)	no[4]/yes[3] (only as bystander)
Data set	conflicts N	120	206	876
	reconcil N	4 (3.3%)	24 (11.6%)	26 (3.0%) / 56 (6.4%)[4]
Latency of bystander PCI	consolation		4 min.	2 min.
Who initiates consolation?	consolation	—	victim's "friend"	aggressor's "friend"
Solicited consolation	functions as post-conflict management?	—	no, independent of conflict	no, independent of conflict

(1) Arnold and Whiten 2001; (2) Kutsukake and Castles 2004; (3) Wittig and Boesch 2003b; (4) this chapter.

(when subadults are included), the rate of "consolation" in the wild appears to be rather small, and is less than half the frequency of reconciliation in captive chimpanzees (compare with table 17.1). Third, although "consolation" is rare in the wild, its latency does not seem different from the latency to reconcile. This may be an indication that consolation follows rules similar to those of reconciliation.

In any case, data on wild chimpanzees do show some support for both hypotheses. Support for the consolation hypothesis may come from chimpanzees in Mahale, where victims receive more "consolation" from partners with whom they associate often (Kutsukake and Castles 2004). This could indicate that a victim's affiliation partner aims to calm and ameliorate the "friend's" stress. Alternatively, the very fact that the affiliation partner affiliates with the victim more means that he or she has more opportunity to "console" the victim.

On the other hand, support for the relationship-repair hypothesis may come from Taï chimpanzees. There, "consolation" is offered in situations where direct interactions among former opponents are likely to lead to renewed aggression rather than reconciliation (Wittig and Boesch 2003b). This could indicate that when reconciliation is too risky, "consolation" may occur instead.

Here I will contrast the two hypotheses and test some of the predictions. The consolation hypothesis predicts that the bystander emphasizes with the victim's distress and wants to reduce it. Therefore, a victim's close bonding partner is expected to offer "consolation" (de Waal and Aureli 1996). The relationship-repair hypothesis predicts that the bystander reconciles with the victim on behalf of the aggressor, since the aggressor is not accepted in the victim's close proximity (Wittig et al. 2007). In this case, therefore, the aggressor's close bonding partner is expected to offer "consolation" to the victim. I will compare the relationship between bystander and victim with the relationship between bystander and aggressor in order to find support for either or none of these hypotheses.

Approach

The chimpanzees of the Taï National Park, Côte d'Ivoire, have been observed continuously since 1979 (for further information: Boesch and Boesch-Achermann 2000). I collected data between October 1996 and April 1999. In October 1996 the community consisted of 4 males, 11 females, and 16 subadults (juveniles and infants).

Aggressive interactions, or conflicts, are not rare events in chimpanzee societies. An aggressor attacks a victim to assert dominance, to gain control over a resource, or to enforce a mating opportunity. If the first interaction of either combatant after the conflict is an affiliative interaction between aggressor and victim, it is called reconciliation. If the victim's first interaction after the conflict is a bystander-initiated affiliative interaction, it is called "consolation". Other post-conflict interactions include renewed aggression, redirected aggression, and offered consolation (see Wittig and Boesch 2003b). Both reconciliation and "consolation" are functional labels rather than operational ones. Reconciliation, however, has been shown to restore friendly relations between former opponents. "Consolation", in contrast, has not yet proven its functional component and therefore will be enclosed within quotation marks.

Relationships in primates are measured using several different behaviors (Silk 2002), and I used four of these behaviors in this set of analyses: association, grooming, agonistic support, and food sharing. I calculated the dyadic association index (DAI), which reflects how often two individuals are associated in the same party. The DAI ranges from 0 to 1, with 0 showing that the individuals are never associated and 1 meaning that they are always associated (Nishida 1968). Grooming rates of dyads were calculated as the amount of time a dyad spent grooming divided by the observation time for both partners. Finally, I calculated a relationship benefit index (RBI) by combining measures of the third and fourth behaviors. The RBI is an indicator of a dyadic relationship's value based on food sharing and agonistic support, both of which are beneficial for chimpanzees. When a pair of chimpanzees shared food and supported each other, they were scored with a high RBI. When they exchanged only one of these behaviors, the dyad was scored with a medium RBI. When they did neither, they were scored with a low RBI (see Wittig and Boesch 2003a). Of the 105 dyads observed, 19 were scored with a high RBI, 48 with a medium RBI, and 38 with a low RBI. For "consolation" events the RBI was calculated for the initiating bystander/aggressor dyad as well as for the initiating bystander/victim dyad. To compare the relationship qualities of one dyad with that of the other, I subtracted the bystander's RBI with the victim from the bystander's RBI with the aggressor. The term was positive when the bystander was a better friend with the aggressor, and negative when the bystander was a better friend with the victim.

Results

Reconciliation

Taï chimpanzees showed aggressive interactions within 90 of the 105 possible dyads. In 48 of these dyads, reconcili-

Table 17.3   Inter-correlation between the four different relationship measurements of 90 dyads.

Correlation-coefficient r	DAI	Grooming	Relationship benefit index	Support
DAI	—	0.128*	0.175*	0.150*
Grooming	0.048	—	0.150*	0.128
Relationship benefit index	0.178*	0.221*	—	0.586***
Support	0.163	0.148	0.701***	—

Significance markers: * = P < 0.1, ** = P < 0.01, *** = P < 0.001
Correlation coefficients above the diagonal represent Kendall Tau correlations of absolute values for DAI, grooming, and support. Correlation coefficients below the diagonal represent Spearman Rank correlations of categorical representation of DAI, grooming, and support.

ation took place at least once. Reconciliation was dependent on the sex of the opponents. All 6 male-male dyads had some reconciliation events, as did most male-female dyads (32 of 40) and 11 of 44 female-female dyads.

Four different relationship measurements and the sex combination of opponents seemed to influence the reconciliation rate in wild chimpanzees (table 17.1). In Taï chimpanzees most of the relationship measurements were intercorrelated (table 17.3).

The sex combination of opponents showed a strong relation with the rate of reconciliation (figure 17.1a). Both male-male and male-female dyads showed a higher rate of reconciliation than female-female dyads. However, no difference was found between male-male and male-female dyads. The dyadic grooming rate of opponents also correlated with the rate of reconciliation in Taï chimpanzees (figure 17.1b). Here, only the dyads with the highest grooming rates (>10 s/h) proved to have significantly higher rates of reconciliation than dyads in the two categories with low grooming rates (<5 s/h) or no grooming at all.

The relationship benefit index also went hand-in-hand with the reconciliation rate (figure 17.1c). Pairs of opponents with a high RBI reconciled at higher rates than pairs with a low medium RBI. However, no difference was found between dyads with low and medium RBIs.

In contrast, the dyadic association index (DAI) was not associated with the rate of reconciliation in Taï chimpanzees (figure 17.1d). Since the difference is close to significance and the DAI had been shown to explain a certain variance in the reconciliation rate of Taï chimpanzees beforehand (see Wittig and Boesch 2003b), the index was tested in the following general linear model (GLM). To

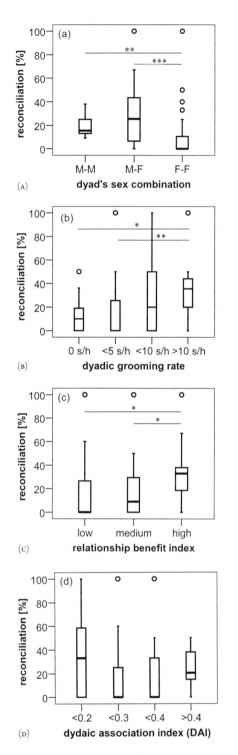

Table 17.4 GLM of predictors of reconciliation in Taï chimpanzees. Model proved significant to predict reconciliation (GLM: F = 1.997, DF = 20, P = 0.018).

Predictor	DF	F	P
RBI (sex)	4	2.838	0.031
DAI (sex)	5	0.678	0.641
Grooming (sex)	6	0.801	0.572

Predictors: Relationship measurements are corrected for sex combination of opponents.

test which of the relationship measurements predicted the occurrence of reconciliation in Taï chimpanzees, I ran a GLM that controlled for intercorrelation of the predictor variables (Tabachnick and Fidell 2007). All three relationship measurements were tested at the same time, in interaction with the sex combination of the opponents. Although the model (table 17.4) proved to predict the rate of reconciliation, only the RBI, corrected for sex of opponents, remained significant. Neither grooming rate nor DAI showed a significant influence on the rate of reconciliation in Taï chimpanzees.

Consolation

Third parties among Taï chimpanzees initiated post-conflict affiliation with the victims of aggression in 56 cases. About half of the "consolation" events (n = 26) were initiated by adult members of the community while the other 30 cases were initiated by subadult group members. Since differences between the age classes were expected, I analyzed subadult and adult bystanders separately.

To test whether the consolation hypothesis or the relationship-repair hypothesis was more likely to describe the "consolation" pattern in Taï chimpanzees, I scored the relationship of each bystander with the aggressor on one side and with the victim of aggression on the other, according to their relationship benefit index (RBI). Subadult individuals were scored with their mothers' RBIs, and mother-offspring dyads were scored with the highest possible value. The consolation hypothesis predicts that bystanders should have the highest RBIs with the victims of aggression they "console." The relationship-repair hypothesis predicts that bystanders should have highest RBIs with the aggressors, since bystanders would "reconcile" with victims in substitution for aggressors.

When the RBI between the bystander and the aggres-

Figure 17.1 Rate of reconciliation dependent on the sex combination of opponents (a) and three measurements of the opponents' relationship (b – d). Reconciliation rate differed depending on the dyad's sex combination (Kruskal Wallis: χ^2 = 19.92, DF = 2, P < 0.001), the dyad's grooming rate (Kruskal Wallis: χ^2 = 7.79, DF = 3, P = 0.020) and the dyad's relationship benefit index (Kruskal Wallis: χ^2 = 6.61, DF = 2, P = 0.037). However, the dyad's association index did not change the reconciliation rate (Kruskal Wallis: χ^2 = 5.36, DF = 3, P = 0.147). Mann-Whitney U tests were conducted to test between categories and significant results are presented using asterisks (* ≡ P < 0.05, ** ≡ P < 0.01, *** ≡ P < 0.001). Boxplots show the median and 25 to 75% of the reconciliation rate, presented for each category. Whiskers show 95% interval and circles represent outliers.

sor is compared to the RBI between bystander and victim (RBI with aggressor – RBI with victim), the difference showed a skewed distribution such that bystanders had better relationships with aggressors (figure 17.2). This skew was similar for both adult and subadult initiators, although adult bystanders showed a clearer skew in the extremes. Thus, bystanders among Taï chimpanzees are more likely to have close relationships to aggressors when they decide to "console" victims—thereby providing support for the relationship-repair hypothesis.

In comparisons within relationship benefit indices, adult bystanders among Taï chimpanzees tended to initiate more "consolation" when they had a high RBI with the aggressor than when they had a high RBI with the victim (figure 17.3a). Similarly, but in the opposite direction, adult bystanders tended to engage more in "consolation" when they had a low RBI with the victim than when they had a low RBI with the victim's opponent. This indicates that adult bystanders are more likely to "console" the victim of their bonding partners than to "console" their own bonding partners when they are victimized—thus providing support for the relationship-repair hypothesis.

Subadult Taï chimpanzees showed a different pattern of relationships as adult bystanders when initiating "consolation" (figure 17.3b). They seemed to initiate "consolation" as often with their own high-benefit partners as with their high-benefits partners' victims. Additionally, three of the six cases of subadults' engagement with high RBI victims were interactions with their own mothers, while

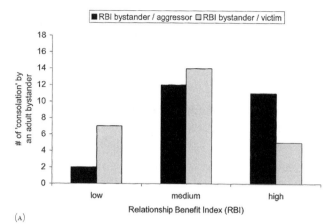

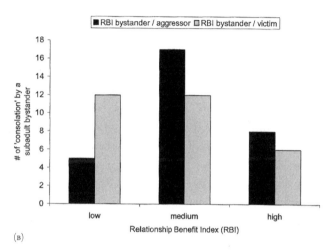

Figure 17.3 Number of "consolation" events initiated by a bystander depending on the bystander's RBI with the aggressor or the victim: (a) represents adult bystanders and (b) represents sub-adult bystanders. High RBI indicates the strongest bonding partners.

sudadults never initiated post-conflict affiliation with the victims of their mothers' aggression. Here, subadult bystanders may be as likely to "console" their own bonding partners as to "console" their bonding partners' victims—thus giving some support to the consolation hypothesis. This effect, however, may be a byproduct of the (close) mother-offspring bond.

Discussion

Both reconciliation and "consolation" in Taï chimpanzees is highly dependent upon the partners' relationship benefit. Opponents of higher relationship benefit are more likely to reconcile their conflicts. In contrast, neither grooming rates nor the dyadic association index proved to predict reconciliation. Similarly, the relationship benefit between bystanders and the victims' former aggressors

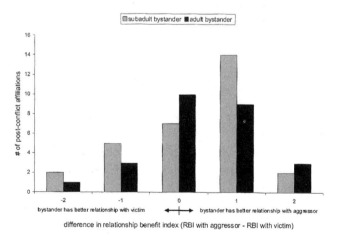

Figure 17.2 Difference in the relationship benefit index between the bystander and the aggressor versus the victim (RBI bystander / aggressor – RBI bystander / victim). Positive values on the right side indicate a higher RBI between bystander and aggressor; negative values on the left side indicate higher RBI values between bystander and victim.

seemed to determine whether or not a bystander was likely to "console" a victim. The relationship between bystanders and victims, however, did not seem to influence this "decision."

The Function and Cognitive Underpinnings of Reconciliation in Wild Chimpanzees

The valuable relationship hypothesis is a good predictor of the pattern of reconciliation in wild chimpanzees. The question remains, however, what kind of benefit best defines the value of relationships. In Taï chimpanzees the relationship benefit index, a combination of food sharing and agonistic support, seems the best explanation of relationship value among community members. These two behaviors may act as commodities that chimpanzees can trade within a biological market to increase their fitness (Noë et al. 2001). Food sharing among Taï chimpanzees occurs mainly during the eating of monopolizable foods such as meat and nuts (Boesch and Boesch-Achermann 2000; Wittig and Boesch 2003c). Agonistic support may help subordinates to keep possession of a resource after a conflict—not an unusual situation in Taï chimpanzees (Wittig and Boesch 2003a). Both behaviors are directly beneficial, and may also be excellent commodities for trading.

Grooming, in contrast, did not predict the rate of reconciliation (table 17.4), although these two measures were significantly related (figure 17.1b). Taï chimpanzees may exchange grooming less for other commodities. Although grooming is undoubtedly a beneficial behavior in terms of ectoparasite removal and the building and maintenance of friendships, grooming in Taï chimpanzees seems to be exchanged for itself rather than for other behaviors, thus suggesting reciprocity (Boesch and Boesch-Achermann 2000). Nonetheless, in other wild populations grooming was found to affect frequencies of temporary support (Watts 2002) or to contribute to the development of alliances (Nishida and Hosaka 1996). Thus a grooming relationship, although beneficial to both partners, may better reflect the establishment, maintenance, or change in intensity of a relationship. On the other hand, different populations with different group compositions, resources, and predation risks may have different needs—and therefore the same beneficial behaviors may be valued differently.

Association, compared with the other behaviors, seems to be a rough measure of relationship. In the fission-fusion communities of chimpanzees, for example, a female may join the same party as another female not because they have a valuable relationship with each other but because both females share a valuable relationship with a third female in the party. Nonetheless, since chimpanzees can choose which party they join, association may reflect "friendships" to a certain degree. Some studies have used association and grooming patterns to describe friendships among apes (Preushoft et al. 2002; Palagi et al. 2004). Applying this approach to Taï chimpanzees is not enough to predict the occurrence of reconciliation between former opponents. Chimpanzees instead take tradable benefits into account when "deciding" whether or not to initiate reconciliation.

In conclusion, chimpanzees appear to have a clear perception of what kind of relationships are valuable to them. While more frequent behaviors, like grooming or associating in the same party, don't seem to play a role in their choices regarding reconciliation, less frequent behaviors like food sharing and agonistic support do seem to play a role in their "decisions." Whether chimpanzees keep score of these events, and whether these behaviors truly reflect the value of their friendships, remains unclear.

The Function and Cognitive Underpinning of "Consolation" in Wild Chimpanzees

The Taï chimpanzees' manner of "consolation" seems to support the relationship-repair hypothesis rather than the consolation hypothesis. In most cases they did not "console" their own closely bonded partners, but instead may have "substituted" for them by reconciling with their victims.

Wild chimpanzees may have good reasons to practice third-party reconciliation in certain cases. To avoid further aggression, a victim may not allow their former aggressor to come into close range soon after a conflict, even when the aggressor's intention is to reconcile. A closely bonded partner of the aggressor, however, may be able to repair the damage without threatening renewed aggression. Similar effects are known from savannah baboons, where close kin can substitute for the aggressor in reconciling with the victim (Wittig et al. 2007). Third-party post-conflict affiliation initiated by kin seems to be rather frequent in female-philopatric monkey species (e.g., vervet monkeys:

Cheney and Seyfarth 1989; pigtail macaques: Judge 1991; longtailed macaques: Das et al. 1997; baboons: Judge and Mullen 2005). In wild chimpanzees, however, kin structures are different. First, chimpanzees are male-philopatric. Second, the vast majority of adults (as reflected in Taï's North community) are not closely related. Here, high-benefit partners rather than kin may be able to substitute for the aggressors' presence during reconciliation.

Even though "consolation" may have rather a reconciliatory effect on a victim, it remains a mechanism requiring complex cognitive abilities. In contrast to direct reconciliation, the victim must first correctly ascertain the value of the relationship between aggressor and bystander. He or she must then connect the recent history of the aggression with the bystander's relationship to the aggressor, and the bystander's affiliative interaction with himself or herself, for the interaction to substitute effectively for reconciliation. Such indirect causal inferences are similar to those in baboons, with one major difference: baboons' close bonds, which are kin-based, do not change over time, while chimpanzee bonds can be stable over years but may also change over time (see chapter 15).

Learning about the affiliative relationships of others appears to be more complex than learning about the dominance relationships of others. A dominant chimpanzee almost never greets a subordinate (Wittig and Boesch 2003c), but affiliative behaviors are exchanged within high- and low-value dyads. Therefore, a dominance relationship may be deduced from one interaction while affiliative relationships need a longer and a more subtle observation effort. This might be the reason why subadult bystanders seem to have a different approach to "consolation" than adults, since subadults tend to "console" their own high-benefit partners as often as they "console" their high-benefit partner's victims. Alternatively, subadults' benefit system may be different. Their main source of commodity exchange may still be with kin, whereas adults' main source is not with kin.

In conclusion, the label of "consolation" may not always reflect the function of the "consolation" behavior (see Koski and Sterck 2007; Fraser et al. 2009). However, the data presented here comprise only one snapshot of one community of wild chimpanzees. It would be too early to dismiss the consolation hypothesis. We need analyses from different wild populations with different relationships and ecological influences to compare both hypotheses and to

better understand chimpanzees' understanding of each other's affiliative relationships. Further analyses might indicate the need to split "consolation" into two types: third-party reconciliation and true consolation.

Implications of Data Gathered from Captive Chimpanzees

In contrast to this study, results from captivity indicate that friendships, measured by association and grooming frequency, determine whether or not to reconcile (Preuschoft et al. 2002; Koski et al. 2007). This may reflect a different range of benefits provided by social partners in captivity. In wild populations, chimpanzees gain benefits from social partners during territory patrols (Mitani and Watts 2005; see also chapter 15), hunting (Boesch 1994; see also chapter 18), the learning of tool use (see chapters 10 and 12), agonistic support (Nishida 1983; see also chapter 16), and even adoption of their offspring by others after death (Boesch and Boesch-Achermann 2000). Although these behaviors may all be present in the repertoire of captive chimpanzees, the benefits are usually not. In captivity, nobody gets killed by neighbors, nobody's diet and energy intake changes dramatically due to use of a certain tool, nobody loses out on high-value foods when not cooperating. Misjudging the intent of group members and the relationships amongst them will have more severe consequences for a wild chimpanzee than for a captive one. This also implies that the values of relationships are different for wild and captive populations. A direct comparison between both types seems inappropriate, and only makes sense when the differing demands on the relationships of wild and captive chimpanzees are also considered.

Future Directions

There is good evidence that social complexity correlates with social intelligence (reviewed in Cheney and Seyfarth 2007). This background suggests that chimpanzees should show levels of third-party knowledge at least similar to that of baboons. Fission-fusion social structure, as seen in wild chimpanzees, may add one or more additional levels of complexity. Nonetheless, there is still little direct understanding of what chimpanzees know about other chimpanzees' relationships. One way to approach this problem is to understand whether "consolation" in chimpanzees is an act of consoling the victim, reconciling the victim with the aggressor, or doing both. True consolation only requires a victim to know his or her relationship with the consoler, while third-party reconciliation requires a victim to understand what kind of bond the consoler has with the former aggressor. Knowing the function of "consolation" will provide an insight into how chimpanzees perceive conspecifics and how they rate their relationships. Although such knowledge of third-party dominance relationships has been observed in many other animals (e.g., fish: Oliveira et al. 1998; birds: Peake et al. 2001; non-primate mammals: Engh et al. 1999), an understanding of the affiliative and transient relationships of others has been documented only in primates (Cheney and Seyfarth 1999; Crockford et al. 2007; Wittig et al. 2007). In chimpanzees and other apes it is assumed, but not yet proven. Knowing what chimpanzees know about their conspecif-

ics' relationships will allow us to better understand how our own cognitive abilities have evolved.

Acknowledgments

I thank the Ministère de la Recherche Scientifique and the Ministère de l'Agriculture et des Ressources Animales of Côte d'Ivoire and the director of the Taï National Park for permission to conduct this study; Kpazahi Honora Néné and Nohon Gregoire Kohon for valuable support in the field; Christophe Boesch, Dorothy Cheney, Catherine Crockford, and Robert Seyfarth for their essential support and constructive comments on different parts of the manuscript; and the Swiss National Foundation, the Max Planck Society, and the German Science Foundation (DFG grant: WI 2637/2–1) for funding during different periods.

Literature Cited

Arnold, K., and A. Whiten. 2001. Post-conflict behaviour of wild chimpanzees (*Pan troglodytes schweinfurthii*) in Budongo Forest, Uganda. *Behaviour* 138:649–90.

———. 2003. Grooming interactions among chimpanzees of Budongo Forest Uganda: Tests of five explanatory models. *Behaviour* 140: 519–52.

Aureli, F. 1997. Post-conflict anxiety in non-human primates: The mediating role of emotion in conflict resolution. *Aggressive Behaviour* 23:315–28.

Aureli, F., and F. B. M. de Waal. 2000. *Natural Conflict Resolution*. Berkeley: University of California Press.

Aureli, F., S. D. Preston, and F. B. M. de Waal. 1999. Heart rate responses to social interactions in free-moving rhesus macaques (*Macaca mulatta*): A pilot study. *Journal of Comparative Psychology* 113:59–65.

Boesch, C. 1994. Cooperative hunting in wild chimpanzees. *Animal Behaviour* 48:653–67.

Boesch, C., and H. Boesch-Achermann. 2000. *The Chimpanzees of the Taï Forest*. Oxford: Oxford University Press.

Call, J., F. Aureli, and F. B. M. de Waal. 2002. Post-conflict third party affiliation in stumptailed macaques. *Animal Behaviour* 63:209–16.

Cheney, D. L., and R. M. Seyfarth. 1989. Redirected aggression and reconciliation among vervet monkeys, *Cercopithecus aethiops*. *Behaviour* 110:258–75.

———. 1997. Reconciliatory grunts by dominant female baboons influence victim's behaviour. *Animal Behaviour* 54:409–18.

———. 1999. Recognition of other individuals' social relationships by female baboons. *Animal Behaviour* 58:67–75.

———. 2007. *Baboon Metaphysics: The Evolution of a Social Mind*. Chicago: University of Chicago Press.

Connor, R. C. 2007. Dolphin social intelligence: Complex alliance relationships in bottlenose dolphins and a consideration for selective environments for extreme brain size evolution in mammals. *Philosophical Transactions of the Royal Society B* 362:587–602.

Cools, A. K. A., A. J.-M. Van Hout, and M. H. J. Nelissen. 2008. Canine reconciliation and third-party-initiated postconflict affiliation: Do

peacemaking social mechanisms in dogs rival those of higher primates? *Ethology* 114:53–63.

Cords, M. 1992. Post-conflict reunions and reconciliation in long-tailed macaques. *Animal Behaviour* 44:57–61.

Crockford, C., R. M. Wittig, R. M. Seyfarth, and D. L. Cheney. 2007. Baboons eavesdrop to deduce mating opportunities. *Animal Behaviour* 73:885–90.

Crockford, C., R. M. Wittig, P. L. Whitten, R. M. Seyfarth, and D. L. Cheney. 2008. Social stressors and coping mechanisms in wild female baboons (*Papio hamadryas ursinus*). *Hormones and Behavior* 53:254–65.

Das, M., Z. Penke, and J. A. R. A. M. van Hooff. 1997. Affiliation between aggressors and third parties following conflicts in long-tailed macaques (*Macaca fascicluaris*). *International Journal of Primatology* 19:53–71.

De Waal, F. B. M. 1996. Conflict as negotiation. In *Great Ape Societies*, ed. by W. C. McGrew, L. F. Marchant, and T. Nishida, 159–172. Cambridge: Cambridge University Press.

De Waal, F. B. M., and F. Aureli. 1996. Consolation, reconciliation and a possible difference between macaques and chimpanzees. In *Reaching into Thought: The Minds of Great Apes*, ed. by A.E. Russon, K. A. Bard, and S. T. Parker, 80–110. Cambridge: Cambridge University Press.

———. 1997. Conflict resolution and distress alleviation in monkeys and apes. *Annals of the New York Academy of Sciences* 807:317–28.

De Waal, F. B. M., and A. van Roosmalen. 1979. Reconciliation and consolation among chimpanzees. *Behavioral Ecology and Sociobiology* 5:55–66.

Emery, N. J. 2006. Cognitive ornithology: The evolution of avian intelligence. *Philosophical Transactions of the Royal Society B* 361:23–43.

Engh, A. L., E. R. Siebert, D. A. Greenberg, and K. E. Holekamp. 1999. Patterns of alliance formation and postconflict aggression indicate spotted hyaenas recognize third-party relationships. *Animal Behaviour* 69:209–17.

Fraser, O. N., S. E. Koski, R. M. Wittig, and F. Aureli. 2009. Why are bystanders friendly to recipients of aggression? *Communicative and Integrative Biology* 2(3): 1–7.

Fraser, O. N., D. Stahl, and F. Aureli. 2008. Stress reduction through consolation in chimpanzees. *Proceedings of the National Academy of Science USA* 105:8557–62.

Holekamp, K. E. 2007. Questioning the social intelligence hypothesis. *Trends in Cognitive Sciences* 11:65–69.

Huntingford, F. A., and A. Turner. 1987 *Animal Conflict*. London: Chapman & Hall.

Judge, P. G. 1991. Dyadic and triadic reconciliation in pigtailed macaques. *American Journal of Primatology* 23:225–37.

Judge, P. G., and S. H. Mullen. 2005. Quadratic postconflict affiliation among bystanders in a hamadryas baboon group. *Animal Behaviour* 69:1345–55.

Koski, S. E., K. Koops, and L. H. M. Sterck. 2007. Reconciliation, relationship quality and postconflict anxiety: Testing the integrative hypothesis in captive chimpanzees. *American Journal of Primatology* 69:158–72.

Koski, S. E., and L. H. M. Sterck. 2007. Triadic post-conflict affiliation in captive chimpanzees: does consolation console? *Animal Behaviour* 73:133–42.

Kutsukake, N., and D. L. Castles. 2004. Reconciliation and post-conflict third-party affiliation among wild chimpanzees in Mahale Mountains, Tanzania. *Primates* 45:157–65.

Mitani, J. C., and D. P. Watts. 2005. Correlates of territorial boundary patrol behaviour in wild chimpanzees. *Animal Behaviour* 70:1079–86.

Neat, F. C., A. C. Taylor, and F. A. Huntingford. 1998. Proximate costs of fighting in male cichlid fish: The role of injuries and energy metabolism. *Animal Behaviour* 55:875–82.

Nishida, T. 1968. The social group of wild chimpanzees in the Mahali Mountains. *Primates* 9:167–224.

———. 1983. Alpha status and agonistic alliance in wild chimpanzees (*Pan troglodytes schweinfurthii*). *Primates* 24:318–36.

Nishida, T., and K. Hosaka. 1996. Coalition strategies among adult male chimpanzees of Mahale Mountains, Tanzania. In *Great Ape Societies*, ed. by W. C. McGrew, L. Marchant, and T. Nishida, 114–34. Cambridge: Cambridge University Press.

Noë, R., J. A. R. A. M. van Hooff, and P. Hammerstein. 2001. *Economics in Nature: Social Dilemmas, Mate Choice, and Biological Markets.* Cambridge: Cambridge University Press.

Oliveira, R. F., P. K. McGregor, and C. Latruffe. 1998. Know thine enemy: Fighting fish gather information from observing conspecific interactions. *Proceedings of the Royal Society B* 265:1045–49.

Palagi, E., T. Paoli, and S. Borgonini. 2004. Reconciliation and consolation in captive bonobos (*Pan paniscus*). *American Journal of Primatology* 62:15–30.

Peake, T. M., A. M. R. Terry, P. K. McGregor, and T. Dabelsteen. 2001. Male great tits eavesdrop on simulated male-to-male vocal interactions. *Proceedings of the Royal Society B* 268:1183–87.

Preston, S. D., and F. B. M. de Waal. 2002. Empathy: Its ultimate and proximate bases. *Behavioral and Brain Sciences* 25:1–72.

Preuschoft, S., X. Wang, F, Aureli, and F. B. M. de Waal. 2002. Reconciliation in captive chimpanzees: A reevaluation with controlled methods. *International Journal of Primatology* 23:29–50.

Sapolsky, R. M. 1995. Social subordinance as a marker of hypercortisolism: Some unexpected subtleties. *Annals of the New York Academy of Sciences* 771:626–39.

Shutt, K., A. MacLarnon., M. Heistermann, and S. Semple. 2007. Grooming in Barbary macaques: Better to give than to receive? *Biology Letters* 3:231–33.

Seed, A. M., N. S. Clayton, and N. J. Emery. (2007). Postconflict third-party affiliation in rooks, *Corvus frugilegus. Current Biology* 17:152–58.

Silk, J. B. 2002. What are friends for? The adaptive value of social bonds. *Behaviour* 139:173–75.

Silk, J. B., D. L. Cheney, and R. M. Seyfarth. 1996. The form and function of post-conflict interactions among female baboons. *Animal Behaviour* 52:259–68.

Tabachnick, B. G., and L. S. Fidell. 2007. *Using Multivariate Statistics.* Boston: Allyn and Bacon.

Wallner, B., E. Moestle, J. Dittami, and H. Prossinger. 1999. Fecal gluccocorticoids document stress in female Barbary macaques (*Macacca sylvanus*). *General and Comparative Endocrinology* 113:80–86.

Watts, D. P. 2000. Grooming between male chimpanzees at Ngogo, Kibale National Park. I. Partner number and diversity and grooming reciprocity. *International Journal of Primatology* 21:189–210.

———. 2002. Reciprocity and interchange in the social relationships of wild male chimpanzees. *Behaviour* 139:343–70.

———. 2006. Conflict resolution in chimpanzees and the valuable-relationships hypothesis. *International Journal of Primatology* 27:1337–64.

Watts, D. P., F. Colmenares, and K. Arnold. 2000. Redirection, consolation, and male policing: How targets of aggression interact with bystanders. In *Natural Conflict Resolution*, ed. by F. Aureli and F. B. M. de Waal, 281–301. Berkeley: University of California Press.

Wittig, R. M., and C. Boesch. 2003a. "Decision-making" in conflicts of wild chimpanzees: An extension of the Relational Model. *Behavioral Ecology and Sociobiology* 54:491–504.

———. 2003b. The choice of post-conflict interactions in wild chimpanzees (*Pan troglodytes*). *Behaviour* 140:1527–59.

———. 2003c. Food competition and linear dominance hierarchy among female chimpanzees in the Taï National Park. *International Journal of Primatology* 24:847–67.

———. 2005. How to repair relationships: Reconciliation in wild chimpanzees (*Pan troglodytes*). *Ethology* 111:736–63.

Wittig, R. M., C. Crockford, E. Wikberg, R. M. Seyfarth, and D. L. Cheney. 2007. Kin-mediated reconciliation substitutes for direct reconciliation in female baboons. *Proceedings of the Royal Society B* 274:1109–15.

18

The Role of Intelligence in Group Hunting: Are Chimpanzees Different from Other Social Predators?

Ian C. Gilby and Richard C. Connor

It is the height of the dry season at Gombe National Park, Tanzania. Dry leaves crunch underfoot as a large party of male chimpanzees from the Kasekela community travels from a grove of Mgwiza trees in Kahama valley toward a woodland ridge where they have recently been feeding on Mhande hande fruit. Suddenly, a crash and a shrill squeak from the canopy alert them to the presence of a troop of red colobus monkeys. The chimpanzees bristle and rush beneath the monkey troop, scanning for vulnerable individuals. Frodo is particularly vigilant, and soon he charges up a tree to chase a female with an infant. Several male colobus drop down toward him, shaking branches and alarm calling. Frodo screams as a large male jumps on his back and sinks its canines into his shoulder. A branch breaks in the scuffle, and Frodo and the monkey fall several meters into a thicket. The monkey escapes, and Frodo resumes the hunt. Other adult male chimpanzees, including Goblin, Freud, Tubi, and Wilkie, advance toward their own targets. Meanwhile, Beethoven and Gimble watch the hunt unfold from the ground, barking and pant-hooting. They do not join their group mates in the hunt. In the chaos above, Freud is repelled by a pair of male colobus, but Tubi and Frodo manage to penetrate the monkeys' beleaguered defenses and, 20 minutes after the hunt has begun, each captures a juvenile. Frodo quickly descends, and Beethoven rushes to him, submissively pant-grunting. He grimaces and whimpers as he tugs on the carcass, and reaches both hands to Frodo's mouth. Frodo bristles and displays, but soon Beethoven resumes begging. His efforts pay off and he eventually obtains several pieces of meat, some directly from Frodo's mouth.

Most chimpanzee populations eat meat (Uehara 1997), and the hunts of their preferred prey, red colobus monkeys, typically involve multiple hunters. At all study sites where hunting has been studied intensively, the probability of hunting upon encountering prey increases with the number of adult male chimpanzees that are present (Gombe National Park, Tanzania: Gilby et al. 2006; Mahale Mountains National Park, Tanzania: Hosaka et al. 2001; Kibale National Park, Uganda (Ngogo): Mitani and Watts 2001; Kibale (Kanyawara): Gilby and Wrangham 2007). Similarly, at Taï National Park, Côte d'Ivoire, 84% of red colobus hunts involve multiple individuals (Boesch and Boesch-Achermann 2000). Why do chimpanzees typically hunt in groups? Group hunting seems susceptible to a classic collective action problem, whereby nonparticipation would appear to be an individual's optimal strategy. Hunting costs include energy expenditure and the risk of injury from falling or being mobbed. Since meat is regularly shared among party members (Gilby 2006; Mitani and Watts 2001), it seems counterintuitive for an individual to hunt if he could obtain meat from others without paying the costs of capturing it. This predicts that no one will hunt.

Therefore, in order for group hunting to have evolved, an individual chimpanzee must benefit by hunting with others, but how? One attractive possibility is that chimpanzees' extraordinary cognitive abilities enable them to solve the collective action problem by making group action beneficial to all participants. First, greater intelligence may increase the range of potential benefits available to a hunter. For example, possessing the mental ability to track transactions of tradable commodities like meat, grooming and coalitionary support may open up a new realm of social benefits. Second, chimpanzees' intelligence may enable them to use complex, coordinated hunting tactics which increase the probability of success.

By contrast, behavioral ecologists rarely invoke cognitively complex processes to explain communal hunting in more distantly related species such as canids or cetaceans. Instead, for those taxa, researchers tend to favor more parsimonious mechanisms. In this chapter we apply a similar approach to chimpanzee predation, considering whether mechanisms that do not require advanced cognitive ability may explain why chimpanzees hunt in groups. As is illustrated in this volume, there is no doubt that chimpanzees have extraordinary brain power and problem-solving ability. However, possessing the mental capacity for a particular task does not prove that there has been selective pressure for intelligence in that context. Therefore, a clearer understanding of how chimpanzees solve

problems in the wild provides important insight into the evolutionary causes and consequences of intelligence (see chapter 15).

We address three main questions (table 18.1). First, how do individual chimpanzees obtain meat at a successful hunt? Group hunting presents a collective action problem only if nonhunters are able to obtain meat. To understand the cognitive processes underlying the decision to hunt, it is therefore critical to identify all possible mechanisms by which meat is distributed. Second, how do individuals benefit by hunting in groups? Here, the "currency" of the payoff is key. It has been proposed that chimpanzees hunt in order to obtain items to trade for social benefits such as grooming or sex (Stanford et al. 1994). Such behavior arguably requires the cognitive ability to plan ahead, base decisions on who is present, and overcome impulsivity. Apart from chimpanzees and humans, this putative social benefit of group hunting has not been proposed for any other species, and is therefore one crucial way in which chimpanzees' cognitive ability may set them apart from other social predators. Third, we evaluate the importance of intelligence in the process of acquiring prey. We argue that group hunting in chimpanzees can be characterized as a simple form of by-product mutualism in which each hunter acts in his own self-interest rather than to increase the group's success. We conclude that ecological factors, rather than cognitive ability, are the primary de-

Table 18.1 Summary of questions and conclusions

Question	Conclusions	Implications
How do chimpanzees obtain meat at a successful hunt?	*Hunters*: Capture, sharing according to hunting effort	There are several mechanisms by which nonhunters acquire meat, creating potential for a collective action problem.
	Hunters and nonhunters: Scrounging, begging, intimidation, reciprocal exchange, kin-based sharing	Reciprocal exchange and sharing according to hunting effort arguably demand greater intelligence.
How do individuals benefit by hunting in groups?	*Nutritional*: No per capita caloric benefit of group hunting at most sites, with the exception of Taï	No striking differences between chimpanzees and social predators.
	Group hunting may increase a chimpanzee's chances of obtaining important micronutrients.	Lack of support for social benefits suggests limited indirect effects of intelligence on hunting.
	Social: Little evidence that chimpanzees hunt for social reasons, including sex or social bonding	
What is the role of intelligence in the process of acquiring prey?	We cannot rule out the possibility that collaborative hunting in chimpanzees can be explained by a simple by-product mutualism in which hunters are trying to maximize their own chances of making a kill.	"Sophisticated" cooperation does not require advanced cognitive ability.
	Like other social predators, chimpanzees hunt cooperatively when ecological conditions require it.	

terminant of social hunting patterns across taxa, including chimpanzees.

Meat Distribution at a Successful Hunt

There are several ways in which a chimpanzee may acquire meat at a successful hunt. First, let us consider the range of possibilities for an individual who is present when others hunt, but who does not actively participate. This nonhunter might be a scrounger, obtaining scraps of meat that others drop from branches (Barnard and Sibly 1981). Whether or not the scrounging is considered parasitic—an extracted benefit, coming at a cost to the meat possessor (Connor 2007)—depends on whether the possessor would have recovered the scraps consumed by the scrounger. A nonhunter might also be able to obtain meat by begging, essentially harassing a meat possessor into relinquishing a portion (Wrangham 1975; Stevens 2004; Gilby 2006). This is another example of an extracted benefit. Similarly, a dominant individual might steal meat or use intimidation, the threat of force, to obtain meat from a subordinate. Finally, a nonhunter might receive meat as a gift from the owner. Such altruism might be favored if the donor and recipient are related (Hamilton 1964), the donor anticipates reciprocation by the recipient in the form of food, sex, or grooming (Trivers 1971; Noë and Hammerstein 1994), or the donor is signaling his status (Zahavi 1975) and/or investing in the relationship—for example, if the recipient is an alliance partner (Wrangham 1982; Connor 1986).

Now consider the ways in which a hunter—a chimpanzee who actively chases prey—might obtain meat. First, he may be the one who actually makes the kill. Or, if another chimpanzee is successful, our hunter might benefit if the captor gives away shares according to degree of participation in the hunt (Boesch 1994b; Boesch and Boesch-Achermann 2000; Boesch 2002). Finally, an unsuccessful hunter may obtain meat by scrounging, begging, or through reciprocal altruism, as discussed above. Mechanisms such as reciprocity and sharing according to hunting effort would push group hunting into a cognitively complex arena—one that could potentially distinguish chimpanzees from other social predators, for which such explanations are rarely if ever invoked (Packer and Ruttan 1988).

These observations demonstrate that there are many ways in which nonhunters may obtain meat, thus indicating that group hunting in chimpanzees does indeed present a collective action problem (table 18.1).

Benefits of Group Hunting

Nutritional Benefits: Social Predators

Hunting is considered to be cooperative if an animal obtains a greater net benefit by hunting with others than by hunting solitarily (Packer and Ruttan 1988; Mesterton-Gibbons and Dugatkin 1992; Clements and Stephens 1995). While costs and benefits should ultimately be measured in relative fitness, calories are a logical (and more practical) currency to use. For social predators, the evidence that group hunting is cooperative is equivocal—in some species, individuals obtain more meat by hunting in groups, while in others they do not (for review see Packer and Ruttan 1988; Creel and Creel 2002). Using theoretical models, Packer and Ruttan (1988) predict that cooperative hunting is most likely to evolve when solo hunting success rates are low and scrounging opportunities are limited. Indeed, hunting success in carnivore species that meet these criteria typically increases with group size (Packer and Ruttan 1988). Additionally, hunting in groups may be advantageous in ways other than simply catching more or larger prey. For example, co-feeding by orca pods increases individual foraging efficiency by preventing large carcasses from sinking (Guinet et al. 2000). Also, larger groups may be more likely to prevent or compensate for kleptoparasitism by scavengers—for example, grey wolves versus ravens (Vucetich et al. 2004) or African wild dogs versus spotted hyenas (Carbone et al. 2005; Fanshawe and Fitzgibbon 1993).

For social predators, ecological factors are a critical determinant of whether there is a per-capita benefit to hunting in groups. Variables such as prey size and abundance, habitat type, and seasonality greatly affect hunting payoffs within and among species and populations. For example, African lions experience a net payoff by hunting in groups in Etosha National Park, Nambia (Stander 1992a; Stander 1992b), where prey density is low. For lions in the Serengeti National Park, Tanzania, large groups enjoyed greater per-capita meat intake than did small groups (Packer et al. 1990), but only when prey was scarce.

Creel and Creel (2002) have argued that the reduced costs of social predation are as important as the increased benefits. While the amount of meat consumed per capita may sometimes be lower after a group hunt than after a solo hunt, an individual may have expended much less energy when hunting with others. As long as the benefit-to-cost ratio of group hunting is higher compared to that of hunting alone, social predation will yield a greater per-capita net benefit. Note, however, that large gains made at a large cost may result in a greater net benefit than small gains at a small cost, even though the benefit-to-cost ratio may be the same (Stephens and Krebs 1986; Packer and Caro 1997). In wild dogs, while observed group sizes yielded suboptimal per-capita meat intake rates (Creel and Creel 2002), chase distances decreased with increasing group size (Creel and Creel 1995). Thus, larger groups resulted in greater net kilojoules per dog per day.

Group hunting may also arise as a by-product of other selective pressures for living in groups, such as territory defense or communal offspring rearing. Here, hunting in groups may not be the optimal foraging strategy, but the cost of sharing meat with others is outweighed by the benefits of a larger territory or increased offspring survivorship. Under these circumstances, group hunting may evolve even if it yields a lower individual payoff than hunting alone (Packer et al. 1990). Additionally, even if hunting in groups does favor sociality, theory predicts that groups will tend to be larger than is optimal for individual food consumption (Sibly 1983).

Nutritional Benefits: Chimpanzees

Given the considerable variation in meat consumption across chimpanzee populations (Uehara 1997), few would argue that hunting is a primary cause of chimpanzee sociality. While there are many opportunities for lone individuals to hunt, chimpanzees tend to hunt in groups, suggesting that there is an incentive for each individual to hunt with others. At Taï, there is evidence that this incentive is caloric. There, social predation yields a net caloric benefit for hunters, but not for nonhunters (Boesch and Boesch-Achermann 2000). Boesch describes a "social mechanism" that prevents nonhunters from obtaining meat (Boesch 1994b; Boesch and Boesch-Achermann 2000; Boesch 2002). However, this is not the case at other study sites. There is no evidence of a caloric benefit of hunting in groups at Gombe (Gilby et al. 2006; Boesch 1994b), Ngogo (Watts and Mitani 2002a), or Kanyawara (Kibale

Chimpanzee Project, unpublished data). It is important to note, however, that with the exception of Boesch (1994b), estimates of per-capita meat consumption are typically obtained by dividing the total amount of meat captured by the party size. More accurate measures are required to test whether the per-capita meat *intake* of hunters is correlated with the number of hunt participants. Such data are extremely challenging to collect, for multiple observers and good visibility are critical for tracking the movements of each meat eater. As males are often able to sneak away with their individual portions, many bouts are likely to be missed by observers.

Nevertheless, are calories the right currency with which to assess hunting profitability in chimpanzees? Unlike social predators, chimpanzees clearly do not rely upon meat for survival, because some communities rarely hunt (e.g., Budongo Forest, Uganda: Newton-Fisher et al. 2002). Even in communities where hunting is common, some females and low-ranking males hardly ever eat meat (Goodall 1986). Additionally, the emerging picture is that hunting frequency increases when ripe fruit is abundant, indicating that hunting is not motivated by a caloric shortfall. At Gombe and Ngogo, this increase is a by-product of the fact that larger parties form when ripe fruit is plentiful (Gombe: Gilby et al. 2006; Ngogo: Mitani and Watts 2005). At Kanyawara, however, the positive effect of diet quality on hunting was independent of party size (Gilby and Wrangham 2007).

If not for calories, then why do east African chimpanzees hunt? Most researchers agree that meat is a concentrated source of important micronutrients (e.g., salt, iron, zinc, vitamins) that are beneficial to chimpanzees in small quantities (Boesch 1994a; Milton 2003; Mitani and Watts 2001; Stanford 1996; Takahata et al. 1984; Teleki 1973; Teleki 1981). Gilby et al. (2008) argue that small amounts of meat may yield a favorable payoff as long as they contain sufficient amounts of these micronutrients. Thus, if hunting with others increases the chances that a given individual eats even a scrap of meat, then we could characterize chimpanzee hunting as cooperative, regardless of its energetic costs. This idea is supported by a recent mathematical model (Tennie et al. 2009) as well as by empirical data. The probability that an individual obtained meat increased with party size both at Kanyawara (Gilby et al. 2008) and at Gombe (Tennie et al. 2009). At Ngogo, a greater proportion of males obtained meat in large hunting parties than in small ones (Watts and Mitani 2002b). At Taï, where large parties are more likely to hunt successfully, and hunters preferentially share with one another (Boesch and Boesch-Achermann 2000), hunting with others should increase a hunter's chances of obtaining meat. Further testing of this "meat-scrap" hypothesis will require nutritional analysis and detailed long-term dietary records. Nevertheless, there is good reason to believe that chimpanzees at all sites obtain valuable micronutrients by hunting in groups. At Taï, there appears to be an additional caloric benefit.

Thus, the nutritional benefits associated with group hunting by chimpanzees are consistent with what is observed in many social predators, in which per-capita caloric benefits vary within and among species. Most chimpanzee communities (with the exception of Taï) do not experience a net caloric benefit. However, if social predation increases the chances of obtaining important micronutrients found in meat, then there arguably is a nutritional benefit to communal hunting at all sites.

Social Benefits: Chimpanzees

In addition to nutritional benefits of hunting in groups, it has also been suggested that chimpanzee hunting decisions are affected by the potential social benefits of acquiring meat (Stanford et al. 1994; Stanford 1998; Mitani and Watts 2001). Obtaining meat provides an opportunity to trade a valuable commodity in return for social favors such as grooming, sex, or coalitionary support (joint aggression or defense against a rival). As party size increases it becomes more likely that a profitable trading partner will be present, and this provides greater incentive to hunt. If chimpanzees base their hunting decisions on the potential for trading social favors, then they must possess the cognitive ability to track relationships, anticipate the consequences of hunting and sharing, and overcome problems associated with impulsivity and temporal discounting. In this way, by opening up a new realm of benefits, intelligence may have an indirect effect on hunting. To our knowledge, this "social incentives" model has been proposed only for chimpanzees. Therefore, it would represent a critical difference between chimpanzees and social predators that hinges upon cognitive ability.

However, there are few data to support the claim that chimpanzees base their hunting decisions on the potential

social benefits of sharing meat. Despite the popular appeal of the idea that males hunt to obtain meat to trade for sex, there is no evidence that they do so. At Ngogo, there was no effect of the presence of sexually receptive (swollen) females on hunting probability (Mitani and Watts 2001). At Gombe, parties containing swollen females were significantly less likely to hunt than those without swollen females, suggesting that males face a tradeoff between hunting and mate guarding (meat-or-sex rather than meat-for-sex; Gilby et al. 2006). There was a similar negative trend at Kanyawara (Gilby and Wrangham 2007). Additionally, Gilby (2006) showed that Gombe males did not share preferentially with swollen females, nor did sharing increase the probability of mating during a meat-eating bout. Similarly, males at Ngogo did not mate more frequently with females they shared with (Mitani and Watts 2001).

There is more evidence that males trade meat reciprocally with each other to establish and maintain social bonds. At Mahale the longtime alpha male, Ntologi, shared more often with frequent grooming partners, possibly to increase rates of coalitionary support (Nishida et al. 1992). Sharing patterns among adult males at Ngogo were consistent with reciprocal exchange of meat, grooming, and support (Mitani 2005). However, Gilby (2006) has argued that such patterns may emerge as a by-product of harassment by persistent beggars. At Gombe, harassment predicted the probability, amount, and mode (active versus passive) of sharing, while grooming and association patterns did not (Gilby 2006). Regardless of the reasons why males share meat with one another, however, males in at least two communities do not base their hunting decisions on the potential for cementing social bonds. At Kanyawara, a male's probability of hunting was unaffected by the presence of preferred associates or grooming partners (Gilby et al. 2008). At Gombe, the probability that a focal male hunted was independent of party size (Gilby et al. 2006), which at other sites is correlated with the presence of preferred social partners (Gilby et al. 2008).

Therefore, after assessing the potential nutritional and social benefits of acquiring meat, we may conclude that there is nothing strikingly different in the reasons why chimpanzees and other social predators hunt in groups. Social predation tends to yield a nutritional benefit, although the currencies (calories versus micronutrients) may be different within and between taxa. Critically, there is no strong evidence to support the idea that chimpanzees hunt for social reasons, a mechanism that arguably would require superior cognitive ability.

The Role of Intelligence in Capturing Prey

We have shown that social predation yields nutritional benefits for both chimpanzees and obligate social predators, thus demonstrating similar incentives for group hunting across taxa. However, we have not yet addressed the behavior of individuals during group hunts. To what extent do hunters modify their behavior based on the actions of others, and what role does intelligence play? We will argue that across taxa (including chimpanzees), group hunting strategies ranging from simultaneous independent efforts to complex collaboration can be explained by a simple by-product mutualism that does not require advanced cognitive ability.

Additive Probability

The simplest explanation for why hunting probability is correlated with party size in chimpanzees is the additive probability hypothesis (Gilby et al. 2006), which assumes that the probability that a given male will hunt remains constant across all party sizes. Therefore, the likelihood that at least one male will hunt is a simple function of the intrinsic hunting propensities of the individuals present— larger groups will be more likely to contain at least one willing hunter. Hunting probability will be positively correlated with party size, but critically, individuals do not base their decision to hunt on the presence or actions of others. The additive probability hypothesis therefore provides an explanation for group hunting that does not require complex cognitive processes.

Chimpanzee studies provide mixed support for the additive probability hypothesis. At Gombe, the likelihood that a given male hunted upon encountering red colobus monkeys was independent of party size (Gilby et al. 2006), a result which is consistent with the additive probability hypothesis. This was not the case at Kanyawara, however, where individual hunting probability was positively correlated with party size, and group hunting rates were significantly higher than those predicted by a mathematical additive probability model (Gilby et al. 2008). While the additive probability hypothesis has not been explicitly tested elsewhere, anecdotal descriptions suggest that rates

of individual participation increase with party size at most sites (Muller and Mitani 2005). However, the fact that at least one population of chimpanzees hunts in a manner consistent with the additive probability hypothesis suggests that social predation in chimpanzees is not rooted in a complex cognitive arena.

To our knowledge, the additive probability hypothesis has not been tested in (or proposed for) other social predators. It would be most likely to occur in fission-fusion species that form groups for benefits other than predation.

A Range of By-Product Mutualisms

Why then do chimpanzees at most sites appear to base their hunting decisions on the presence or actions of others? One possibility is that collaboration increases the probability that the group will make a kill. Boesch and Boesch (1989) define collaboration as when individuals perform "different but complementary actions, all directed toward the same prey". Individuals observe (or anticipate) the actions of others and adjust their behavior accordingly, increasing the likelihood that someone in the group will be successful. Note that this definition excludes scenarios in which two hunters simultaneously pursue a single target from the same direction, or independently pursue different prey.

The degree to which chimpanzees collaborate varies across sites. At Gombe, most hunts have been described as "simultaneous, solitary hunts" in which collaboration is rare (Boesch 1994b). Similarly, although collaborative hunting has not been explicitly tested at Ngogo or Mahale, anecdotal evidence from both sites indicates that while collaboration does occur, it is rare (Ngogo: Watts and Mitani 2002b; Mahale: Hosaka et al. 2001). As collaboration is dependent upon the actions of others, a hunter must be confident that his partners won't quit in the middle of a hunt, which would result in a loss of benefits for all. Therefore, we would expect males with a close social bond to hunt together. Strong, mutual association is often indicative of a cooperative alliance, and allies often reciprocally take risks with one another—for example, providing coalitionary support or participating in territorial patrols (Mitani et al. 2000; Watts and Mitani 2001). Therefore, a close social partner should be less likely to defect (cheat) during a joint activity. At Kanyawara, however, males were equally likely to hunt whether or not their preferred social partners hunted (Gilby et al. 2008), suggesting that hunts are not collaborative.

By contrast, one population of chimpanzees reportedly collaborates regularly in the context of predation. As described by Boesch (1994b, 2002) and by Boesch and Boesch-Achermann (2000), hunts at Taï typically involve one or more "drivers" who chase the prey toward "ambushers." "Blockers" climb to fill critical escape routes, and "chasers" move quickly after the prey in an attempt to catch them. Such a division of labor has also been described in several species of social predator. For example, hunts by lions in Etosha National Park, are similar to those described for the Taï chimpanzees. "Centers" drive quarry toward "wings" which anticipate the movements of both predator and prey, and all hunters share in the spoils (Stander 1992a). Unlike chimpanzees, however, the Etosha lions tend to adopt the same roles (including a preferred side) from hunt to hunt. An individual is less likely to be successful if it adopts a "non-preferred" position.

Similar collaboration appears to be customary among wild dogs in the Selous Game Reserve, Tanzania. Although the roles of individuals have not been systematically identified, Creel and Creel (2002) describe individual dogs driving prey toward waiting ambushers, and certain dogs distracting the prey while others attack from behind. In the Serengeti there are anecdotal descriptions of a division of labor among wild dogs after they have captured an animal but before they have killed it. One animal typically immobilizes the prey by holding its nose while others disembowel it (van Lawick-Goodall and van Lawick-Goodall 1971). Bottlenose dolphins in Cedar Key, Florida, exhibit a division of labor with strict (100%) role specialization while hunting fish (Gazda et al. 2005). A single dolphin drives a school of fish toward several other dolphins that line up to form a barrier and trap them. In the two groups in which this has been regularly observed, the driver is always the same individual. Collaborative hunting has also been described in grey wolves (Peterson and Ciucci 2003), spotted hyenas (Kruuk 1972), and even a raptor, Harris' hawk (*Parabuteo unicinctus*; Bednarz 1988), but most studies do not attempt to quantify specific roles.

Division of labor need not be limited to intraspecies cooperation. For example, groupers and giant moray eels collaborate to capture prey that move between types of habitat that require different hunting strategies (Bshary et al. 2006). Groupers hunt in open water while moray eels hunt in reefs, moving quickly through crevices and cornering prey in holes. These species-specific complementary hunting skills have led to interspecies collaboration in which a grouper initiates a joint hunt by visiting an eel's resting place and shaking its head vigorously from side to side. Such signaling often occurs after an unsuccessful hunt in which the grouper's potential prey has escaped into a reef crevice, where hunting by eels is favored. The eel then pursues the fish in the reef, either catching it, or driving it back out into open water where it is caught by the grouper. Even though the predators do not share the prey, both groupers and eels were more likely to be successful while hunting jointly than when hunting solitarily (Bshary et al. 2006). Similarly, coyotes and badgers sometimes collaborate to capture ground-dwelling rodents (Minta et al. 1992). It is surprising that such collaboration has not been described in more species with similar habitats and different hunting strategies. Such species would seem to be primed for collaboration because there is a "built-in" (species-specific) division of labor.

Anderson and Franks (2001) identify certain types of collaboration as "team tasks." A team task is a cooperative endeavor in which individuals adopt roles (perform subtasks) that increase group success *even if doing so reduces the individuals' own chances of making a kill.* Certain cases, such as the eel/grouper example, are clearly not team tasks. Each participant is simply trying to maximize its own hunting success; only one of them gets to eat if a fish is caught. The logic is the same as that given for male savanna baboons that form a coalition to disrupt a sexual consortship; each has a better chance of mating with the female if both attack at the same time (Bercovitch 1988). Collaboration among the Etosha lions, however, may be a true team task since individuals take on roles that improve other individuals' chances of obtaining meat.

In chimpanzees, and likely in some of the other cases of collaboration in other species, we are not confident that the different behaviors observed during a hunt are part of a true team task. Is a chimpanzee that moves into a "blocking" position attempting to maximize the chance that another chimpanzee will make a kill (a subtask in a team task) or is he simply trying to capture the monkey himself (a selfish task)? In the descriptions of collaborative hunts at Taï (Boesch 1994b; Boesch and Boesch-Achermann 2000; Boesch 2002), it is assumed that hunters act for the benefit of the group—that the decision to adopt a certain role is based on maximizing the probability that the hunting party as a whole is successful. In order to support this claim, one must demonstrate that individuals aren't simply attempting to maximize their own chances of making a kill (note that this applies to studies of group hunting in all taxa).

If it can be shown conclusively that "driving," "blocking," and "chasing" are not simply strategies that maximize individual hunting success, then it may be argued that chimpanzee group hunting requires a certain level of intelligence. Individuals would need to be able to overcome the temptation to switch roles, even when doing so would provide a small increase in their own probability of making a kill (but a decrease in the chance of group success). Additionally, at Taï, a putative "social mechanism" of preferential sharing ensures that an individual's payoff depends upon his hunting role (Boesch 1994b). Hunters obtain more net calories than bystanders or latecomers (Boesch 1994b), and those with cognitively demanding roles (e.g., anticipating prey movements and ambushing) obtain more than hunters with "less decisive roles" (Boesch 2002). This implies that individuals have the cognitive ability to assess and remember each individual's relative contribution and share accordingly. However, the same pattern could arise if hunters are simply more motivated than nonhunters to acquire meat, and therefore beg more persistently if they fail to capture their own monkeys. Finally, Boesch (2002) suggests that collaboration requires the ability to understand and "consider the perspectives of others." We suspect, however, that cognitively simpler mechanisms might suffice. For example, a chimpanzee might employ simple associative learning to predict that a fleeing monkey will change direction upon encountering a blocking chimpanzee or an uncrossable canopy gap.

While the reports of complex collaboration at Taï are intriguing, we suggest that observed patterns there and elsewhere may have a simpler explanation. Furthermore, some researchers have challenged the feasibility of identifying specific hunting roles. Fast action, the number of participants, and reduced visibility all contribute to considerable uncertainty regarding precise individual movements (Mitani and Watts 2001; Gilby et al. 2006), which are critical for identifying the existence of true "subtasks" (sensu Anderson and Franks 2001). For these reasons, we now focus upon whether chimpanzee group hunts can be explained by a simple form of by-product mutualism. A by-product mutualism occurs when an individual's selfish

actions incidentally benefit others (West-Eberhard 1975; Brown 1983; Connor 1995). Therefore, if two or more individuals act at the same time, each participant obtains a greater benefit than by hunting alone. By-product mutualism is a very broad category of cooperation that includes collaboration. It may range from simple, uncoordinated behavior, such as vigilance in groups, to much more complex coordinated behavior, such as hunting with a division of labor (Connor 1995). Below, we argue that a simple form of by-product mutualism that does not require advanced intelligence can explain group hunting across chimpanzee sites.

To put chimpanzee hunting into perspective, let us first consider the range of by-product mutualisms found in group-hunting cetaceans. Bottlenose dolphins are commonly reported to feed in groups on schools of fish, trapping them against a barrier, encircling them, or even chasing them out of the water (reviewed in Connor 2000, Duffy-Echevarria et al. 2008). As discussed earlier, some groups exhibit 100% role specialization (Gazda et al. 2005). However, there is no evidence to suggest that such collaboration represents a true team task; each participant appears to experience an increase in individual capture probability. Orca have been observed simultaneously creating waves to wash seals off ice floes (Smith et al. 1981). Larger waves increase each whale's chances of catching a seal. Baird and Dill (1996) reported that the average group size found in "transient" orca was optimal for group hunts of harbor seals. Targeted seals sometimes took refuge in underwater caves and crevices, but the orca waited them out by coordinating diving and breathing patterns. This, and examples of collaboration during hunts of large whales (Baird 2000; Pitman et al. 2001), may indeed be true team tasks.

We suggest that social predation in chimpanzees can be characterized as a by-product mutualism in which each hunter experiences an increased chance of success by hunting in a group. Each chimpanzee tries to capture a monkey, and the consequence of several simultaneous individual hunts is that colobus defenses are diluted, thereby increasing the probability of a kill for *each* participant. In some ecological circumstances (see the ecology of cooperation, below), this strategy may result in a division of labor (Connor 1995), but not a true team task in which hunters are acting to increase the probability of a group kill at the expense of their own success.

If a hunter benefits from the chaos created by multiple simultaneous chases, how are hunts initiated? One possibility is that, like the baboon males (Bercovitch 1988; Packer 1977), individuals use a signal to recruit partners, however there is no evidence of this in chimpanzees. Alternatively, Gilby et al. (2008) have proposed that certain individuals act as hunting "catalysts" by being the first to hunt. In this manner, others could simply take advantage of the fact that another chimpanzee is already hunting. Recent data from Kanyawara support this idea. There, a hunt was significantly more likely to occur if either of two particular males, Imoso and Johnny, were present when a party encountered red colobus than if both were absent (Gilby et al. 2008). In fact, parties missing these "impact" males almost never hunted (figure 18.1). Additionally, "non-impact" males hunted on only 4 of 333 occasions when one or both of the impact males were present but did not hunt. Informal observations suggest that Imoso and Johnny commonly climb higher than other males and are usually the first to be confronted by adult male colobus (R.W. Wrangham, personal communication). By doing so, they engage red colobus defenders, thereby reducing the number of monkeys available to attack other hunters. Anecdotal evidence reveals similar patterns in other populations. At Gombe, upon encountering red colobus, males often cluster around Frodo, apparently waiting for him to hunt (Gilby, personal observation). At Ngogo, Monk is usually one of the first to hunt, apparently prompting others to follow (D.P. Watts, personal communication). Boesch and Boesch (1989) maintain that the maturation

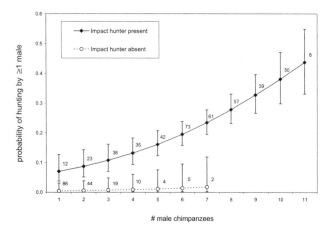

Figure 18.1   Party-level hunting probability as a function of the presence of "impact" hunters at Kanyawara. A hunt was significantly more likely to occur if Johnny and/or Imoso were present (solid diamonds) than if both were absent (open circles), even after statistically controlling for party size. Error bars represent 95% confidence intervals, and numbers indicate the number of red colobus encounters for a given party size. From Gilby et al. (2008).

of Snoopy, a particularly persistent hunter, significantly changed hunting patterns at Taï.

The factors responsible for variation in intrinsic hunting rates among males have not been rigorously examined. However, it is unlikely that age—once hunting skill has been developed—is a major factor, as certain individuals have been recognized as impact hunters for many years. Dominance rank may also be important, but given that some of the qualities that promote hunting, such as strength and agility, are often correlated with high rank, it will prove difficult to test whether hunting enhances a male's status (Kortlandt 1972). Hunting skill is likely to be critical, as hunting costs should be lower for those who are able to avoid falling and are capable of thwarting the defensive efforts of male colobus. Finally, personality is clearly important, although challenging to quantify. Some males quickly give up after being challenged by colobus, while others persist. Boesch and Boesch-Achermann (2000) describe how an adult male at Taï (Ulysse) typically seemed to invite mobbing, during which he would seize and kill one of his attackers. At Gombe, Frodo rarely backs down when faced with male colobus (Gilby, personal observation).

These data suggest that variation in individual motivation or skill creates opportunities for others to hunt when they normally wouldn't. It is presumably less costly to join a hunt than it is to initiate one. Therefore, once a hunt is underway, more individuals are likely to participate, and their combined but independent efforts increase their own chances of success. Why then would an individual ever refrain from hunting? It would seem that each additional hunter would contribute to a positive feedback loop of reduced hunting costs and increased success. However, in some cases, if there are a sufficient number of hunters, an individual may be able to obtain meat (by scrounging or begging) without actively participating in the hunt (Gilby et al. 2008). For example, it might therefore benefit a fourth or fifth chimpanzee to join a hunt, but a sixth might remain on the ground because of the opportunity to obtain meat without experiencing the costs of hunting. This threshold will depend upon several factors, including the size and composition of the monkey troop and the vegetation type. Note that this model assumes that failure to participate does not exclude a chimpanzee from access to meat (cf. Boesch 1994b). Empirical data from Kanyawara match this model exactly. There, an individual's probabil-

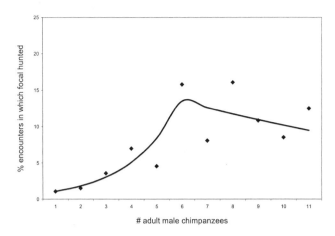

Figure 18.2   Probability that a randomly chosen focal male at Kanyawara hunted as a function of the number of adult male chimpanzees present at a colobus encounter. A male was more likely to hunt in larger parties, but hunting probability reached an asymptote in large parties. This is consistent with Packer and Ruttan's (1988) mathematical model of cooperative hunting. From Gilby et al. (2008).

ity of hunting reaches an asymptote as party size increases (Gilby et al. 2008; figure 18.2), and the party size at which an individual's hunting probability reaches a maximum is the one at which hunters and nonhunters are equally likely to obtain meat.

Here we have provided evidence that individual variation in hunting skill and motivation is critical for determining social hunting patterns in chimpanzees. We argue that hunting in groups can be explained by a simple by-product mutualism whereby individuals take advantage of the actions of others in an effort to maximize their own chances of success. Such behavior invokes simple behavioral rules (e.g., hunt when two others are hunting), instead of cognitively complex mechanisms that a true team task may require.

The Ecology of Cooperation

We have seen that the frequency and mode of group hunting varies across chimpanzee study sites. At Gombe, lone hunters often succeed, and group hunts are best characterized as "simultaneous individual hunts" (Boesch 1994b). By contrast, solitary hunts at Taï are rare, and 77% of group hunts feature differentiated roles that are coordinated in space and time (Boesch 2002). What explains this variation? Again, we turn to studies of social predators for clues. Using a mathematical model, Packer and Ruttan (1988) showed that cooperative hunting is most likely to evolve when additional hunters greatly increase the

chances of success. This is most likely to be the case when ecological conditions make solitary hunting difficult. This prediction is supported by empirical data. For example, the hunting success rate of solitary lions is low in the semi-arid environment of Etosha (Stander 1992a, 1992b), where the terrain is open and flat and prey occur at low densities (East 1984). There, 71% of hunts are collaborative (Stander 1992b). In contrast, in the Serengeti, where prey is abundant and daily feeding rates of solitary individuals are high (Packer et al. 1990), such division of labor is rare (Schaller 1972). Similarly, collaboration among wild dogs in the Selous is frequent (Creel and Creel 2002), while in the relatively easy hunting grounds of the Serengeti, collaboration is rare (Fanshawe and Fitzgibbon 1993).

We see a similar pattern of ecology-driven cooperation in the context of chimpanzee hunting. Reports of collaboration are linked to habitats where it is intrinsically more difficult to catch prey. At Taï, the forest canopy is high and uninterrupted, and red colobus are usually found in trees that reach above the canopy to heights of 40 to 50 meters (Boesch and Boesch-Achermann 2000). Red colobus therefore have more avenues by which to escape, and more time to do so, thus making it extremely difficult for a solitary hunter to succeed (Boesch 1994b). This indicates that there is strong selective pressure for group hunting with a division of labor at Taï. At the other extreme, Gombe is a mosaic of evergreen forest and deciduous woodland (Clutton-Brock and Gillett 1979), and the canopy is frequently low and discontinuous, and hence more favorable toward solitary hunters. This suggests that at Gombe group hunting (and especially collaboration) does not greatly increase the chances of success, at least in woodland or semi-deciduous forest. It remains to be seen, but is certainly possible, that the Gombe chimpanzees are more likely to adopt a division of labor when hunting in evergreen forest. The same may be true at Ngogo, where hunts are most likely to succeed where the canopy is broken (Watts and Mitani 2002b).

In sum, current evidence does not reveal any major differences between chimpanzees and other social predators in the frequency, causal factors, or mechanisms of cooperation in the context of social predation. Cooperation is rare across taxa, but when present, it appears to be highly sensitive to ecological factors, generating variation within and among species and populations. Similar to other predators, chimpanzees demonstrate varying degrees of cooperation which seem to be related to habitat differences that affect the probability of capture. Currently the relationship between intelligence and cooperation and collaboration is equivocal, as we have discussed strict role specialization in both highly intelligent (dolphins) and less intelligent (groupers, moray eels) species. Therefore, we have argued that a simple form of by-product mutualism, rather than differences in intelligence, may explain the distribution of cooperation and collaboration across taxa.

Implications and Future Directions

In this chapter we have discussed the role of intelligence in chimpanzee predatory behavior. We have asked whether the cognitive abilities of chimpanzees have generated unique, "sophisticated" solutions to the collective action problem posed by hunting in groups (table 18.1). In our examination of the benefits and mechanisms of group hunting, we identified no major factor separating chimpanzees from other social predators. Critically, there is no strong evidence to support the claim that chimpanzees hunt for social benefits such as sex or grooming. Therefore, it is unlikely that increased intelligence indirectly affects chimpanzee hunting behavior by widening the range of possible benefits that meat provides.

There is also little evidence that chimpanzees directly apply their advanced intelligence to the process of acquiring prey. First, like social predators (including those with limited intelligence), they appear to collaborate to catch prey only when ecological conditions make solitary hunting unprofitable. Second, collaboration may not be as cognitively demanding as it may seem. We argue that a simple by-product mutualism, in which individuals take advantage of the actions of others, may generate a hunting pattern consistent with that of a true team task. To support this claim, one must prove that individuals aren't simply attempting to maximize their own chances of making a kill, but rather are participating in a cooperative group effort. This presents a significant challenge for researchers on two fronts. First, it requires knowledge of all individuals' precise movements in space and time. The most convincing study would require multiple observers, each responsible for recording the behavior of a single individual during a hunt. A second-by-second account of the movements of each individual would allow researchers to assess the extent to which hunting strategies are predetermined, and whether they unfold as selfish responses to the actions of others. Second, a clearer understanding of the nutritional benefits of meat eating is required. Specifically, do chimpanzees benefit by obtaining small scraps of meat and the rare micronutrients they contain? Or should net calories be the ultimate currency with which the costs and benefits of hunting are measured? Answers to these questions require a combination of long-term data, detailed field observations, and laboratory experiments. We are confident that as such multidisciplinary approaches become more common, we will gain a more complete understanding of the role intelligence plays in chimpanzee cooperation.

Acknowledgments

We are very grateful to Richard Wrangham for encouraging us to tackle this topic, and to Zarin Machanda and three anonymous reviewers for their comments on early manuscript versions of this chapter. Finally, this chapter would not have been possible without extensive theoretical and empirical contributions to our understanding of chimpanzee predation by Christophe Boesch, John Mitani, David Watts, and Craig Stanford. We look forward to many future discussions.

Literature Cited

Anderson, C., and N. R. Franks. 2001. Teams in animal societies. *Behavioral Ecology* 12:534–40.

Baird, R. W. 2000. The killer whale: Foraging specializations and group hunting. In *Cetacean Societies: Field Studies of Whales and Dolphins*, edited by J. Mann, R. C. Connor, P. Tyack, and H. Whitehead, 127–53. Chicago: University of Chicago Press.

Baird, R. W., and L. M. Dill. 1996. Ecological and social determinants of group size in transient killer whales. *Behavioral Ecology* 7: 408–16.

Barnard, C. J., and R. M. Sibly. 1981. Producers and scroungers: A general model and its application to captive flocks of house sparrows. *Animal Behaviour* 29:543–50.

Bednarz, J. C. 1988. Cooperative hunting in Harris' hawk. *Science* 239:1525–27.

Boesch, C. 1994a. Chimpanzees–red colobus monkeys: A predator-prey system. *Animal Behaviour* 47:1135–48.

———. 1994b. Cooperative hunting in wild chimpanzees. *Animal Behaviour* 48:653–67.

———. 2002. Cooperative hunting roles among Taï chimpanzees. *Human Nature: An Interdisciplinary Biosocial Perspective* 13:27–46.

Boesch, C., and H. Boesch. 1989. Hunting behavior of wild chimpanzees in the Taï National Park. *American Journal of Physical Anthropology* 78:547–73.

Boesch, C., and H. Boesch-Achermann. 2000. *The Chimpanzees of the Taï Forest: Behavioural Ecology and Evolution*. Oxford: Oxford University Press.

Brown, J. L. 1983. Cooperation: A biologist's dilemma. In *Advances in Behaviour*, edited by J. S. Rosenblatt, 1–37. New York: Academic Press.

Bshary, R., A. Hohner, K. Ait-el-Djoudi, and H. Fricke. 2006. Interspecific communicative and coordinated hunting between groupers and giant moray eels in the Red Sea. *PLoS Biology* 4:2393–98.

Carbone, C., L. Frame, G. Frame, J. Malcolm, J. Fanshawe, C. FitzGibbon, G. Schaller, I. J. Gordon, J. M. Rowcliffe, and J. T. Du Toit. 2005. Feeding success of African wild dogs (*Lycaon pictus*) in the Serengeti: The effects of group size and kleptoparasitism. *Journal of Zoology* 266:153–61.

Clements, K. C., and D. Stephens. 1995. Testing non-kin cooperation: Mutualism and the Prisoner's Dilemma. *Animal Behaviour* 50:527–35.

Clutton-Brock, T. H., and J. B. Gillett. 1979. A survey of forest composition in the Gombe National Park, Tanzania. *African Journal of Ecology* 17:131–58.

Connor, R. C. 1986. Pseudo-reciprocity: Investing in mutualism. *Animal Behaviour* 34:1562–84.

———. 1995. The benefits of mutualism: A conceptual framework. *Biological Reviews of the Cambridge Philosophical Society* 70:427–57.

———. 2000. Group living in whales and dolphins. In *Cetacean Societies: Field Studies of Whales and Dolphins*, edited by J. Mann, R. C. Connor, P. Tyack, and H. Whitehead, 199–218. Chicago: University of Chicago Press.

———. 2007. Invested, extracted & byproduct benefits: A modified scheme for the evolution of cooperation. *Behavioral Processes*. 76:109–113.

Creel, S., and N. M. Creel. 1995. Communal hunting and pack size in African wild dogs, *Lycaon pictus*. *Animal Behaviour* 50:1325–39.

———. 2002. *The African Wild Dog: Behavior, Ecology, and Conservation*. Princeton, NJ: Princeton University Press.

Duffy-Echevarria, E., R. C. Connor, and D. St. Aubin. 2008. Observations of strand feeding by bottlenose dolphins (*Tursiops truncatus*) in Bull Creek, South Carolina. *Marine Mammal Science* 24: 202–6.

East, R. 1984. Rainfall, soil nutrient status and biomass of large Africa savanna mammals. *African Journal of Ecology* 22:245–70.

Fanshawe, J. H., and C. D. Fitzgibbon. 1993. Factors influencing the hunting success of an African wild dog pack. *Animal Behaviour* 45:479–90.

Gazda, S. K., R. C. Connor, R. K. Edgar, and F. Cox. 2005. A division of labour with role specialization in group-hunting bottlenose dolphins (*Tursiops truncatus*) off Cedar Key, Florida. *Proceedings of the Royal Society B: Biological Sciences* 272:135–40.

Gilby, I. C. 2006. Meat sharing among the Gombe chimpanzees: Harassment and reciprocal exchange. *Animal Behaviour* 71:953–63.

Gilby, I. C., L. E. Eberly, L. Pintea, and A. E. Pusey. 2006. Ecological and social influences on the hunting behaviour of wild chimpanzees (*Pan troglodytes schweinfurthii*). *Animal Behaviour* 72:169–80.

Gilby, I. C., L. E. Eberly, and R. W. Wrangham. 2008. Economic profitability of social predation among wild chimpanzees: Individual variation promotes cooperation. *Animal Behaviour* 75:351–60.

Gilby, I. C., and R. W. Wrangham. 2007. Risk-prone hunting by chimpanzees (*Pan troglodytes schweinfurthii*) increases during periods of high diet quality. *Behavioral Ecology and Sociobiology* 61:1771–79.

Goodall, J. 1986. *The Chimpanzees of Gombe: Patterns of Behavior*. Cambridge, MA: Harvard University Press.

Guinet, C., L. G. Barrett-Lennard, and B. Loyer. 2000. Co-ordinated attack behavior and prey sharing by killer whales at Crozet Archipelago: Strategies for feeding on negatively-buoyant prey. *Marine Mammal Science* 16:829–34.

Hamilton, W. D. 1964. The genetical evolution of social behavior. *Journal of Theoretical Biology* 7:1–52.

Hosaka, K., T. Nishida, M. Hamai, A. Matsumoto-Oda, and S. Uehara. 2001. Predation of mammals by the chimpanzees of the Mahale Mountains, Tanzania. In *All Apes Great and Small, Volume I: African Apes*, edited by B. Galdikas, N. Briggs, L. Sheeran, G. Shapiro, and J. Goodall, 107–30. New York: Klewer Academic Publishers.

Kortlandt, A. 1972. *New Perspectives on Ape and Human Evolution*. Amsterdam: Stichting voor psychobiologie.

Kruuk, H. 1972. *The Spotted Hyena: A Study of Predation and Social Behavior*. Chicago, IL: University of Chicago Press.

Mesterton-Gibbons, M., and L. A. Dugatkin. 1992. Cooperation among unrelated individuals: Evolutionary factors. *Quarterly Review of Biology* 67:267–81.

Milton, K. 2003. The critical role played by animal source foods in human (*Homo*) evolution. *Journal of Nutrition* 133:3886S–92S.

Minta, S. C., K. A. Minta, and D. F. Lott. 1992. Hunting associations between badgers (*Taxidea taxus*) and coyotes (*Canis latrans*). *Journal of Mammalogy* 73:814–20.

Mitani, J. C. 2005. Reciprocal exchange in chimpanzees and other primates. In *Cooperation in Primates: Mechanisms and evolution*, ed. by Kappeler, P. M. & van Schaik, C. P., 101–13. Heidelberg: Springer-Verlag.

Mitani, J. C., D. A. Merriwether, and C. Zhang. 2000. Male affiliation, cooperation and kinship in wild chimpanzees. *Animal Behaviour* 59:885–93.

Mitani, J. C., and D. P. Watts. 2001. Why do chimpanzees hunt and share meat? *Animal Behaviour* 61:915–24.

———. 2005. Seasonality in hunting by non-human primates. In *Seasonality in Primates: Studies of Living and Extinct Human and Non-human Primates*, edited by Brockman, D. K. & van Schaik, C. P., 215–40. Cambridge: Cambridge University Press.

Muller, M. N., and J. C. Mitani. 2005. Conflict and cooperation in wild chimpanzees. *Advances in the Study of Behavior* 35:275–331.

Newton-Fisher, N. E., H. Notman, and V. Reynolds. 2002. Hunting of mammalian prey by Budongo Forest chimpanzees. *Folia Primatologica* 73:281–83.

Nishida, T., T. Hasegawa, H. Hayaki, Y. Takahata, and S. Uehara. 1992. Meat-sharing as a coalition strategy by an alpha male chimpanzee? In *Topics in Primatology*, edited by T. Nishida, W. C. McGrew, P. Marler, M. Pickford, and F. B. M. de Waal, 159–74. Tokyo: University of Tokyo Press.

Noë, R., and P. Hammerstein. 1994. Biological markets: Supply and demand determine the effect of partner choice in cooperation, mutualism and mating. *Behavioral Ecology and Sociobiology* 35: 1–11.

Packer, C. 1977. Reciprocal altruism in *Papio anubis*. *Nature* 265:441–43.

Packer, C., and T. M. Caro. 1997. Foraging costs in social carnivores. *Animal Behaviour* 54:1317–18.

Packer, C., and L. Ruttan. 1988. The evolution of cooperative hunting. *American Naturalist* 132: 159–98.

Packer, C., D. Scheel, and A. E. Pusey. 1990. Why lions form groups: Food is not enough. *American Naturalist* 136:1–19.

Peterson, R. O., and P. Ciucci. 2003. The wolf as a carnivore. In *Wolves: Behavior, Ecology and Conservation*, edited by L. D. Mech and L. Boitani. Chicago: University of Chicago Press.

Pitman, R. L., L. T. Ballance, S. L. Mesnick, and S. J. Chivers. 2001. Killer whale predation on sperm whales: Observations and implications. *Marine Mammal Science* 17:494–507.

Schaller, G. 1972. *The Serengeti Lion*. Chicago: University of Chicago Press.

Sibly, R. M. 1983. Optimal group size is unstable. *Animal Behaviour* 31:947–48.

Smith, T. G., D. B. Siniff, R. Reichle, and S. Stone. 1981. Coordinated behavior of killer whales, (*Orcinus orca*), hunting a crabeater seal (*Lobodon carcinophagus*). *Canadian Journal of Zoology* 59:1185–89.

Stander, P. E. 1992a. Cooperative hunting in lions: The role of the individual. *Behavioral Ecology and Sociobiology* 29:445–54.

———. 1992b. Foraging dynamics of lions in a semiarid environment. *Canadian Journal of Zoology–Revue Canadienne De Zoologie* 70:8–21.

Stanford, C. 1996. The hunting ecology of wild chimpanzees: Implications for the evolutionary ecology of Pliocene hominids. *American Anthropologist* 98:96–113.

Stanford, C. B. 1998. *Chimpanzee and Red Colobus*. Cambridge, MA: Harvard University Press.

Stanford, C. B., J. Wallis, E. Mpongo, and J. Goodall. 1994. Hunting decisions in wild chimpanzees. *Behaviour* 131:1–18.

Stephens, D. W., and J. R. Krebs. 1986. *Foraging Theory*. Princeton, NJ: Princeton University Press.

Stevens, J. R. 2004. The selfish nature of generosity: Harassment and food sharing in primates. *Proceedings of the Royal Society of London Series B: Biological Sciences* 271:451–56.

Takahata, Y., T. Hasegawa, and T. Nishida. 1984. Chimpanzee predation in the Mahale Mountains from August 1979 to May 1982. *International Journal of Primatology* 5:213–33.

Teleki, G. 1973. *The Predatory Behavior of Wild Chimpanzees*. Lewisburg, PA: Bucknell University Press.

———. 1981. The omnivorous diet and eclectic feeding habits of chimpanzees in Gombe National Park, Tanzania. In *Omnivorous Primates: Gathering and Hunting in Human Evolution*, edited by R. S. O. Harding and G. Teleki. 303–43. New York: Columbia University Press.

Tennie, C., I. C. Gilby, and R. Mundry. 2009. The meat-scrap hypothesis: Small quantities of meat may promote cooperation in wild chimpanzees (*Pan troglodytes*). *Behavioral Ecology and Sociobiology* 63:421–31.

Trivers, R. L. 1971. The evolution of reciprocal altruism. *Quarterly Review of Biology* 46: 35–57.

Uehara, S. 1997. Predation on mammals by the chimpanzee (*Pan troglodytes*). *Primates* 38:193–214.

Van Lawick-Goodall, H., and J. van Lawick-Goodall. 1971. *Innocent Killers*. Boston, MA: Houghton Mifflin.

Vucetich, J. A., R. O. Peterson, and T. A. Waite. 2004. Raven scavenging favours group foraging in wolves. *Animal Behaviour* 67:1117–26.

Watts, D. P., and J. C. Mitani. 2001. Boundary patrols and intergroup encounters in wild chimpanzees. *Behaviour* 138:299–327.

———. 2002a. Hunting and meat sharing by chimpanzees at Ngogo, Kibale National Park, Uganda. In *Behavioural Diversity in Chimpanzees and Bonobos*, edited by C. Boesch, G. Hohmann, and L. Marchant, 244–55. Cambridge: Cambridge University Press.

———. 2002b. Hunting behavior of chimpanzees at Ngogo, Kibale National Park, Uganda. *International Journal of Primatology* 23:1–28.

West-Eberhard, M. J. 1975. The evolution of social behavior by kin selection. *Quarterly Review of Biology* 50:1–33.

Wrangham, R. W. 1975. *The Behavioural Ecology of Chimpanzees in Gombe National Park, Tanzania*. PhD thesis, Cambridge University.

———. 1982. Mutualism, kinship and social evolution. In *Current Problems in Sociobiology*, edited by Kings College Sociobiology Group., 269–89. Cambridge: Cambridge University Press.

Zahavi, A. 1975. Mate selection: A selection for a handicap. *Journal of Theoretical Biology* 53:205–14.

PART IV

Social Minds:
Empirical Perspectives

19

Chimpanzee Social Cognition

Michael Tomasello and Josep Call

A researcher sits at a table with Alexandra, a young chimpanzee, who intently watches him scrub the table's surface with a sponge. The human moves the sponge back and forth until it suddenly tumbles from his grip, falling across the table and to the ground on the other side. Seemingly unable to move from his current position, he reaches across the table, trying unsuccessfully to reach the fallen object. Alexandra glances down at the sponge on the ground, up at the straining human, and back down at the sponge. She then deftly leaps to the ground and retrieves the sponge, handing it back to the human without a sound. Did she understand that the researcher couldn't reach the sponge—or that he wanted to? What level of comprehension might she and other chimpanzees have in terms of the goals and intentions of other beings?

Like many primate species, chimpanzees live in a complex "social field" that is perhaps especially complex due to its fission-fusion structure. Virtually every decision individuals make on a daily basis is affected by which other individuals are present and what they are likely to do. Judgments about what others may do next are based not only on a variety of immediate situational factors, but also on past interactions and relationships that those others may have had with still others who might be present (see Muller and Mitani 2005 for a recent review of fieldwork documenting this social complexity).

The fundamental question of chimpanzee social cog-

nition is that of how individuals understand the actions of others. One extreme position is that they understand others in virtually the same way as humans do; that is, chimpanzees have basically a full-blown, human-like "theory of mind." The opposite extreme is that chimpanzees are simply behavior readers; that is, they learn which behavioral and contextual cues predict which future behaviors of others, with no mentalistic understanding of those others at all. In between these two extremes, our own view is that chimpanzees understand others as intentional agents in basically the same way as humans do—in terms of their goals and perceptions—but unlike humans, they do not understand others in terms of their beliefs, especially false beliefs. This makes chimpanzees' social cognition fundamentally the same as that of humans—but at the same time subtly different.

In this chapter we lay out our case for this "third way" position—cognitivist but not anthropocentric—by summarizing and integrating recent experimental research on the social-cognitive skills of chimpanzees, with a focus on our own work. We begin with a brief history.

A Brief History

Premack and Woodruff's (1978) original groundbreaking study—aimed at determining whether chimpanzees have a "theory of mind"—was actually about chimpanzees' understanding of human goals. They presented the language-trained chimpanzee Sarah with videos of human actors coming upon obstacles in problem-solving situations—for example, a human trying to exit a locked door. Sarah then saw two pictures, one of which was irrelevant and one of which represented in some way a solution to the problem—for example, a key. Sarah was reasonably reliable in choosing the correct picture, but the basis of her choice was immediately questioned. Thus, Savage-Rumbaugh et al. (1978) presented data from their chimpanzees suggesting that these tasks could be solved simply by associating keys with doors, and so forth. Supporting this skepticism, Povinelli and Perilloux (1998) had six juvenile chimpanzees observe both a "clumsy" experimenter spilling their juice by accident and a different, "mean" experimenter pouring out their juice on purpose before they could get any. In a later choice test, the chimpanzees showed no preference for either the "clumsy" or the "mean" human, suggesting that they did not discriminate (or care about) the intentionality involved. On the basis of a number of experimental studies, Tomasello (1996) concluded that in social learning situations chimpanzees show no understanding of the goals or intentions of others.

Negative experimental evidence also accrued during the 1990s about chimpanzees' understanding of visual perception. Most prominently, in a series of 15 experiments Povinelli and Eddy (1996) presented six juvenile chimpanzees with a choice of two humans from whom they could beg food (which was situated on a table between the two). The surprising finding was that the chimpanzees begged from both humans indiscriminately, even when one was wearing a blindfold or had a bucket on his head. Similarly, Povinelli et al. (1994)—in a more controlled version of Povinelli et al. (1990)—found that the same chimpanzees failed to differentiate between a knowledgeable and an ignorant human experimenter. Specifically, when the chimpanzees saw two humans pointing to different locations to indicate the location of a single piece of hidden food—and they had seen one of those humans watch the original hiding process and the other not—they followed the different humans' pointing gestures indiscriminately.

But beyond understanding goals, perceptions, and knowledge, in the debates following Premack and Woodruff's (1978) paper a consensus emerged that mature hu-

man beings understood others in terms of their beliefs about the world. Beliefs are mental representations that guide action—I search for my keys in the drawer because I believe they are there. The key diagnostic is an understanding of false beliefs—that is, an understanding that even if my keys are really in the car, I will nevertheless search in the drawer if that's where I believe they are. Young children from around four to five years of age understand that people's actions may be guided by false beliefs (Wimmer and Perner 1983). Call and Tomasello (1999) devised a nonverbal false-belief test, and found that although five-year-old children passed it readily, chimpanzees failed it miserably.

All of these data led Tomasello and Call (1997) to the general conclusion that chimpanzees and other nonhuman primates do not understand the psychological states of others. That is, they understand that others act, and they can predict those actions in many situations based on their own experience (and perhaps some specialized cognitive adaptations), but they do not go beneath the surface to an understanding of the goals, perceptions, knowledge, and beliefs that guide action. Not all of the data existing at the time of that study were consistent in suggesting this conclusion, however, and chimpanzees had been observed doing many things in their natural habitats that would seem to require more than just an understanding of surface-level behavior. For example, Byrne and Whiten (1990) collected many informal observations of nonhuman primates, including chimpanzees, doing things that seemed to involve some form of deception. In prototypical forms of deception, one individual conceals things from another or misleads them in ways that clearly require an understanding that the other sees and knows things.

The story since the late-1990s has been one of experimenters finding better ways to tap into what chimpanzees and other nonhuman primates know about the psychological states of others, and thus getting many more positive results. In most cases this has been guided by attempts to model the experiments more closely on situations that chimpanzees routinely encounter in their natural environments—for example, presenting them with problems less often in situations in which they must cooperate and/or communicate with humans, and more often in situations in which they must compete with either conspecifics or humans. Skeptics still abound, as represented most prominently by Povinelli and colleagues (e.g., Povinelli and Vonk 2006). They continue to cling to the hypothesis that chimpanzees only understand surface-level behavior, and indeed this explanation is almost always possible for any single experiment. But the approach we have taken in our laboratory over the past decade or so is to devise multiple experimental paradigms—in many cases representing highly novel problem situations for the chimpanzees—that are all aimed at a single psychological state. Thus, in the next section we present seven different experimental paradigms involving five different response measures, all indicating that chimpanzees understand the goals underlying the actions of others. In the section following that, we present five different experimental paradigms involving four different response measures, suggesting that chimpanzees understand the visual perception and knowledge of others (but no evidence yet that they also understand the beliefs of others). This converging-operations approach forces proponents of the behavior-reading hypothesis to create many different behavioral explanations, sometimes differing wildly from one another and almost never having an empirical basis, to explain all of the different results. At some point, this kind of explanation simply sinks of its own weight.

Understanding Goals and Intentions

In order to compete and cooperate effectively with others in their group, highly social animals such as chimpanzees must be able not only to react to what others are doing, but also to anticipate what they will do. One way they do this is by observing what others do in particular situations and deriving "behavior rules" (or, in some cases, already possessing them innately) that will enable them to predict the behavior of others when the same situation, or a very similar one, arises again. But another, better way to do it is to discern directly what the other is trying to do—what state of the environment he is trying to bring about. This enables one individual to predict another's behavior not only in previously observed situations but in novel situations as well.

The methodological problem is that when an actor is successful in achieving a goal, then what he is trying to do and what he actually does are the same thing. Thus, distinguishing whether the observer is reading the actor's behavior or just reading his goals is very difficult or impossible. But there are also situations in which what the actor is

trying to do—his goal—does not match what he actually does. This mismatch occurs in unsuccessful attempts and in accidents. The best evidence that an observer is reacting to goals and not to behavior, therefore, is when she reacts to the actor's goal and not to his actual behavior when he is trying unsuccessfully or having an accident. Such behavioral prediction will be enhanced even further if the observer can assess the physical and other constraints on the actor's actions, and so figure out which behavioral plan he will choose to achieve his goal in the given situation (see Tomasello et al. 2005 for more detailed definitions and discussions).

In this section we present evidence that chimpanzees read and react not only to the surface behavior of others but also to their underlying goals, and that in some cases they can understand why the actor has chosen a particular course of action (i.e., what constraints it is reacting to) in a given situation. They demonstrate such understanding in a number of different contexts, some quite novel, both in straight behavior prediction situations and in social learning situations.

Interpreting Actions on the Basis of Goals

Call and Tomasello (1998) taught chimpanzees that a marker on top of a container indicated hidden food inside. During this training, the subjects were prevented from seeing the experimenter place the marker on the baited container—instead, the subjects found the marker already there. They were then tested in a situation in which two containers had each been marked before the subject could choose one of them. The trick was that in one case the marker had been placed accidentally—either the experimenter had dropped it on top of the container accidentally ("Whoops!") or it had fallen by itself—whereas in the other case it had been placed on purpose by the experimenter ("There!"). When the chimpanzees were then allowed to choose between the two containers, they mostly chose the one marked intentionally, especially in the early trials. It is easy to specify for this experiment a behavior rule leading to successful performance, because the intentional and accidental markings were behaviorally distinct. But it is important to note that the chimpanzees made this distinction from the earliest trials even though the placing of markers, especially accidentally, should have been fairly novel behaviors for them to observe.

Call et al. (2004) took a different approach to the problem of behavior reading. In this study a human experimenter gave food to the chimpanzee repeatedly through a glass panel. Then, on some trials, he did not give any food. The experimental manipulation was that sometimes he did not give the food because he was *unwilling* to do so, whereas at other times he did not give it because he was *unable* to do so. The methodological advance was that the experimenter's failure to give the food was actually accomplished in many different ways. Three of those ways instantiated unwillingness: the experimenter either stared at the food on the table in front of him without giving it, ate the food himself, or teased the ape with it. Yoked to each of these three actions of unwillingness were two actions of inability, each of which resembled its counterpart fairly closely behaviorally, including in the way the food was moved and in exactly where the experimenter looked. For example, yoked to the experimenter's teasing action were a "clumsy" action, in which while bringing the food to the chimpanzee the human dropped it "accidentally" and it rolled back to him, and a "trying" action in which the human attempted "unsuccessfully" to force the food through a small hole in the glass and then brought it back to himself. Each of the three different behavioral sets had these "unwilling" and "unable" instantiations that superficially differed from one another in very different ways. And yet, in terms of goals, what the three "unwilling" actions had in common was that the experimenter did not want to give the food to the ape, while what the six "unable" actions had in common was that the experimenter did want to give the food to the ape and was "trying" to do so. The main result was that the chimpanzees reacted similarly to the "unwilling" actions by expressing frustration and impatience, and also reacted similarly to the "unable" actions by being patient. The obvious conclusion is that the chimpanzees' similar reactions in the two different experimental conditions were due to their understanding of the two different goals involved in the two different sets of conditions, no matter how they were expressed behaviorally.

A related finding comes from two studies actually aimed at a different question. Warneken and Tomasello (2006) wanted to know whether young chimpanzees would help a human to fetch an object positioned out of reach. A human experimenter thus strained and reached toward an object (actually several different objects in several different situations) in the presence of chimpanzees, who tended to fetch it for him. Significantly, they did not fetch it for him in various control conditions in which he threw the object away or otherwise indicated lack of interest. The difference in the chimpanzees' behavior between the experimental and control conditions could be interpreted as indicating their understanding of the experimenter's different goals in the two situations. The problem was that this behavior may have derived from some history of interaction that the chimpanzees had previously had with humans. Warneken et al. (2007) therefore set up a much more novel situation in which one chimpanzee might help another. In this study, one chimpanzee attempted to open a door into an adjoining room, often shaking it in his attempt. Another chimpanzee then quite often, from her advantageous location, pulled a chain that unlocked the door so that the first chimpanzee could open it. They did this more than in a control condition in which the first chimpanzee was trying to get out another door.

Together, then, these two studies of so-called "instrumental helping" suggest that a chimpanzee can tell when another chimpanzee needs help in achieving a goal.

Any one of the studies cited so far could be explained in terms of behavioral rules that chimpanzees are either born with or learn. In our view, however, such explanations are inadequate to explain the diversity of behaviors investigated across the different studies, as well as the different response measures required (choosing a container, reacting appropriately to the human's actions, helping another achieve his goal). Nevertheless, in all of these studies there are behavioral differences between experimental and control conditions that could conceivably be input for chimpanzee behavioral rules. It is not clear what would be the adaptive history or significance of these rules, or how they would be connected to such different responses—and indeed, to repeat an earlier point, researchers who raise the possibility of such rules never supply empirical evidence that the apes actually employ them, or evidence of how they could have acquired them. But still, behavioral rules are possible in all these cases.

In two recent experiments, however, we have completely ruled out the possibility of behavioral rules because the experimenter's actions in the experimental and control conditions of the study's test phase were identical;

the only difference was in the context leading up to those actions—which, for organisms capable of understanding goals, leads to two different interpretations of what the actor was doing. Specifically, Buttelmann et al. (submitted) ran two experiments of this type, each with a different response measure. In the first, the experimenter—identically in the experimental and control conditions—twirled a piece of metal on top of a box that the chimpanzee knew contained food—an action that in some contexts could be seen by the chimpanzee as an attempt to open the box. In the experimental condition, the chimpanzees had previously seen the experimenter manipulate locks and latches on the tops of other boxes containing food, and opening them and offering the food to the chimpanzees. This set up the chimpanzees' expectation that the experimenter would also try to open the final target box as well. This was different from the control condition, in which they had watched the experimenter simply manipulate locks and latches on the tops of boxes without opening them, and then draw food from his pocket and offer it to them. Using the same response measures as in Call et al. (2004), it was found that chimpanzees tended to wait longer in the experimental condition than in the control condition, presumably because only in the experimental condition did they see the experimenter's action as trying to open the box.

In a second study, Buttelmann et al. used a very different paradigm in which a human sat on a stool, drawing food from a bucket and giving it to the chimpanzee through a glass panel. A second bucket also containing food sat in front of another glass panel several meters away, and the chimpanzee had previously received food from it as well. In both the experimental and control conditions of the experiment's test phase, the experimenter stood up from his stool and turned his body toward the second bucket. In situations like this, and in the absence of other information, the chimpanzees naturally anticipate that the human is headed for the second bucket, and so they go there first themselves. That is what they did in this experiment. There was also an experimental condition, however, in which something happened before the experimenter stood up—for example, a call came from a walkie-talkie in the same direction as the second bucket, or another human threw a clipboard at the experimenter and it landed short (again, in the same direction as the second bucket). In this case, the chimpanzees waited longer to move, presumably because they understood that the experimenter's goal in standing up was not to go to the second bucket but rather to answer the walkie-talkie or fetch the clipboard. They interpreted the exact same behavior differently depending on their understanding of the experimenter's goal in standing up.

Because cognitive experiments with nonverbal organisms may typically be interpreted in many different ways, our strategy has been to use converging operations. In this case, we have investigated whether chimpanzees understand the goals underlying an actor's behavior by seeing if they (1) discriminate his intentional from his accidental actions, (2) discriminate when he is trying but failing from when he is refusing to do something, (3) determine the goal another is pursuing in the context of helping; (4) interpret a human's actions differently based on the goal he is seemingly pursuing (determined by previous context), and (5) predict what a human will do next based on his apparent goal, given the context. Behavioral-rules explanations are possible for many of the experiments, though these cannot include rules learned during the experiment, since in many cases the chimpanzees performed well from the outset. But there will have to be different behavioral rules for each individual experiment, and they will have to be connected to the observing individual's behavioral responses in various ways, since there are four different response measures involved in these different studies. And behavioral rules are not possible—at least not in a straightforward way—for the studies in which the behavior being reacted to is exactly the same in different conditions. The alternative is to simply propose that in many situations, chimpanzees are able to understand and discern the goals that others are pursuing.

Selective Imitation of Goal-Directed Actions

We have also investigated chimpanzees' understanding of goal-directed action in several social learning paradigms. Beyond just reacting to or anticipating the behavior of others, an individual who attempts to imitate the actions of another gives us a fairly direct expression of what she understands the other to be doing; she is actually acting out her understanding. Social learning paradigms may be especially informative in this context, because it is not

Figure 19.1   Human-raised chimpanzees in the Tomasello and Carpenter (2005) study

at all clear how the behavioral-rules explanation—focused as it is on the organism reacting to or predicting the behavior of others—might work in the case of social learning and imitation. The problem is that chimpanzees in general are not such skillful imitators of the actions of others; they tend to focus much more on the results produced by actions (with some disagreement among researchers about the strength of this tendency; see Call and Carpenter 2002, Tomasello 1996; see also chapter 8). But there is fairly strong evidence that chimpanzees raised extensively by humans focus much more on the actual actions involved (e.g., Tomasello et al. 1993). The social learning studies that provide evidence for an understanding of goal-directed actions all come, therefore, from chimpanzees raised from a fairly young age by humans.

Tomasello and Carpenter (2005) presented two different imitation tasks (previously used with human infants) to three young chimpanzees, two to four years of age at the time of testing, who had been raised from birth mostly by humans (figure 19.1). In the first test, based on Meltzoff (1995), in the key condition the human demonstrator attempted unsuccessfully in various ways to effect some change of state on an apparatus—for example, she attempted unsuccessfully to separate the two parts of a dumbbell. Like human infants, these three young human-raised chimpanzees did not copy the surface actions of the demonstrator (i.e., the actual behavioral attempts), but instead reproduced the outcome he intended to produce (i.e., the pulling apart of the dumbbell)—doing this just as often as when they had seen a full demonstration of the action leading to its full result. Significantly, they did not

do this in two different control conditions in which the experimenter either manipulated the apparatus idly or else just showed them the end state (e.g., the separated dumbbell) without any intervening behavioral attempts. If we interpret this experiment the way it is normally interpreted for human infants, the conclusion is that the chimpanzees understood what the experimenter was trying to do—his goal.

In the second test, based on Carpenter et al. (1998), the same three human-raised chimpanzee juveniles watched a human demonstrator perform two actions on a single apparatus and then offer each of them a turn at doing the same. The demonstrator performed one of the actions intentionally ("There!") and performed the other "accidentally" ("Whoops!") in varying order on different apparatuses. Like human infants, across all the trials the chimpanzees selectively imitated the intentional action over the "accidental" action, regardless of the order in which they had seen them performed. If we interpret this experiment as it is normally interpreted for human infants, the conclusion is that the chimpanzees understood which action the experimenter was doing "on purpose" and which action was only an "accident" not representing his goal.

Finally, in a recent study we have gone beyond even these fairly direct expressions of human-raised chimpanzees' understanding of goal-directed action in social learning contexts. Using the basic method of Gergeley et al. (2002), Buttelmann et al. (2007) tested six human-raised chimpanzees in the so-called rational imitation paradigm. The chimpanzees were shown how to operate an apparatus to produce an interesting result (there were many different apparatuses), and then they were given a turn at operating it. The most natural behavior for them in all cases was to operate it with their hands. But this obvious behavior was never demonstrated for them. Instead, they always saw a human manipulate the apparatus in a novel way with some other body part. The idea was that in some cases the physical constraints of the situation dictated that the human had to use that unusual body part. For example, he had to turn on a light with his head because his hands were occupied holding a blanket, or he had to operate a light with his foot because his hands were occupied with a heavy bucket (figure 19.2). After seeing this forced use of the unusual body part, the chimpanzees mostly discounted it and used their hands as they normally would, since the same constraints were not present for them. However, when they saw the human use the unusual body part without any physical constraint requiring it, they quite often copied the unusual behavior. If we interpret this experiment the way it is normally interpreted for human infants, the conclusion is that the chimpanzees understood not only what the experimenter was trying to do—his goal—but also why he was doing it in the way he was doing it—the rationality behind his choice of action toward the goal. According to Tomasello et al. (2005), an understanding of the action plan chosen toward a goal constitutes an understanding of the intention.

It is difficult to see how behavioral rules apply to the imitation paradigm at all. Behavioral rules are for how to react to others and perhaps predict their behavior. There are no rules for how to reproduce the goal-directed actions of others, especially when we are not talking about the surface actions involved but rather their intended outcome, and especially when the means for producing that outcome are selectively imitated depending on the contextual circumstances for the demonstrator and learner. It is true that all of these results in the social learning paradigm are from human-raised chimpanzees. But these chimpanzees were not specifically trained to imitate or to discern goals, nor had they any previous experience with any of the tasks presented to them. Various attempts to perform similar experiments with non–human-raised chimpanzees have led to mixed or negative results (e.g., Buttelmann et al., submitted). These results, in our opinion, are negative not because non–human-raised chimpanzees do not understand goals or intentions, but only because they lack the imitation skills needed to perform in the experiments.

Conclusion

We believe that there is only one reasonable conclusion to be drawn from the totality of the studies reviewed here. Chimpanzees, like humans, understand the actions of others not only in terms of surface behaviors but also in terms

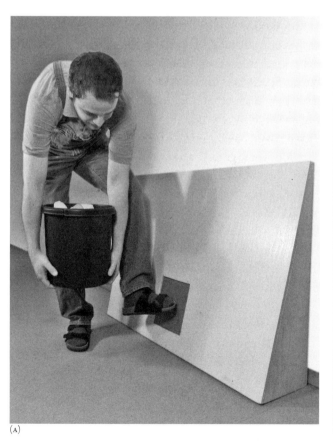

(A)

(B)

Figure 19.2 Turning on the light with the foot (a) because E wanted to, or (b) because E had to (since his hands were occupied) in the Buttelmann et al. (2007) experiments

of the underlying goals, and possibly intentions, involved. Behavioral rules may be created to explain the results of each of the various studies in which the chimpanzees reacted to or predicted the behavior of others based on an understanding not of their surface actions but of their goals. However, this will require many different ad hoc behavioral rules for which there is absolutely no positive evidence—and indeed, consistent use of this explanatory strategy would deny the ability of human children to understand goals and intentions, as most of the ape studies are modeled on child studies. Moreover, the studies from the social learning paradigm would not seem amenable to behavioral-rules explanations at all. The role of human rearing in chimpanzees' performance in the social learning paradigm is unclear, and so this is an important question for future research. Another important question is whether other nonhuman primate species have some of these same skills. In some of the studies cited above (e.g., Call and Tomasello 1998; Buttelmann et al., submitted), we obtained similar results with other great ape species, though in all cases the number of subjects investigated was small and so the results for each species individually were not statistically reliable.

Understanding Perception and Knowledge

To understand how others work as goal-directed agents, one must understand not just their goals but also their perceptions, since what organisms see and know helps to determine what they do. Here we examine what a chimpanzee understands about what another individual sees—not just what he is oriented to, but what he registers from the environment in ways that affect his actions, and also about what others know in the sense of information they have registered from the past that still affects their current actions (e.g., their knowledge of where food is, even though they cannot see it now). Understanding false beliefs is the special case in which an observer predicts or explains the behavior of an actor based on what that actor believes to be the case, not what really is the case as the observer knows it (e.g., the actor believes the food is in one place when the observer knows that it is really in another).

Again in the current case, the main alternative hypothesis we must consider is that chimpanzees either are born with or learn certain behavioral rules that determine how they respond to others' surface orienting behaviors without any understanding of their perception or knowledge. And as in the experiment described above, involving an individual's understanding of another individual's goals, the best evidence that an individual understands another's perception is when the perceptions of two individuals differ (e.g., they see different sides of a barrier) and the observer acts on the basis not only of what he himself sees or knows, but also on the basis of what the other sees or knows.

Detecting and Following Gaze Direction

Many mammals follow the gaze direction of conspecifics to outside targets. By itself, gaze following does not require an understanding of what the other sees. But a number of experimental variations have been tested with chimpanzees, and these, especially in combination with the research to be reviewed in the two subsections to follow, do suggest something in this direction.

First, chimpanzees do not just follow the gaze direction of a conspecific or human in a given direction (Povinelli and Eddy 1996b; Tomasello et al. 1998), which could represent some kind of simple orienting reflex or learned behavioral rule—they follow it to a particular external target. Specifically, they follow the direction of a human's gaze past distractors and to locations behind barriers, even when it means they must locomote some distance to change their viewing angle (Tomasello et al. 1999). When they follow the gaze direction and see nothing interesting, older but not younger chimpanzees check back to make sure they have followed the gaze accurately (Bräuer et al. 2005; Call et al. 1998). When they look repeatedly and still see nothing interesting, older chimpanzees eventually stop looking but younger ones do not (Tomasello et al. 2001). These variations on basic gaze following suggest that chimpanzees are not orienting themselves using the head direction of the other; they really want to know what the other sees, as it might be important for them.

Second, chimpanzees can detect when others can and cannot see them. With one another, for example, they use visually-based gestures mostly when the potential recipient is attending to them already (Tomasello et al. 1994, 1997). Indeed, if the potential recipient is not attending to them, they will sometimes walk around in front of her before gesturing (Liebal et al. 2004). In support of these findings, Liebal et al. (2004) found in an experimental study that when a human who was facing a chimpanzee and giving him food then turned his back to the chimpanzee, the chimpanzee subject walked around him to face him again before gesturing.

As noted above, Povinelli and Eddy (1996a) found that chimpanzees do not preferentially beg food from a human with uncovered eyes over begging it from one wearing a blindfold. This suggests that they do not know when others can and cannot see them. But Kaminski et al. (2004) modified this paradigm slightly to reflect more natural communicative situations, and they found different results. Their modification was that the chimpanzees did not have to choose between two human communicators—a very unnatural situation (and indeed the young human children tested by Povinelli and Eddy, 1996a, struggled in this paradigm)—but were always faced with only one communicator who was oriented in different ways in different experimental conditions. For example, in one condition the human faced the subject. In another, his back and head were both facing away. In yet another, his body was turned away as he looked back over his shoulder at the subject. The main finding was that chimpanzees gestured differently to the human depending on whether the human's face was oriented toward them, but only if the human was facing them bodily as well. Kaminski et al. (2004) argued, therefore, that body orientation and face orientation indicate two different things to an ape when it begs food from a human. Whereas body orientation indicates the human's disposition to give the subject food (i.e., when he is oriented so as to transfer food effectively), face orientation indicates whether the human is able to see the subject's begging gesture. This two-factor account helps to explain Povinelli and Eddy's (1996) negative findings.

Interestingly and significantly, Kaminski et al. (2004) found that as long as the human was facing them, chimpanzees did not differentiate between the human's eyes being open and closed. This accords with the findings of Tomasello et al. (2007) that in gaze-following situations chimpanzees follow mainly the direction of the head as a whole, and to a much lesser extent the direction of the eyes. It also helps to explain most of the remaining negative findings of Povinelli and Eddy (1996a). As originally suggested by Tomasello (1996) in his commentary on those studies, it is possible to understand that others see things

without knowing that the eyes specifically are the mechanisms involved. Based on the evidence reviewed here and in the coming subsections, this is apparently the case with chimpanzees.

Competition: Knowing What Others See

As noted, simply following the gaze of others or gesturing to their faces are not by themselves unequivocal evidence that chimpanzees know what others can see, because each of these behaviors could also be governed by some natural tendency or learned rule. But a number of other recent studies have provided much stronger evidence that chimpanzees know what others see, since they required the subject to act not just on the basis of what she saw but on the basis of what the other saw (which was different). And many of the situations involved in these studies have been novel to the degree that learning explanations are extremely implausible. Importantly, a common factor in almost all of the studies is that they engage chimpanzees competitively, whereas many of the experiments with negative results from the 1990s (noted in the introduction) required chimpanzees to engage with humans cooperatively. By many accounts, competition is chimpanzees' most natural form of interaction, especially when food is involved (Hare 2001).

In a series of five studies, Hare et al. (2000) placed a subordinate and a dominant chimpanzee into two rooms situated on either side of a third room (figure 19.3). Each of the outer rooms had a guillotine door leading into the middle room which, when opened at the bottom, allowed them to see two pieces of food placed at different locations in the middle—and also to see another chimpanzee

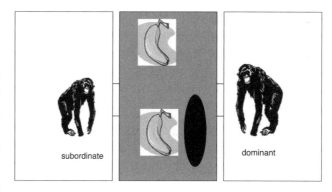

Figure 19.3 Experimental setup for the Hare et al. (2000) experiments. Each chimpanzee's entrance into the middle cage is regulated by an experimenter-controlled guillotine door.

looking into the middle room from under her door on the other side. After the food was placed in the middle room, the two side doors were opened and both chimpanzees were allowed to enter the middle room. The problem for the subordinate chimpanzee in this situation was that the other, dominant chimpanzee would take all of the food she could see. In some cases, however, the pieces of food were arranged so that the subordinate could see one that the dominant could not see. For example, one piece of food might be placed on the subordinate's side of a small barrier, with the other piece of food positioned out in the open. The question was, therefore, whether the subordinate knew that the dominant could not see a particular piece of food, and thus that it was safe to reach for it. The basic finding was that the subordinate did indeed reach for the food that only they could see much more often than they reached for the food that both they and the dominant could see. Importantly, there were no differences between the early and later trials, thus ruling out the possibility that individuals were learning during the experiment the most effective way to get the food.

One possibility is that the subordinates in these studies may have been monitoring, and reacting to, the dominants' behavior rather than their perceived access to the food. But this possibility was ruled out in some of the studies in this series by giving the subordinates a small head start and forcing them to make their choice between the two pieces of food before the dominant was released into the middle cage. Moreover, in one additional control condition the dominant's door was lowered before the two competitors were let into the room (and again the subordinate got a small head start), so that the subordinate could not see the dominant at all at the moment of choice—and so could not react to her behavior. In these cases, the subordinates still targeted the piece of food the dominant could not see. In still another control condition, food was placed on the subordinate's side of a transparent barrier, and subordinates, understanding that the dominant's view of the food was not blocked, chose equally between the two pieces of food. Finally, to demonstrate the flexibility of the skills involved here, dominant individuals were placed in the position of the subordinate—that is, with one piece of food visible to both themselves and the dominant, and the other piece visible only to them. In this case, the strategy of the dominants—who were the same individuals who in other pairings had been subordinates—was to go first for the

food that lay out in the open, and only then to the hidden food. The strategy was presumably to use their dominance to first take the piece of food that was "at risk"—the one the subordinate could also see—and save the food about which they had privileged knowledge for later, when the subordinate was already in the middle room.

Karin-D'Arcy and Povinelli (2002) failed to replicate the study of Hare et al. (2000). They found that subjects did not preferentially pursue the hidden piece of food as they had in the Hare et al. studies. But there is a very good reason for this discrepancy that is attributable to a crucial difference in methodology: the size of the testing arena in the Karin-D'Arcy and Povinelli study was about half the size of that in the Hare et al. (2000) studies and, most crucially, the distance between the food pieces was 1.25 meters instead of 2 meters. This spatial difference almost certainly changed the nature of the competition between the chimpanzees. If a subordinate thought that the dominant would go for the piece out in the open, this would bring her extremely close to the hidden piece (by accident, as it were)—and indeed Karin-D'Arcy and Povinelli report that in their study, subordinates often refrained from entering the cage at all. Bräuer et al. (2008) thus attempted to replicate the original finding, and at the same time tried to demonstrate changes in the competitive structure of the interaction as a function of space. They succeeded in the first aim by replicating the original Hare et al. (2000) findings, in this case with a completely new set of chimpanzees (and again with no evidence of learning during the experiment). They succeeded in the second aim by replicating Karin-D'Arcy and Povinelli's failed study, varying the spatial arrangement of the food in the middle cage, which affected the nature of the competition. When both pieces of food were closer to the subordinate's door, the subordinate just moved quickly and did not preferentially target the food that only she could see. But when both pieces of food were closer to the dominant's door, again, as in Hare et al. (2000), they targeted preferentially the food that only they could see since the competitive structure of the situation demanded this strategy. The original Hare et al. (2000) finding thus stands.

In a recent set of studies, Hare et al. (2006) used a similar competitive paradigm, but with a human competitor inside a glass booth (figure 19.4). In the first study, one piece of food was placed behind a large transparent barrier and the other behind an opaque barrier of the same size

Figure 19.4 Experimental set-up for the studies of Hare et al. (2006)

while the human competitor stared straight ahead. Chimpanzees preferentially went for the piece that was behind the opaque barrier, the one that the competitor could not see, thus providing still further confirmatory evidence for the findings of the food competition studies of Hare et al. (2000). Moreover, and importantly, chimpanzees in this competition with a human sometimes attempted to hide the beginning of their approach to the food by using circuitous routes that hid them from the human's view (behind the large opaque barrier) from as early in their approach as possible—which would seem to indicate a fairly sophisticated understanding of the viewing angle of the human experimenter.

The conservative hypothesis for these results is that chimpanzees do not compute what the human competitor can and cannot see, but instead simply avoid approaching food if they themselves can see a competitor's face. Hare et al. (2006) ruled this out in another experiment in which subjects had to approach food that lay either behind (1) an opaque barrier that obstructed the human competitor's view of the subject and the food, or (2) a split barrier—each part half the size of the barrier on the other side, so that overall area covered was the same—that oc-

cluded the subject's view of everything at eye level (including the human competitor's face) and also at the lower body level—but with a gap allowing the human competitor to see the chimpanzee subject's body approaching. This condition thus compared the hypothesis that subjects were simply going to the side where they themselves could not see the human competitor—in which case they would show no preference—with the hypothesis that they were going to the side where the competitor could not see *them*—in which case they would approach behind the opaque barrier. The finding was that chimpanzees preferred approaching the food that was behind the opaque barrier, indicating their knowledge that the competitor could see them approaching behind the split barrier.

It is possible that in this condition chimpanzees could see the competitor's torso or feet through the split in the barrier, and so simply tried to avoid being seen by their competitor at all. But Melis et al. (2006) took care of that hypothesis by having chimpanzees always approach the food from behind an opaque barrier but then be made to choose between reaching through a clear tunnel, in which the human competitor could see her arm, and an opaque tunnel. In this task, in which the chimpanzees could not see the human under any conditions, they reached more often through the opaque tunnel than through the clear tunnel—presumably imagining the perspective of the human. In a further extension, Melis et al. (2006) used this same booth arrangement, but with two clear tunnels leading to the food. Here the experimenter lowered her head between her knees so that she was unable to see the chimpanzee. The trick was that in this case, each tunnel had a small door midway through its length. One door made a loud noise when opened and the other could be opened silently (the chimpanzees had learned about the doors' properties earlier in a different context). If the human heard a door opening, she would raise her head and grab the food before the chimpanzee had a chance—but not if she could hear no door opening. The chimpanzees in this case preferentially reached through the silent door. In both the visual and auditory versions of this tunnel study, the chimpanzees performed the task well from the very earliest trials, once again ruling out the possibility that they were learning during the experiment.

These findings are especially important in our current inquiry, as they suggest that chimpanzees not only know what others can and cannot see in certain situations, but also actively manipulate that by concealing their actions from others when it is to their advantage to do so (see also Whiten and Byrne 1990 for natural observations to the same effect). The fact that they do this in both the visual and auditory domains again suggests a very flexible skill based on social cognition rather than on reading surface behavioral cues. Significantly, there is no evidence in this study or any others that chimpanzees actively mislead others by attempting actively to instill in them some false belief that, for example, the food is located over there and not over here.

Competition: Knowing What Others Know

Hare et al. (2001) used the basic chimpanzee-versus-chimpanzee competition paradigm as Hare et al. (2000), but with only one piece of food, which was always hidden from the dominant behind one of two barriers. But in order to investigate the chimpanzees' understanding not only of perception but of knowledge, they varied whether or not the dominant chimpanzee saw the food being hidden. Thus, in some trials the dominant did not see the food being hidden because her door was closed, whereas in other trials she witnessed the hiding process from her partially open doorway (subordinates always saw the entire baiting procedure, and could monitor the visual access of the dominant competitor as well). Subordinates in this case preferentially went for the food that dominants had not seen being hidden. This indicates that the subordinates know not only what others can and cannot see at the moment, but also what others have and have not just seen in the immediate past. In another condition, just before the moment of choice the dominant individual was removed, and replaced by another dominant individual who had seen nothing. In this case, subordinates now felt free to go for the food no matter what had transpired earlier, presumably based on their knowledge of what the particular individuals involved (the two dominants) had seen and not seen. This is important, because it rules out the possibility that the subordinates merely used the mere presence or absence of a dominant, any dominant, during the baiting process as a behavioral cue.

One final conservative interpretation of these food competition experiments is the so-called "evil eye" hypothesis. Perhaps subordinates believe that any piece of food observed by a dominant is "contaminated"—that it is for-

bidden once the dominant has put the "evil eye" on it— and so the only safe food is food that the dominant cannot see and has never seen. In one final study of Hare et al. (2001), both the dominant and the subordinate watched the food being hidden behind one of the two barriers. The dominant's "evil eye" was thus placed on it, and so by this interpretation the subordinate should have avoided it at all costs. But then in one experimental condition, only the subordinate watched the food being moved to a new location (the dominant's door was closed), whereas in another condition both the subordinate and the dominant watched it being moved. Subordinates went for the food when they alone had watched the moving process, but not when the dominant had also watched it. It is thus clear that they did not believe in any dominant "evil eye," since they went for the food whose movement to a new location the dominant had not witnessed (even though he had put his "evil eye" on it earlier).

Nevertheless, one version of the "evil eye" hypothesis is still viable. It could be that chimpanzees have learned a behavioral rule: if a dominant orients himself toward a piece of food in a particular location—which does not, in this view of things, lead to his knowing where it is—then that food must be avoided. Kaminski et al. (2008) attempted to rule out this final version of the "evil eye" hypothesis by developing a new methodology, which also serves to replicate the original findings in a new experimental arrangement. The experimental task was a back-and-forth game in which a chimpanzee subject and a chimpanzee competitor took turns choosing from a row of three opaque buckets, two of which contained food (figure 19.5). The game began with a hiding event, which established one piece of food as a "known" to the competitor—since both chimpanzees saw it being hidden in one of the buckets, and each saw the other watching—and another piece of food as "unknown" to the competitor, since only the subject saw it being hidden in another bucket. The third bucket stayed empty.

In the "competitor first" condition, the competitor then got to make the first choice by selecting a bucket without the subject witnessing it. The subject made her choice only after the competitor had done so. If the subject had noted which of the two pieces of food the competitor had watched being hidden (the "known" piece), then she should avoid that one, since by the time of her choice it should be gone. In the second condition, the

Figure 19.5   Experimental set-up for the studies of Kaminski et al. (2008)

"subject first" condition, the subject chose first. In this case, even if the subject noted which piece of food the partner had watched being hidden, she still might choose either piece of food. This new version of the "evil eye" hypothesis predicts that the subject should avoid the piece of food that both participants saw being hidden, irrespective of whether she chooses first or second, since it was seen in its current location by the dominant in both cases. In contrast, our own hypothesis was that when the subject chooses second she should indeed avoid the piece of food that the competitor saw being hidden (because it is likely that he has already taken it) and instead choose the piece that only she saw being hidden (which should still be available). But this should not be the case when the subject chooses first; the "evil eye" put on this food earlier will have no effect because the subject understands that when she chooses first, what the competitor saw before is irrelevant.

The main finding was that subjects in the "competitor first" condition avoided the known food and chose the unknown food, whereas in the "subject first" condition they chose randomly between the known and unknown pieces of food. The chimpanzee subjects did not change the nature of their choices from the first to the last half of the trials, thus again indicating no significant learning during the course of the experiment. These results effectively rule out the "evil eye" hypothesis in all its versions, and so represent the strongest evidence to date that, at least in some situations, a chimpanzee knows what another individual knows and has seen in the recent past, and can use that information to infer what she will do. Indeed, in this experiment they knew what she did some moments before, based on what she knew at that time, which in turn was based on what she had seen even earlier.

In a final study, Kaminski et al. also devised a false-belief version of this same basic back-and-forth test. The new twist was that in this study, a piece of food was held up above the occluder so that both could see it, and then it was lowered toward one of two buckets. In one condition it stayed in the bucket, so that now the competitor knew its location and the subject might then suppose that the competitor would choose it when he went first, so that it would be gone by her turn (in which case she should choose a safe, less desirable alternative food made available on the side). In the key false-belief condition, however, the piece of food was lowered toward one bucket but then, behind the occluder, moved surreptitiously to the other bucket. In this case the subject, if she understood the existence of false beliefs, should suppose that the competitor would choose wrongly and that the food should therefore still be available when it was her turn (and so she should take it). The results were very clear. Although human children passed this false belief test at about the same age that they normally pass the standard false belief test, the chimpanzees failed it. In combination with the negative findings of Call and Tomasello (1999), well as negative findings from a new competitive version of that task (Krachun et al., submitted), all evidence points to the conclusion that although chimpanzees understand knowledge versus ignorance, they do not understand that someone's actions can be driven by knowledge that differs from reality—that is, they do not understand the existence of false beliefs. The current negative result is particularly compelling because it was obtained in exactly the same experimental paradigm, requiring exactly the same supporting skills, as other positive results concerning knowledge versus ignorance.

Summary

As in the case of goals and intentions, reviewed above, we believe that in the current case there is only one reasonable conclusion to be drawn from the totality of these studies: chimpanzees, like humans, understand that others see, hear, and know things. Again, behavioral rules may be created to explain the results of each of the various studies, but that would require a host of post hoc explanations to be created on the basis of basically no evidence. And, again, if one were to use the behavioral-rules critique rigorously and fairly across the board, one would have to conclude that human infants and young children also have no social cognitive skills, but only surface behavioral rules. Nevertheless, we have been unable so far to obtain any positive evidence for chimpanzees understanding false beliefs, even using a conspecific competition paradigm in which they previously have been successful in understanding knowledge versus ignorance.

In several of the studies cited above, we obtained similar results with other great ape species, though again in all cases the number of subjects investigated was small, and so the results for each species individually were not statistically reliable. Perhaps more tellingly, Flombaum and Santos (2005) replicated the Hare et al. (2006) competition-with-human concealing studies, using a modified methodology, with rhesus monkeys. Santos et al. (2006) replicated the auditory version of this task (Melis et al., 2006) with rhesus monkeys as well. This suggests the possibility that many nonhuman primates understand something of the visual perception, and even auditory perception, of others.

Conclusions

Human cognition is clearly different from that of other apes; nothing we are saying here threatens this unassailable claim. At the moment, in the domain of social cognition we believe that one key difference is that humans understand that people's actions are driven by their beliefs, even false beliefs, whereas chimpanzees and other apes do not understand this. For some theorists, this means they do not have a representational theory of mind.

But chimpanzees and other apes are not only behavior readers (Call 2001). Even if they do not understand beliefs, they also do not just perceive the surface behaviors of others and learn behavioral rules for effective action. Instead, all of the evidence reviewed here suggests that chimpanzees understand others as goal-directed agents who mentally represent desired states of affairs (i.e., goals) and who also perceive the world in order to devise behavioral strategies for meeting those goals; they understand something of the interrelations among perception, goals, and action. Obviously, this understanding begins with observations of others' behavior, as it does in the case of humans, but it does not have to end there. The food competition studies (e.g., Hare et al. 2000, 2001, 2006; Melis et al. 2006) are especially important because in those studies, subordinate chimpanzees made inferences about the dominants' impending actions depending on their knowledge of the dominants' perceptions and goals. Thus, if a dominant could see the food or had just seen it, the subordinate could infer that she would go for it—because chimpanzees in normal circumstances have the goal of obtaining especially delicious food. Note that they would not make this inference if what the dominant could see or had just seen was a rock, because that would not be a goal. Chimpanzees may thus be said to have a perception-goal psychology in which they understand that another individual *acts* in a certain way because she *perceives* the world in a certain way and has certain *goals* of how she wants the world to be. Chimpanzees may not have a more human-like belief-desire psychology in which they understand that others actually have mental representations of the world which drive their actions, even when those do not correspond to reality—but why should beliefs be the only mental state worthy to be called theory of mind?

Our view is thus that, despite the proposals of Tomasello and Call (1997) and of Povinelli and Giambrone (2001), the totality of current evidence strongly suggests that chimpanzees do indeed have a coherent understanding of the psychological functioning of others. The fundamentals are the same for both chimpanzees and humans:

a perception-goal psychology enabling the understanding of important aspects of intentional, rational action and perception. Humans may go beyond this in also understanding false beliefs, but that additional social-cognitive skill is built upon the general *Pan* foundation, perhaps even a more ancient *Anthropoidea* foundation. It is likely that at least some other nonhuman primates operate with this same perception-goal psychology as well, but at the moment there are few experimental studies investigating this question with other primate species. This should be a major goal of future research.

We ourselves think that most of human cognitive uniqueness derives less from their "theory of mind," narrowly defined, than from some species-unique social-cognitive skills and motivations for sharing intentional states with others in special types of cooperative and communicative activities, since these are responsible for humans' unique forms of cultural organization and even language (Tomasello et al. 2005; Tomasello 2008). These too are built on the general *Pan* understanding of others as goal-directed agents, but the extent to which chimpanzees and other nonhuman primates can share intentions and attention in human-like fashion is not clear. Determining similarities and differences between chimpanzees and humans in terms of cooperative, communicative, and cultural activities—and their cognitive consequences—should be a major goal of future research as well.

Literature Cited

Bräuer, J., J. Call, and M. Tomasello 2008. Chimpanzees do not take into account what another can hear in a competitive situation. *Animal Cognition* 11:175–78.

Bräuer, J., J. Call, and M. Tomasello. 2005. All four great ape species follow gaze around barriers. *Journal of Comparative Psychology* 119:145–54.

Buttelmann, D., M. Carpenter, J. Call, and M. Tomasello. 2007. Enculturated chimpanzees imitate rationally. *Developmental Science* 10:F31–38.

Buttelmann, D., S. Schüette, J. Call, M. Carpenter, and M. Tomasello, submitted. Chimpanzees understand prior intentions.

Byrne, R. W., and A. Whiten. 1990. Tactical deception in primates: The 1990 data-base. *Primate Report* 27:1–101.

Call, J. 2001. Chimpanzee social cognition. *Trends in Cognitive Sciences* 5:369–405.

Call, J., and M. Carpenter, 2002. Three sources of information in social learning. In K. Dautenhahn and C. Nehaniv, eds., *Imitation in Animals and Artifacts*, 211–28. Cambridge, MA: MIT Press.

Call, J., B. Hare, M. Carpenter, and M. Tomasello. 2004. Unwilling or unable? Chimpanzees' understanding of intentional action. *Developmental Science* 7:488–98.

Call, J., B. Hare, and M. Tomasello. 1998. Chimpanzee gaze following in an object choice task. *Animal Cognition* 1:89–100.

Call, J., and M. Tomasello. (1998). Distinguishing intentional from accidental actions in orangutans (*Pongo pygmaeus*), chimpanzees (*Pan troglodytes*), and human children (*Homo sapiens*). *Journal of Comparative Psychology* 112(2): 192–206.

———. 1999. A nonverbal false belief task: The performance of children and great apes. *Child Development* 70:381–95.

Carpenter, M., N. Akhtar, and M. Tomasello. (1998). 14- through 18-month-old infants differentially imitate intentional and accidental actions. *Infant Behavior and Development* 21:315–30.

Karin-D'Arcy, M., and D. J. Povinelli. 2002. Do chimpanzees know what each other see? A closer look. *International Journal of Comparative Psychology* 15:21–54.

Flombaum, J. I., and L. R. Santos. 2005. Rhesus monkeys attribute perceptions to others. *Current Biology* 15:447–52.

Gergely, G., H. Bekkering, and I. Király. (2002) Rational imitation in preverbal infants. *Nature* 415:755.

Hare, B. 2001. Can competitive paradigms increase the validity of social cognitive experiments on primates? *Animal Cognition* 4:269–80.

Hare, B., J. Call, B. Agnetta, and M. Tomasello. 2000. Chimpanzees know what conspecifics do and do not see. *Animal Behaviour* 59:771–85.

Hare, B., J. Call, and M. Tomasello. 2001. Do chimpanzees know what conspecifics know? *Animal Behavior* 61:139–51.

———. 2006. Chimpanzees deceive a human by hiding. *Cognition* 101:495–514.

Kaminski, J., J. Call, and M. Tomasello. 2004. Body orientation and face orientation: Two factors controlling apes' begging behavior from humans. *Animal Cognition* 7:216–23.

———. 2008. Chimpanzees understand what others know but not what they believe. *Cognition* 109:224–34.

Krachun, C., M. Carpenter, J. Call, and M. Tomasello, submitted. Children read minds but apes read reaches in a competitive nonverbal false belief task.

Liebal, K., J. Call, J., and M. Tomasello. 2004. The use of gesture sequences by chimpanzees. *American Journal of Primatology* 64:377–96.

Liebal, K., S. Pika, J. Call, and M. Tomasello. 2004. To move or not to move: How apes adjust to the attentional state of others. *Interaction Studies* 5:199–219.

Melis, A., J. Call, and M. Tomasello. 2006. Chimpanzees conceal visual and auditory information from others. *Journal of Comparative Psychology* 120:154–62.

Meltzoff, A. 1995. Understanding the intentions of others: Re-enactment of intended acts by 18-month-old children. *Developmental Psychology* 31:1–16.

Muller, M., and J. C. Mitani. 2005. Conflict and cooperation in wild chimpanzees. In P. Slater, J. Rosenblatt, C. Snowdon, T. Roper, and M. Naguib, eds., *Advances in the Study of Behavior*, 275–331. New York: Elsevier.

Povinelli, D., and T. J. Eddy. 1996. What young chimpanzees know about seeing. *Monographs of the Society for Research in Child Development* 61(3).

Povinelli, D., and S. Giambrone. 2001. Reasoning about beliefs: A human specialization? *Child Development* 72:691–95.

Povinelli, D. J., K. E. Nelson, and S. T. Boysen. 1990. Inferences about guessing and knowing by chimpanzees (*Pan troglodytes*). *Journal of Comparative Psychology* 104:203–10.

Povinelli, D., H. Perilloux, J. Reaux, amd D. Bierschwale. 1998. Young chimpanzees' reactions to intentional versus accidental and inadvertent actions. *Behavioral Processes* 42:205–18.

Povinelli, D. J., A. B. Rulf, and D. Bierschwale. 1994. Absence of knowledge attribution and self-recognition in young chimpanzees (*Pan troglodytes*). *Journal of Comparative Psychology* 180:74–80.

Povinelli, D. J., and J. Vonk. 2006. We don't need a microscope to explore the chimpanzee's mind. In S. Hurley, ed., *Rational Animals*, 385–412. Oxford: Oxford University Press.

Premack, D., and G. Woodruff. 1978. Does the chimpanzee have a theory of mind? *Behavioral and Brain Sciences* 1:515–26.

Santos, L. R., A. G. Nissen, and J. Ferrugia. 2006. Rhesus monkeys (*Macaca mulatta*) know what others can and cannot hear. *Animal Behaviour* 71(5): 1175–81.

Savage-Rumbaugh, E. S., D. M. Rumbaugh, and S. Boysen. 1978. Linguistically mediated tool use and exchange by chimpanzees (*Pan troglodytes*). *Behavioral and Brain Sciences* 4:539–54.

———. 1996. Chimpanzee social cognition. Commentary for *Society for Research in Child Development Monographs* 61:161–73.

Tomasello, M. 2008. *Origins of Human Communication*. MIT Press.

Tomasello, M., and J. Call. 1997. *Primate Cognition*. Oxford University Press.

Tomasello, M., J. Call, and B. Hare. 1998. Five primate species follow the visual gaze of conspecifics. *Animal Behaviour* 55:1063–69.

Tomasello, M., J. Call, K. Nagell, R. Olguin, and M. Carpenter. 1994. The learning and use of gestural signals by young chimpanzees: A transgenerational study. *Primates* 37:137–54.

Tomasello, M., J. Call, J. Warren, T. Frost, M. Carpenter, and K. Nagell. 1997. The ontogeny of chimpanzee gestural signals: A comparison across groups and generations. *Evolution of Communication* 1:223–53.

Tomasello, M., and M. Carpenter. 2005. Intention-reading and imitative learning. In S. Hurley and N. Chater, eds., *New Perspectives on Imitation*. Oxford: Oxford University Press.

Tomasello, M., M. Carpenter, J. Call, T. Behne, and H. Moll. 2005. Understanding and sharing intentions: The origins of cultural cognition. *Behavioral and Brain Sciences* 28:675–91.

Tomasello, M., B. Hare, and B. Agnetta. 1999. Chimpanzees follow gaze direction geometrically. *Animal Behaviour* 58:769–77.

Tomasello, M., B. Hare, and T. Fogleman. 2001. The ontogeny of gaze following in chimpanzees and rhesus macaques. *Animal Behaviour* 61:335–43.

Tomasello, M., B. Hare, H. Lehmann, and J. Call. 2007. Reliance on head versus eyes in the gaze following of great apes and human infants: The cooperative eye hypothesis. *Journal of Human Evolution* 52(3): 514–20.

Tomasello, M., S. Savage-Rumbaugh, and A. Kruger. 1993. Imitative learning of actions on objects by children, chimpanzees and enculturated chimpanzees. *Child Development* 64:1688–1705.

Warneken, F., B. Hare, A. P. Melis, D. Hanus, and M. Tomasello. 2007. Spontaneous altruism by chimpanzees and young children. *PLoS Biology* 5(7): 1414–20.

Warneken, F., and M. Tomasello. 2006. Altruistic helping in human infants and young chimpanzees. *Science* 31:1301–03.

Wimmer, H., and J. Perner. 1983. Beliefs about beliefs: Representation and constraining function of wrong beliefs in young children's understanding of deception. *Cognition* 13:103–28.

20

Intentional Communication and Comprehension of the Partner's Role in Experimental Cooperative Tasks

Satoshi Hirata, Naruki Morimura, and Koki Fuwa

A group of five chimpanzees was asked to move from a small enclosure to larger one at Hayashibara Great Ape Research Institute (GARI) as one of their daily procedures. Zamba, Tsubaki, and Mizuki responded to our request and moved smoothly to the large enclosure through a gate. Loi, the alpha male of the group, appeared slightly excited for some reason; his hair bristled and he kicked the nearby wall as he went through the gate. The last chimpanzee, a small female named Misaki, remained in the small enclosure. We asked her to move by calling to her loudly, but she hesitated and simply watched Loi, who is occasionally rough with the females, getting excited outside the doorway. Loi sat in front of the gate, watching her. We continued calling Misaki to move, but she refused. Then Loi came back through the gate into the small enclosure and approached Misaki with his arm extended. Misaki stepped back a little, but Loi continued approaching her with his arm out. Finally he reached and touched her softly, embraced her, and the two chimpanzees went out to the large enclosure together.

We wonder if we are allowed to interpret the above anecdote as Loi's attempt to cooperate with Masaki in moving together. He may instead have been simply attempting to reconstruct a friendly relationship with her. Cooperation is an interesting behavior to look for in the behavior of chimpanzees, but it is difficult to determine its goal and to verify whether an episode fulfills the definition of coop-

eration during natural interactions. We have several years of experience with captive chimpanzees, but we find it hard to imagine a situation in which the chimpanzees would work together to create something—namely, a clearly distinguishable goal—in their daily lives. It is possible that they do not lack the ability to cooperate, but instead that the situations they are confronted with in captivity provide insufficient motivation for cooperation. It also seems to be difficult to find cooperation in wild chimpanzees, as two individuals rarely need to work together to obtain food. Their main diets are fruits and other vegetation, which they can obtain by themselves. But wild chimpanzees do appear to cooperate in some specific instances, such as in reports of cooperative hunting (but see chapter 18) and cooperative traveling (see chapter 27).

In this chapter, we describe the behavior of a captive female chimpanzee in two types of cooperative tasks in which she was paired with a human and with a conspecific partner. Cooperation in the common sense may refer to a behavior in which an individual actively assists or supports another, with benefits to the receiver and often with costs to the actor (van Schaik and Kappeler 2006). In the tasks reported in this chapter, the chimpanzee actor benefited by obtaining food instead of having to pay the cost of assisting the partner. From this perspective, the chimpanzee's behavior is not cooperative. However, even when a human participates in apparently costly cooperation, he or she may do so to obtain reputation or social approval in exchange for helping another person (Gächter and Herrmann 2006). Human cooperation has various types of cost/benefit distribution among its participants, along with various types of motivation. We therefore would like to take a more behavioral perspective, rather than emphasizing the outcome in terms of cost and benefit or preceding motivation. This allows cooperation to be defined as two individuals acting together to reach a common goal (Boesch and Boesch 1989). From this perspective, the behavior of chimpanzees described in this chapter can be considered cooperative.

Background

Prior to our work, several studies investigated cooperation in nonhuman primates. Köhler (1925) was a pioneer in studying the behavior and intelligence of chimpanzees in experimental situations, and he described his interest in the cooperation among chimpanzees at his laboratory. Köhler created one of his experiments as a variation of a tool-use test in which the chimpanzees had to pile one box upon another, step on it, and obtain food that was hanging from the ceiling. After the chimpanzees became familiar with piling the boxes, they were allowed to attempt the task while together in the playground. They gathered underneath the hanging food, and each tried to make a box pile to climb in his or her own way. Several of the chimpanzees wanted to climb at the same time and to build their piles unaided. If one was close to finishing a pile, another often came to pilfer the boxes, resulting in the pile being destroyed in the struggle. Thus, there was generally no systematic collaboration or strict division of labor among individuals. However, one of the chimpanzees occasionally helped another. This chimpanzee was better at piling boxes than the others, and when he watched another chimpanzee piling boxes unsuccessfully, he could not keep from lending a hand and supporting a box that threatened to fall. Köhler did not consider this to be helping in a true sense, but rather interpreted it as the skilled chimpanzee being interested in the process of piling the boxes.

Köhler described another example that more strongly resembles cooperation. In this instance, he again provided some food that was tethered from the ceiling and hanging out of reach. The chimpanzees made repeated efforts to reach the food, but without success. A heavy cage was located some distance away; one of the chimpanzees no-

ticed it, shook it back and forth, but could not move it. Another chimpanzee then also went over to hold the cage, and the two chimpanzees acted together to lift and roll it. A third chimpanzee joined them, taking hold of one side of the cage and helping to move it. The three moved the cage to a position under the food and eventually one of them climbed up on it and obtained the food, leaving the remaining two unrewarded. The chimpanzees showed no trace of altruism, but Köhler wrote that all three had the same aim (i.e., to move the heavy cage) and understood one another's intentions. A more experimental approach to cooperation was developed by Crawford (1937), in which he presented chimpanzees a task that required them to pull a pair of ropes to access a box containing food that was too heavy for one chimpanzee to pull alone. In the initial trials, the two chimpanzees did not cooperate. The human experimenter then actively taught them to pull the rope when they heard a verbal cue. Once their cooperation was established in this way, the chimpanzees worked together to pull the box, and they continued to do so even after the human experimenter stopped giving cues. One of the chimpanzees began to solicit the other by touching her, placing an arm over her body, or vocalizing when she was not motivated to pull the string (figure 20.1). When other pairs were tested, the process was somewhat similar. Some years after Köhler's observation, Menzel (1972) observed young chimpanzees cooperating to use ladders. More recently, the method used by Crawford was reintroduced to chimpanzees by Povinelli and O'Neill (2000), who studied the possible use of gestures by an experienced individual to instruct a naïve partner, but found no evidence for such gestures.

Chalmeau (1994; see also Chalmeau and Gallo 1996a, 1996b) also carried out an experimental study of cooperation in chimpanzees. A specially constructed fruit distributor was presented to a group of captive chimpanzees. Two individuals had to simultaneously pull a handle connected to the device to make the fruit fall into the enclosure. A dominant male and an infant produced most of the pull-

Figure 20.1   A chimpanzee solicits the partner to pull the rope. Picture taken from a video clip of Crawford's (1937) study. © Yerkes National Primate Research Center, Emory University.

ing responses, and the male obtained nearly all of the fruit. This male displayed an increasing number of glances toward the infant partner, suggesting that he had learned to cooperate with the infant. In contrast, the infant did not reliably check the partner, and the authors interpreted the infant's behavior as partially linked to play activity.

More recently, Tomasello and colleagues conducted a comparative study of cooperation between chimpanzees and humans in various tasks (Herrmann and Tomasello 2006; Warneken et al. 2006; Warneken and Tomasello 2006). The chimpanzees had difficulty understanding the cooperative communicative motive of a human experimenter, and did not try to maintain joint collaborative activities with the human. However, when the human reached for objects but failed to grasp them, they helped by fetching the object for the human.

Several other studies have probed cooperation in nonhuman primates; some have failed to see cooperation while others have succeeded (see Tomasello and Call 1997, Noë 2006 for review). For example, capuchin monkeys have been studied in various settings. Chalmeau et al. (1997), using the same method as in Chalmeau's (1994) chimpanzee study, found that capuchin monkeys succeeded in accomplishing the task, but had only a limited understanding of the task requirements and did not take the role of the partner into account. The two individuals pulled the handle randomly. However, Mendres and de Waal (2000) used the paradigm pioneered by Crawford (1937) and showed that the capuchin monkeys did understand the role of the partner; they pulled more frequently when the partner was present than when the partner was absent. De Waal and Davis (2003) extended this paradigm and compared cooperation in pairs of monkeys with different dominance or kinship relationships. The results suggested that expectations about the behavior of a partner played a role in the decision to cooperate. Hattori et al. (2005) introduced another type of intuitive cooperative task to capuchin monkeys, and showed that they could successfully cooperate and divide labor.

Questions Addressed

We address three primary issues in this chapter: (1) understanding of the role of others in cooperation, (2) intentional communication between potential cooperative partners, and (3) the role of eye contact in these sorts of interactions. We also use a unique comparative approach in which identical tasks are performed with the opportunity to cooperate with human or conspecific partners.

Earlier, we described studies with chimpanzees in which Crawford (1937) and Chalmeau (1994) both succeeded in creating situations in which two individuals worked together. However, Crawford's initial training phase included human cues—and in Chalmeau's study, one of the two individuals appeared not to comprehend the situation. We attempted to determine how the chimpanzees would begin to coordinate their behavior without external human cues, and to what extent they understood their partner's role in solving the task. More precisely, we were looking for behavioral evidence that the chimpanzees understood their partner's role, such as watching the partner carefully and coordinating their behavior with that of the partner. In addition, we were interested in replicating the result of Crawford's (1937) experiment and Menzel's (1972) observation of chimpanzees, showing soliciting behavior. If the chimpanzees in our studies understood the necessity of the partner and had the ability to communicate to change the partner's behavior, they would show soliciting behavior when the partner was not cooperative.

We combined this question with another line of research about intentional communication. Several studies have investigated this kind of communication in chimpanzees and other apes, with reference to their understanding of the attention of others. They have shown that chimpanzees use communicative signals when a recipient is oriented toward them (Gómez, 1996a, 1996b; Hostetter et al. 2001; Leavens et al. 2004), although there is no clear evidence of their understanding of whether the recipient can actually see (e.g., Povinelli et al. 2000). Orangutans repeated communicative signals when the humans they were signaling did not respond to a request, and in other instances they modified their gestures according to a human response (Cartmill and Byrne 2007). Studying the use of gestures among conspecifics, Tomasello et al. (1994) found that chimpanzees used more visual gestures when recipients were looking at them and more tactile signals when recipients were not looking at them. In the stone-pulling task described below, we modified the orientation of the human partner as well as his responsiveness to the chimpanzee's gesture to further examine the nature of the chimpanzee's soliciting behavior in the cooperative task.

The third focus of our study is the occurrence of eye contact, which is related to the issue of intentional communication and understanding of the attention of others. Gómez (1996a) stated that eye contact in humans is a case of ostensive behavior—a way to express and assess communicative intent—and suggested that in the great apes, it has evolved into a similarly ostensive behavior. For one thing, eye contact is within the repertoire of spontaneous friendly interaction in chimpanzees and other great apes (de Waal 1982, 1989; Goodall 1986), in contrast to the use of eye contact as threats by monkeys. Gómez (1996a) described how the chimpanzees in his experiment established eye contact when they requested food. He reported that the chimpanzees had been waiting for the human partner to direct her gaze at them. Gómez (1990) also described a gorilla making eye contact with a human when requesting the human to open the latch of a door. The great apes thus seem to be capable of some sort of ostensive function by means of eye contact. However, there is conflicting evidence indicating that chimpanzees did not differentiate between a human's eyes being open or closed when they gestured toward the human (Kaminski et al. 2004). To elucidate the possibility of chimpanzees using eye contact to express and assess communicative intent, our study examined whether they made eye contact with cooperative partners. The sample size we report here is small, but we hope that these additional examples contribute to an understanding of the cooperative nature of chimpanzees.

Approach and Results

Participants and Study Site

The subjects were two young female chimpanzees, Tsubaki and Mizuki, housed at the Great Ape Research Institute (GARI) of Hayashibara Biochemical Laboratories, Inc., established in 1999 (Idani and Hirata 2006). These subjects had been moved to GARI together with two young male chimpanzees in January 1999, when Tsubaki was approximately three years old and Mizuki was two. The four chimpanzees lived as a group in a facility consisting of a large outdoor compound of 7400 m² that contained natural forest, a pond, and a climbing structure 13 m high. The outdoor compound was attached to a smaller compound and to an indoor shelter. The study took place when Tsubaki and Mizuki were six and five years old respectively. Tsubaki had been mother-reared until a few months before she was moved to GARI, and Mizuki had been hand-reared from

a few days after her birth. Both chimpanzees had participated in several types of cognitive tasks, such as tool use and sequential learning using computer-controlled touch screens (see Morimura 2006). The human experimenters had extensive direct contact with the chimpanzees, including feeding, playing, body checks, and training for studies. The chimpanzees typically spent a few hours each day interacting with humans indoors for study or husbandry purposes, and spent their remaining hours with other chimpanzees in the outdoor enclosure or the indoor sleeping areas.

Stone-Pulling Task

TEST WITH A CONSPECIFIC PARTNER. We dug a hole in the ground inside the chimpanzee enclosure, placed a piece of food in it, and covered it with a set of stones wrapped in netting, with attached metal rings that could be used to pull the stones off the hole. The chimpanzees first learned to pull the stones from the hole to obtain the food. Then additional stones were added, gradually increasing the weight of the set until a single chimpanzee could no longer pull it off the hole. When the maximum load, approximately 120 kg, was introduced, we brought in the Tsubaki and Mizuki to see if they would move the set of stones together. A session started when they were released into the enclosure from an adjacent waiting area, and it ended when 3 min passed without either individual manipulating the stones.

When Tsubaki and Mizuki were released to the enclosure, they approached the set of stones and pulled one of the attached rings. In the first session Mizuki approached the stones first, and pulled one of the rings by herself, but the stones did not move. She then sat beside them. Soon after this, Tsubaki approached and also pulled one of the rings by herself while Mizuki watched from nearby, but the stones still did not move. In this way in the first session, Tsubaki made 14 attempts to pull the stones and Mizuki made 13. On three attempts, both chimpanzees pulled the stones at the same time: on two of those attempts they pulled in opposite directions, while on the other they pulled in the same direction but stopped before the stones moved. They never succeeded in moving the stones during this first session. In the second session Tsubaki made three attempts to pull the stones and Mizuki made one attempt; they never both pulled the stones at the same time. In the third session Tsubaki made one attempt

Figure 20.2   Tsubaki pulls the set of stones while Mizuki sits nearby.

to pull the stones and Mizuki made three attempts; they never pulled at the same time. In all three of these sessions, they never succeeded in moving the stones. Because they were beginning to lose interest, we terminated the test after the third session. The frequency of Tsuabki and Mizuki pulling the stones at the same time was significantly lower than would be expected if they had proportioned their pulling efforts randomly throughout the session. Indeed, it seemed that each individual avoided pulling the stones when her partner was pulling them (figure 20.2).

TEST WITH A HUMAN PARTNER. As the next step, we wanted to see whether cooperation in an identical setting would occur between a chimpanzee and a human. SH worked as a cooperative partner with Mizuki. Tsubaki also participated in the test, but since her motivation was inconsistent, her participation was terminated and those results are not reported hereafter.

In the initial test with Mizuki we tested whether she understood the need to adjust the timing of her pulling to match that of her human partner. The human alternately pulled and stopped pulling every 10 s for a total of 3 min during each session. While the human was doing this, Mizuki was released to the enclosure. Across the three sessions conducted, she pulled the stones four times while the human was not pulling. She also pulled four times while the human was simultaneously pulling, but only for a very short time. The stones did not move, and all three sessions ended without success.

We then began training sessions in which less weight was used. Mizuki could pull the weight alone, but the human also intervened and they pulled the stones together. The weight of the stones was gradually increased, and the human adjusted the timing and direction of pulling to

match that of the chimpanzee. When the maximum load was again introduced, the chimpanzee successfully coordinated her pulling with that of the human. After she began to move the stones with the human partner, we introduced a test situation in which the stones were pulled in a predetermined direction to see whether Mizuki understood that to move them she had to pull them in the same direction as the partner. The human started pulling in a predetermined direction before the chimpanzee was released into the enclosure. Three trials were conducted in each session. In early trials the chimpanzee did not appear to adjust her pulling direction on her first attempt. She sometimes pulled the stones in a completely opposite direction to that of the human, of course in vain (figure 20.3). In 4 of the first 10 trials, she pulled the stones in the same direction as the human on her first attempt. However, she changed the direction of her pulling when she could not move the stones, and in all of the trails she pulled the stones in the same direction as the human sooner or later, which led to successful displacement of the stones. After approximately 60 trials (20 sessions), the chimpanzee began pulling the stones in the same direction as the human partner on her first attempt in the majority of trials (figure 20.4).

Figure 20.3   Mizuki pulls the set of stones in the direction opposite to that of the human partner.

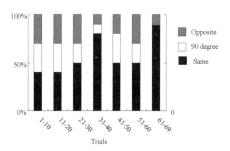

Figure 20.4   Change in the direction in which the chimpanzee initially chose to pull in each trial across blocks of 10 trials. Opposite: the chimpanzee pulled the stone in the opposite direction of the pulling direction of the partner. 90 degree: she pulled at 90 degrees to the direction of the partner. Same: her pulling direction was the same as that of her partner.

Figure 20.5   Mizuki (a) pulls the stones alone, (b) takes the human's hand, and (c) brings him to the stones.

Finally we introduced a test situation to determine whether the chimpanzee would actively solicit assistance in the task. The human experimenter stood at a distance of 1.5 m beyond the stones from the viewpoint of the door through which the chimpanzee was released into the enclosure. There were two variations on this position: the human either faced toward the door or away from it. In the first trial of this test condition, the chimpanzee first tried unsuccessfully to pull the stones alone for a total of 12 s. She then approached the human, took his hand, led him to the stones, and took one of the rings by herself to pull. The human responded and they pulled the stones together. During the six of the first eight trials of this test condition, the chimpanzee first tried to pull the stone by herself and then approached the human to lead him to the stones to pull together (figure 20.5). During the rest of the 40 trials

of these test conditions, directly after entering the enclosure without making any effort to pull alone, the chimpanzee approached the human, took his hand, and led him to the stones.

Mizuki never attempted eye contact when she solicited the human and took his hand. The condition in which the human stood with his back turned was used to see whether the chimpanzee would move around the human to make eye contact when she solicited him. No such behavior was observed. When the human's back was turned, the chimpanzee always solicited him by taking his hand from behind his back. She did look up to the area around the human's face most clearly in the first trial, when his back was turned (figure 20.6), although eye contact was not actually established because Mizuki shifted her attention before the human turned around to face her.

To further investigate the chimpanzee's soliciting behavior, we implemented another variation of the test in which the human did not respond to the chimpanzee's soliciting behavior (i.e., the taking of his hand) for 5 s, to see whether this would induce any further communicative behavior by the chimpanzee. We conducted a total of 12 such trials. In half of them the human stood facing the door through which Mizuki was released, and in the other half he stood with his back to the door. The result was that Mizuki repeated the same soliciting behavior—pulling the human's hand—but when the human did not respond, Mizuki pulled his hand again in the same manner. Overt eye contact did not occur in any of these trials.

Because the chimpanzee's soliciting behavior always consisted of taking the human's hand, we made it more difficult for her in the next trials by keeping the human's hands up out of her reach, to see whether that would induce any further communicative behavior on her part. We conducted a total of 24 trials. The human faced the door

Figure 20.6   Mizuki takes the hand of the human, who is facing away from her.

in half of the trials and away from the door in the other half. Mizuki did not use any new behavior patterns, such as going around to stand in front of the human when he had his back turned. She did, however, make eye contact in 3 of 12 trials in which the human was facing her when she took his hand; but this might be explained by her looking for his hand, which he was holding up close to his face.

String-Pulling Task

TEST WITH A CONSPECIFIC PARTNER. We conducted another type of cooperative task with the same chimpanzees using an indoor experimental room (Hirata and Fuwa 2007; Hirata 2007). This experiment began during the same time period as the aforementioned stone-pulling task. The two chimpanzees, Tsubaki and Mizuki, were required to pull both ends of a string simultaneously to drag food within reach. Two blocks, each with a piece of food on top, were placed on the floor outside an experimental room where the two chimpanzees were located. The blocks were connected by a plastic rod and a single string passed through a hole in each block. Both ends of the string extended into the experimental room through openings in the lower wall. The distance between the two ends was greater than a chimpanzee's arm span. Although a chimpanzee could reach a hand through the opening in the wall, the blocks were out of reach. By pulling on both ends of the string, however, the chimpanzee could draw the blocks within reach to obtain the food. If a chimpanzee pulled only one end of the string, she would get only the string while the block remained out of reach. Before the tests, we trained each of the two chimpanzees separately, each with a single block, until they learned to pull both ends of the string by themselves to draw it toward them. Once Tsubaki and Mizuki had each learned to pull both ends of the string by themselves, we began the test situation. First they were brought to a waiting area about 2 m away from the string while the apparatus was prepared. When the setup was complete, they were allowed to behave freely. Ten trials were conducted in each session.

In initial tests, the length of each end of the string extending into the experimental room was short (10 cm), thus requiring two chimpanzees to pull both ends simultaneously. Soon after the start of the first trial, Tsubaki approached one end of the string and pulled it without paying attention to the other end. The string slipped through the blocks and out of the apparatus. She briefly glanced at

the other end, which had gone out of the room, and released the end she was holding. Mizuki remained in the waiting area. In the second trial the result was similar. By the third trial Mizuki also began to approach and pull the string; but the two chimpanzees never cooperated, nor did they succeed in pulling the blocks within reach during any of the three sessions conducted. In all 30 trials, only one end of the string was pulled by either of the two chimpanzees.

We then made the situation a little easier by increasing the length of the two ends of string extending into the experimental room to 130 cm (this was called the long-string treatment). In this treatment, both ends of the string would remain inside the room even if one was pulled before the other. Thus, the chimpanzees did not have to pull them simultaneously. Moreover, it was possible for one chimpanzee to pull both ends of the string and succeed by herself in pulling the food within reach. The result was that Mizuki pulled both ends herself, and succeeded in the first trial. Tsubaki did not approach, staying instead at the waiting zone. In the second trial the result was the same; Mizuki succeeded alone. When the third trial began, Tsubaki moved from the waiting zone, approached one end of the string, and started pulling the string, Mizuki then arrived at the other end of the string and pulled. This led to the chimpanzees' first success, in which they drew both blocks within reach and both obtained the food.

During this condition, the chimpanzees' frequency of success gradually increased (figure 20.7). Mizuki began frequently looking at Tsubaki, waiting to see if she was holding the string, and then pulling the string in synchrony with her (figure 20.8). Tsubaki began to behave similarly,

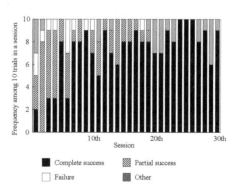

Figure 20.7 Progress across sessions consisting of 10 trials. Complete success: the two chimpanzees pulled the string together and each reached the food after joint pulling behavior. Partial success: the two chimpanzees pulled the string together but only one of them reached the food after joint pulling behavior; when the first chimpanzee reached the food and released the string, the block containing food for the second chimpanzee was still beyond her reach. Failure: the chimpanzees failed to draw the blocks. Other: other result, such as one chimpanzee pulling both ends.

Figure 20.8 Mizuki holds one end of the string and waits for her partner Tsubaki to take the other end. From Hirata and Fuwa 2007. Reprinted with permission of Japan Monkey Center and Springer Japan.

glancing at Mizuki and waiting if necessary. Thus, after some trial and error, the two chimpanzees learned to coordinate their behavior. The string was gradually shortened and they succeeded in pulling both ends simultaneously when the original condition was reintroduced. Of note was that they did not use interactive behavior or eye contact to synchronize their behavior. Both chimpanzees became experienced at this task, and each could coordinate her behavior to that of the other, but this coordination behavior consisted only of glancing and waiting.

We also observed that neither chimpanzee ever waited until her partner drew the block within reach. As illustrated in figure 20.7, there were several cases with partial success, in which both chimpanzees pulled the string together at first, but only one drew the block within reach and then released the string before her partner had finished doing the same. In such cases the remaining individual continued pulling the other end of the string, or succeeded by grabbing the end released by her partner and pulling both ends by herself. We observed no clear evidence of an individual continuing to hold her end of the string and actually waiting until her partner had succeeded in drawing the block within reach. Each individual pulled her end as quickly as possible once she noticed her partner holding the other end, and this resulted in their drawing the blocks within reach together at the same time.

TEST WITH A HUMAN PARTNER. To further investigate the potential for soliciting behavior, we paired Mizuki with a human partner (SH) in the same situation, in which there were two conditions. In the first, the human adjusted his timing to pull the string simultaneously with the chimpanzee. In the second, the human delayed his approach,

remaining still for 2 s. This resulted in failure in almost all of the first eight trials, because Mizuki then pulled the string alone. The exception was the seventh trial, in which Mizuki waited for the human; it resulted in success. On the ninth trial, when the human did not approach the string at all, Mizuki looked up at his face, whimpered, and took his hand (figure 20.9). The human was unaware that Mizuki was looking at his face because he always looked forward to avoid cueing her. When Mizuki took his hand he approached the string; this resulted in success (figure 20.10). Mizuki almost always took the human's hand in later trials.

Given that Mizuki had looked up into the human's face during the first trial, we reanalyzed the videotape of later trials and found that she had done so in 6 of 24 of them. The human looked forward to avoid cueing her, and she did not wait for him to look down at her, so eye contact was not established. Taking the human's hand became her routine in this test condition, and she did so regardless of whether he delayed his approach. However, this same behavior was never observed when she was paired with a conspecific partner.

Figure 20.9 Mizuki looks up at the face of the human partner, whimpers, and takes the partner's hand. From Hirata and Fuwa 2007. Reprinted with permission of Japan Monkey Center and Springer Japan.

Figure 20.10 Mizuki and her human partner pull the string together. From Hirata 2007. Reprinted with permission of the Japanese Society for Animal Psychology.

Discussion

Cooperation in Successful Trials

The chimpanzees did not show signs of cooperation in early trials of either the stone-pulling or string-pulling tasks. Furthermore, in the stone-pulling experiment it seemed that the two chimpanzees actually avoided working at the same time. One started to pull the stones after the other had stopped pulling, or stopped when the partner approached the stone. Previous attempts to probe chimpanzee social understanding in cooperative situations have also not been successful (Hare 2001). Hare and Tomasello (2004) showed that chimpanzees behave more skillfully in competitive tasks than in cooperative tasks, and suggested that might be cognitively hardwired to outperform rivals in competitive situations. Our results can be interpreted from a similar perspective, but to make a slightly different point. That is, the tendency to avoid working at the same time may imply that the chimpanzees are trying to avoid conflict in a possibly competitive situation. In addition to the possibility that they are skilled at outperforming rivals in competitive situations, they may also be careful to avoid conflict situations that could be damaging to existing social relationships.

Melis et al. (2006a) conducted experiments of cooperation in chimpanzees using a method fundamentally identical to the string-pulling task (Melis et al. 2006b; see also chapter 21), and they point out the importance of tolerant relationships for the successful performance of a cooperative task. They report that chimpanzees were not more cooperative when they faced a rival; their tendency to avoid working at the same time may have constrained success in the tasks, both in Melis's experiment and in our studies. A similar tendency was noted by Hare et al. (2007), who used the same method to study cooperation in bonobos for comparison with chimpanzees. In that study, the bonobos were more tolerant and also more successful at solving the cooperative task when the food could be monopolized, as they shared food while the chimpanzees did not. Petit et al. (1992) also suggested that tolerance was a critical factor in the different performances of rhesus and Tonkean macaques in a cooperative task; while the Tonkean macaques sometimes engaged in coordinated activity with others, the rhesus macaques did not show such coordination. In summary, a tolerant relationship may have been

the basis for the emergence of cooperative behavior in primates.

The fact that Tsubaki and Mizuki did not show an immediate understanding of cooperation is in line with the results of several earlier studies, including the pioneering work by Crawford (1937). It is notable that Mizuki was introduced to the string-pulling task with a conspecific partner after having learned to pull a stone with a human partner (after she had adjusted the direction of pulling, but before having undergone the tests of solicitation), but she did not work with the conspecific partner in the initial stages of the string-pulling task. This is consistent with Crawford's (1941) study in which chimpanzees who mastered one cooperative task failed to show generalization to another kind of cooperative task.

Shared Action and Shared Goal

The chimpanzees in our studies became successful at solving the tasks after some experience. In the string-pulling task, they checked the behavior of the conspecific partner, waited for the partner to hold the string, then pulled the other end themselves. In the stone-pulling task, the chimpanzee adjusted her direction of pulling to match that of the human partner. Thus, the chimpanzees understood the partner's role in these tasks. Such an understanding has been suggested by other studies of several primate species such as chimpanzees (Chalmeau et al. 1994), capuchins (Mendres and de Waal 2000; Hattori et al. 2005), and tamarins (Cronin et al. 2005). In brief, they have shown that an individual performed a necessary behavior more often when the partner was present than when he or she was absent. Therefore, an understanding of the partner's role in cooperative tasks is not a special capability of chimpanzees (see also chapter 21). One may say that adjusting one's own behavior—waiting in the partner's absence and pulling in the partner's presence—may be achieved by mechanical learning, like pulling when a green light is on and not pulling when a red light is on. However, the study by de Waal and Berger (2000), which showed that capuchin monkeys shared more food with the partner with whom they solved a cooperative task, indicates that capuchins regard a partner as more than a red or green light. Note that the string-pulling task described in this chapter is more complex than the simple pulling tasks used in other studies. In our task, the two individuals needed to pull in a very precisely coordinated way, and when their timing was not coordinated at the beginning of a trial, the trial ended in failure. Such bad timing in other simple pulling tasks does not result in failure; the individuals have a chance to pull again in the same trial, and if they keep pulling, the trial will eventually end in success. In the short-string condition of our string-pulling task, the chimpanzees must coordinate the timing of their pulling precisely; whether other species have the ability to do this is an interesting question for future research.

When the chimpanzee's partner was human, in both the string- and stone-pulling tasks, the chimpanzee solicited the partner to work with her. In other studies with monkeys, such soliciting behavior has never been observed. Together with the examples of Crawford's (1937) study and Menzel's (1972) observation, the emergence of soliciting behavior may reflect chimpanzees' deeper understanding about other individuals as agents, and their greater ability to communicate to alter other individuals' attentional states and behavior.

The chimpanzees' motivation for establishing such behaviors appeared to be the desire to obtain food for themselves. In other words, they may have been using the partner as a tool to achieve their own goal. They never appeared to care about whether the partner achieved his or her goal. This is not surprising in the case of the chimpanzee-human pair, as the human experimenter gives food to the chimpanzee but does not try to obtain food in front of the chimpanzee during their daily interactions. When the two chimpanzees succeeded in the string-pulling task, both chimpanzees attempted to obtain food, but we never observed either chimpanzee waiting until their partner achieved the goal.

Tomasello et al. (2005) argued that chimpanzees lack shared intentionality. We are not certain that the chimpanzees in our tasks had no understanding of each other's individual's intentions. However, it is clear that rather than helping their partner to achieve a goal, they engaged in shared action with the partner to achieve a self-oriented goal. Few studies have explicitly investigated whether an individual would assist a partner to achieve a goal, but Hattori et al. (2005) presented a related situation. They tested pairs of capuchin monkeys in a cooperative task in which two individuals had to perform a sequence of two actions—pulling a tab at one location in the experimental area and then pushing a block in another area—to obtain food. The pairs solved this task by dividing their roles, and they maintained this cooperation even when only one of the two obtained a reward in each trial and their roles were reversed in alternate trials. The authors concluded that the monkeys engaged in attitudinal reciprocity (Brosnan and de Waal 2002), in which a positive attitude is mirrored by the partner. It would be interesting to examine whether these monkeys understood that their behavior assisted the partner in achieving the goal. In addition, one study with chimpanzees tested helping behavior. Warnenken and Tomasello (2006) observed whether chimpanzees would help a human achieve a goal, and found positive evidence in one of the conditions tested. The difference between our result and that of Warnenken and Tomasello may be partly explained by the different goals of the two experiments; the goal in our study was to obtain food while the goal in their study was to obtain objects. As Moll and Tomasello (2007) noted, food is a resource for which apes and monkeys compete with conspecifics; thus a nonfood goal may be better in this context.

In wild chimpanzees, the case of Bossou chimpanzees crossing the road can be considered an example of a shared goal or assisting the goal of another (see chapter 27). The goal in this example is to cross the road, and the role of adult males scanning the road while group members cross it together may be to help others achieve their goal or to facilitate the achievement of a shared goal. A chimpanzee in the Taï forest who appears to be driving a target monkey in the direction of ambusher chimpanzees may also be helping others to achieve a goal (Boesch and Boesch-Achermann 2000). Another possibility that we cannot reject is that the collective efforts of individuals to achieve their own goals could appear like individuals helping each other achieve their goals. More observations in the wild and further experimental study in the laboratory are necessary to understand to what extent the chimpanzees understand the goals of other individuals and whether and how they would assist them in achieving them.

Intentional Communication in Cooperative Tasks

Mizuki's soliciting behavior can be regarded as imperative intentional communication in that she used a communicative signal to get another individual to help her attain a goal (Bates 1976). Several studies have shown that chimpanzees make visual communicative signals to a recipient who is facing them but not to a person turning away (e.g., Hostetter et al. 2001). The fact that Mizuki did not differentiate her communicative behavior according to whether her human partner was facing toward her or away from her does not conflict with such studies. Her strategy was to pull the partner's hand, and this could be categorized as tactile communication that should work whether or not the partner was looking at her.

We examined not only the occurrence of imperative intentional communication, but also whether it included eye contact. Mizuki did not make eye contact with her conspecific partner. This was, in a sense, a matter of course because she also did not use any other behavior to solicit that partner in the tasks described here. In contrast, she did use soliciting behavior when her partner was a human, but even then eye contact was rare. In the string-pulling task she looked up at the human experimenter, but the experimenter was not attending to her at the moment and Mizuki did not wait for him to look at her. In the stone-pulling task, she never went around in front of the human when he was facing away. Although the chimpanzee repeated her communicative behavior when the human did not respond, eye contact was still rare in these situations. In general, Mizuki solved the situation by establishing a routine (see Hirata and Fuwa 2007 for more description) of repeating a tactic that had worked once and then treating the human as if he were a social tool. However, it should be noted again that the chimpanzee looked up at the human partner in both of the two tasks when she solicited him (figures 20.6 and 20.9). The face, but perhaps not specifically the eyes, might have some ostensive role for chimpanzees, and this role may have emerged as a precursor to eye contact in humans.

Gómez (1996a) noted individual differences in the occurrence of intentional communication among chimpanzees; the chimpanzees that performed better in those experiments had a more extensive hand-rearing history with humans. He wrote that these individual differences could be related to Tomasello et al.'s (1993) concept of enculturation. Call and Tomasello (1996) claimed that intentional communication was one of the domains in which humans seemed to have the greatest effect on apes, and they hypothesized that the experience of being treated intentionally by others in home-raised environments may lead to a fundamental change in their social cognition. Home-raised apes may acquire a deeper understanding of others in terms

of their intentions. The results presented here came from a single subject, so making general comments on such a phenomenon is not our aim. However, the results of these studies call careful attention to this issue because the same chimpanzee, Mizuki, showed different communicative behavior with conspecific and human partners. Our tentative interpretation is that experience with others, probably through trial and error in various types of interactions including play, may lead a chimpanzee to an understanding of how specific individuals respond to his or her own behaviors. That is, in the course of daily interactions, Mizuki may have learned that it was fruitless to show soliciting behavior toward Tsubaki, who as a conspecific would mostly be a competitor in the presence of food and not help her to obtain it. On the other hand, she may also have learned that soliciting behavior worked with humans, as her human partners did not compete over food but instead gave it to her or shared it with her in their daily interactions.

We also suspect that younger chimpanzees may tend more than adults to show soliciting behavior toward a conspecific partner in a cooperative task to obtain food. Younger individuals have had less opportunity to learn that such behavior is fruitless, and our observations of the chimpanzees in our facility suggest that competition over food is less severe when they are young. The classic experiments mentioned earlier in this chapter also provide evidence for this idea. When Crawford (1937) found that chimpanzees showed soliciting behavior in a cooperative task, the chimpanzees in his study were juveniles. Another line of support comes from Menzel's (1974) study in a different setting. He created a game for chimpanzees in which a piece of food was hidden in their play yard and one of them knew its location while the others did not. When young individuals who preferred to travel together were tested, the individual who knew the food's location solicited peers by tapping on their shoulders or taking their hand to lead them to it. When older individuals were likewise tested, the situation became more competitive; another individual tried to steal the food from the one who knew its location, who in turn attempted to deceive the competitor by taking a detour to uncover the food in an unguarded moment.

Further interpretation of Mizuki's differentiated behavior will be that she may be able to judge a partner's cooperative intention in advance—that is, she may understand that the human partner is willing to cooperate while Tsubaki is not (see chapter 19, about chimpanzees' understanding of the intentions of others). Unfortunately, we do not have enough evidence to determine that this is true. We do at least consider, however, that she has an understanding of the effect her behavior has on a particular individual—thus suggesting a sophisticated understanding about others.

Implications and Future Directions

Unlike Crawford, who trained chimpanzees to respond to a human voice saying "pull," we did not use an intensive training phase to teach chimpanzees to respond to a cue given by a third party, but instead allowed them to succeed at the cooperative tasks gradually over time. They spontaneously began to glance frequently at their partners and wait until they were ready. This suggests that they have the ability to adjust their own behavior toward a partner. When wild chimpanzees hunt in groups or cross a road in a systematic progression, they may be using this ability to engage in shared activities.

Intentional communication in a cooperative situation is an important topic to pursue with reference to ostensive behavior (Gómez 1996a) and shared intentionality (Tomasello et al. 2005). The results so far indicate that rearing history affects the production of communicative behavior, and even that the same chimpanzee can behave differently depending on a partner's identity. Comparative cognitive approaches to the chimpanzee generally consider the similarities and differences between species (i.e., chimpanzees and humans), but differences within species (i.e., between chimpanzees with different histories) are another useful source for research on the evolution of intelligence or other features unique to humans.

The chimpanzees in our study did not appear to care whether their partner achieved a goal, which may have been due to our use of food as a reward. Segerdahl et al. (2005) described a process by which a male bonobo acquired language, and explained that using food as a reward sometimes inhibits rather than stimulates the spontaneous behavior of apes. It may also inhibit the cooperative nature of chimpanzees. Hirata (2008) has noted the helping behavior of chimpanzee mothers toward their immature offspring when mother and offspring travel together. As in the case of Bossou chimpanzees crossing the road, they may cooperate and assist each other more readily in achieving a social goal than in a context involving food. Researchers should consider and further investigate the context and situations under which cooperative events occur or fail to occur, as well as the characteristics of the goal for which partners might cooperate.

Acknowledgments

This study was financially supported by the Ministry of Education, Culture, Sports, Science and Technology, Japan (Grant 1870266 and 20680015 to S. Hirata) and the Core-to-Core Program HOPE by the Japan Society for the Promotion of Science. We thank S. Sekine, K. Sugama, K. Kusunoki, and C. Houki for assistance with data collection, and G. Idani for support and suggestions. Thanks are also due to N. Sato, F. Kawashima, and T. Nanba for support in conducting the experiment and caring for the chimpanzees. The chimpanzees were cared for according to "Guide for the Care and Use of Laboratory Animals" of Hayashibara Biochemical Laboratories, Inc., and the guidelines formulated by the Primate Society of Japan.

Literature Cited

Bates, E. 1976. *Language and Context*. New York: Academic Press.

Boesch, C., and H. Boesch, H. 1989. Hunting behavior of wild chimpanzees in the Taï National Park. *American Journal of Physical Anthropology* 78:547–73.

Boesch, C., and H. Boesch-Achermann, H. 2000. *The Chimpanzees of the Taï Forest: Behavioural Ecology and Evolution*. New York: Oxford University Press.

Brosnan, S. F., and F. B. M. de Waal. 2002. A proximate perspective on reciprocal altruism. *Hum. Nat.* 13:129–52.

Cartmill, E. A., and R. W. Byrne. 2007. Orangutans modify their gestural signaling according to their audience's comprehension. *Curr. Biol.* 17:1345–48.

Call, J., and M. Tomasello. 1996. The effect of humans on the cognitive development of apes. In *Reaching into Thought: The Minds of the Great Apes*, edited by A. E. Russon, K. A. Bard, and S. T. Parker, 371–403. Cambridge, UK: Cambridge University Press.

Chalmeau, R. 1994. Do chimpanzees cooperate in a learning task? *Primates* 35:385–92.

Chalmeau, R., and A. Gallo. 1996a. What chimpanzees (*Pan troglodytes*) learn in a cooperative task. *Primates* 37:39–47.

Chalmeau, R., and Gallo, A. 1996b. Cooperation in primates: Critical analysis of behavioral criteria. *Behav. Proc.* 35:101–11.

Chalmeau, R., E. Visalberghi, and A. Gallo. 1997. Capuchin monkeys, *Cebus apella*, fail to understand a cooperation task. *Anim. Behav.* 54:1215–25.

Crawford, M. P. 1937. The cooperative solving of problems by young chimpanzees. *Comp. Psychol. Monogr.* 14:1–88.

———. 1941. The cooperative solving by chimpanzees of problems requiring serial responses to color cues. *J. Soc. Psychol.* 13:259–80.

Cronin, K., A. V. Kurian, and C. T. Snowdon. 2005. Cooperative problem solving in a cooperatively breeding primate (*Saguinus oedipus*). *Anim. Behav.* 69:133–42.

De Waal, F. B. M. 1982. *Chimpanzee Politics: Power and Sex among Apes.* London: Jonathan Cape.

———. 1989. *Peacemaking among Primates.* Cambridge, MA: Harvard University Press.

De Waal, F. B. M., and M. L. Berger. 2000. Payment for labor in monkeys. *Nature* 404:563.

De Waal, F. B. M., and J. M. Davis. 2003. Capuchin cognitive ecology: Cooperation based on projected returns. *Neuropsychologia* 41:221–28.

Gächter, S. and B. Herrmann. 2006. Human cooperation from an economic perspective. In *Cooperation in Primates and Humans: Mechanisms and Evolution*, edited by C. P. van Schaik and P. M. Kappeler, 279–301. Berlin: Springer-Verlag.

Gilby, I. C., L. E. Eberly, L. Pintea, and A. E. Pusey. 2006. Ecological and social influences on the hunting behaviour of wild chimpanzees, *Pan troglodytes schweinfurthii*. *Anim. Behav.* 72:169–80.

Gómez, J. C. 1990. The emergence of intentional communication as a problem-solving strategy in the gorilla. In *"Language" and Intelligence in Monkeys and Apes: Comparative Developmental Perspectives*, edited by S. T. Parker and K. R. Gibson, 333–55. New York: Cambridge University Press.

Gómez, J. C. 1996a. Ostensive behavior in great apes: The role of eye contact. In *Reaching into Thought: The Minds of the Great Apes*, edited by A. E. Russon, K. A. Bard, and S. T. Parker, 131–51. Cambridge, UK: Cambridge University Press.

Gómez, J. C. 1996b. Non-human primate theories of non-human primate minds: some issues concerning the origins of mind-reading. In *Theories of Theories of Mind*, edited by P. Carruthers and P. K. Smith, 330–43. Cambridge, UK: Cambridge University Press.

Goodall, J. 1986. *The Chimpanzees of Gombe: Patterns of Behavior.* Cambridge, MA: Harvard University Press.

Hare, B. 2001. Can competitive paradigms increase the validity of experiments on primate social cognition? *Anim. Cogn.* 4:269–80.

Hare, B., A. P. Melis, V. Woods, S. Hastings, and R. Wrangham. 2007. Tolerance allows bonobos to outperform chimpanzees on a cooperative task. *Curr. Biol.* 17:619–23.

Hare, B., and M. Tomasello. 2004. Chimpanzees are more skillful in competitive than in cooperative tasks. *Anim. Behav.* 68:571–81.

Hattori, Y., H. Kuroshima, K. and Fujita. 2005. Cooperative problem solving by tufted capuchin monkeys (*Cebus apella*): Spontaneous division of labor, communication, and reciprocal altruism. *J. Comp. Psychol.* 119:335–42.

Herrmann, E., and M. Tomasello. 2006. Apes' and children's understanding of cooperative and competitive motives in a communicative situation. *Dev. Sci.* 9:518–29.

Hirata, S. 2007. Competitive and cooperative aspects of social intelligence in chimpanzees. *Jpn. J. Anim. Psychol.* 57:29–40.

———. 2008. Communication between mother and infant chimpanzees and its role in the evolution of social intelligence. In *Origins of the So-cial Mind: Evolutionary and Developmental Views*, edited by K. Fujita and S. Itakura, 21–38. Tokyo: Springer-Verlag.

Hirata, S., and K. Fuwa. 2007. Chimpanzees (*Pan troglodytes*) learn to act with other individuals in a cooperative task. *Primates* 48:13–21.

Hockings, K. J., J. R. Anderson, and T. Matsuzawa. 2006. Road-crossing in chimpanzees: A risky business. *Curr. Biol.* 16:668–70.

Hostetter, A. B., M. Cantero, and W. D. Hopkins. 2001. Differential use of vocal and gestural communication by chimpanzees (*Pan troglodytes*) in response to the attentional status of a human (*Homo sapiens*). *J. Comp. Psychol.* 115:337–43.

Idani, G., and Hirata, S. 2006. Studies at the Great Ape Research Institute. In *Primate Perspectives on Behavior and Cognition*, edited by D. A. Washburn, 29–36. Washington, DC: American Psychological Association.

Kaminski, J., J. Call, and M. Tomasello. 2004. Body orientation and face orientation: Two factors controlling apes' begging behavior from humans. *Anim. Cogn.* 7:216–23.

Köhler, W. 1925. *The Mentalities of Apes.* New York: Harcourt Brace.

Leavens, D. A., A. B. Hostetter, M. J. Wesley, and W. D. Hopkins. 2004. Tactical use of unimodal and bimodal communication by chimpanzees (*Pan troglodytes*). *Anim. Behav.* 67:467–76.

Melis, A. P., B. Hare, and M. Tomasello. 2006a. Engineering cooperation in chimpanzees: Tolerance constraints on cooperation. *Anim. Behav.* 72:275–86.

Melis, A. P., B. Hare, and M. Tomasello. 2006b. Chimpanzees recruit the best collaborators. *Science* 311:1297–1300.

Mendres, K. A., and F. B. M. de Waal. 2000. Capuchins do cooperate: The advantage of an intuitive task. *Anim. Behav.* 60:523–29.

Menzel, E. W. 1972. Spontaneous invention of ladders in a group of young chimpanzees. *Folia Primatol.* 17:87–106.

Moll, H., and M. Tomasello. 2007. Cooperation and human cognition: The Vygotskian intelligence hypothesis. *Phil. Trans. R. Soc. B.* 362:639–48.

Noë, R. 2006. Cooperation experiments: Coordination through communication versus acting apart together. *Anim. Behav.* 71:1–18.

Petit, O., C. Desportes, and B. Thierry. 1992. Differential probability of "coproduction" in two species of macaque (*Macaca tonkeana, M. mulatta*). *Ethol.* 90:107–20.

Povinelli, D. J., J. M. Bering, and S. Giambrone. 2000. Toward a science of other minds: Escaping the argument by analogy. *Cogn. Sci.* 24:509–41.

Povinelli, D. J., and D. O'Neill. 2000. Do chimpanzees use their gestures to instruct each other? In *Understanding Other Minds, 2nd Edition: Perspectives from Developmental Cognitive Neuroscience*, edited by S. Baron-Cohen, H. Tager-Flusberg, and D. Cohen, 459–87. Oxford: Oxford University Press.

Segerdahl, P., W. Fields, and S. Savage-Rumbaugh. 2005. *Kanzi's Primal Language: The Cultural Initiation of Primates into Language.* Hampshire, UK: Palgrave Macmillan.

Tomasello, M., and J. Call. 1997. *Primate cognition.* New York: Oxford University Press.

Tomasello, M., J. Call, K. Nagell, R. Olguin, and M. Carpenter. 1994. The learning and use of gestural signals by young chimpanzees: A transgenerational study. *Primates* 35:137–54.

Tomasello, M., M. Carpenter, J. Call, T. Behne, and H. Moll. 2005. Understanding and sharing intentions: The origins of cultural cognition. *Behav. Brain Sci.* 28:675–90.

Tomasello, M., A. C. Kruger, and H. H. Ratner. 1993. Cultural learning. *Behav. Brain Sci.* 16: 495–552.

Van Schaik, C. P., and P. M. Kappeler. 2006. Cooperation in primates and humans: Closing the gap. In *Cooperation in Primates and Humans:*

Mechanisms and Evolution, edited by C. P. van Schaik and P. M. Kappeler, 3–21. Berlin: Springer-Verlag.

Warneken, F., F. Chen, and M. Tomasello. 2006. Cooperative activities in young children and chimpanzees. *Child Dev.* 77:640–63.

Warneken, F., and M. Tomasello. 2006. Altruistic helping in human infants and young chimpanzees. *Science* 311:1301–3.

Watts, D. P., and J. C. Mitani. 2002. Hunting and meat sharing by chimpanzees at Ngogo, Kibale National Park, Uganda. In *Behavioural Diversity in Chimpanzees and Bonobos*, edited by C. Boesch, G. Hohmann, and L. F. Marchant, 244–58. Cambridge, UK: Cambridge University Press.

21

Collaboration and Helping in Chimpanzees

Alicia P. Melis, Felix Warneken, and Brian Hare

In September 1979, all but one of the Kasakela adult males were traveling in Lower Mkenke Valley when they came upon a female baboon with a tiny black infant. She was feeding in an oil-nut palm and appeared to be quite on her own. Goblin grinned, squeaked softly, and reached to touch Satan. All six males had their hair erect. When the baboon noticed the chimpanzees, she stopped feeding and gazed towards them. . . . Jomeo, moving very slowly, left the other males and climbed a tree close to the palm. At this point the female baboon began to scream but she did not run off. When he had climbed to a branch level with her and about 5 meters away, Jomeo stopped. He stared at her, then began to shake a branch— possibly to try to make her run. She screamed even louder, but apparently no other baboons were within earshot. Two minutes later Figan and Sherry, also moving slowly, climbed two other trees. A male chimpanzee was now in each of the trees to which the baboon could have leaped from her palm; the other three, still looking up, waited on the ground. At this point Jomeo leaped over onto the palm. The baboon made a large jump into Figan's tree, where she was easily grabbed and her infant seized. The mother ran off 6 meters or so, where she remained, screaming, then uttering waa-hoo calls for the next fifteen minutes while the chimpanzees consumed her infant. (Goodall 1986, p. 286)

In many ways the life of a chimpanzee is like that of most primates—a delicate balancing act between out-competing one's competitors within the group and needing to cooperate with some of those same individuals in defense of oneself and one's kin and allies against threats from other group members, neighboring groups, and predators (Muller and Mitani 2005). What becomes apparent from field research, however, is that male cooperation in chimpanzees is particularly unusual among primates. The males form coalitions and longer-term alliances to overpower group members when competing for food and mating opportunities (e.g., de Waal 1982, Goodall 1986). In some cases they can be completely dependent on one another, since in large groups no single male is able to dominate the group in all contexts. For example, pairs or even trios of males in the Ngogo community have been seen to mate-guard ovulating females, trying to prevent them from copulating with all other adult males. These coalition males have been found to gain higher shares of copulations than could have expected from solo mate-guarding (Watts 1998). Male chimpanzees also reciprocally exchange grooming, meat, and support within and across currencies, and these relationships are still highly significant after controlling for proximity, similarity in rank, age, and kinship within the group (Watts 2002; Mitani 2006).

Chimpanzee males stand out most prominently from other primates, though, in terms of the breadth of forms in which they express these cooperative partnerships. Not only do they protect a territory as a group, but they also frequently hunt in groups. They support each other during agonistic interactions with other groups and patrol the borders of their territory together. Patrolling chimpanzees, upon detecting members of another group, reportedly react most strongly when discovering isolated individuals. In such cases they will occasionally enter the neighboring territory and coordinate attacks on these strangers that can result in the death of infants or even adults (Boesch and Boesch-Achermann 2000; Watts and Mitani 2001, Wilson and Wrangham 2003). Second, across all sites where groups of multiple males are found living sympatrically with monkeys, chimpanzees are extremely successful arboreal predators; they frequently hunt in groups—primarily red colobus monkeys—and share the meat afterwards (Boesch and Boesch-Achermann 2000; Watts and Mitani, 2002; see also chapter 18). Finally, like other primates they also spend a considerable amount of time grooming kin and other preferred partners (Goodall 1986; Mitani 2006).

While it is male cooperation that makes chimpanzees seem to stand out among primates, it is not the case that females never work together with others. Coalitionary behavior of female chimpanzees does occur—particularly in sites where the costs of sociality are reduced and females can spend time together in larger groups (i.e., Tai: Boesch and Boesch-Achermann 2000). Moreover, in captive situations females have been observed to form coalitions with one another, and in special circumstances they can even dominate males together (de Waal 1982; Wrangham and Peterson 1996).

What is also becoming increasingly apparent with the growing data from field scientists is that chimpanzees likely adjust the form of their cooperative behavior depending on socioecological factors (Boesch et al. 2002). This suggests the possibility that they develop cooperative strategies relatively flexibly. For example, the ways in which chimpanzees form coalitions and alliances can vary depending on the stability of the male hierarchy (see Muller and Mitani 2005 for review). Likewise, it has been suggested that their hunting strategies may vary depending on the canopy structure of the forest, since that affects the level of difficulty of catching prey, and thus it may create different pressures for hunters to develop cooperative strategies (Boesch 1994a, 1994b; see also chapter 18). At Taï, three-quarters of the hunts have been described

as collaborative, in which different hunters perform different complementary actions towards the same prey (Boesch and Boesch 1989, Boesch and Boesch-Achermann 2000). Boesch describes how during collaborative hunts individuals coordinate their actions, performing different roles (chaser, blocker) that aim to drive the prey towards locations where they can be caught (by the ambusher). In Gombe, however, collaboration seems to be rare, and group hunts appear to be mostly simultaneous solitary hunts during which hunters do not coordinate their actions and even pursue different animals from the same group of prey (Goodall 1986; Boesch 1994a; see also chapter 18). Boesch (1994b) argued that the canopy structure at Taï makes the cost of pursuing red colobus higher than at Gombe where the trees are shorter and the canopy is broken. Thus, hunting at Taï requires coordinated action between individuals, whereas at Gombe the hunting success of single hunters is already high enough. Corroborating this idea, Gilby et al. (2006) found evidence that even within a particular site (Gombe), the probability and success of hunting are higher in woodland forest (where the canopy is low and broken) than in evergreen forest.

From a cost-benefit analysis perspective, some of these cooperative interactions entail immediate benefits for all participants, whereas in many other cases it is less clear how (directly or indirectly) or when the actor of the behavior will benefit. Perhaps most surprising is that contrary to previous models of male behavior based on kinship, at Ngogo (with the largest male community to be studied to date) coalition and alliance partners have exceptionally low degrees of relatedness. Therefore, instead of benefiting from such cooperative relationships through indirect reproductive benefits, it seems that male chimpanzees must benefit directly through mutualistic or reciprocal rewards—both of which require a careful choice of cooperative partners to maximize benefit and minimize cheating (Langergraber et al. 2007).

The frequency, variety, and adaptability of chimpanzee cooperation suggest that the underlying motivational and cognitive processes could be similar to those of humans. Yet it recently has been proposed that the range and types of human cooperative behavior are unique in the animal kingdom, and that this diversity relies on uniquely derived human cognitive and emotional adaptations (Fehr and Fischbacher 2003; Tomasello et al. 2005; Stevens and Hauser 2004). Therefore, in order to understand the evolutionary roots of human cooperation, it is crucial to investigate the similarities and differences in the proximate mechanisms underlying the cooperative behavior of our closest living relatives, chimpanzees and bonobos. Only then can we identify the traits that are derived and unique in humans, and which might account for our unusual levels of cooperation.

Proximate versus Ultimate Accounts of Cooperation

Experimental studies with captive chimpanzees can make an important contribution to our understanding of the proximate mechanisms underlying chimpanzee cooperation. We use the term "cooperation" in its broadest sense to mean behavior through which both actor and recipient or only the recipient can benefit (van Schaik and Kappeler 2006). In mutualistic cooperative interactions, in which both actor and recipient benefit, the interest from a cognitive point of view is the extent to which the participants might understand the role of the others in the interaction. Do they understand how the different roles interrelate with each other? Do they actively coordinate their actions, and what types of social and communicative strategies can they use to help them do so? Observations like the baboon hunting episode described above illustrate how different levels of understanding about any joint activity can be put forward to explain the behavior. For example, when Goblin squeaked and touched Satan, it is possible that he understood how success depended on Satan's help and was therefore trying to recruit him. But it is equally possible that he was aroused and excited at the sight of a potential prey and that touching Satan simply calmed him down. When Jomeo, Figan, and Sherry climbed three different trees blocking the baboon's potential escape routes, it is possible that they understood how these different escape routes had to be blocked so that they could succeed. Possibly the chimpanzees were taking into consideration the positions and most likely behavior of their partners. However, it is also possible that each of them simply took a position in a free tree anticipating a possible escape route of the baboon, without any understanding of how the others' actions were related to their own chances of success. It is only with careful and controlled experiments in captivity that we can distinguish between these alternative explanations. We use the term *collaboration* to indicate

joint activities that involve active coordination between partners who understand something about the role of the others in the interaction (this would comprise coordination and collaboration as defined by Boesch and Boesch 1989). This is in contrast to *by-product cooperation* in the case of simultaneous actions that are directed towards the same goal, but without coordination or participants understanding how the others' actions relate to one's own actions and potential success (similarity and synchrony sensu Boesch and Boesch 1989).

It is important to emphasize that the motivation behind these mutualistic interactions (including collaboration) may be purely selfish. Collaborating partners understand that by coordinating their actions with each other, they increase their own probability of success and their likelihood of increasing their own individual payoffs. This leads to the prediction that collaborating chimpanzees will differentiate between different partners and, if given the choice, preferentially choose to collaborate with those who are more skilled, or with whom they have a more tolerant relationship—and thus the highest chances of obtaining the spoils afterwards.

But is chimpanzee cooperation strictly limited to mutualistic endeavors? Or are chimpanzees also motivated and skilled to help others when it does not entail immediate benefits? In other words, can their behavior be more altruistic in nature? Anecdotes like the following would seem to suggest so.

Washoe spent some time on an island ringed with an electric fence. One day a three-year-old female, Cindy, somehow jumped this fence. She fell into the moat, splashed wildly, and sank. As she reappeared, Washoe leaped over the fence, landed on the narrow strip of ground at the water's edge, and, clinging tightly to a clump of grass, stepped into the water and managed to seize one of Cindy's arms as the infant surfaced again. Washoe was about nine years old at the time; she was not related to Cindy and had not known her for very long. (R. Fouts and D. Fouts, personal communication in Goodall 1986)

Many of the cooperative exchanges reviewed in the previous section (food sharing, support in agonistic encounters, grooming) potentially represent examples of altruistic behavior since they do not seem to bring immediate benefits for the actors (although for alternative mutualistic explanations see Stevens and Gilby 2004 and Gilby 2006, on meat sharing; Mitani 2006 and Bercovitch 1988 on co-alitionary behavior; and Shutt et al. 2007 on grooming). This creates an evolutionary puzzle for anyone attempting to understand these social behaviors at the ultimate level of explanation, since the behaviors potentially increase the recipients' fitness at a cost to the performers (Hamilton 1964). How then, could such behaviors evolve? Ultimate accounts of altruism focus on the mechanisms that have selected for such social behavior because of its inclusive fitness effects on the actor. For example, kin selection explains altruistic behavior that increases the actor's indirect fitness (e.g., Hamilton 1964), and reciprocal altruism (Trivers 1971) explains altruistic behavior directed towards unrelated members that increases the actor's direct fitness in the long term (see Bshary and Bergmueller 2007 and de Waal 2008 for reviews on the topic). However, the major goal of our experiments is to test hypotheses about the proximate processes that make such behavior possible. Specifically, our goal is to empirically investigate the motivational and cognitive mechanisms of behaviors that do not result in short-term benefits (and possibly entail some costs) for the actor. Our research is guided by the following questions: Do chimpanzees know when someone else needs help? In what situations are they willing to help? If so, then how are these prosocial tendencies maintained, and what psychological mechanisms allow individuals to detect and avoid cheaters? For example, it has been argued that although the idea of contingency-based reciprocal altruism is theoretically very compelling, there is scarce evidence for it in nonhuman animals (Hammerstein 2003; Stevens and Hauser 2004). For reciprocal altruism, individuals need to keep track of past interactions with different partners, potentially over long periods of time. This requires memory and learning capacities as well as low temporal discounting rates for delayed benefits. Since cognitive constraints could make the establishment and maintenance of reciprocal altruistic strategies difficult, this could account of the relative rarity of evidence for reciprocal altruism in nonhuman animals.

In the current chapter, we will first review the most instructive examples of experimental work on chimpanzee cooperation which has focused on an individual's ability to solve mutualistic instrumental tasks requiring coordinated actions. We then focus on our own recent work, which also concentrates on the skills of captive chimpanzees in collaborative mutualistic tasks in addition to altruistic

helping. Finally, we reflect on how these findings might inform observations in the wild and contribute to a better understanding of the differences between cooperation in humans and cooperation in nonhuman apes.

Cooperative Problem Solving in the Lab

More than seventy years ago, Crawford (1937) presented pairs of chimpanzees with three different instrumental problems requiring simultaneous joint action. The tasks were all very similar, requiring the chimpanzees to pull two ropes or levers at the same time in order to gain access to food. Although some of the chimpanzees eventually learned to solve each problem, no chimpanzee showed spontaneous skill at solving them. The first task required both chimpanzees to pull a rope in order to obtain a heavy baited box that neither individual could pull alone. The subjects did not succeed until the experimenters intervened and trained them to pull on command. This result was replicated with each new cooperative problem, thus suggesting that the subjects did not understand their partners' role in their success. Such an understanding would have facilitated problem solving, enabling the chimpanzees to coordinate their efforts in very similar mutualistic tasks. The author summarizes his subjects' behavior as follows.

When Ross and Bulla were introduced into the situation immediately after individual training with a single rope, their efforts toward moving the box were entirely uncoordinated. There was no suggestion in the behavior of either which would indicate any attempt of one to coordinate its activity with that of the other. While their efforts were directed toward the food-baited box, and while they both made strenuous efforts to draw it in alone, they never, even for an instant, pulled simultaneously. Typically, one animal would pull while the other watched and played about the cage. When the first finished pulling, the other would take the rope, of the same one, and pull while the first animal did something else. Frequently they took a succession of turns at the ropes in this manner (Crawford 1937, p. 21).

Eventually Crawford did see behaviors in one individual that he interpreted as communication used to recruit a cooperative partner, but this interpretation was moderated by the fact that the behavior was rare and occurred only after hundreds of trials. Crawford did see the potential limitations in his experimental approach, however, and he suggested the need for further research with a larger sample of individuals and with tasks perhaps more appropriate for use with chimpanzees. In another attempt several years later with more individuals, Crawford did find that one pair of chimpanzees could learn to coordinate their behavior very quickly (Crawford 1941).

Using a method highly similar to that used by Crawford (1937), Povinelli & O'Neill (2000) trained a pair of chimpanzees to simultaneously pull two adjacent ropes in order to retrieve food from a heavy baited box. Each of the successful cooperators was then paired with a series of naïve subjects. Of the ten resulting pair combinations of experienced and naïve chimpanzees, only three were ever successful in obtaining the food by pulling the ropes simultaneously. Although they were equally constrained by sample size (n = 7) and had essentially used the same method that had not worked sixty years before, Povinelli and O'Neill (2000) quickly concluded that the chimpanzees' failure to cooperate was largely due to cognitive constraints that both prevented a naïve individual from learning socially from an experienced partner, and also prevented an experienced partner from teaching a naïve individual (e.g., with encouraging gestures).

Chalmeau and colleagues (Chalmeau 1994; Chalmeau and Gallo 1996) also presented a task that required joint actions (i.e., pull two handles simultaneously to obtain a small amount of food) to another small group of chimpanzees (n = 6). Like previous investigations, their found very low levels of cooperation, with only two individuals consistently solving the task. The successful cases involved one dominant male coordinating his pulling so that it occurred when a younger female was near the apparatus. On several occasions this dominant individual even coerced the young female into pulling the handle by herding her into the correct location, and in all cases the male was the one to obtain the small fruit rewards. Here the authors were able to draw an important conclusion that would be the basis for future success: they noted that the chimpanzees' cooperative behavior was limited by social constraints on the subordinates. In almost every trial the most dominant individual monopolized the apparatus, thereby preventing others from cooperating, while also eating any food retrieved through joint effort.

Recently Hirata and colleagues (see chapter 20) tested two chimpanzees in a cooperation task similar to the one previously used by Petit et al. (1992) with two species of macaques. Food was placed in a hole in the ground and covered by a set of heavy stones. Pairs of chimpanzees were required to pull simultaneously on the rings attached to the stones in order to uncover the hole. Although both chimpanzees pulled the stones, they almost never did so at the same time, and therefore never succeeded in moving them. In observations similar to those made by Crawford (1937), Hirata et al. report that each chimpanzee rather seemed to avoid pulling at the stones when the partner was pulling them.

These experiments on cooperation, together with other repeated findings that chimpanzees fail tasks in cooperative-communicative contexts (see Tomasello et al. 2003; Hare and Tomasello 2004; Hare 2007) have led researchers to propose that the sophisticated cognitive abilities of nonhuman primates might be best demonstrated in competitive rather than cooperative contexts (e.g., Hare 2001; Tomasello et al. 2005). It is possible, however, that just as competitive paradigms engage chimpanzees particularly well in certain cognitive tasks, there are cooperative paradigms that can engage them as well. In fact, it is possible that in previous problem-solving experiments requiring cooperation chimpanzees interpreted the situation not as potentially cooperative, but as competitive. In order to tease apart the psychology that underlies what seems to be flexible chimpanzee cooperation in the field, we sought to elicit spontaneous cooperation in captivity and investigate key factors that may cause this phenomenon to be relatively scarce in captive settings.

Tolerance as a Prerequisite

One possible explanation for the poor performance of chimpanzees in previous experimental studies on cooperation is that their low levels of social tolerance constrained their ability to cooperate in food-retrieval tasks (e.g., Chalmeau 1994). Experiments with other primate species have demonstrated that the ability to cooperate to produce mutually beneficial outcomes is likely not just a product of complex cognitive abilities, but also a result of tolerance between potential cooperative partners. For example, species with the highest level of tolerance (low dominance asymmetries) are generally more reliable cooperative problem solvers (Tonkean macaques: Petit et al. 1992; capuchins: Mendres and de Waal 2000; marmosets: Werdenich and Huber 2002; cotton-top tamarins, Cronin et al. 2005). Moreover, capuchin monkeys have been found to be most cooperative in a joint food-retrieval task when their partner is closely related (i.e., one with whom they have more a tolerant relationship) and the food reward is sharable (de Waal and Davis 2003).

To systematically investigate whether or not social tolerance plays a significant role in chimpanzee cooperation, we conducted a set of studies in which we measured both the tolerance level and the problem-solving performance of chimpanzee dyads (Melis et al. 2006a). Subjects were 32 unrelated chimpanzees from Ngamba Island Chimpanzee Sanctuary in Uganda. We found that levels of social tolerance (measured as the tendency of a pair of individuals to share food) predicted the dyad's likelihood of solving a cooperative problem. In the tolerance test that was conducted prior to the cooperation test, we allowed the two subjects to enter the testing room and feed from a baited tray that was placed outside the room but within their reach. The baited tray was placed against the metal bars of the subjects' test room (figure 21.1). We measured the level of sharing within and across several trials in which we also varied the amount and distribution of the food. Each dyad obtained a sharing score and participated after this phase in the cooperation test. As in previous studies, the cooperation problem was a food-retrieval task in which two individuals were required to pull simultaneously on two ends of a rope in order to obtain a tray of food that had been placed outside the testing room and out of their reach. The pulling mechanism (originally conceived by Hirata and Fuwa 2007; see chapter 20) consisted of a single rope threaded through loops fixed on top and across the length of the tray (figure 21.1a). Both ends of the rope were extended into the testing room through the metal bars. The rope's two ends were too far from each other for one chimpanzee to pull both ends simultaneously. If one chimpanzee pulled only one end of the rope, it would simply slide out of the loops on the tray. Thus, to pull the tray and food within reach, subjects had to cooperate to pull both ends of the rope simultaneously. This cooperation test consisted of six trials in which, to facilitate the emergence of cooperation and reduce possible situational tolerance constraints, we made the food rewards extremely sharable: an abundance of sliced banana pieces on two separate dishes.

All 16 dyads (32 individuals) showed a positive correlation between the tendency to share food in the tolerance

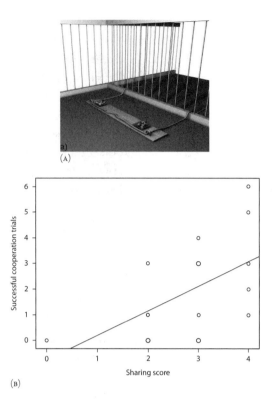

(A)

(B)

Figure 21.1 (a) The cooperation apparatus from the Melis et al. 2006a and 2006b. experiments. Two food dishes 2.7 m apart were baited with fruit pieces. The two ends of the rope were 3.4 m apart. Metal bars separated the testing room, where the subjects were located, from the room where the tray was located. (b) Correlation between the score obtained in the food-sharing tests and the number of successful trials in the cooperation tests ($r_s = 0.56$, N = 16, P = 0.02). Larger circles represent two data points. From Melis et al. 2006a.

test and the level of success in the cooperation test (figure 21.1b). That is, levels of food sharing between members of a dyad predicted spontaneous success or failure in the cooperation task. We even found that social tolerance was a better predictor of success than was the subjects' understanding of the task's physical properties. However, an alternative explanation for these findings could be that it was not necessarily the dyad's level of inter-individual tolerance but rather some characteristic of the individuals themselves that determined their success at cooperating. Therefore, to provide further evidence that the level of tolerance was indeed the critical factor enabling cooperation, we conducted another experiment in which we attempted to turn cooperation on and off by systematically varying the level of tolerance between members of a pair. We tested whether a previously unsuccessful subject would succeed in the cooperation test when paired with a more tolerant partner and, conversely, whether a previously successful subject would fail when paired with a less tolerant partner. Confirming the earlier results, we found that previously unsuccessful subjects were immediately successful when paired with partners with whom they had a more tolerant relationship—but even when subjects knew how to solve the pulling task, they stopped working together when paired with less tolerant partners. This appears to be clear evidence that the level of tolerance was in fact the causal factor in these situations (Melis et al. 2006a, experiment 3).

Low levels of inter-individual tolerance constrained the chimpanzees' tendency to cooperate or act simultaneously on the apparatus even when the tray had been baited with dispersed and shareable food and the working space between the partners had been maximized. Interestingly, unsuccessful dyads typically did not show any aggressive behavior in the cooperation test. Instead, individuals seemed to avoid approaching or manipulating the apparatus in the presence of certain partners. This suggests that the individuals were inhibited by some partners since they probably knew from their experience outside the experimental setting that approaching and/or manipulating the baited tray in the presence of that more dominant partner could lead to a negative response (aggression, or lack of sharing afterwards). Not only do these results demonstrate that chimpanzees can spontaneously cooperate when their tolerance level is high, but they also help to interpret the failure of chimpanzees in previous studies (Crawford 1937; Povinelli and O'Neill 2000; Chalmeau 1994; see also chapter 20) in which the social tolerance between individuals was not controlled and dyads were presented with highly monopolizable food rewards (often in a small reduced working space). These results suggest that high levels of social tolerance within a social group may be an important prerequisite for certain forms of cooperation to appear. Only if individuals are tolerant enough to approach and manipulate objects or food items that are in possession of others are they likely to find a cooperative solution to a problem and share the rewards of the joint effort afterwards.

Collaborative Chimpanzees

Although lifting tolerance constraints in chimpanzees allowed us to elicit spontaneous cooperation between individuals, it was still not possible to distinguish between *collaboration*, in terms of individuals actively coordinating their actions to those of their partners, and *by-product cooperation*, in which two individuals act simultaneously without each having any understanding of, or behavioral coordination with, their partner's actions. Since in the previous experiment both subjects entered the room simultaneously to participate and both were attracted to the baited tray, it was possible that their success was due to their simultaneous but independent actions taken toward the same tray (by-product cooperation)—and thus we were left with the question of what each chimpanzee understood about their partner's contribution to the task, and to what extent they coordinated their actions with those of their partner.

To investigate these questions, we presented a subset of chimpanzees from the previous study with several variations of the same cooperation task described above. After an individual introduction to the pulling task, in which each subject could learn about the physical properties of the pulling mechanism (see Melis et al. 2006b; online material), subjects participated in the partner delay test, which consisted of allowing one individual in each pair to enter the testing room first while the other waited in an adjacent room. The baited tray could not be pulled within reach by a single subject since, as described above, the length of the rope ends inside the testing room were too short and too far apart for one chimpanzee to reach both simultaneously. Therefore, the subject who entered first had to wait for the partner (thus inhibiting pulling the rope out of the loops), who was allowed to enter after increasingly long delays of 5, 10, 20, and 30 seconds. Once subjects were able to wait for the partner without pulling the rope out, and were successful in obtaining the food on two consecutive trials at a given delay, we increased the delay. This test ended when all subjects were able to wait 30 seconds for their partner. In order to move to a longer delay, subjects first had to wait for the partner and successfully obtain the food on two consecutive trials at a given delay. Two test subjects made no mistakes, two subjects made three to four mistakes respectively, and the other four subjects made between 12 and 28 mistakes before completing two consecutive successful trials at all of the delays. These results suggest that half of the subjects understood from the very beginning the need to wait for their partner, whereas the other half needed additional experience to learn to inhibit the impulse to manipulate the rope before the partner's arrival.

Our results contrast with the slower learning process described by Hirata and colleagues (Hirata and Fuwa 2007; chapter 20) using the same cooperative task. In that study, subjects began to show waiting behavior only after 90 to 170 trials, even though they had already participated in a long learning phase of 600 trials in which the rope ends had been increasingly shortened. It is possible that this difference between studies is due to individual differences (as evidenced by the fact that some Ngamba subjects made no mistakes whereas others made up to 28 mistakes in the partner delay test). Alternatively, it is possible that this difference is due to the different methods we used to familiarize the subjects with the task. Hirata and Fuwa (2007) report several treatments in which subjects actually did not need to cooperate since both ends of the rope could be reached simultaneously by a single individual (and, in fact, that is what subjects often did). Thus, when confronted with the crucial tests, subjects might not have fully understood the impossibility of solving the task without a partner, and instead might have been very motivated to continue pulling by themselves (thus trying to find an individual solution to the task). Furthermore, it is possible that subjects were not highly tolerant of each other and therefore tended to avoid pulling simultaneously from the rope. In our introduction to the pulling task, on the other hand, subjects experienced that they could not solve the problem alone, since they could not reach both ends of the rope—and if they pulled one end of the rope, they experienced how the rope came out of the loops. After that, they were paired with highly tolerant partners, with whom they experienced successful cooperation. This different method may have accelerated the learning process in our test.

Although our results from the partner-delay test were evidence of the chimpanzees' ability to synchronize their actions with that of their partners in order to obtain a common goal, we reasoned that a further test of their level of understanding about the partner's role would be if they were allowed to recruit a collaborative partner when they either did or did not need help in solving the problem. Specifically, they could choose to let another chimpanzee enter the testing room to collaborate on the task with them. For this purpose, we showed subjects how to remove a "key" (wooden peg) that locked a sliding door between the testing room and an adjacent room (see online material in Melis et al. 2006b). All subjects learned very quickly that upon removing the key, they could open the door and

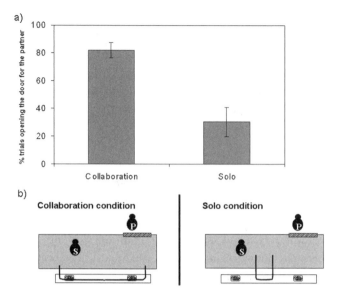

Figure 21.2 (a) Mean percentage of trials (± SEM) in which subjects opened a door for the potential partner. Subjects unlocked the door to recruit their partner significantly more often in the collaboration condition, when they needed assistance to obtain the food, than in the solo condition, when they did not (t = 7.27, df = 7, P < 0.001, paired t test). As individuals, seven of eight subjects recruited significantly more in the collaboration condition than in the solo condition. From Melis et al. 2006b. (b) Set-up of the collaboration and solo conditions: the subject was released from an adjacent room into the testing room, while the partner was "locked" in another adjacent room that only the subject could open with a key from inside the testing room. The rooms were separated by metal bars so that the subjects could see each other.

enter the adjacent room. In the collaboration condition of this test (figure 21.2) the tray of food could not be pulled within reach unless two individuals pulled simultaneously, as the ends of the rope were too far apart. Therefore, the first subject had to open the door to the adjacent room to recruit their partner and initiate the collaborative activity. In the control (solo) condition, however, the two ends of the rope were placed close to each other so that a single individual could simultaneously reach both ends and did not need to recruit a partner. The results showed that individuals recruited a partner significantly more often when solving the problem required collaboration (figure 21.2). Furthermore, this preference for recruiting a collaborator when needed appeared relatively fast, with five of eight subjects doing so within the first twelve trials.

Given the finding that chimpanzees recruit collaborators when needed, a second experiment was designed to test whether they can also learn to choose the more effective of two possible partners to recruit. The testing procedure was identical to that in the collaborative condition from the previous experiment, with the exception that now two potential collaborators were in the separate rooms adjacent to the testing room and each of the doors that led to the testing room was locked with a key. The subject could open either door from inside the testing room (figure 21.3b). The two potential collaborators—both males—had previously demonstrated very different levels of success in pulling the food tray with others, and on this basis they were designated by the experimenters as the "more effective" and the "less effective" partner. Since the subjects had not interacted with these potential collaborators in this context, they first participated in an introductory session in which they were allowed to recruit either collaborator—by choosing whose door to open—in six consecutive trials. In the test session, occurring on a later day, subjects were again released into the testing room for another six trials and allowed to choose which of the two partners to recruit for the collaboration task (the positions of the two potential partners were counterbalanced across trials). In the introductory session the subjects had

chosen both partners with equal frequency, but in the test session they almost exclusively chose the more effective partner (figure 21.3a). A trial-by-trial analysis revealed that the subjects were basing their recruitment choices in a given trial on the outcome of the preceding trial. If subjects succeeded in a trial, they tended to stay with the same partner on the next trial, whereas when they failed they shifted partners on the next trial (known as a "win-stay, lose-shift" strategy). Overall, subjects were significantly more successful in retrieving the baited tray with the "effective" partner than with the "less effective" partner, which explains the development of a preference for the latter: subjects preferred the partner with whom they were more successful in retrieving the tray. This experiment revealed that after minimal experience, chimpanzees could identify and remember their more skillful and less skillful collaborators, choosing to collaborate preferentially with the partner who led them to higher benefits and avoiding the individual with whom they experienced more failures.

Taken together, these results show that chimpanzees can learn extremely quickly, and without any extensive training, to collaborate with others in reaching a common goal. Furthermore, they show that chimpanzees not only coordinate their actions with those of their partners, but also understand the role a partner plays in joint activity, as evidenced by the fact that they actively recruited the most effective partners for solving a problem that required collaboration. However, several prerequisites are necessary if any potential collaboration is to emerge between chimpanzees in a food-retrieval task. First, the two individuals in a dyad need to be very tolerant towards each other. Second, the food rewards for the task must be sufficiently sharable, and, if possible, distributed widely enough in space, to reduce the potential for competition between the partners. Finally, subjects first need to experience and understand that they cannot solve the problem by themselves. During all of our tests, and often before recruiting their partner, subjects repeatedly checked the apparatus to make certain that the tray of food could really not be pulled successfully by a single chimpanzee. Nearly all subjects employed various means searching for nonsocial solutions to the problem (e.g., tool use, or attempts to fix one end of the rope to the bars). Thus, it seems as though they tried their best to solve the problem alone and were not particularly motivated toward collaboration (most likely

a)

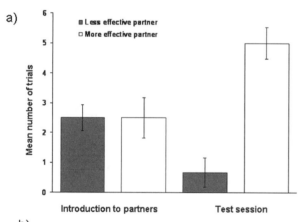

b)

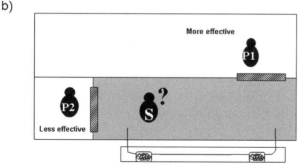

Figure 21.3   (a) The mean number of trials (± SEM) in which the subjects chose the less or more effective partner in the introductory and test sessions of the second recruitment experiment of Melis et al. 2006b. Subjects preferred the most effective partner in the test session only (F (1.5) = 9, P = 0.027). From Melis et al. 2006b. (b) The two potential collaborators were locked in two rooms adjacent to the testing room. Each door was locked with a key that only the subject could remove from inside the testing room. All three rooms were separated by metal bars so that subjects could see the partner behind the door before opening it. The position of the potential collaborators was counterbalanced across trials.

because it would involve sharing the food afterwards), choosing collaboration only as a last resort.

This leads us to the question about the motivation for cooperative behavior in chimpanzees. Is it solely based on egocentric and immediate self-interest, and therefore confined to mutualism and situations in which individuals can immediately benefit? Or are chimpanzees in some cases motivated to help and collaborate with others when it does not directly benefit them, at least in the short term?

Helping in Chimpanzees

As previously reviewed observational studies and several anecdotes suggest, chimpanzees might possess a certain tendency to act prosocially, maybe even altruistically, towards unrelated individuals. However, recent experiments have demonstrated that when chimpanzees had the opportunity to deliver food to a conspecific, they showed no altruistic tendencies (Silk et al. 2005; Jensen et al. 2006; Vonk et al. 2008). In particular, chimpanzees in the Jensen et al. study did not reliably pull a tray with food within reach of a conspecific if they themselves would not benefit from the act. In the Silk et al. experiment, one chimpanzee (the actor) could choose between one handle delivering food only to herself or another handle delivering the same amount of food to herself with an additional piece of food to another chimpanzee (the recipient). Chimpanzees chose randomly, and the results have been taken as evidence that chimpanzees are self-regarding and do not care about the welfare of others.

However, this food-retrieval context might not be representative of all potential helping situations. In particular, chimpanzees can be very competitive over monopolizable food, perhaps with the exception of meat. Thus, when chimpanzees are tested in food-retrieval contexts, in which the rewards often are strongly preferred fruit items, it might preclude expression of altruistic tendencies, especially when the actor is preoccupied with retrieving food for herself. Moreover, the recipient's need for assistance might not have been obvious to the actor, since the recipient was not actively struggling with a problem. The situation was thus a test more of the chimpanzee's generosity than of his willingness to help. Taken together, these circumstances allowed for the possibility that when the constraints related to food were lifted and the problem

situation was made more salient to the potential helper, a different picture might emerge.

We tested this possibility in an initial study with three nursery-reared chimpanzees who interacted with their primary human caregiver (Warneken and Tomasello 2006). The caregiver enacted different tasks in which she was unable to achieve her goal and needed help from the chimpanzee. For example, in an "out-of-reach" task, the caregiver accidentally dropped an object on the floor and reached for it unsuccessfully. In a "physical obstacle" task, she had problems putting objects into a container because her hands were full and she could not open the lid. Importantly, these tasks involved objects other than food and the caregiver never rewarded the chimpanzees for helping, giving them neither a food reward nor praise. The intriguing finding was that all three chimpanzees helped in different "out-of-reach" tasks: They reliably handed her objects she was unsuccessfully reaching for, while not doing so in the control condition in which she was not reaching for them. These chimpanzees were able to determine her goals and had the motivation to help her in the absence of an external reward. However, in contrast to the "out-of-reach" tasks, the same chimpanzees did not reliably help the caregiver in the other types of tasks, like opening a container or stacking objects. This may reflect a difference between types of tasks in the complexity of their goal structure. In the out-of-reach tasks, the recipient's goal was in principle easier to identify (an arm outstretched towards a visible object) and the intervention followed straightforwardly. The goals of the other types of tasks might not have been obvious to them, or perhaps they did not know how to intervene. This shows the importance of taking into account the social-cognitive demands that tests of altruism entail. These human-reared chimpanzees were, in principle, willing to help, but did so only in contexts in which the caregiver's goal was easy to identify.

This was the first experimental demonstration of altruistic helping in chimpanzees. It must be kept in mind, however, that these were human-reared chimpanzees helping a human with whom they interacted regularly. Thus, it is possible that human-reared chimpanzees present a special case, since they often develop behaviors not found in individuals with less human contact (Bering 2004; Call and Tomasello 1996). Moreover, although the helping situations were novel for these chimpanzees, the caregiver had previously reinforced other types of compliant behavior. Therefore, from this initial experiment it remains unclear whether helping behavior in chimpanzees is restricted to interactions with highly familiar individuals who have rewarded them before, or extends also to unfamiliar individuals.

We investigated this question by testing a sample of wild-born chimpanzees who live on Ngamba Island in Uganda. These semi–free-ranging chimpanzees spend the day in the island's forest and come to a human-built shelter for feeding and sleeping in the evening. While they have regular contact with humans, they have not been exposed to human rearing practices comparable to those experienced by the three human-raised chimpanzees from the initial helping study. Of particular importance for the current purposes was that these chimpanzees were tested by a human with whom they had not interacted before the experiment (no training, no feeding, no previous testing). In a first experiment with 36 chimpanzees, we examined whether their helping behavior was indeed motivated by an altruistic tendency to assist other individuals with their problems or by the selfish goal of receiving an immediate benefit for themselves (Warneken et al. 2007, experiment 1). Therefore we varied between subjects whether the human experimenter attempted unsuccessfully to get at an out-of-reach object (conditions of reaching for it versus not reaching for it) and whether he rewarded the subjects with pieces of food in exchange for the object (conditions of reward versus no reward). The rationale was that if subjects were responsive to another's goal, they should hand the object to the researcher more often in the reaching condition than in the no-reaching condition. If they are primarily interested in their own immediate benefit, they should help more often in the reward condition than in the no-reward condition. Chimpanzees helped the human more often when he was reaching for the object than when he was not reaching for it, and they did so irrespective of being rewarded. Rewarding their helping was unnecessary and it did not even raise the rate of helping. This indicates that the chimpanzees were primarily motivated by the other person's unachieved goal, and not by the prospect of a reward for themselves.

In a follow-up experiment, we again probed the chimpanzees' helpfulness by making the act of helping more costly (Warneken et al. 2007; experiment 2). In previous experiments the costs for helping had been low, requiring not much more than picking up an object or pulling a handle. In this experiment, however, the cost of helping was increased by locating the target object in a raceway 2.5 meters above the ground. Chimpanzees continued to help the human over repeated trials, even if they first had to climb up to retrieve the object for him. Taken together, these experiments show that helping extends also to chimpanzees who act on behalf of unfamiliar humans, even if it requires some effort and they receive no reward in return.

Despite the fact that chimpanzees helped an unfamiliar human who had never rewarded them previously and did not reward them in the test situation, it is conceivable that they had been rewarded in the past for the general behavior of handing objects to humans. Moreover, all these positive instances of helping involved chimpanzees helping humans, and not chimpanzees helping chimpanzees. The use of a human as the recipient has the advantage that particular factors can be controlled and manipulated systematically (such as the cues given by the recipient, or the offer or withholding of a reward). It can reveal what chimpanzees are in principle able to do, but the crucial test case remains that of chimpanzees interacting with conspecifics. Thus, it still remains an open question whether chimpanzees would help other chimpanzees. Previous experiments with chimpanzee conspecifics yielded negative results (Silk et al. 2005, Jensen et al. 2006) but those tests involved the delivering of food to a partner who was not actively trying to retrieve it, so that the need for help might not have been obvious to the actor. In addition, in the study by Silk et al. (2005), there was no direct evidence that chimpanzees actually understood the physical setup of the helping task.

To address these issues, we created a novel situation in which the recipient's problem was made more salient for the chimpanzees (Warneken et al. 2007, experiment 3): the recipient chimpanzee was trying to open a door leading to a room containing a piece of food, and the door was fixed with a chain that the chimpanzee could not unlock (figure 21.4). Only if the actor, who was located in an adjacent room, released the chain could the recipient enter. All chimpanzees were genetically unrelated group members. To exclude the possibility of short-term reciprocation, the roles of actor and recipient were not switched within a dyad. In individual pretests, the chimpanzees had demonstrated that they could manipulate the chain-door mechanism successfully to open the door for themselves. Results showed that in the majority of cases when the re-

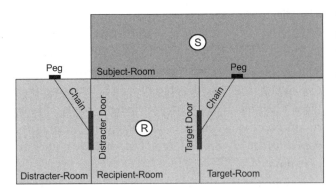

Figure 21.4 Test area and setup in the third helping experiment of Warneken et al. 2007. All rooms were separated by metal bars. In the experimental condition, food was placed in the target room so that the recipient would try to open the target door. The target door was shut by a chain that could only be released by the subject. In the control condition, food was placed in the distracter room so that the recipient would try to open the distracter door, but releasing the target chain would be useless with respect to the recipient's goal of opening the distracter door. From Warneken et al. 2007.

cipient was trying to open the door, chimpanzees helped by releasing the chain. Importantly, they performed this behavior significantly less often (1) in a control condition, in which releasing the chain would not have helped the recipient because he was trying to open a different door leading to the distracter room where the food had been placed, and (2) in a baseline condition in which no other individual was present. This indicated that the chimpanzees were attentive to each other's goals and provided help when needed. It also showed that they could help conspecifics without any expectation of immediate returns.

Taken together, these experiments show that chimpanzees perform acts of helping. This finding sheds light on both the motivation and cognition of chimpanzees. They appear to have an altruistic motivation to act on behalf of other individuals, even at some cost to themselves and when no immediate benefit is expected. They also have the cognitive capacity to intervene with some flexibility on behalf of others who pursue goals different from their own—even using novel skills to help in novel situations. Chimpanzees therefore possess some of the core capacities to engage in altruistic helping.

Discussion

We have been able to draw a number of conclusions about the psychological mechanisms underlying chimpanzee cooperation. Let us now reflect back on what these findings might reveal about the cooperation we see in the wild, and what the critical differences might be between coopera-

tion in chimpanzees and in humans. This can provide hints about the possible species-unique evolutionary trajectory that our own lineage took after it split from our closest living relatives, the chimpanzees and bonobos.

Chimpanzee Collaboration

Our results support the idea that chimpanzees are capable of active collaboration. The spontaneity with which the chimpanzees in our experiments solved problems suggests that wild chimpanzees likely understand the potential role of their partners when they are cooperating during hunting or during coalitionary behavior. It is important to differentiate the motivational and cognitive components underlying such behavior. For example, although the motivation underlying group hunting might be completely selfish, it is possible that in sites where ecological conditions make solitary hunting difficult, individuals can learn that they have increased chances of success (and potential payoffs—see Boesch and Boesch-Achermann 2000, Watts and Mitani 2002; see also chapter 18) if they coordinate their behavior with that of their partners. In our opinion, these results rule out the hypothesis that chimpanzees lack the capacity to engage in intentional coordination and that social predation in chimpanzees is just by-product cooperation involving the simultaneous but uncoordinated actions of multiple individuals towards the same prey (see chapter 18). However, this level of understanding does not rule out the hypothesis that chimpanzees view their cooperative partners as social tools to reach their own individual goals, rather than as collaborative partners with shared goals—a feature that is characteristic of human collaboration already in early ontogeny (Tomasello et al. 2005; Warneken et al. 2006). Specifically, collaborative activities that are based on shared intentions entail a shared goal (and not just a common goal), including a mutual understanding among partners that they have this goal and a joint commitment to subsume their actions in its pursuit. Shared intentions often entail (and can be identified by) some form of communication in which partners attempt to influence each other's goals and intentions (Tomasello et al. 2005). In fact, it is in regard to these communicative abilities (or tendencies) during cooperative interactions that our ongoing research has found the greatest differences between chimpanzees and even the very youngest human children. In a study by Warneken et al. (2006), a human experimenter collaborated in several problem-solving activities and social games with one-year-old human children and three human-raised chimpanzees. When the adult partner stopped playing his role during certain predetermined interruption periods, children attempted to reactivate him using several communicative means, whereas chimpanzees attempted to solve the problem without him or disengaged from it completely. Chimpanzees never produced any communicative signals to influence their partner's behavior (although the same chimpanzees did produce such signals in other noncollaborative contexts; Tomasello and Carpenter 2005). This is also suggested by the results of another of our studies (Melis et al. 2009), in which the same skillful and cooperative chimpanzees who participated in our previous studies (Melis et al. 2006a, 2006b) were presented with a variation of the cooperative task that required them to agree on pulling from one, and only one, of two food trays. The study created a conflict of interests between the partners, with the dominant individual in each pair aiming to obtain one tray and the subordinate partner the other tray. Although the subjects were finally very successful at cooperating, and ended up agreeing on one tray or the other (a result which further supports the idea that they have a sophisticated level of understanding about the social nature of the collaborative interaction), there were surprisingly few instances of intentional communication between partners, and none of the communication was as overt as that previously observed by Crawford (1937) between chimpanzee partners, or that observed by Hirata and colleagues (Hirata and Fuwa 2007; see also chapter 20) between a young female chimpanzee and a human partner. This was surprising because the negotiation process often lasted as long as two to three minutes, with partners unable to agree on which tray to pull, and each waiting for the other to help pulling a different tray. It seems that in these situations, any type of communicative effort could have facilitated the negotiation process. In the same way that chimpanzees readily gesture for support in coalitionary contexts (side-directed communication: de Waal and van Hoof 1981; see also Call and Tomasello 2007 for a review of gestural communication in apes), one could expect the use of similar communicative attempts in new cooperative contexts. Therefore, the lack of communication among the chimpanzees in our study clearly marks a difference from humans in their ability to influence and regulate a cooperative activity. Whether this actually reflects a fundamental difference in how chimpanzees and humans conceive the joint action is a question that deserves further investigation (see chapter 20). Importantly, we will need first to establish which variables constrain chimpanzees' use of intentional communication during cooperative interactions. One could imagine, for example, that the lack of communication is related to their low levels of tolerance and their high competition over food. Maybe communicating the need for support to a potential competitor in a food-related context only decreases the likelihood of obtaining help, or increases the competition between individuals. This would also fit with the observations made by Hirata and colleagues (chapter 20) regarding the young age of those chimpanzees who have been observed to engage in soliciting behavior. A new study by Wobber, Hare, and Wrangham (unpublished data) shows that young chimpanzees are more tolerant than adult ones.

Future experimental studies should also investigate whether chimpanzees remain as skillful when confronted with collaborative tasks that involve more than two partners who potentially need to fill different but complementary roles. This would be a closer parallel to cooperative hunting, and it would also shed light on the mechanisms that might underlie the maintenance of such group-level cooperation in the wild. Collaborative hunts present a collective-action problem, since free-riders (nonhunters) can benefit at the hunters' expense unless hunters have a non-costly way of controlling their behavior (see chapter 18). Boesch (1994a) describes how in the Taï forest there is a fair distribution of meat between hunters and nonhunters: hunters obtain more meat than do bystanders and latecomers, and the good hunters receive the most meat. This is an observation that deserves further investigation in the laboratory and at other field sites, since it could shed light on how individuals conceive the joint action. If confirmed, it would suggest a sense of fairness that rewards individual contributions to the collaborative act, and would also imply that partners are viewed not as mere social tools, but as true collaborative partners with whom they have formed a shared goal (sensu Tomasello et al. 2005). At the same time, it would also suggest that chimpanzees possess another non-costly punishing mechanism (in addition to shunning) by which hunters might be controlling the behavior of nonhunters (i.e., cheaters; see van Schaik and Kappeler 2006).

Chimpanzee Helping

Although it seems that chimpanzees may not be helpful to conspecifics universally across all contexts, they can in some situations be helpful toward others when there is no immediate reward for their effort. The conflicting results between Silk et al. (2005), Jensen et al. (2006), and Vonk et al. (2008), on the one hand and Warneken and colleagues (2006, 2007) on the other can be due to different reasons. One possibility is that helping is especially constrained in food-retrieval situations, as is suggested by the results of Silk et al. (2005), Jensen et al. (2006) and Vonk et al. (2008). As mentioned above, chimpanzees are very competitive over monopolizable food, making active forms of food sharing (in which the donor initiates the transition and actively gives food away) a rare event. Natural observations show that most instances of food sharing are of a passive nature: possessors of food tolerate beggars obtaining scraps from them (de Waal 1989), or food is shared as the result of harassment (Wrangham 1975; Gilby 2006). This is also in line with the results from our previous experiments on mutualistic collaboration, which have shown that to find flexible collaboration, the potential for competition between partners must be extremely reduced (they must be tolerant of each other, and the food rewards must be highly sharable). A second possibility is that helping behavior in chimpanzees is constrained to situations in which actors can use very salient cues to infer each other's goals and needs. In all the experimental procedures in which we obtained positive results, the recipients were clearly indicating the goal they were trying to achieve (e.g., by an arm outstretched towards a visible object, by an alternation of gaze, or by the manipulation of a door). In the future it will be important to test these two hypotheses against each other. Are chimpanzees unable or unwilling to actively provide food to others? Or are they limited in their ability to infer recipients' goals and needs in a way that constrains them to help? Future tests should also carefully control for chimpanzees' level of understanding of the food-delivering apparatus, and rule out the possibility that the actors' interest in obtaining the food for themselves interferes with their potential helping tendencies.

But what are the mechanisms that allow such altruism to be maintained as evolutionarily stable strategies? First, kin-selection mechanisms are an unlikely candidate, as our chimpanzee subjects helped genetically unrelated group members as well as both familiar and unfamiliar humans. Moreover, as previously mentioned, recent field studies provide evidence that kin relationships play a negligible role in such other types of cooperative behaviors as alliance formation or group hunting (Muller and Mitani 2005; Langergraber et al. 2007). A second candidate would be reciprocal altruism (Trivers 1971). Social relationships among chimpanzees often appear to be characterized by some form of reciprocity—even across different commodities, such as meat sharing and alliance formation (e.g., de Waal 1997; Koyama et al. 2006; Mitani 2006; but see Gilby 2006). However, with the exception of de Waal (1997) and Koyama et al. (2006), these results are mostly correlational in nature, so further studies must address the question of whether the provision of services to others is conditional upon receiving them as well. If these interactions are based upon reciprocal altruism, bouts of exchange should collapse if one individual does not return the other's incurred costs at later encounters. Alternatively, this effect could also be facilitated by partner switching—preferentially choosing to interact with individuals who reciprocate the services provided (Noë 2001). Our previous results (Melis et al. 2006b) in which chimpanzees readily identified and chose to collaborate with the most competent partners, suggest that they likely have the cognitive capacity to recognize and remember favors they have previously received from others, being thus able to avoid cheaters. Using a similar paradigm, future studies could directly test how chimpanzees terminate relationships with noncooperators, and how they decide to switch partners (Melis et al. 2008). A third proposed mechanism for stabilizing altruistic behaviors is social punishment. Its importance is emphasized both as part of evolutionary models relying on reciprocal altruism (as moralistic punishment, Trivers 1971) and as part of models relying on cultural group selection (Boyd et al. 2003). Through punishment, individuals inflict harm on cheaters, thus increasing the cost of not cooperating and reducing the probability that individuals will repeat the action or refuse to perform an altruistic one (Clutton-Brock and Parker 1995). Recent experimental work shows that chimpanzees are vengeful as a form of active punishment towards individuals who have stolen food from them (Jensen et al. 2007a), and there is evidence of retaliation in the patterns of conflict intervention among chimpanzees (de Waal and Luttrell 1988). Therefore, it appears that chimpanzees possess a basic tendency to punish others who directly harm them. In conclusion, reciprocity and some form of punishment are among the most likely stabilizers of altruism in chimpanzees, although empirical validation of this remains a goal for future research.

Our current results challenge the view that only humans perform acts of altruism towards non-kin in situations without an obvious return benefit—as a result of a human-unique psychology and cultural practices such as the transmission and enforcement of social norms (Richerson and Boyd 2005; Silk et al. 2005). It appears instead that some of the crucial components of human altruism can already be found in chimpanzees. Although chimpanzees seem with humans to share some of the core motivations and cognitive skills for altruistic helping, humans show helping behaviors more readily and in a wider variety of contexts. One possibility for this difference between chimpanzees and humans is that human culture has created unique social mechanisms to preserve and foster these prosocial tendencies, ultimately resulting in forms of altruism not found beyond the human species. This is where culture and socialization come into play, as humans rear their children in ways that might promote these altruistic tendencies in species-unique ways. In addition, the emotional, cognitive (theory-of-mind and perspective-taking abilities), and motivational systems of humans may have undergone significant changes which have contributed to the increase in altruistic behaviors and the readiness to show them.

Future Directions

While we have taken some initial steps in uncovering the social constraints and cognitive abilities that underlie chimpanzee cooperation, many questions remain unanswered. However, we finally have a number of methodologies with which we can probe the depths of the cooperative abilities of chimpanzees and other primates. In the same way we found that inter-individual tolerance levels affect chimpanzees' ability to cooperate, we have also found that inter-species differences in levels of social tolerance affect the species' tendency to cooperate (Petit et al. 1992;

Hare et al. 2007). Even though male bonobos have never been observed to cooperate in many of the ways observed in male chimpanzees—for example, they have never been seen to hunt in groups—we have found that due to their increased levels of social tolerance, bonobos can outperform chimpanzees in a collaborative task when the food rewards are easily monopolizable (Hare et al. 2007). Although chimpanzees understand the role their partners play in a collaborative activity, their low levels of food tolerance act as a constraint in a collaborative context. These findings raise important questions about the conditions that may have played a role in the evolution of human cooperation. Higher levels of social tolerance or certain emotional adaptations might have been a prerequisite for the evolution of human-like cooperation (see also van Schaik et al. 1999; and see also chapter 9 for a similar argument with regard to the influence of tolerance in social learning). Further studies with bonobos and chimpanzees will be fundamental for our understanding of the interplay of tolerance (or temperament) and cognition (e.g., the question of whether high levels of food tolerance in bonobos allow for altruistic food sharing in the same experimental setups where such sharing is not seen in chimpanzees). It is only by combining our knowledge of wild chimpanzee and bonobo cooperation with our knowledge of their behavior in captive settings that we will reach a full understanding of the flexibility, motivation, and constraints that drive chimpanzee and bonobo cooperation—and thus gain a better understanding of the evolutionary roots of cooperative behavior in our own species.

Acknowledgments

We would like to thank Ian Gilby, Mike Tomasello, and the reviewers for their helpful comments on an earlier version of this chapter. The research of B.H. and A.P.M. is supported by a Sofja Kovalevskaja award to B.H. from the Alexander von Humboldt Foundation and the German Federal Ministry for Education and Research.

Literature Cited

Bering, J. M. 2004. A critical review of the "enculturation hypothesis": The effects of human rearing on great ape social cognition. *Animal Cognition* 7(4): 201–12.

Berkovitch, F. B. 1988. Coalitions, cooperation and reproductive tactics among adult male baboons. *Animal Behavior* 36:1198–1209.

Boesch, C. 1994a. Cooperative hunting in wild chimpanzees. *Animal Behavior* 48(3): 653–67.

———. 1994b. Chimpanzees–red colobus monkeys: A predator-prey system. *Animal Behavior* 47(5): 1135–48.

Boesch, C., and H. Boesch. 1989. Hunting behavior of wild chimpanzees in the Taï National Park. *American Journal of Physical Anthropology* 78:547–73.

Boesch, C., and H. Boesch-Achermann. 2000. *The Chimpanzees of the Taï Forest*. Oxford: Oxford University Press.

Boesch, C., G. Hohmann, and L. Marchant. 2002. *The Behavioral Diversity of Chimpanzees*. Cambridge: Cambridge University Press.

Boyd, R., H. Gintis, S. Bowles, and P. J. Richerson. 2003. The evolution of altruistic punishment. *Proc Natl Acad Sci USA* 100:3531–35.

Bshary, R., and Bergmüller. 2007. Distinguishing four fundamental approaches to the evolution of helping. *J. Evol. Biol.* 21:405–20.

Call, J., and M. Tomasello. 1996. The effect of humans on the cognitive development of apes. In *Reaching Into Thought: The Minds of the Great Apes*, A. E. Russon, K. A. Bard, and S. T. Parker, eds. New York: Cambridge University Press, 371–403.

Call, J., and M. Tomasello, eds. 2007. *The Gestural Communication of Apes and Monkeys*. Mahwah, NJ: Lawrence Erlbaum.

Chalmeau, R. 1994. Do chimpanzees cooperate in a learning task? *Primates* 35:385–92.

Chalmeau, R., and A. Gallo. 1996. Cooperation in primates: Critical analysis of behavioural criteria. *Behavioural Processes* 35:101–11.

Clutton-Brock, T. H., and G. A. Parker. 1995. Punishment in animal societies. *Nature* 373:209–16.

Crawford, M. P. 1937. The cooperative solving of problems by young chimpanzees. *Comparative Psychology Monographs* 14:1–88.

———. 1941. The cooperative solving by chimpanzees of problems requiring serial responses to color cues. *Journal of Social Psychology* 13:259–80.

Cronin, K. A., A. V. Kurian, and C. T. Snowdon. 2005. Cooperative problem solving in a cooperatively breeding primate (*Saguinus oedipus*). *Animal Behavior* 69:133–42.

De Waal, F. B. M. 1982. *Chimpanzee Politics*. New York: Harper and Row.

———. 1989. Food sharing and reciprocal obligations among chimpanzees. *Journal of Human Evolution* 18(5): 433–60.

———. 1997. The chimpanzee's service economy: Food for grooming. *Evol. Hum. Behav* 18:375–86.

———. 2008. Putting the altruism back into altruism: The evolution of empathy. *Annu. Rev. Psychol.* 59:279–300.

De Waal, F. B. M., and J. M. Davis. 2003. Capuchin cognitive ecology: Cooperation based on projected returns. *Neuropsychologia* 41:221–28.

De Waal, F. B. M., and L. Luttrell. 1988. Mechanisms of social reciprocity in three primate species: Symmetrical relationship characteristics or cognition? *Ethol. Sociobiol.* 9:101–18.

De Waal, F. B. M., and J. A. R. A. M. van Hooff. 1981. Side-directed communication and agonistic interactions in chimpanzees. *Behavior* 77:164–98.

Fehr, E., and U. Fischbacher. 2003. The nature of human altruism. *Nature* 425:785–91.

Gilby, I. C. 2006. Meat sharing among the Gombe chimpanzees: Harassment and reciprocal exchange. *Animal Behavior* 71:953–63.

Gilby, I. C., L. E. Eberly, L. Pintea, and A. E. Pusey. 2006. Ecological and social influences on the hunting behavior of wild chimpanzees (*Pan troglodytes schweinfurthii*). *Animal Behavior* 72:169–80.

Goodall, J. 1986. *The Chimpanzees of Gombe: Patterns of Behavior*. Cambridge, MA: Belknap Press.

Hammerstein, P. 2003. Why is reciprocity so rare in social animals? A protestant appeal. In *Genetic and Cultural Evolution of Cooperation*, P. Hammerstein, ed. Cambridge, MA: MIT Press.

Hamilton, W. D. 1964. The genetical evolution of social behavior. *Journal of Theoretical Biology* 7:1–52.

Hare, B. 2001. Can competitive paradigms increase the validity of social cognitive experiments on primates? *Animal Cognition* 4:269–80.

———. 2007. From nonhuman to human mind: what changed and why. *Current Directions in Psychological Science* 16:60–64.

Hare, B., and M. Tomasello. 2004. Chimpanzees are more skilful in competitive than in cooperative cognitive tasks. *Animal Behavior* 68:571–81.

Hare, B., A. P. Melis, V. Woods, S. Hastings, and R. Wrangham. 2007. Tolerance allows bonobos to outperform chimpanzees on a cooperative task. *Current Biology* 17:619–23.

Hirata, S., and K. Fuwa. 2007. Chimpanzees (*Pan troglodytes*) learn to act with other individuals in a cooperative task. *Primates* 48:13–21.

Jensen, K., B. Hare, J. Call, and M. Tomasello. 2006. What's in it for me? Self-regard precludes altruism and spite in chimpanzees. *Proc R Soc B* 273:1013–21.

Jensen, K. J. Call, and M. Tomasello. 2007a. Chimpanzees are vengeful but not spiteful. *PNAS* 104:13046–50.

Koyama, N. F., C. Caws, and F. Aureli. 2006. Interchange of grooming and agonistic support in chimpanzees. *International Journal of Primatology* 27:1293–1309.

Langergraber, K. E., J. C. Mitani, and L. Vigilant. 2007. The limited impact of kinship in cooperation in wild chimpanzees. *Proc Natl Acad Sci USA* 104(19): 7786–90.

Melis, A. P., B. Hare, and M. Tomasello. 2006a. Engineering cooperation in chimpanzees: Tolerance constraints on cooperation. *Animal Behavior* 72:275–86.

———. 2006b. Chimpanzees recruit the best collaborators. *Science* 311:1297–1300.

———. 2009. Chimpanzees coordinate in a negotiation game. *Evolution and Behavior* 30:381–92.

Mendres, K. A., and F. B. M. de Waal. 2000. Capuchins do cooperate: The advantage of an intuitive task. *Animal Behavior* 60:523–29.

Mitani, J. 2006. Reciprocal exchange in chimpanzees and other primates. In *Cooperation in Primates: Mechanisms and Evolution*, P. Kappeler and C. van Schaik, eds. Berlin: Springer-Verlag.

Muller, M. N., and J. C. Mitani. 2005 Conflict and cooperation in wild chimpanzees. In *Advances in the Study of Behavior*, P. J. B. Slater, J. Rosenblatt, C. Snowdon, T. Roper, and M. Naguib, eds. New York: Elsevier, 275–331.

Noë, R. 2001. Biological markets: Partner choice as the driving force behind the evolution of cooperation. In *Economics in Nature: Social Dilemmas, Mate Choice and Biological Markets*, R. Noë, J. A. R. A. M van Hooff, and P. Hammerstein, eds. Cambridge: Cambridge University Press, 92–118.

Petit, O., C. Desportes, and B. Thierry. 1992. Differential probability of coproduction in two species of macaque (*Macaque tonkeana, M. mulatta*). *Ethology* 90:107–20.

Povinelli, D. J., and D. K. O'Neill. 2000. Do chimpanzees use their gestures to instruct each other? In *Understanding Other Minds: Perspectives from Developmental Cognitive Neuroscience*. S. Baron-Cohen, H. Tager-Flusberg, and D. J. Cohen, eds., 459–87. Oxford: Oxford University Press.

Richerson, P. J., and R. Boyd. 2005. *Not by Genes Alone*. Chicago: University of Chicago Press.

Shutt, K., MacLarnon, A., Heistermann, M., and Semple, S. (2007) Grooming in Barbary macaques: Better to give than to receive? *Animal Behavior* 3:231–33.

Silk, J., Brosnan, S., Vonk, J., Henrich, J., Povinelli, D., Richardson, A.S., Lambeth, S.P., Mascaro, J., and Schapiro, S.J. (2005) Chimpanzees are indifferent to the welfare of unrelated group members. *Nature* 437:1357–59.

Stevens, J. R., and M. D. Hauser (2004) Why be nice? Psychological constraints on the evolution of cooperation. *Trends in Cognitive Sciences* 8(2): 60–65.

Stevens, J., and I. Gilby. 2004. A conceptual framework for nonkin food sharing: Timing and currency of benefits. *Anim. Behav.* 67:603–14.

Svetlova, M. L., S. R. Nichols, L. Shuck, and C. Brownell. 2007. What's mine is mine and what's yours is . . .": Development of sharing in toddlers. Poster presented in SRCD, Boston.

Trivers, R. 1971. The evolution of reciprocal altruism. *Quarterly Review of Biology* 46:189–226.

Tomasello, M., and M. Carpenter. 2005. *The Emergence of Social Cognition in Three Young Chimpanzees.* Monographs of the Society for Research in Child Development 70(1): 1–132.

Tomasello, M., J. Call, and B. Hare. 2003. Chimpanzees understand psychological states: The question is which ones and to what extent. *Trends in Cognitive Sciences* 7(4): 153–56.

Tomasello, M., M. Carpenter, J. Call, T. Behne, and H. Moll. 2005. Understanding and sharing intentions: The origins of cultural cognition. *Behavioral and Brain Sciences* 28:675–735.

Van Schaik, C. P., and P. Kappeler. 2006. Cooperation in primates and humans: Closing the gap. In *Cooperation in Primates: Mechanisms and Evolution,* P. Kappeler and C. van Schaik, eds. Berlin: Springer-Verlag.

Van Schaik, C. P., R. O. Deaner, and M. Y. Merril. 1999. The conditions for tool use in primates: Implications for the evolution of material culture. *Journal of Human Evolution* 36:719–41.

Vonk, J., S. F. Brosnan, J. B. Silk, J. Henrich, A. S. Richardson, S. P. Lambeth, S. Schapiro, and D. J. Povinelli. 2008. Chimpanzees do not take advantage of very low cost opportunities to deliver food rewards. *Animal Behavior* 75:1757–70.

Warneken, F., and M. Tomasello. 2006. Altruistic helping in human infants and young chimpanzees. *Science* 311:1301–3.

———. 2007. Helping and cooperation at 14 months of age. *Infancy* 11(3):271–94.

Warneken, F., F. Chen, and M. Tomasello. 2006. Cooperative activities in young children and chimpanzees. *Child Development* 77(3): 640–63.

Warneken, F., B. Hare, A. Melis, D. Hanus, and M. Tomasello. 2007. Spontaneous altruism in chimpanzees and children. *Public Library of Science* 5, e184.

Watts, D. 1998. Coalitionary mate-guarding by male chimpanzees at Ngogo, Kibale National Park, Uganda. *Behav. Ecol. Sociobiol.* 44:43–55.

———. 2002. Reciprocity and interchange in the social relationships of wild male chimpanzees. *Behavior* 139:343–70.

Watts, D., and J. Mitani. 2001. Boundary patrols and intergroup encounters in wild chimpanzees. *Behavior* 138:299–327.

Watts, D. P., and J. C. Mitani. 2002a. Hunting and meat sharing by chimpanzees at Ngogo, Kibale National Park, Uganda. In *Behavioural Diversity in Chimpanzees and Bbonobos,* C. Boesch, G. Hohmann, and L. Marchant, eds., 244–255. Cambridge: Cambridge University Press.

Werdenich, D., and L. Huber. 2002. Social factors determine cooperation in marmosets. *Animal Behavior* 64:771–81.

Wilson, M., and R. Wrangham. 2003. Intergroup relations in chimpanzees. *Ann. Rev. Anthro.* 32:363–92.

Wrangham, R. 1975. *Behavioural Ecology of Chimpanzees in Gombe National Park.* PhD thesis, Cambridge University.

Wrangham, R. W., and D. Peterson. 1996. *Demonic Males.* New York: Houghton Mifflin.

22

Inequity and Prosocial Behavior in Chimpanzees

Sarah F. Brosnan

Emily, an adult female chimpanzee, had just obtained a large piece of frozen fruit juice, a very valuable treat on a hot summer's day. Tony, a high-ranking male, soon noticed her juice and quickly came over and threatened her, trying to grab it from her hands. Emily screamed loudly and gestured towards the alpha male, Punch, reaching her arm out towards him while continuing to scream at Tony. Punch responded, joining Emily and threatening Tony. Soon Emily was threatening Tony too, with her arm draped across Punch's shoulders. Within minutes, three more adult females stood around Punch and Emily, all threatening Tony. Emily kept her treat and Tony left without any frozen juice. Perhaps most interestingly, none of Emily's defenders received any of the food—not even Punch, whom she had recruited to help her. The chimpanzees' defense of Emily could have been a political maneuver meant to increase their chances of favorable interactions later, but it might also have been a reaction to Tony's violation of the social norm against taking food which belongs to another.

Social species face difficult challenges as they must make decisions based on not only their own interests but also those of others to succeed over the long term. Individuals in these species must keep track of their relationships with others and determine whether their actions will serve not only their current interests, but also their future ones. Failure to do so may mean a better outcome in the short

term, but over the long term it may be disadvantageous as more opportunities for beneficial interaction (e.g., cooperation) are lost. Of course, individuals must also ensure that they are receiving their fair share from others. To maintain such a balance, these individuals must keep track of what is given and received and ultimately react appropriately both when they receive less than they should and when they receive more than they should. Incidents like the one described above highlight the complex social network that chimpanzees must navigate on a daily basis. Understanding how they do this opens a window into the inner workings of the chimpanzee's mind.

One avenue for investigating these tradeoffs between short-term costs and long-term outcomes is in the context of cooperation. In a cooperative interaction, multiple individuals work together to achieve some end, and they must assess outcomes, or payoffs, as well as each individual's contribution to determine whether the endeavor is beneficial. Although differences in payoffs are present in many situations, in this chapter I am concerned only with social situations that involve more than one individual. For instance, an ape foraging for food may realize from watching another ape that it is possible to obtain superior food elsewhere, as when young individuals learn foraging skills from their mothers. However, learning to get better food by watching another individual is quite different than interacting with that individual. In the case of foraging, a difference in outcomes informs the first individual that she should alter her strategy, but provides no information about the other as a social partner. In social enterprises, in which individuals work towards a common goal and gain payoffs as a direct result of others' behavior, a difference in payoffs provides information which should guide an individual's future interactions with a given social partner. Thus, the individual who receives less should be motivated to stop participating and the individual who receives more should be motivated to equalize the benefits in order to maintain their partner's cooperation.

Chimpanzees are an ideal species for studying such complex social interactions. These apes live in a rich social environment in which they routinely interact with a large number of individuals, ranging from frequent companions to those with whom they rarely interact, while managing to keep track of their relationships. They appear to be cooperative, working together on many tasks, and they perform well on cooperation experiments in the lab (Boesch 1994; Crawford 1937; see also chapters 19–21). Even in competitive situations, chimpanzees are often highly cooperative, working together in cooperative groups—alliances and coalitions—to defend territorial boundaries or females who are in estrus (see Mitani 2006 and chapter 15).

Until recently, most of our understanding of chimpanzee cooperation and competition came from observational studies. Observations of these interactions—and their aftermaths—led to the conclusion that these apes were cognizant of both their own and others' payoffs, or outcomes, and were proactive in making decisions affecting them (de Waal 1991). However, observational studies, while essential and informative in their own right, typically cannot provide an understanding of causation because there are no manipulations of the environment or controls to verify which factors are the cause and which are the effect. Thus, recent work has studied these behaviors in captive settings, where carefully controlled experiments aim to disentangle the mechanisms underlying the behaviors and the conditions that lead to cooperation. Studies on both chimpanzees and capuchin monkeys have investigated the reactions and behaviors of individuals who are both advantaged and disadvantaged as compared to their partners, as well as the responses to situations in which chimpanzees could potentially assist another. Thus, we can begin to understand how chimpanzees handle tradeoffs between their own interests and those of others. Ultimately these studies help us un-

derstand how the chimpanzee is able to maintain complex social networks, thus shedding light both on their minds and on the evolution of our own.

In this chapter I review the current state of knowledge regarding inequity responses in chimpanzees. I begin by defining and differentiating a potentially confusing set of terminologies which exist in this field. I then describe the existing research on a range of nonhuman species, including chimpanzees, in terms of their responses to inequity and the factors that may affect those responses. Finally, I shift to an examination of the related phenomenon of prosocial behavior, and attempt to tease out the proximate and ultimate explanations for these cognitive mechanisms.

Inequity Responses and Prosocial Behavior

Responding to Inequity

Observations of chimpanzees paint a picture of a species concerned about equity. For instance, chimpanzees will attack individuals who fail to respond with support in a fight for which their help has been solicited (de Waal 1982) and will scream in apparent frustration if their begging goes unrewarded (de Waal 2006). They are not only interested in their own outcomes, though. Chimpanzees show behaviors which appear empathetic, or concerned with the wellbeing of others (de Waal 1996; figure 22.1; chapter 23). Although these two categories of behavior typically are handled separately in animal studies, in humans they are often treated as two sides of the same coin, typically termed fairness. While it is instructive to address each category separately, as I will do below, I ultimately combine these two approaches to the behavior into a more unified theory of inequity in primate society.

Most people can agree on a definition for inequity (Brosnan 2006b; Hatfield et al. 1978). Inequity occurs when one individual receives a payoff or outcome which is in some dimension inferior to that which was expected based on another's outcome. Often equity is assumed to be synonymous with equality, although an outcome may be equitable and not equal (or equal and not equitable). Moreover, inequity must be differentiated from "greed," or from wanting more for its own sake rather than because another individual got more. Thus, it is a negative reaction to inequity if you want a larger piece of cake because everyone else at the party got a larger slice, but not if you want a larger piece because you really like cake. Both may be perfectly legitimate desires, but a reaction to inequity must be based on a comparison of one's payoffs with another's. Hence, it is inherently social.

The flip side of inequity is responding when you are the advantaged individual or when you observe another individual being treated inequitably. This goes by several names in different disciplines, including advantageous inequity aversion in economics or overcompensation in psychology. In all fields, such overcompensation is typically addressed either by behavior (that is, equalizing outcomes by offering a part of one's rewards) or by justification (that is, the use of psychological leveling mechanisms in which one's own surfeit is explained by one's greater deservedness; Fehr and Schmidt 1999; Hatfield et al. 1978). Of course, in a nonverbal organism justification is extraordinarily difficult, if not impossible, to uncover. Thus we are left with determining reactions to overcompensation only by the subjects' behavior; if they act to redistribute in favor of a less endowed conspecific, then they have demonstrated an aversion to overcompensation.

In human societies, "fairness" typically encompasses both of the above characteristics. Generally, too, there is an appeal to an objective level of fairness, almost a Platonic ideal, which is the standard by which all actions (or at least actions of a given kind) are to be judged. I do not mean to imply that there is only one such standard—there certainly are many, and each person's standard may shift with the circumstances—but that there is a relating of the current situation to a separate "objective" standard that is removed from the current situation. Again, since it is difficult or impossible to determine an ape's ideals, I prefer not to use the term fairness with regard to them.

Prosocial Behavior

Behavioral demonstration of an aversion to overcompensation is closely related to the current literature on prosocial behavior in chimpanzees. These studies, meant to complement the empathy literature (which is primarily based on observations), address how the apes respond to a situation in which they can provide a benefit to another individual, generally without any potential for recompense (e.g., reciprocity, mutualism, or kin selection).

Prosocial behavior itself means many different things,

depending upon the discipline. In social psychology, prosocial behavior is defined by intention (Eisenberg and Mussen 1989). Thus, an act is prosocial if it is performed with the intention of providing benefit to another. However, focusing on intentions ignores the distinctions between various evolutionary mechanisms which have selected for prosocial behavior. Thus, at least for students of animal behavior, an evolutionary approach is more useful.

From an evolutionary perspective, the question of interest is that of the conditions which have led to a given behavior or trait being selected over other possible behaviors. Thus, behaviors are distinguished not on the basis of their outcome, but on the basis of the evolutionary (ultimate) mechanisms that have led to them being selected. Behavior that benefits others can be selected because (1) the individual bestows benefits upon those who are closely related, thereby indirectly benefiting their own genes (kin selection; Hamilton 1964a, 1964b), (2) the individual, in the act of providing a benefit to himself, simultaneously provides a benefit for another individual (mutualism; Dugatkin 1997), and (3) the individual enhances her long-term fitness by alternating the receipt and donation of favors with another individual (reciprocal altruism; Trivers 1971). Cooperative behaviors not obviously explained by any of these three mechanisms are termed prosocial behaviors (Gintis 2000). Thus, examining whether other species also exhibit cooperative behaviors that do not fit neatly into these three explanations, and under which conditions these cooperative actions occur, will help delineate this new class of cooperative interaction and clarify the ultimate mechanisms that lead to it.

Experimental Approaches to Inequity

Responses to Inequity in Nonhuman Species

Although much work across a wide variety of disciplines has been done to investigate responses to inequity in humans, little has been done in nonhuman species. One of the first researchers to explicitly investigate the effects of fairness considerations on animal societies was Marc Bekoff in his work on social canids (Bekoff 2004). In social canids' play behavior, individuals who are much stronger or older typically self-handicap during play, and excessive aggression is not tolerated. This behavior apparently represents a group norm. Individuals who fail to exhibit such leveling behavior are excluded from play sessions. As a result, they do not interact nearly as much with other pups their age. Play seems to be important for these social animals as an opportunity to learn how to cooperate and negotiate social agreements in a situation in which mistakes or transgressions will be forgiven.

Moreover, individuals apparently require play behavior to integrate properly into the group. As coyotes mature, individuals ostracized from play sessions are much more likely to leave the pack than those who are included. The mortality rate for individuals who stay in the pack is 20%, while mortality for those who leave is 55%—almost three times higher (Bekoff 2004). Thus, there is apparently a direct fitness link between playing fairly and early survival, rendering this behavior very susceptible to alteration through natural selection. Bekoff suggests that in learning the art of social negotiation, conflict resolution, compromise, and cooperation, young canids (and other species) can learn the basics of the social norms that will govern their adult lives in the pack (Bekoff 2001). This is likely true in many group-living species. Ravens, too, show third-party enforcement of social norms (Heinrich 1999). Ravens have a social norm that a food possessor can maintain possession of the food regardless of which other ravens are present. Any individual who violates this by attempting to grab a piece will be attacked by another raven who is neither related to the food possessor nor stands to lose any food of her own (i.e., there is no self-serving intent). Thus, it appears that there is an expectation of equitable behavior (in the sense of following social norms) that is of so much importance to ravens that they will attack the offending party *even though* they themselves were not wronged. Thus, aggression is used to enforce the equitable treatment not only of oneself but of others as well.

Distributional Inequity in Nonhuman Primates

The earliest work on inequity in nonhuman primates established the presence of distributional inequity in two species, capuchin monkeys and chimpanzees (Brosnan and de Waal 2003; Brosnan et al. 2005; for more information on capuchin monkeys' responses see Brosnan 2006, van Wolkenten et al. 2007, and Brosnan 2009). In initial studies we investigated whether the primates would be willing to complete an exchange interaction (return a token and accept a food reward) if their partners received a better

food reward for the same "price" in a similar exchange. We tested subjects in pairs; each subject had a member of his or her own social group as a partner. In each case, the partner received a good reward (in this case, a grape) for completing an exchange with the experimenter while the subject received a less valuable but nonetheless typically desired reward (a piece of cucumber) for the same exchange. These results were compared to a control in which both individuals had received a cucumber, as well as several other conditions discussed below. We measured the subjects' willingness to complete the exchange interaction and to consume the cucumber once they had received it. In fact, they responded in each way about half of the time, so for the following analyses these categories are pooled and I report the subjects' willingness to complete the interaction (give the token and consume the food). Chimpanzees responded negatively to the situation in which a partner got a better reward as compared to the situation in which both individuals received the same reward. Their reactions included refusing to participate in the exchange, refusing to accept the reward of lesser value, and throwing away the rewards or the tokens.

Of course, to demonstrate inequity, it is not enough that the subject responds in the above situation. They must respond more strongly in this situation than in a different situation in which a better reward is available but is given to no one. This is the separation of "greed" from "envy." To control for this separation, another condition was included in which the experimenter showed each subject a grape immediately prior to that individual's exchange interaction. After the chimpanzee gestured towards the grape, the experimenter then hid it by placing it back in the container; the chimpanzee then exchanged a token and received a cucumber. This was done before each exchange, so that both subjects saw the better reward, a grape, prior to completing each of their exchanges for the cucumber. Critically, neither primate ever received the grape, so neither saw their partner receive a grape—an inequitable situation. Chimpanzees were much less likely to respond negatively in this situation than after seeing their partner get the grape (Brosnan et al. 2005).

Another possibility, however, is that subjects were simply frustrated about not receiving the grape. This would essentially contrast their current outcome with the outcome of a previous session or test (Brosnan and de Waal 2006; Roma et al. 2006). Roma et al. offered food items to capuchin monkeys in a similar situation (except without their having to complete a task, such as an exchange) and found that the subjects were more likely to refuse a cucumber in trials immediately following those in which they had received grapes than in trials following those in which they had received cucumber. This indicates that subjects may form expectations based on previous rewards, as was previously found by Tinkelpaugh (1928). However, Roma and colleagues failed to run the critical comparison in which the same subjects were compared for frustration and for inequity (Brosnan and de Waal 2006). In other words, although there was a control in which subjects were given a cucumber while their partners received a grape, these were not the same individuals in the two tests, and so an individual subject's behavior in the two situations could not be compared. Moreover, no task was required of the capuchin monkeys; instead they were simply handed food, a difference that could alter the results, as is discussed below.

Additional studies which do include this critical comparison find that inequity exerts a stronger force than frustration. In a follow-up study with capuchin monkeys, we compared their reactions in sessions immediately preceded by sessions in which they had received a grape to their reactions in sessions immediately preceded in which they had received cucumber rewards. If subjects are experiencing frustration, they would be expected to respond much more strongly in the first case, when comparing their cucumbers to grapes received in the previous session. However, the subjects showed no change in response, thus indicating that previous rewards did not affect results (van Wolkenten et al. 2007). Using a similar comparison, Bräuer and colleagues found that great apes also do not respond to frustration (Bräuer et al. 2006).

Social Factors Affecting Inequity

Among chimpanzees, we also found that the social environment plays a strong role in subjects' reactions (Brosnan et al. 2005). The chimpanzees used in this study came from two different social groups housed at the Yerkes National Primate Research Center Field Station, plus a few individuals from pair-housed situations at the Yerkes Main Center. All of the individuals lived in indoor-outdoor enclosures and received the same food and enrichment; however, their social situations were dramatically different. Chim-

panzees in the pair-housed groups each lived with only one other individual, although they also had visual and vocal contact with other chimpanzees. The two groups at the Field Station, on the other hand, lived in large outdoor corrals of more than 20 individuals, with multiple adult males and females as well as juveniles.

There were also other differences between these two groups. One, the "long-term" group, had been together for almost 30 years at the time of the study. All of its members had been born into the group except for one adult female, who had been present as a juvenile at the group's formation and was currently the alpha female. Thus, these individuals had grown up together and had long-established relationships. The second group, the "short-term" group, had been put together approximately eight years prior to the study, when all of the members were adults; most had not known each other before that time. Thus, although these individuals were now in a stable social group, they had not grown up with their partners as had the individuals from the long-term group.

These differences appeared in their responses to the test. Individuals from the pair-housed situation and the short-term group showed very strong reactions to distributional inequity. Chimpanzees from the long-term group, however, showed virtually no reaction to inequity, hardly ever refusing to exchange and accept a cucumber piece under any of the conditions. These results have several interesting implications. It appears that social environment may affect responses to inequity. Perhaps the close relationships of the long-term group were kin-like, and thus did not elicit the same sort of reactions as would have occurred with individuals who knew each other less well. Human individuals who know each other well and have very close relationships (termed communal) tend not to respond to inequity within those relationships, while those who have more superficial relationships (termed contingent) tend to respond strongly to inequity (Clark and Grote 2003). We were unable to test the same individual chimpanzees with partners they knew less well, and the length of time during which the partners had known each other was confounded with potential differences in personality or group character. Nonetheless, future testing may further establish the role of relationship quality in chimpanzees' responses to inequity. Since the responses in our study also indicated strong differences in the behavior of chimpanzees from different social groups, even those with similar environments, future chimpanzee studies should include multiple groups whenever possible.

The Context of Inequity

One important aspect of inequity responses is the context in which the inequity occurs. In several studies which have investigated inequity responses and failed to find a reaction, the lack of response may be explained by other factors. Given the variety of potential responses, it is necessary to understand which behaviors are the best measures of response to inequity. It is also important to understand the context in which the inequity occurs. For instance, as discussed above, social context and relationships are critical (Brosnan et al. 2005). One study failed to find any response to inequity among capuchin monkeys that had been preselected for tolerance in partnerships (Dubreuil et al. 2006). Perhaps these tolerant partners also had close relationships with each other, thus mitigating their response to inequity as in the earlier chimpanzee study. Another study that found that chimpanzees did not respond negatively to inequity also used individuals from a stable social group (Jensen et al. 2007), which may have curtailed their negative response.

Inequity can present itself in many different ways, and different methodologies and measurements may affect the results of a study. Our initial studies focused on the subject's willingness to complete an interaction, defined as both returning the token *and* accepting the food reward. Later studies have also included a measure of whether the interaction varied from a standard exchange in either subjects' behavior or response latency (van Wolkenten et al. 2007). No other study has used these same criteria, partly because few other studies have involved tasks, thus rendering the measure of subjects' behavior irrelevant for comparison. The majority of studies have compared latencies, which typically do not differ between conditions of inequity and equity (Bräuer et al. 2006; Dindo and de Waal 2006; Dubreuil et al. 2006; Roma et al. 2006). Likewise, refusal to accept food alone (in situations which do not involve a task) also does not vary between the two conditions (Bräuer et al. 2006; Dindo and de Waal 2006; Dubreuil et al. 2006). Social facilitation plays a role as well: when a partner is present, a subject is less likely to ignore food (Bräuer et al. 2006) and eats more rapidly (Dindo and de Waal 2006). A chimpanzee is also more likely to

spend time away from the experimenter when no competitor is present (Bräuer et al. 2006).

Perhaps the most critical difference between these various studies has been the presence or absence of a task for which subjects receive equitable or inequitable rewards. The subjects' perception of inequity likely arises as they compare their own efforts and payoffs to those of others in cooperative and other joint tasks. Outside of this context, their behavior may be quite different. In our earlier studies the requirement of effort was added with the inclusion of a simple exchange task, chosen for being easily visible by the partner and a correlate to social exchange, in which chimpanzees and capuchin monkeys are known to engage. Following these initial studies, several others were conducted on both chimpanzees and capuchin monkeys in which no task was employed; in these later studies no difference in response was found between equitable and inequitable conditions (Bräuer et al. 2006; Dindo and de Waal 2006; Dubreuil et al. 2006). Although such differences can also be due to extraneous factors, such as social or environmental context, in one case the same capuchin monkey subjects were also used in inequity studies both with and without a task; they responded to inequity only in the two studies with a task (Brosnan and de Waal 2003; van Wolkenten et al. 2007) but not in the one without a task, which took place between the other two (Dindo and de Waal 2006).

Thus the presence of a task seems to be crucial to activate a subject's reaction to inequity. Further work needs to be done to determine the situations in which equity is important and the situations in which subjects are willing to tolerate inequitable outcomes. If the hypothesis that inequity responses evolved in the context of cooperative enterprise is correct, then responses to inequity are more likely to occur in situations that evoke cooperation.

Cooperation and Inequity

While previous work on distributional inequity focused on the distribution of rewards between various subjects within a study, more recent work on cooperation involving unequal rewards shows that subjects may be more attuned to their conspecific partners' behavior than to the rewards themselves. Previous work using a mutualistic cooperation task in capuchin monkeys had already demonstrated how sensitive the monkeys are to both social and

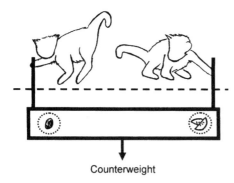

Counterweight

Figure 22.1 Two monkeys, who are not separated from each other, must work together to pull the counterweighted tray that has two rewards on it. If they pull it toward themselves successfully, they can access the rewards. Sometimes the rewards are the same (two apple slices or a grape for each monkey) and sometimes the rewards are different (two apple slices for one, one grape for the other; pictured). All subjects preferred the grapes to the apple slices. Pairs of monkeys succeeded more than twice as often when they took turns receiving the grape than in instances when the rewards for each monkey were different.

physical aspects of cooperation (de Waal and Berger 2000; Mendres and de Waal 2000; de Waal and Davis 2002). We examined how they would respond if the rewards were present and not monopolizable, but were unequal (Brosnan et al. 2006). In the previous exchange experiments, the monkeys' unequal rewards had been allocated by the experimenter; the partners did not affect the outcome and neither individual monkey could do anything to affect the other's reward. This is likely not the case in natural situations. Thus, this later experiment was constructed so that the subjects controlled the distribution of rewards.

This was done using a bar-pull task, which has been used extensively to test cooperation in capuchin monkeys and chimpanzees (figure 22.1). A tray with two long handles was set in front of a pair of monkeys. On the tray were two rewards, one at each end of the tray, with one for each monkey. When the subjects successfully pulled the tray toward themselves, both could access their rewards. But the tray was counterweighted such that no individual capuchin could pull it in successfully without help; only two individuals working together could do so. Sometimes both rewards on the tray were the same (equal condition) and sometimes they were different (unequal condition). Because they were spaced so far apart, each monkey could access only one of the rewards. The subjects were not separated for this task, so they had to work out for themselves which monkey would receive which reward, as determined by which handle each of them pulled. Moreover, if either individual refused to participate, the tray

could not be pulled in at all, and neither individual got a reward.

Unlike in the studies discussed above, the distribution of rewards in this situation did not affect a pair's cooperative success. However, the behavior of each partner with respect to the other did affect success (Brosnan et al. 2006). In the case of unequal rewards (unequal condition), some partners dominated the better reward, pulling the handle on that side in virtually every case. Other partners were more egalitarian, choosing the side with the better reward only about half of the time. Pairs in which the individuals shared in the better rewards over the long term (e.g., the more egalitarian pairs) were more than twice as successful at the cooperative task overall as those pairs in which one member consistently claimed the better reward. Moreover, an analysis of rewards obtained shows that the willingness to share the better reward resulted in far more rewards for both monkeys. Because cooperation succeeded so much more often, both individuals received substantially more rewards, and substantially more of the better reward. It may seem counterintuitive that individuals can increase their outcomes by sharing, but it is a robust strategy for reward maximization.

This is also likely to be the case in chimpanzees. Although the same study has not been completed with chimpanzees, work with cooperative tasks shows that they, too, are sensitive to partner behavior. They are much more likely to cooperate with individuals who tolerantly share food with them (Melis et al. 2006b) and they actively recruit tolerant partners (Melis et al. 2006a).

Responses to Being the Advantaged Partner

Finally, although none of these studies was explicitly designed to investigate how subjects respond when they are the advantaged partner, they do shed some light on that question. In the distributional inequity (exchange) studies, the advantaged partner, who received the grape, could have equalized the situation by sharing the grape. In fact neither species showed this behavior, although in a few cases capuchin monkeys did share their (less desirable) cucumber. In five instances (of 2000), chimpanzees shared their grape—however, this is much less sharing than is typical of these groups.

The cooperative bar-pull study indicates that, at least in some situations, monkeys are willing to let their partner have a good reward. In this situation, unlike in the distributional inequity task, sharing the better reward may directly impact an individual's short-term payoff, thus supporting the idea that a subject's response to inequity is tied to his or her evaluation of whether the joint outcomes accurately reflect their joint efforts. Neither of these studies, however, addresses prosocial behavior in great detail.

Experimental Approaches to Prosocial Behavior

Some cooperative behavior cannot easily be explained by mutualism, kinship, or reciprocity. My colleagues and I recently completed a series of studies explicitly designed to exclude those possibilities as much as possible. We did this by examining whether chimpanzees would bring food to unrelated members of their group when given the opportunity to do so (Silk et al. 2005; Vonk et al. 2008).

This set of studies was done with two very different populations of chimpanzees, allowing us to compare responses across different environments and social situations. One group collected data at the Michale E. Keeling Center for Comparative Medicine and Research of the University of Texas M.D. Anderson Cancer Center, in Bastrop, Texas (hereafter Bastrop) from six different corral-housed social groups of chimpanzees. These groups were all multi-male and multi-female, and many contained infants and juveniles. The chimpanzees had had very little previous experience with behavioral or cognitive testing. The sample size was sufficiently large that we were able to test 11 pairs without reusing subjects (i.e., each subject was paired with only one partner) and without testing pairs in both directions (i.e., the roles of partner and subject in each pair never switched). Thus there was no opportunity for reciprocity within the experimental situation. The second group collected data on a group of chimpanzees at the Cognitive Evolution Group of New Iberia Primate Center (hereafter NI), in Lafayette, Louisiana. This group, which has a single adult male, has experienced extensive cognitive and behavioral testing since the chimpanzees were very young. As there were only seven potential subjects, each chimpanzees was tested with each available partner. Thus, the potential for reciprocity within the experiment did exist.

In the first study, each subject was paired with a partner and given the opportunity to choose between an option that brought a piece of food only to them and an option that brought pieces of the same quantity and type of food

to both chimpanzees. At Bastrop this was accomplished with a two-tiered bar-pull. Food was placed in front of one or both chimpanzees, and one chimpanzee—the subject—could pull a handle to bring the food on that tier within reach. Subjects were positioned next to their partners, from whom they were separated by mesh fencing. At NI, subjects and their partners were seated across from one another in separate rooms, and a "scissor" apparatus simultaneously brought one option to the subject and one to the partner.

In order to verify that the subjects were not simply confused, choosing the option with two pieces of food because it contained more food (even though each subject could access only one piece), we compared their responses in sessions with a partner present to sessions with no partner present in the other cage. Critically, none of the pairs was composed of related individuals, there were no mutualistic benefits for bringing the partner a reward, and reciprocity was not possible within a session (or, in Bastrop, between sessions; see above). Thus, any subject's decision to bring rewards to the partner would represent prosocial behavior.

None of the 18 subjects varied his or her responses dependent upon whether another chimpanzee was present. In fact, most subjects chose the "prosocial" option, which brought both individuals food about half of the time, or at chance levels. Shortly after this study was published, a very similar study showed the same results for a third group of captive chimpanzees (Jensen et al. 2006). Taken together, this is very good evidence that chimpanzees are not inclined to benefit their partners in this sort of task. However, it is possible that the subjects were simply so excited by the prospect of receiving food that they made a choice for themselves without even attending to the option of rewarding a partner. To address this possibility, we ran a second study with the same subjects. Again the chimpanzees had two options, but this time they could choose one or both of them. One option brought a food reward only to the subject, and the other brought an identical food reward only to the partner. Thus, a subject could obtain a reward and then, after no longer being focused on the outcome, could also choose to obtain a reward for the partner.

At Bastrop this was done using the same apparatus as in the other test, except that in this case the subject could pull both options (in fact, at Bastrop this study was actually run first, so that subjects did not need to learn to pull both handles after having been taught to pull only one). At NI (where this study was run second), the chimpanzees were again positioned across from each other, but the apparatus used was different from before. Food rewards were placed at the top of an assembly with two ramps, each of which sloped towards one of the chimpanzees. The subject could use a tool to dislodge a food reward from the top and make it roll either only towards the subject or only towards the partner. Again, each subject was tested both adjacent to (or across from) a partner and adjacent to (or across from) an empty cage.

The majority of subjects showed no tendency to obtain a reward for their partner more often than they obtained a reward for the empty cage, thus implying again that they do not choose to reward others when given an opportunity to bring them food. However, there was one female from the Bastrop colony who chose the prosocial option at a significantly higher rate when a partner was actually present in the adjoining cage than when that cage was empty (p = 0.002), and several other subjects also showed a tendency in this direction. Moreover, in their initial session the subjects chose the prosocial option somewhat more often when a partner was present. These data certainly do not indicate a strong tendency to equalize rewards between oneself and a partner. However, they do supply evidence that in some situations, some chimpanzees may have the capacity to make prosocial choices. In which situations and with which individuals these prosocial choices may occur is still unknown, although recent studies provide evidence that begins to address this question (see below for more detail; Warneken and Tomasello 2006; Warneken et al. 2007; see also chapter 19).

Reconciling Experimental and Observational Data

There are many anecdotes of empathetic behavior in the chimpanzee and great ape literature (e.g., de Waal 1996; Goodall 1986), but no consensus on whether apes may actually have empathy. In the majority of the animal literature, "empathy" is used to denote behavior that seems to indicate the presence of the underlying psychological mechanism of empathy. However, behavior is not the same as mechanism, so I prefer the term "empathetic behavior" for denoting actions and reserve "empathy" for the psychological mechanism. Current definitions of empathy are slippery, ranging from emotional contagion to cogni-

tive empathy, and it is not always clear how empathy would manifest in behavior. However, it is generally assumed that an individual exhibits empathetic behavior if they alter their behavior in a way that affects another individual's outcome, and receives no individual gain for doing so. For instance, chimpanzees will interfere when a male's attack becomes too harsh, comfort a human caregiver, or assist another chimpanzee in a too-difficult task (de Waal 2006). Moreover, such empathetic behavior is not restricted to great apes. Mice, rats, and monkeys have all been shown to desist in behavior which caused another individual pain, even if it meant forgoing food themselves (Church 1959; Langford et al. 2006; Masserman et al. 1964). Although most of these studies did not examine the selective pressures that would have led to such behavior, the assumption is that they cannot be easily explained by individual fitness and thus represent prosocial behavior.

However, these examples are open to interpretation, which is made more treacherous by the attempt to elucidate the subjects' motivations. Consider consolation, a behavior in which chimpanzees or other animals affiliate with individuals who have recently been involved in a fight (see chapter 17). These acts of affiliation are much more common immediately after a fight than in a time-matched sample from a different day. This has been interpreted as consolation of the victim (de Waal and van Roosmalen 1979), yet there is evidence that consolation also provides solace to the consoler (Koski and Sterk 2007). Is it empathetic behavior if the consoler is also receiving benefit?

A second problem is that much interpretation depends on the definitions one uses. For instance, a behavior can be functionally defined as consolation if it relieves some of the victim's stress, whether or not the consoler also receives some benefit. Another definition, however, could rely on motivation: behavior would only be consolation if the consoler intended it to result in alleviating the victim's stress. Thus, an act that resulted in some benefit to both individuals would not be consolation from a motivational approach but would be consolation from a functional one. As nonhuman primates are not verbal, of course, it is difficult to discern their motivations.

I propose that it is possible to reconcile the two approaches. First and foremost, many situations in which apes have been argued to show empathetic behavior revolve around what has been termed helping behavior—such as rescuing birds, humans, or other chimpanzees. In experimental situations, chimpanzees will assist both humans

Figure 22.2 A juvenile chimpanzee begs for food from an unrelated adult female. Photograph by the author.

and other chimpanzees in "helping" tasks (e.g., retrieving an object that is out of another's reach; Warneken et al. 2007; see also chapters 19–21). This assistance behavior is apparently prosocial (there do not appear to be explanations based on kinship, mutualism, or reciprocity) but not very costly; thus it may be a much more likely scenario for the emergence of prosocial behavior.

Second, many observations of empathetic behavior take place in the absence of food. It may be that preferred food is such a salient cue for chimpanzees, and one that is so difficult to procure in their natural setting, that they do not readily distribute it (see figure 22.2). In the wild, examples of food-sharing between unrelated adults are relatively rare outside the context of hunting; the sharing of plant material is particularly rare (Slocombe and Newton-Fisher 2005; Hockings et al. 2007). There is also evidence that chimpanzees may share food after experiencing harassment, which perhaps leads to less sharing in situations in which harassment is not possible (Gilby 2006; Stevens 2004).

Finally, we need more data on how humans behave in similar situations in order to make valid comparisons between the two species. More data collected from controlled experiments will help to clarify the mechanisms involved in prosocial behaviors, and observations from the wild will help clarify the situations in which prosocial behaviors are likely to occur.

The Evolution of Inequity and Prosocial Behavior

Responses to inequity and prosocial behavior are inextricably linked. Both require individuals to compare their own rewards to those given to others, and to make judgments as to whether those rewards are inappropriately different. Yet there are also differences between these two kinds of responses. In the case of inequity, the response is inescapably self-oriented with many decisions made for short-term, self-benefiting reasons. In the case of prosocial behavior, the other-regarding response may ultimately provide a benefit to the self, but it also requires that individual to assess that another individual has been underbenefitted, and to make a decision to act to change *that other* individual's outcome. This decision requires the same assessment ability as in the response to inequity, but it is applied to another individual.

Both the response to inequity and the prosocial response undoubtedly evolved in the context of complex societies (these responses are not necessarily limited to great apes or even primates; any highly social species may evolve similar propensities; Range et al. 2008; see also figure 22.3). Once complex societal structures had developed, natural selection may have driven the development of behaviors that increased fitness in this new social environment which included many other individuals with whom one interacted on a regular basis. Through such mechanisms as kin selection, mutualism, and reciprocal altruism, there arose behaviors that benefited social partners—i.e., were cooperative. Prosocial behaviors are a class of cooperative behavior that is not satisfactorily explained by these existing mechanisms. It has been argued that these behaviors are special, as their evolution was driven not through individual selection but through mechanisms such as multilevel selection or gene-culture coevolution. Thus it is possible that reliance on other individuals or the group may reach the point where individuals began to perform low-cost acts of assistance, such as grooming or support in a minor (i.e., relatively risk-free) altercation. These assistance behaviors may have evolved into costly prosocial behavior.

Complex sociality is also the foundation for responses to inequity. The evolutionary origins of such responses were undoubtedly couched in cooperative interactions with other individuals, and likely proceeded through six stages of development. By judging their own outcomes against those of others and, at more complex levels, evaluating how those outcomes compare to the effort put into the interaction, individuals can achieve more efficient cooperation (van Wolkenten et al. 2007).

Initially, individuals had to notice that their rewards differed from those of others (Brosnan 2006a; Brosnan and de Waal 2004a). This required that an animal be sufficiently cognizant of its surroundings to notice that others were being rewarded better. An individual of any species that learns socially must pay attention to the rewards received by others. Given that many species of primate do learn socially (Brosnan and de Waal 2004b; chapters 8, 9,

14), we can assume that the individuals already possess the capability to recognize each other's rewards.

In the second stage, individuals had to react negatively to discrepancies between the rewards received by themselves and other individuals. This required them to notice another's rewards, process the understanding that there is a difference in quality or quantity, and react to this difference. In fact, this ability alone may have provided a fitness benefit that would drive the inequity response. If this negative response to inequity caused individuals to find new partners with whom to cooperate, they might have increased their fitness. For instance, chimpanzees cooperatively hunt and then share the prey after the kill (Boesch 1994). If one individual consistently receives little or no meat after hunts with a certain individual or individuals, they might increase their meat consumption (and fitness) by hunting instead with other individuals who may share more. Individuals need not intentionally seek out better sharers, nor understand their own motives for switching partners, for this to be fitness-enhancing. A similar mechanism may help individuals avoid partners who shirk in dangerous activities and thus increase the risk for others (Packer 1988).

The third step was to take action to rectify inequity. This is something we see in humans, and it involves paying a cost of some sort to reduce the relative level of inequity between oneself and another. In experimental games, individuals will pay to reduce another's earnings (Fehr and Gachter 2002; Zizzo and Oswald 2001). This is the most complex level of the inequity response. It requires an understanding of how one's actions will alter the partner's outcome and an ability to inhibit, as punishment in this sense requires giving up an immediate reward. While primates can inhibit their own behavior, even in the face of an immediate reward (e.g., Beran 2002), it is a difficult task.

Chimpanzees exhibit behavior consistent with the first two of these stages as discussed above, although no work has yet been done to see whether they will sacrifice gains to rectify inequity. However, capuchin monkeys, who respond similarly to chimpanzees in distributional inequity tasks, will only cooperate with a partner who shares rewards, even if both individuals will receive the same rewards on a particular trial (Brosnan et al. 2006). This response requires the individual to give up good rewards to avoid interacting with a partner, although it is unclear

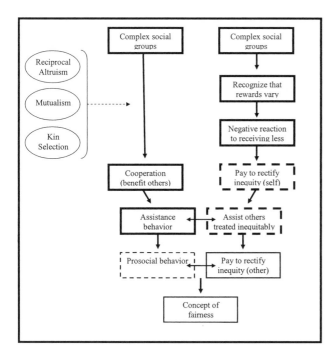

Figure 22.3   Schematic diagram of the evolution of inequity reactions and prosocial behaviors. Arrows indicate the direction of selection in the primate lineage. Ovals indicate ultimate mechanisms that lead to cooperation (behaviors not immediately explicable through individual selection). Thick lines indicate behaviors for which there is good observational and experimental support in chimpanzees; thin lines indicate behaviors for which there is neither. Dotted lines indicate circumstantial evidence, but no strong support. Horizontal arrows indicate some equivalence.

whether the individual intends to keep the partner from being rewarded or is just responding negatively to the partner. Future work may clarify whether chimpanzees behave in this way, and help to pinpoint the underlying cause of the decision.

In the fourth stage, individuals responded to assist those who had been treated inequitably—a response similar to low-cost prosocial behavior, and the place at which our two evolutionary trajectories converge. This behavior, already discussed above, may include such actions as consolation towards an individual who has been wrongly attacked by another (de Waal and van Roosmalen 1979). However, although low-cost prosocial behavior has been documented experimentally in chimpanzees (Warneken et al. 2007; Warneken and Tomasello 2006; see also chapter 19), it has not been demonstrated whether individuals specifically help others who have been treated inequitably.

The fifth stage was high-cost prosocial behavior, or material compensation for inequitable treatment. In the case of chimpanzees, this most likely would involve high-cost commodities such as food, or costly behaviors such

as loser support. Experimentally, there is little evidence for this. In general, chimpanzees do not appear to provide material benefit to each other when given the opportunity to share foods (Jensen et al. 2006; Silk et al. 2005; Vonk et al. 2008). Moreover, there is no behavioral evidence that overcompensated chimpanzees notice those instances in which they are overcompensated in a distributional inequity task (Brosnan et al. 2005), and few of the observed instances of empathic or prosocial behavior are costly (although some do meet this criterion: de Waal 2006).

The final stage is the development of a system of fairness that is outside of the individual and objectively applied to all situations. Such a "Platonic ideal" of fairness is a stage that not even many humans may reach, and there is currently no evidence that any nonhuman species has this sense.

Future Directions

The evolution of responses to inequity and prosocial behaviors is central to our understanding of cooperation and other complex social phenomena. Chimpanzees are a highly cooperative species and an understanding of how they respond to inequity, both as part of cooperative interactions and divorced from them, will help us understand more about both chimpanzees and the role of inequity responses in complex social behavior.

Comparative work typically allows for a better understanding of phenomena, and there is a need for such studies in other species of primates as well as non-primates. Other species with complex societies—for instance cetaceans, social canids, or elephants—may also respond negatively to inequity, but may also show differences from the primate data that are illuminating. Moreover, investigations into species with simpler societies or solitary species will help uncover situations in which complex sociality is not a prerequisite for responding to inequity. Such comparisons will shed greater light on the evolution of both inequity responses and chimpanzee social behavior and cognition.

Acknowledgments

I thank R. Kurzban for helpful discussion and comments on an earlier version of this chapter. Support was provided by a Human and Social Dynamics grant from the National Science Foundation (SES 0729244) and an NIH IRACDA grant to Emory University.

Literature Cited

Beran, M. J. 2002. Maintenance of self-imposed delay of gratification by four chimpanzees (*Pan troglodytes*) and an orangutan (*Pongo pygmaeus*). *Journal of General Psychology* 129(1): 49–66.

Boesch, C. 1994. Cooperative hunting in wild chimpanzees. *Animal Behavior* 48:653–67.

Bräuer, J., J. Call, and M. Tomasello. 2006. Are apes really inequity averse? *Proc. R. Soc. Lond. B* 273:3123–28.

Brosnan, S. F. 2006a. At a crossroads of disciplines. *Social Justice Research* 19:218–27.

Brosnan, S. F. 2006b. Nonhuman species' reactions to inequity and their implications for fairness. *Social Justice Research* 19:153–85.

Brosnan, S. F. 2009. Responses to inequity in nonhuman primates. In P. Glimcher, C. Camerer, E. Fehr, and R. Poldrack, eds., *Neuroeconomics: Decision Making and the Brain*. London: Elsevier, 285–302.

Brosnan, S. F., and F. B. M. de Waal. 2002. A proximate perspective on reciprocal altruism. *Human Nature* 13(1): 129–52.

———. 2003. Monkeys reject unequal pay. *Nature* 425:297–99.

———. 2004a. Reply to Henrich and Wynne. *Nature* 428:140.

———. 2004b. Socially learned preferences for differentially rewarded tokens in the brown capuchin monkey, Cebus apella. *Journal of Comparative Psychology* 118(2): 133–39.

———. 2006. Partial support from a non-replication: Comment on Roma, Silberberg, Ruggiero, and Suomi (2006). *Journal of Comparative Psychology* 120(1): 74–75.

Brosnan, S. F., C. Freeman, and FF. B. M. de Waal. 2006. Partner's behavior, not reward distribution, determines success in an unequal cooperative task in capuchin monkeys. *American Journal of Primatology* 68:713–24.

Brosnan, S. F., H. C. Schiff, and F. B. M. de Waal. 2005. Tolerance for inequity may increase with social closeness in chimpanzees. *Proc. R. Soc. Lond. B* 1560:253–58.

Church, R. M. 1959. Emotional reactions of rats to the pain of others. *Journal of Comparative and Physiological Psychology* 52:132–34.

Clark, M. S., and N. K. Grote. 2003. Close Relationships. In T. Millon and M. J. Lerner, eds., *Handbook of Psychology: Personality and Social Psychology*, vol. 5, 447–61. New York: John Wiley & Sons.

Crawford, M. 1937. The cooperative solving of problems by young chimpanzees. *Comparative Psychology Monographs* 14(2): 1–88.

De Waal, F. B. M. 1982. *Chimpanzee Politics*. Baltimore: Johns Hopkins University Press.

———. 1991. The chimpanzee's sense of social regularity and its relation to the human sense of justice. *American Behavioral Scientist* 34(3): 335–49.

———. 1996. *Good Natured: The Origins of Right and Wrong in Humans and Other Animals*. Cambridge, MA: Harvard University Press.

———. 2006. *Primates and Philosophers*. Princeton, NJ: Princeton University Press.

De Waal, F. B. M., and M. L. Berger. 2000. Payment for labour in monkeys. *Nature* 404:563.

———. 2002. Capuchin cognitive ecology: Cooperation based on projected returns. *Neuropsychologia* 1492:1–8.

De Waal, F. B. M., and A. van Roosmalen. 1979. Reconciliation and consolation among chimpanzees. *Behavioral Ecology and Sociobiology* 5:55–66.

Dindo, M., and F. B. M. de Waal. 2006. Partner effects on food consumption in brown capuchin monkeys. *American Journal of Primatology* 69:1–6.

Dubreuil, D., M. S. Gentile, and E. Visalberghi. 2006. Are capuchin monkeys (Cebus apella) inequity averse? *Proc. R. Soc. Lond. B* 273:1223–28.

Dugatkin, L. A. 1997. *Cooperation among Animals: An Evolutionary Perspective*. New York: Oxford University Press.

Eisenberg, N., and P. H. Mussen. 1989. *The Roots of Prosocial Behavior in Children*. New York: Cambridge University Press.

Fehr, E., and S. Gachter. 2002. Altruistic punishment in humans. *Nature* 415:137–40.

Fehr, E., and K. M. Schmidt. 1999. A theory of fairness, competition, and cooperation. *The Quarterly Journal of Economics* 114:817–68.

Gilby, I. C. 2006. Meat sharing among the Gombe chimpanzees: Harassment and reciprocal exchange. *Animal Behavior* 71:953–63.

Gintis, H. 2000. Strong reciprocity and human sociality. *Journal of Theoretical Biology* 206:169–79.

Goodall, J. 1986. *The Chimpanzees of Gombe*. Cambridge, MA: Belknap Press of Harvard University Press.

Hamilton, W. D. 1964a. The genetical evolution of social behavior, I & II. *Journal of Theoretical Biology* 7:1–52.

Hatfield, E., G. W. Walster, and E. Berscheid. 1978. *Equity: Theory and Research*. Boston: Allyn and Bacon.

Hockings, K. J., T. Humle, J. R. Anderson, D. Biro, C. Sousa, G. Ohashi, and T. Matsuzawa. (2007). Chimpanzees share forbidden fruit. *PLoS One* 2(9).

Jensen, K., J. Call, and M. Tomasello. 2007. Chimpanzees are rational maximizers in an Ultimatum Game. *Science* 107–9.

Jensen, K., B. Hare, J. Call, and M. Tomasello. 2006. *Proc. R. Soc. Lond. B*, published online (10.1098/rspb.2005.3417).

Koski, S., and E. H. M. Sterk. 2007. Triadic postconflict affiliation in captive chimpanzees: Does consolation console? *Animal Behaviour* 73:133–42.

Langford, D. J., S. E. Crager, Z. Shehzad, S. B. Smith, S. G. Sotocinal, J. S. Levenstadt, et al. 2006. Social modulation of pain as evidence for empathy in mice. *Science* 312:1967–70.

Masserman, J., Wechkin, M. S., and W. Terris. 1964. Altruistic behavior in rhesus monkeys. *American Journal of Psychiatry* 121:584–85.

Melis, A. P., B. Hare, and M. Tomasello. 2006a. Chimpanzees recruit the best collaborators. *Science* 311:1297–1300.

———. 2006b. Engineering cooperation in chimpanzees: Tolerance constraints on cooperation. *Animal Behavior* 72:275–86.

Mendres, K. A., and F. B. M. de Waal. 2000. Capuchins do cooperate: the advantage of an intuitive task. *Animal Behaviour* 60(4):523–29.

Mitani, J. C. 2006. Reciprocal exchange in chimpanzees and other primates. In P. Kapeller and C. P. van Schaik, eds., *Cooperation in Primates and Humans: Evolution and Mechanisms*. Berlin: Springer.

Packer, C. 1988. Constraints on the evolution of reciprocity: lessons from cooperative hunting. *Ethology and Sociobiology* 9:137–47.

Range, F., L. Horn, Z. Viranyi, and L. Huber. 2008. The absence of reward induces inequity aversion in dogs. *PNAS* 106(1):340–45.

Roma, P. G., A. Silberberg, A. M. Ruggiero, and S. J. Suomi. 2006. Capuchin monkeys, inequity aversion, and the frustration effect. *Journal of Comparative Psychology* 120(1):67–73.

Silk, J. B., S. F. Brosnan, J. Vonk, J. Henrich, D. J. Povinelli, A. S. Richardson, S. Lambeth, J. Mascaro, and S. Schapiro. 2005. Chimpanzees are indifferent to the welfare of unrelated group members. *Nature* 437:1357–59.

Slocombe, K. E., and N. E. Newton-Fisher. 2005. Fruit sharing between wild adult chimpanzees (*Pan troglodytes schweinfurthii*): A socially significant event? *American Journal of Primatology* 65(4): 385–91.

Sober, E., and D. S. Wilson. 1998. *Unto Others: The Evolution and Psychology of Unselfish Behavior*. Cambridge, MA: Harvard University Press.

Stevens, J. R. 2004. *The Selfish Nature of Generosity: Harassment and Food Sharing in Primates*. 7:451–56.

Tinklepaugh, O. L. 1928. An experimental study of representative factors in monkeys. *Journal of Comparative Psychology* 8:197–236.

Trivers, R. L. 1971. The evolution of reciprocal altrusim. *The Quarterly Review of Biology* 46:35–57.

Van Wolkenten, M., S. F. Brosnan, and F. B. M. de Waal. 2007. Inequity responses in monkeys modified by effort. *Proceedings of the National Academy of Sciences* 104:18854–59.

Vonk, J., S. F. Brosnan, J. B. Silk, J. Henrich, A. S. Richardson, S. Lambeth, S. Schapiro, D. J. Povinelli. 2008. Chimpanzees do not take advantage of very low cost opportunities to deliver food to unrelated group members. *Animal Behaviour* 75(5): 1757–70.

Warneken, F., B. Hare, A. P. Melis, D. Hanus, and M. Tomasello. 2007. Spontaneous altruism by chimpanzees and young children. *PLoS Biology* 5(7): e184.

Warneken, F., and M. Tomasello. 2006. Altruistic helping in human infants and young chimpanzees. *Science* 311:1301–03.

Zar, J. H. 1996. *Biostatistical Analysis*, 3rd Edition. Saddle River, NJ: Prentice Hall.

Zizzo, D. J., and A. Oswald. 2001. Are people willing to pay to reduce others' incomes? *Annales d'Economie et de Statistique*, 63–64:39–62.

23

The Need for a Bottom-Up Approach to Chimpanzee Cognition

Frans B. M. de Waal

In our chimpanzee colony at the Yerkes Primate Center infants sometimes get a finger stuck in the compound's fence. Their finger has been hooked the wrong way into the mesh and cannot be extracted by force. The adults have learned not to pull at the infant; victims always manage to free themselves eventually. In the meantime, however, the entire colony has become agitated: this is a dramatic event analogous to a wild chimpanzee getting caught in a poacher snare.

On several occasions, we have seen other apes mimic the victim's desperate situation. The last time, for example, I approached to assist but received threat barks from both the mother and the alpha male. As a result, I just stood next to the fence watching. One older juvenile came over to reconstruct the event. Looking me in the eyes, she inserted her finger into the mesh, slowly and deliberately hooking it around, and then pulling as if she, too, had gotten caught. Then two other juveniles did the same at a different location, pushing each other aside to get their fingers in the same tight spot they had selected for this game. These juveniles themselves may, long ago, have experienced the situation for real, but here their charade was prompted by what had happened to the infant.

I wonder where this behavior would fall under the usual classifications of imitation: no problem was being solved, no goal was being copied, and no reward was procured." (de Waal 1998, p. 689).

The field of primate cognition has a long history of going for the top. Only a few decades ago ape language research

was at the forefront: science took the most human of human characteristics—language—to see how far apes could go with it. At first, attempts to teach speech to apes failed miserably, but when researchers moved from the vocal to the gestural domain, the apes' performance suddenly exceeded expectations. Partly as a result, the way linguists defined language changed to include greater emphasis on syntax than symbolic reference.

This history contains two pertinent lessons. First, positive results often make us forget previous negative ones. Second, the human/animal divide remains a major focus outside of biology. Social scientists, psychologists, linguists, and philosophers continue to obsess over it while clinging to the illusion that one can embrace evolutionary theory without its implied continuity.

The evolutionary framework permits no saltationist arguments, however. Charles Darwin could not have been clearer: ". . . the difference in mind between man and the higher animals, great as it is, certainly is one of degree and not of kind" (Darwin 1871, p. 105). Like Darwin, I am not claiming here that humans possess absolutely no unique mental capacities; I am sure they do have such capacities, but they are merely the tip of the iceberg. We need to look at the whole "*berg*" (mountain). I would much rather see science sample the entire array of cognitive capacities—without worrying too much about which ones are uniquely human—to see how they manifest themselves across life forms. I will call this the bottom-up approach to cognition.

This bottom-up approach is sometimes written off as a pursuit of mere "analogies," since the only similarities that matter—"homologies"—are hard to prove (Povinelli 2000). The reason homologies are hard to prove, however, is that the concept does not suit the cognitive domain very well. Developed by anatomists, the homology-versus-analogy distinction remains most useful in relation to highly defined morphological traits. It has been successfully extended to well-defined behavioral displays (Lorenz 1941) as well as to the coordination of specific facial muscles (Preuschoft and van Hooff 1995), but no biologist would comfortably speak of homology in relation to something as hard to pinpoint as cognitive capacities. In this regard the fierce definitional debates in our field speak volumes. Furthermore, to establish homology requires the tracing of traits across evolutionary time, which is rather hopeless in relation to the hominoids. We don't have the twenty duck species Konrad Lorenz could work with, and are facing a seriously impoverished phylogenetic tree of which most side branches have disappeared.

Given the difficulty of establishing homology with regards to complex cognitive abilities, the safest procedure is to opt for parsimony in our assumptions. Here I do not mean the parsimony of behaviorists, and its attendant double standard with regards to human and animal cognition, but rather Darwinian parsimony. Given the recent divergence between humans and apes and the uncontestable homology between their brains, our null hypothesis ought to be that if two hominoid species solve similar problems in similar ways, the cognition behind their solutions is probably similar too. This null hypothesis can be extended to a much wider range of vertebrate species, as it would be highly uneconomic to assume that the impressive intelligence of, say, cetaceans, corvids, and primates arose independently, without reliance on a shared neural substrate. The question then becomes: What are the basic cognitive processes on which all of these life forms rely?

In this essay on bottom-up cognition, the areas to be treated are cooperation, imitation, altruism, and perspective-taking—all of which remain of interest today, as is illustrated by the contributions to this volume. I will explore how we have arrived at our current understanding of the chimpanzee mind, and how its elements can also be found outside of the hominoids. At the end, I will discuss how a top-down focus on cognitive complexity has led to all-or-nothing questions—such as whether apes have imitation, theory of mind, culture, or intentional altruism—

which do little to advance our understanding of how things actually work. Negative answers to these questions are to be treated with the utmost care, both because of the impossibility to prove the null hypothesis and because of the fact that whenever larger capacities are broken down into their constituents, many animals show some of them. An organism cannot imitate, for example, without action understanding and motor mirroring, both of which rely on neurons and brain circuits that predate the hominoids (Iacoboni 2005). A bottom-up approach is more in line with modern neuroscience an d also with evolutionary theory, according to which complex features have simple origins.

Cooperation and Reciprocity

Crawford's (1937) classic experiment in which two juvenile chimpanzees sat side by side, pulling ropes attached to a heavy box outside their cage, has been captured on black and white film.[1] The film nicely illustrates the level of understanding achieved. The apes not only pulled in synchrony to bring the food in but also watched each other carefully, activated each other with a backslap whenever one of them slowed down, and physically pushed or pulled the other to the task if he had temporarily stepped away.

Recent ape studies have more systematically tested these abilities, which indicate an understanding that a partner is needed, a preference for cooperative partners, and the facilitating effect of tolerance during the reward phase (see chapters 20 and 21). For example, bonobos seem better cooperators than chimpanzees because their well-known sociosexual ways of fostering tolerance (de Waal 1987) make them less competitive when it comes to dividing food (Hare et al. 2007).

Similar cooperation experiments with brown capuchin monkeys initially failed, leading to the conclusion that these monkeys lacked intentional cooperation (Chalmeau et al. 1997; Visalberghi et al. 2000). Subsequent adoption of Crawford's (1937) testing paradigm, however, led to immediate success with the same species. This paradigm was more intuitive: the monkeys could see how their pulling brought food closer, and could also feel the effect of their partners' pulling. As in the ape studies, capuchin monkeys seem to understand the need for a partner (that is, they wait until their partner is at the bars before they pull; Mendres and de Waal 2000), and the greater their tolerance during reward division, the more they succeed (de Waal and Davis 2003; Brosnan et al. 2006). Since capuchin monkeys are

Table 23.1 Four proximate mechanisms to explain reciprocal helping. The mechanisms are arranged from the least to the most cognitively demanding. Modified from de Waal and Brosnan (2006).

Mechanism	Catch phrase	Definition
Generalized reciprocity	"Thank goodness!"	Increased tendency to assist *any* others after having received assistance: no partner-specific contingency
Symmetry-based reciprocity	"We're buddies."	Symmetrical relationship characteristics prompt similar behavior in both dyadic directions: low degree of contingency in close relationships
Attitudinal reciprocity	"If you're nice, I'll be nice."	Parties mirror each other's social attitudes: high degree of immediate contingency
Calculated reciprocity	"What have you done for me lately?"	Scorekeeping of given and received benefits: high degree of delayed contingency

also capable of role division during cooperation (Hattori et al. 2005), it is unclear whether any major differences exist between monkeys and apes with regard to cooperation. The argument has even been made that the issue is not so much cognition but lifestyle, and that social carnivores such as spotted hyenas may in fact be better suited for cooperative tasks (Drea and Frank 2003). Many animals survive through coordinated action, so it would be strange indeed if primates were exceptional in this regard.

There is, for example, the well-documented case of joint hunting between groupers and giant moray eels in the Red Sea. Moray eels can enter crevices in the coral reef, whereas groupers hunt in open waters around the reef. Prey can escape from the grouper by hiding in a crevice and from the moray eel by leaving the reef, but it has nowhere to go if hunted by both predators simultaneously. It has been demonstrated that the two predators seek each other's company, and that groupers actively recruit moray eels through a curious head shake, to which the eels respond by leaving their crevices and joining the hunt (Bshary et al. 2006).

Since the benefits of cooperation can be learned by many species—even fish are capable of partner recruitment—any claims that cooperation has reached unique heights in humans, so that human society represents a "huge anomaly" in the animal kingdom (Fehr and Fischbauer 2003), need to be taken with a grain of salt. The common dismissal of chimpanzee cooperation as almost entirely kin-based, for example, was recently countered with actual DNA data from the field (Langergraber et al. 2007)—and it is of course a rather puzzling claim anyway given compelling evidence, both observational and experimental, for well-developed social reciprocity in this species (de Waal 1982, 1997; Koyama et al. 2006).

It has been argued that reciprocal altruism (RA) is cognitively demanding since it requires scorekeeping, and hence that it may be limited to very few species. We should remember, though, that the definition of RA postulates only that the cost of help given is offset by the benefits of future help received. Precisely how return benefits find their way back to the original altruist remains unspecified by the theory. It can be achieved in multiple ways, all of which fall under the general RA rubric. Whereas some proximate mechanisms of RA are cognitively demanding, others are not (de Waal and Brosnan 2006; see table 23.1). Thus, chimpanzees seem to show a "calculated" long-term type of RA, but the reciprocal assistance demonstrated in capuchin monkeys (de Waal and Berger 2000) and cotton-top tamarins (Hauser et al. 2003) may be of a simpler "attitudinal" type. A recent study on rats suggests an even simpler "generalized" form of reciprocity (Rutte and Taborsky 2007).

Perspective-Taking

Intersubjectivity research on primates began with Menzel (1974), who released juvenile chimpanzees into a large outdoor enclosure where food or a toy snake was hidden. Only one chimpanzee knew the food or snake's location, but the others seemed perfectly capable of "guessing" what to expect based on his or her behavior. Menzel's classic experiment, combined with Humphrey's (1978) notion of "natural psychologists" and Premack and Woodruff's (1978) "theory of mind," inspired the guesser-versus-knower paradigm still popular today in research on both apes and children.

Since then, belief in the ability of apes to take another's perspective has had its ups and downs. Failed demonstrations led some to conclude that apes must lack this particular capacity (e.g., Tomasello 1999; Povinelli 2000; see also chapter 19), but since these studies compared the apes' responses in reaction to human experimenters, the obvious problem was that the apes faced a species barrier whereas the children did not (de Waal 1996, 2001; Boesch 2007). When the human experimenter is cut out of the picture, chimpanzees seem to realize that if another individual has seen hidden food, he or she knows where it is (Hare et al. 2001).

This is not to say that human experimenters can never serve as models, but it warns that the human role in these tests may be one way to explain negative findings. Positive findings with human experimenters do not have this problem, and they have since been produced by Shillito et al. (2005) and Bräuer et al. (2005). These studies have thrown the issue of intersubjectivity wide open again. Possibly, perspective-taking and knowledge attribution are not even limited to the apes, since there is now also evidence for these capacities in monkeys (Kuroshima et al. 2003; Flombaum and Santos 2005), dogs (Virányi et al. 2005), and corvids (Bugnyar and Kotrschal 2002; Dally et al. 2006).

One of the most striking naturalistic manifestations of perspective-taking is intentional or tactical deception. A rich body of qualitative accounts has never left much doubt about the deceptive capacities of apes (de Waal 1982; Whiten and Byrne 1988), which indeed have been confirmed experimentally (Hirata 2006; Hare et al. 2006).

Imitation

Imitation used to be an uncontroversial topic, and it was assumed to be widespread in the primates (figure 23.1). Without denying the cognitive limits to ape imitation, Yerkes (1943, p. 142) asserted that "literally scores of times I have seen our subjects acquire useful acts by watching apes or men."

There are so many different ways by which one individual can come to act like another that we now face a confusing plethora of definitions of imitative processes (e.g. Whiten and Ham 1992). Regardless of how we classify

these processes, though, they must be highly developed in the primates given the abundance of population-specific traditions in the wild that likely owe their existence to social transmission (e.g. Whiten et al. 1999; van Schaik et al. 2003; Perry et al. 2003; see also chapter 14).

I proposed "identification" with the other as an important precondition for imitation (de Waal 1998). Identification entails bodily mapping the self onto the other (or the other onto the self), and as such it relates to neural capacities for shared representation that are thought to underlie imitation and empathy (Decety and Chaminade 2003). Bodily similarity—as with members of the same sex and species—and social closeness between subject and object likely enhance identification. This may, in fact, explain much of the variation in the outcome of imitation experiments. After initial skepticism about imitation in nonhuman primates, based on their failure to copy complex human actions, two kinds of studies have yielded more promising results, namely those concerning human-raised apes watching a human model (Tomasello et al. 1993; Bjorklund et al. 2000), or apes raised by their own kind watching a conspecific model (see below). In both cases, identification with and attention to the model species is facilitated by the subject's rearing history.

Identification with others seems so central that I proposed a theory of bonding- and identification-based observational learning (BIOL) (de Waal 2001). Accordingly, cultural transmission of habits and skills is based on conformity to those to whom one feels closest (see also Matsuzawa et al. 2001). Instead of being driven by reward-based learning, it may be motivated by a spontaneous desire to act like others, which then in turn may provide intrinsic rewards. Indeed, if chimpanzees are tested with members of their own species, they faithfully copy others' tool use and foraging techniques (Horner et al. 2006; Whiten et al., 2005, 2007; see also chapters 8 and 9), as well as their arbitrary means to achieve rewards (Bonnie et al. 2006). Tested with human models with whom they had previously developed close bonds, chimpanzees copied necessary actions toward a goal while ignoring unnecessary actions (Horner and Whiten 2007). In light of this role of identification, it is unsurprising that when young chimpanzees learn to use a wand to fish for ants, daughters copy their mothers more precisely than do sons (Lonsdorf et al. 2004).

Monkeys have long been thought to lack imitative capacities, but in experiments with conspecific models they

Figure 23.1 Prince Chim, a young bonobo, imitates the poise of a student. Robert Yerkes wrote that again and again Chim was seen to take a book and turn the pages carefully and neatly one by one, as if he wanted to discover what humans found so interesting about this activity. Photograph by Robert Yerkes, 1923, courtesy of the Yerkes National Primate Research Center.

do copy motor actions (Custance et al. 1999; Dindo et al. 2007; Voelkl and Huber 2007) as well as solutions to a serialization problem (Subiaul et al. 2004). In accordance with the BIOL, marmosets bias their attention in a social learning context towards close relatives (Range and Huber 2006), capuchin monkeys do not need to receive or even see any rewards to be influenced by the foraging choices of their conspecifics (Bonnie and de Waal 2006; figure 23.2), and object manipulation by ravens is biased more by watching siblings than by watching non-siblings (Schwab et al. 2007).

Not all of these studies concern imitation as currently defined, but they do suggest continuity in the social learn-ing capacities of monkeys, apes, and humans. Other animals may need to be included—such as parrots, which show remarkable bodily mimicry (Moore 1992), as well as ravens (Fritz and Kotrschal 1999) and dogs (Range et al. 2007).

Altruistic Behavior

An aging female named Peony spends her days with other chimpanzees in a large outdoor enclosure near Atlanta, Georgia. On bad days, when her arthritis is acting up, she has great trouble walking and climbing. But other females help her out. One day, for example, Peony is huffing and puffing to pull herself up into the climbing frame where

Figure 23.2   A capuchin monkey model opens one of three boxes covered with different colors and markings while the test subject, behind mesh, stands upright to get a better look. Later the subject will be presented with a rearrangement of the same three boxes. Throughout the entire test series, monkeys copied the model's choices even if none of the boxes contained any rewards (Bonnie and de Waal 2007). Drawing from video still by Frans de Waal.

several chimpanzees have gathered for a grooming session. An unrelated younger female moves behind her, places both hands on her ample behind, and with quite a bit of effort pushes Peony up to join the others.

Examples of spontaneous helping among primates are abundant (e.g., de Waal 1996, 1997). This, however, is not the impression one gets from the modern literature according to which humans are the only truly altruistic species, since animals care only about return-benefits (e.g., Fehr and Fischbacher 2003). In any case, the evolutionary reasons for altruistic behavior are not necessarily the animals' reasons. Do animals really help each other in the knowledge that it will ultimately benefit themselves? To assume so is incredibly cognitively demanding; it requires animals to have expectations about the future behavior of others and to keep track of what they did for others versus what others did for them. Thus far, there is little evidence for such expectations. Helpful acts for immediate self-gain are indeed common, but it seems safe to assume that future return-benefits remain beyond most animals' cognitive horizon.

Once evolved, behavior often operates with motivational autonomy: its motivation is relatively independent of evolutionary goals (de Waal 2008). An example is sexual behavior, which arose to serve reproduction. Since animals are, so far as we know, unaware of the link between sex and reproduction, they must be engaging in sex (as do humans much of the time) without progeny in mind. Just as sex cannot be motivated by unforeseen consequences, altruistic behavior cannot be motivated by unforeseen pay-offs such as inclusive fitness or return-benefits in the distant future.

The motivation to help must therefore stem from immediate factors, such as a sensitivity to the emotions and/or needs of others. Such sensitivity would by no means contradict self-serving reasons for the evolution of behavior, so long as it steers altruistic behavior in the direction predicted by theories of kin selection and reciprocal altruism. In humans, the most commonly assumed motivation behind altruism is empathy. We identify with another in need, pain, or distress, which induces emotional arousal that may translate into sympathy and helping (Batson 1991). Inasmuch as there are ample signs of empathy in other animals (de Waal 1996; Preston and de Waal 2002), the same hypothesis may apply. This can be tested by evaluating how animals perceive another's situation, and under which circumstances they try to ameliorate this situation.

Apart from assisting an aging female in her climbing efforts, chimpanzees occasionally perform extremely costly helping actions. For example, when a female reacts to the screams of her closest associate by defending her against a dominant male, she takes enormous risks on her behalf.

She may very well get injured. Note the following description of two longtime chimpanzee friends in a zoo colony: "Not only do they often act together against attackers, they also seek comfort and reassurance from each other. When one of them has been involved in a painful conflict, she goes to the other to be embraced. They then literally scream in each other's arms" (de Waal 1982, p. 67). This kind of cooperation, expressed in alliances and coalitions, is among the best-documented in primatology (Harcourt and de Waal 1992).

Given the observational evidence for ape altruism, one might have expected experimental confirmation by now, but initial results have been negative. Apes seem to ignore the good of others while pursuing immediate gains for themselves (Silk et al. 2005; Jensen et al. 2006). Interpreted as proof of unmitigated selfishness, the results of these studies were overinterpreted in one title as "Chimpanzees are indifferent to the welfare of unrelated group members." All that these experiments have shown, however, is that humans can create situations in which apes focus on their own interests. With regards to our own species, too, it will not be hard to create such situations. Take the way people often trample each other to get to the merchandise as soon as a department store opens its doors for a major sale. Would anyone conclude from these scenes that humans, as a species, are indifferent to each other's welfare?

The above studies have since been followed by one that set out to determine the precise circumstances under which chimpanzees are willing to assist either humans or each other. The investigators tried to rule out reciprocity by having the apes interact with humans they barely knew, and on whom they had never depended for food or other favors. They also tried to rule out the role of immediate return-benefits by manipulating the availability of rewards. The chimpanzees spontaneously assisted persons regardless of whether it yielded a reward, and were also willing to help fellow chimpanzees reach a room with food. One would think that rewards, even if not strictly necessary, at least would stimulate helping behavior, but in fact they seemed to play no role at all. Since the decision to help did not seem to be based on a cost/benefit calculation, it may have been genuinely other-oriented (Warneken et al. 2007).

A recent study demonstrated spontaneous helping in marmosets (Burkart et al. 2007), while setting it apart by stressing its "unsolicited" nature. In Warneken et al. (2007), however, chimpanzees did help each other without solicitation, and the same species is also known to spontaneously reassure distressed parties (e.g., de Waal and van Roosmalen 1979). Indeed, spontaneous helping can be expected in a wide range of species (de Waal 2008), and confirmations from monkeys are on the rise (e.g., de Waal et al. 2008; Lakshminarayanan and Santos 2008).

Trends in Cognitive Science

The top-down approach to ape cognition has brought us a focus on complex capacities and a tendency to compare degrees of complexity (e.g.: Can apes do what humans do? Are apes smarter than monkeys? How do apes compare with cetaceans or corvids?). But to establish a capacity's presence does not amount to understanding it, and to establish a capacity's absence is neither entirely possible nor very informative. Our focus should be on why things exist (ultimate explanation) and how they work (proximate explanation) rather than on where each species ranks on "the" cognitive complexity scale (which on closer inspection probably looks more like a bush than a scale).

Let me start with the issue of negative evidence. If I walk through a forest here in Georgia and fail to hear or see the pileated woodpecker, am I permitted to conclude that this bird is absent? Of course not. I may just have missed it. Even if after one year of daily walks through the forest I still have not heard or seen this bird, all I can say is that I have never seen it. Perhaps it is not around, but I cannot be sure.

What has happened to the field of primate cognition that we see so many negative claims based on a single walk through the forest? It is not hard to recognize this trend in the review above: apes cannot learn symbolic communication; monkeys fail to grasp cooperation; apes lack altruism; apes cannot learn by imitation; only humans take another's perspective, and so on. In other areas of research, not reviewed here, we find similar negative claims, such as that so-called "consolation" fails to calm others, that apes are insensitive to inequity, or that they do not comprehend the laws of physics.

The problem with negative claims is twofold. First, the old mantra "Absence of evidence is not evidence of absence" remains absolutely valid. Negative results are notoriously hard to interpret. They usually have a multitude of possible explanations, only one of which is that the sought-for capacity is absent. Second, negative claims sometimes concern behavior or capacities for which there is actually quite a bit of evidence from observations in the wild or naturalistic captive settings. Experimenters should take such information into account. If chimpanzees, for example, have a rich variety of cultural traditions in the wild, experimenters should think twice before concluding that they lack complex social learning skills. Or, if apes can invent novel tool solutions on the spot, experimenters should think twice before concluding that they have no grasp of the physical world. Claims of absence have been followed over and over by positive evidence, which generally confirms what observers have been saying all along, which is that apes do imitate, do understand cooperation, do help each other altruistically, and do adopt one another's perspective. A slight change in methodology has often sufficed to correct previous negative findings.

There surely is a place for negative findings in science (see chapter 7), but the past decade has seen far too many articles touting such findings without much accompanying skepticism. Ideally, of course, experiments should be guided by observations of naturalistic behavior, so that they address tendencies or capacities for which there is already some ecological context (see chapter 15). There is no better source of inspiration than nature (de Waal 1991; Boesch 2007). Moreover, given that social complexity similar to that of the primates is being reported for other taxa, there is every reason to expand the research on social cognition outside of the primates.

This is exactly what has been happening. Often a given capacity was first demonstrated in apes and then also found in monkeys. Research increasingly includes nonprimates as well. I have given examples above, but for reviews of advanced cognition in corvids, see Emery and Clayton (2004), and in canines, see Csányi (2000). Typical recent developments are the possibility of self-recognition in bottlenose dolphins (Reiss and Marino 2001) and Asian elephants (Plotnik et al. 2006), and consolation in rooks (Seed et al. 2007).

Let me sum up by listing three historical trends in primate cognition research:

1. Cognitive capacities first implied by the qualitative accounts ("anecdotes") of experienced observers have typically been confirmed by systematic observations and experiments.
2. The absence of a given cognitive capacity is impos-

sible to prove, and negative experimental outcomes have more often than not been followed by positive ones after modification of the experimental design.
3. Capacities assumed to be uniquely human or uniquely hominoid are generally found later, in whole or in part, in monkeys and nonprimates.

None of the above is to say that all we need to do is gather qualitative observations and leave it at that, or that negative results can be safely ignored. The problem with qualitative accounts is that they do not afford the careful comparison between alternative hypotheses that is essential to determine which cognitive processes underlie the observed behavior. Ideally, qualitative accounts are followed by systematic data on spontaneous behavior as well as by controlled experiments (de Waal 1991). We also need to heed negative findings; if they accumulate in a certain cognitive area or with regards to a certain animal family, one possible explanation is indeed that a particular cognitive capacity is little-developed or missing. At the same time we need to remain careful in our interpretations of negative results, and understand that every species needs to be tested in a way appropriate to its ecology and perceptual world. If a child and an ape do not react the same way to the same situation, that may tell us something about their respective cognitive capacities, but it may also tell us something about how differently they perceive situations that look alike to us. If this holds for the comparison between human and ape, one can imagine how much more it does for comparisons between, say, human and dog, or monkey and bird.

Bottom-Up Perspective: Implications and Future Directions

Human capacities reach dizzying heights, such as when I understand that you understand that I understand, et cetera. But we are not born with what phenomenologists call "reiterated empathy." Both developmentally and evolutionarily, advanced forms of cognition are preceded by and grow out of more elementary ones. Bottom-up accounts are the opposite of Big Bang theories. They assume continuity between past and present, child and adult, human and animal—even between humans and the earliest mammals.

This is why we would do well to pay attention to the

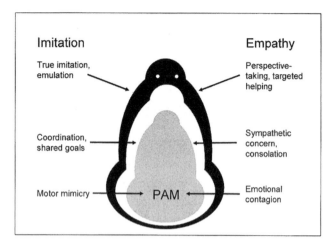

Figure 23.3 The Russian doll model of empathy and imitation. Empathy (right) induces a similar emotional state in the subject as the object, with at its core the perception-action mechanism (PAM). The doll's outer layers, such as sympathetic concern and perspective-taking, build upon this hardwired socio-affective basis. Sharing the same mechanism, the doll's imitation side (left) correlates with the empathy side. Here, the PAM underlies motor mimicry, coordination, shared goals, and true imitation. Even though the doll's outer layers depend on prefrontal functioning, they remain fundamentally linked to its inner core.

simpler, often more mundane processes underlying complex capacities. Instead of asking all-or-none questions regarding complex capacities, it would seem more fruitful to break these capacities down and study their constituent parts. Just as no animal is devoid of intelligence—all we can say is that some animals have more or less of it—there are no animals lacking all the capacities that allow for theory of mind, imitation, or intentional altruism. Many animals possess some of the constituent capacities.

For example, one of the basic capacities underlying empathy is emotional contagion, and one of the basic capacities underlying imitation is motor mimicry. If both go back to a single perception-action mechanism, as proposed by Preston and de Waal (2002), we could envision this as the core of the Russian doll around which cognitive evolution has constructed layers of ever greater complexity (de Waal 2008; figure 23.3). However advanced some of the outer layers may be, they remain connected to the doll's core and probably cannot function without it. Following this bottom-up argument, any proposed boundar-

ies will be arbitrary: the entire doll is evolutionarily and neurologically continuous.

Evolution rarely throws anything out. Instead, structures are transformed, modified, co-opted for other functions, or "tweaked" in another direction. Thus, the frontal fins of fish became the front limbs of land animals, which over time turned into hoofs, wings, and hands. This is why to the biologist, a Russian doll is such a satisfying plaything, especially if it has a historical dimension. I own a doll that shows Russian President Vladimir Putin on the outside, within which we discover, in this order, Yeltsin, Gorbachev, Brezhnev, Khrushchev, Stalin, and Lenin. So, within Putin we find a little Lenin. The same holds for all biological traits: the old always remains present in the new.

Even the most advanced empathy probably relies on emotional mechanisms already present in mice (Langford et al. 2006). The same applies to imitation, which seems neurologically linked to motor mimicry. If mirror neurons, for example, play a central role in these capacities, their presence in monkeys, apes, and humans offers more compelling evidence for homology than any cognitive analysis could potentially provide (Iacoboni 2005; Rizzolati 2005). Motor mimicry is indeed a basic capacity demonstrated in the neonates of both monkeys and apes (Ferrari et al. 2006; Bard 2007).

It seems to me that our first task is to understand the core mechanisms we share with all mammals and birds. This would be a truly evolutionary and comparative approach to primate cognition—one that does not draw lines between capacities or species, but tries to apply a unitary framework. Out of such understanding, the Russian doll may be reconstructed from the inside out, which will be far more productive than the other way around.

Acknowledgments

The author is grateful for constructive feedback on an early draft of this chapter by Nicola Clayton, Nathan Emery, Victoria Horner, and Ludwig Huber.

Literature Cited

Bard, K. A. 2007. Neonatal imitation in chimpanzees (*Pan troglodytes*) tested with two paradigms. *Animal Cognition* 10:233–42.

Batson, C. D. 1991. *The Altruism Question: Toward a Social-Psychological Answer*. Hillsdale, NJ: Erlbaum.

Bjorklund, D. F., J. M. Bering, and P. Ragan. 2000. A two-year longitudinal study of deferred imitation of object manipulation in a juvenile chimpanzee (*Pan troglodytes*) and orangutan (*Pongo pygmaeus*). *Developmental Psychobiology* 37:229–37.

Boesch, C. 2007. What makes us human (*Homo sapiens*)? The challenge of cognitive cross-species comparison. *Journal of Comparative Psychology* 121:227–40.

Bonnie, K. E., and F. B. M. de Waal. 2007. Copying without rewards: Socially influenced foraging decisions among brown capuchin monkeys. *Animal Cognition* 10:283–92.

Bonnie, K. E., V. Horner, A. Whiten, and F. B. M. de Waal. 2006. Spread of arbitrary conventions among chimpanzees: A controlled experiment. *Proceedings of the Royal Society B* 274:367–72.

Bräuer, J., J. Call, and M. Tomasello. 2005. All great ape species follow gaze to distant locations and around barriers. *Journal of Comparative Psychology* 119:145–54.

Brosnan, S. F., C. Freeman, and F. B. M. de Waal. 2006. Partner's behavior, not reward distribution, determines success in an unequal cooperative task in capuchin monkeys. *American Journal of Primatology* 68:713–24.

Bshary, R., A. Hohner, K. Ait-el-Djoudi, and H. Fricke. 2006. Interspecific communicative and coordinated hunting between groupers and giant moray eels in the Red Sea. *PLoS-Biology* 4:2393–98.

Bugnyar, T., and K. Kotrschal. 2002. Observational learning and the raiding of food caches in ravens, *Corvus corax*: Is it "tactical" deception? *Animal Behaviour* 64:185–95.

Burkart, J. M., E. Fehr, C. Efferson, and C. P. van Schaik. 2007. Other-regarding preferences in a non-human primate: Common marmosets provision food altruistically. *Proceedings of the National Academy of Sciences* 104:19762–66.

Chalmeau, R., E. Visalberghi, and A. Gallo. 1997. Capuchin monkeys (*Cebus apella*) fail to understand a cooperative task. *Animal Behaviour* 54:1215–25.

Csányi, V. 2000. *If Dogs Could Talk: Exploring the Canine Mind*. New York: North Point Press.

Custance, S., A. Whiten, and T. Fredman. 1999. Social learning of artificial fruit task in capuchin monkeys (*Cebus apella*). *Journal of Comparative Psychology* 113:13–23.

Dally, J. M., N. J. Emery, and N. S. Clayton. 2006. Food-caching western scrub-jays keep track of who was watching when. *Science* 312:1662–65.

Darwin, C. 1982 (orig. 1871). *The Descent of Man, and Selection in Relation to Sex*. Princeton, NJ: Princeton University Press.

De Waal, F. B. M. 1987. Tension regulation and nonreproductive functions of sex in captive bonobos (*Pan paniscus*). *National Geographic Research* 3:318–35.

———. 1991. Complementary methods and convergent evidence in the study of primate social cognition. *Behaviour* 118:297–320.

———. 1996. *Good Natured: The Origins of Right and Wrong in Humans and Other Animals*. Cambridge, MA: Harvard University Press.

———. 1997. *Bonobo: The Forgotten Ape*. Berkeley: University of California Press.

———. 1998. No imitation without identification. *Behavioral and Brain Sciences* 21:689.

———. 2001. *The Ape and the Sushi Master*. New York: Basic Books.

———. 2007 (orig. 1982). *Chimpanzee Politics: Power and Sex among Apes*. Baltimore: Johns Hopkins University Press.

———. 2008. Putting the altruism back into altruism: The evolution of empathy. *Annual Review of Psychology* 59:279–300.

De Waal, F. B. M., and M. L. Berger. 2000. Payment for labour in monkeys. *Nature* 404:563.

De Waal, F. B. M., and S. F. Brosnan. 2006. Simple and complex reciprocity in primates. In *Cooperation in Primates and Humans: Mechanisms and Evolution*, eds. P. M. Kappeler and C. P. van Schaik, 85–105. Berlin: Springer.

De Waal, F. B. M., and J. M. Davis. 2003. Capuchin cognitive ecology: Cooperation based on projected returns. *Neuropsychologia* 41:221–28.

De Waal, F. B. M., K. Leimgruber, and A. R. Greenberg. 2008. Giving is self-rewarding for monkeys. *Proceedings of the National Academy of Sciences, USA* 105:13685–89.

De Waal, F. B. M., and A. van Roosmalen. 1979. Reconciliation and consolation among chimpanzees. *Behavioral Ecology and Sociobiology* 5:55–66.

Decety, J., and T. Chaminade. 2003. When the self represents the other: A new cognitive neuroscience view on psychological identification. *Conscious Cognition* 12:577–96.

Dindo, M., B. Thierry, and A. Whiten. 2007. Social diffusion of novel foraging methods in brown capuchin monkeys (*Cebus apella*). *Proceedings of the Royal Society B* 275:187–93.

Drea, C. M., and L. G. Frank. 2003. The social complexity of spotted hyenas. In *Animal Social Complexity*, ed. F. B. M. de Waal and P. L. Tyack, 121–48. Cambridge, MA: Harvard University Press.

Emery, N. J., and N. S. Clayton. 2004. The mentality of crows: Convergent evolution of intelligence in corvids and apes. *Science* 306:1903–7.

Fehr, E., and U. Fischbacher. 2003. The nature of human altruism. *Nature* 425:785–91.

Ferrari, P. F., E. Visalbergi, A. Paukner, L. Gogassi, A. Ruggiero, and S. J. Suomi. 2006. Neonatal imitation in Rhesus Macaques. *PLoS-Biology* 4:1501–8.

Flombaum, J. I., and L. R. Santos. 2005. Rhesus monkeys attribute perceptions to others. *Current Biology* 15:447–52.

Fritz, J. and Kotrschal, K. 1999. Social learning in common ravens, *Corvus corax*. *Animal Behaviour* 57:785–93.

Harcourt, A. H., and F. B. M. de Waal. 1992. *Coalitions and Alliances in Humans and Other Animals*. Oxford: Oxford University Press.

Hare, B., J. Call, and M. Tomasello. 2001. Do chimpanzees know what conspecifics know? *Animal Behaviour* 61:139–51.

———. 2006. Chimpanzees deceive a human competitor by hiding. *Cognition* 101:495–514.

Hare, B., A. P. Melis, S. Hastings, V. Woods, and R. Wrangham. 2007. Tolerance allows bonobos to outperform chimpanzees on a cooperative task. *Current Biology* 17:1–5.

Hattori, Y., H. Uroshima, and K. Fujita. 2005. Cooperative problem solving by tufted capuchin monkeys (*Cebus apella*): Spontaneous division of labor, communication, and reciprocal altruism. *Journal of Comparative Psychology* 119:335–42.

Hauser, M. D., M. Chen, M.K. Chen, and E. Chuang. 2003. Give unto others: Genetically unrelated cotton-top tamarin monkeys preferentially give food to those who altruistically give food back. *Proceedings of the Royal Society B* 270:2363–70.

Hirata, S. 2006. Tactical deception and understanding of others in chimpanzees. In *Cognitive Development in Chimpanzees*, ed. T. Matsuzawa, M. Tomanaga, and M. Tanaka, 265–76. Tokyo: Springer Verlag.

Horner, V., and A. Whiten. 2007. Learning from others' mistakes? Limits on understanding a trap-tube task by young chimpanzees (*Pan troglodytes*) and children (*Homo sapiens*). *Journal of Comparative Psychology* 121:12–21.

Horner, V., A. Whiten, E. Flynn, and F. B. M. de Waal. 2006. Faithful replication of foraging techniques along cultural transmission chains by chimpanzees and children. *Proceedings of the National Academy of Sciences* 103:13878–83.

Humphrey, N. 1978. Nature's psychologists. *New Scientist* 29:900–904.

Iacoboni, M. 2005. Neural mechanisms of imitation. *Current Opinion in Neurobiology* 15:632–37.

Jensen, K., B. Hare, J. Call, and M. Tomasello. 2006. What's in it for me? Self-regard precludes altruism and spite in chimpanzees. *Proceedings of the Royal Society B* 273:1013–21.

Koyama, N. F., C. Caws, and F. Aureli. 2006. Interchange of grooming and agonistic support in chimpanzees. *International Journal of Primatology* 27:1293–1309.

Kuroshima, H., K. Fujita, I. Adachi, K. Iwata, and A. Fuyuki. 2003. A capuchin monkey (*Cebus apella*) recognizes when people do and do not know the location of food. *Animal Cognition* 6:283–91.

Lakshminarayanan, V. R., and L. R. Santos (2008). Capuchin monkeys are sensitive to others' welfare. *Current Biology* 18: R999–R1000.

Langergraber, K. E., J. C. Mitani, and L. Vigilant. 2007. The limited impact of kinship on cooperation in wild chimpanzees. *Proceedings of the National Academy of Sciences* 104:7786–90.

Langford, D. J., S. E. Crager, Z. Shehzad, S. B. Smith, S. G. Sotocinal, J. S. Levenstadt, M. L. Chanda, D. J. Levitin, and J. S. Mogil. 2006. Social modulation of pain as evidence for empathy in mice. *Science* 312:1967–70.

Lonsdorf, E. V., L. E. Eberly, and A. E. Pusey. 2004. Sex differences in learning in chimpanzees. *Nature* 428:715–16.

Lorenz, K. 1941. Vergleichende Bewegungsstudien an Anatiden. *Journal of Ornithology* 89:194–293.

Matsuzawa, T., D. Biro, T. Humle, N. Inoue-Nakamura, R. Tonooka, and G. Yamakoshi. 2001. Emergence of a culture in wild chimpanzees: Education by master-apprenticeship. In *Primate Origins of Human Cognition and Behavior*, ed. T. Matsuzawa, 557–74. Tokyo: Springer Verlag.

Mendres, K. A., and F. B. M de Waal. 2000. Capuchins do cooperate: The advantage of an intuitive task. *Animal Behaviour* 60:523–29.

Menzel, E. W. 1974. A group of young chimpanzees in a one-acre field. In *Behavior of Non-human Primates*, ed. A. M. Schrier and F. Stollnitz, 83–153. New York: Academic Press.

Moore, B. R. 1992. Avian movement imitation and a new form of mimicry: Tracing the evoluting of a complex form of learning. *Behaviour* 122:231–63.

Perry, S., et al. 2003. Social conventions in wild white-faced capuchin monkeys: Evidence for traditions in a neotropical primate. *Current Anthropology* 44:241–68.

Plotnik, J., F. B. M. de Waal, and D. Reiss. 2006. Self-recognition in an Asian elephant. *Proceedings of the National Academy of Science* 103:17053–57.

Povinelli, D. J. 2000. *Folk Physics for Apes*. Oxford: Oxford University Press.

Premack, D., and G. Woodruff. 1978. Does the chimpanzee have a theory of mind? *Behavioral and Brain Sciences* 4:515–26.

Preston, S. D., and F. B. M. de Waal. 2002. Empathy: Its ultimate and proximate bases. *Behavioral and Brain Sciences* 25:1–72.

Preuschoft, S., and J. A. R. A. M. van Hooff. 1995. Homologizing primate facial displays: A critical review of methods. *Folia primatologica* 65: 121–37.

Range, F., and L. Huber. 2007. Attention in common marmosets: implications for social-learning experiments. *Animal Behaviour* 73:1033–41.

Range, F., Z. Viranyil, and L. Huber. 2007. Selective imitation in domestic dogs. *Current Biology* 17:868–72.

Reiss, D., and L. Marino. 2001. Mirror self-recognition in the bottlenose dolphin: A case of cognitive convergence. *Proceedings of the National Academy of Sciences* 98:5937–42.

Rizzolati, G. 2005. The mirror neuron system and imitation. In *Perspectives on Imitation: From Neuroscience to Social Science*, eds. S. Hurley and N. Chater, 55–76. Cambridge, MA: MIT Press.

Rutte, C., and M. Taborsky. 2007. Generalized reciprocity in rats. *PLoS-Biology* 5:1421–25.

Schwab, C., T. Bugnyar, C. Schloegl, and K. Kotrschal. 2007. Enhanced social learning between siblings in common ravens, *Corvus corax*. *Animal Behaviour* 75:501–8.

Seed A. M., N. S. Clayton, and N. J. Emery. 2007. Post-conflict third-party affiliation by rooks (*Corvus frugilegus*). *Current Biology* 17:152–58.

Shillito, D. J., R. W. Shumaker, G. G. Gallup, and B. B. Beck. 2005. Understanding visual barriers: Evidence for Level 1 perspective taking in an orang-utan, *Pongo pygmaeus*. *Animal Behaviour* 69:679–87.

Silk, J.B., S.F. Brosnan, J. Vonk, J. Henrich, D. Povinelli, S. Lambeth, A. Richardson, J. Mascaro, and S. Shapiro. 2005. Chimpanzees are indifferent to the welfare of unrelated group members. *Nature* 437:1357–59.

Subiaul, F., J. F. Cantion, R. L. Holloway, and H. S. Terrace. 2004. Cognitive imitation in rhesus macaques. *Science* 305:407–10.

Tomasello, M. 1999. *The Cultural Origins of Human Cognition*. Cambridge, MA: Harvard University Press.

Tomasello, M., E. S. Savage-Rumbaugh, and A. C. Kruger. 1993. Imitative learning of actions on objects by children, chimpanzees, and enculturated chimpanzees. *Child Development* 64:1688–1705.

Van Schaik, C. P., et al. 2003. Orangutan cultures and the evolution of material culture. *Science* 299, 102–5.

Virányi, Z., J. Topál, Á. Miklósi, and V. Csányi. 2005. A nonverbal test of knowledge attribution: A comparative study on dogs and human infants. *Animal Cognition* 9:13–26.

Visalberghi, E., B. P. Quarantotti, and F. Tranchida. 2000. Solving a cooperation task without taking into account the partner's behavior: The case of capuchin monkeys (*Cebus apella*). *Journal of Comparative Psychology* 114:297–301.

Voelkl, B., and Huber, L. 2007. Imitation as faithful copying of a novel technique in marmoset monkeys. *PLoS ONE* 7:e611.

Warneken, F., B. Hare, A. P. Melis, D. Hanus, and M. Tomasello. 2007. Spontaneous altruism by chimpanzees and young children. *PLoS Biology* 5:e184.

Whiten, A., et al. 1999. Cultures in chimpanzees. *Nature* 399:682–85.

Whiten, A., and R. W. Byrne. 1988. Tactical deception in primates. *Behavioral and Brain Sciences* 11:233–73.

Whiten, A, and R. Ham. 1992. On the nature and evolution of imitation in the animal kingdom. In *Advances in the Study of Behavior*, vol. 21, eds. Slater et al., 239–83. San Diego: Academic Press.

Whiten, A., V. Horner, and F. B. M. de Waal. 2005. Conformity to cultural norms of tool use in chimpanzees. *Nature* 437:737–40.

Whiten, A., A. Spiteri, V. Horner, K. E. Bonnie, S. P. Lambeth, S. J. Shapiro, and F. B. M. de Waal. 2007. High-fidelity transmission of multiple traditions within and between groups of chimpanzees. *Current Biology* 17:1038–43.

Yerkes, R. M. 1943. *Chimpanzees: A Laboratory Colony*. New Haven, CT: Yale University Press.

---

[1]Yerkes National Primate Research Center Archives. Watch footage at: http://www.emory.edu/LIVING_LINKS/av/nissencrawford_cut .mov

PART V

Ethics, Care, and Conservation

24

How Cognitive Studies Help Shape Our Obligation for the Ethical Care of Chimpanzees

Stephen R. Ross

I wondered if he would remember me. Years had passed since he and I worked together on the joystick-based computer task research, and now the young chimpanzee I had known then was all grown up. The sounds coming from around the corner were big, burly, and bristling. I pulled the visor down over my face and announced my presence by calling his name: "Drew! I'm back, Drew!" Sudden silence. I opened the door to the vestibule but before I could even turn the corner, new sounds punctured the hot air. With excitement only barely restrained, sweet chirps of anticipation rose and pierced easily through five years of separation. And there he waited, his hand held aloft in exactly the place for the joystick, cued only by the sound of an old friend's voice . . . his once-valued crescents of apple lying abandoned beside him, forgotten for the moment as he greeted his friend.

Throughout this volume you have read accounts of chimpanzees in many settings engaging in cognitive tasks. For the most part, the hours of experience these chimpanzees accrued have been appropriately distilled into data that quantifies the broad-ranging mental abilities of our closest genetic relative. But what has perhaps been less evident in these accounts is the subjective experience of the chimpanzees participating in these studies. Scientists studying chimpanzees who voluntarily participate in cognitive tasks are in consensus regarding the immediate and positive impact these studies have on the chimpanzees themselves. Drew, one of the chimpanzees I worked with at the Yerkes National Primate Research Center, taught me this valuable lesson every day I worked with him on studies of computer-assisted enrichment, and especially that day five years after our final session together.

The question remains as to whether these qualitatively positive experiences are real and/or measurable. Can the cognitive tasks have effects that extend beyond the session itself and provide lasting behavioral benefits for the chimpanzee subjects? What can we learn about the needs and preferences of apes by exploring their cognitive capabilities? These and other questions seed a fertile landscape for discussion by those interested in how chimpanzee minds work and how we care for these complex animals. In this chapter I describe the many ways in which cognitive research and care of captive chimpanzees overlap. By examining how advancements in chimpanzee psychology have influenced husbandry and management and vice versa, I will argue that the link between cognitive research and captive care is not only complimentary but essential to the advancement of both fields.

Connecting Cognition and Care

The connection between cognitive research and captive care practices is a natural one. Improving the way in which animals are housed requires an appreciation and interest in the mental states of animals—a course of study known by many terms, including "comparative cognition" and "animal psychology." Working to improve captive animal care without an understanding of how animal minds work, or the scope in which they perceive and interact with their social and physical environments, is akin to drawing a map without knowledge of any landmarks or bearings. You need to know the characteristics of the destination before you can point yourself in the right direction. It's not unlike the dreaded office Christmas party tradition in which we're forced to buy presents for coworkers whom we barely know or interact with. Without an understanding of individual preferences and needs, it's nearly impossible to hit the mark. Hence the invention of the gift certificate, which provides the important concept of choice and control to the recipient—a point I will return to later.

Of course the need to look carefully at the care of all captive animals is important, but here I will focus primarily, though not exclusively, on the link between cognition and care for chimpanzees. The link between social organization and cognition is well documented and reflects the proximate and ultimate demands placed on animal minds to quantify, analyze, and react to complex social groupings. As such, we might expect that animals that have evolved in complex, fluid social structures (such as chimpanzees) may have cognitive abilities and social and environmental needs very different from those of species with more simple and stable systems. My point is that while using cognitive research to guide the improvement of captive care may be a useful tactic for a variety of species, it is especially compelling in the case of chimpanzees.

Investigations into chimpanzee psychology have been underway for about a century, and since the earliest days of those studies it has been clear that their mental abilities far surpass traditional expectations of nonhuman animals. Evidence of their capacity to reason, project, deceive, communicate, calculate, plan, and learn has grown with our aptitude to design appropriate methods with which to elucidate these abilities. With these data has come a wider understanding and appreciation of the breadth of mental states that chimpanzees possess. For instance, characterizing the emotions of chimpanzees has moved from a misguided anthropomorphic practice to a careful examination of subtle behaviors, facial expressions, and brain function (see chapter 5). As such, there is a natural connection between the study of the chimpanzees' psychological mechanisms and the growing appreciation of their emotional state.

Animals (are) clocks . . . the cries they emit when struck (are) only the noise of little springs that had been touched. . . ."
— René Descartes (1596–1650)

When the sixteenth-century French philosopher, mathematician, and scientist René Descartes claimed that the sounds emitted by animals when struck were nothing more than the mechanic springs of a clock, he was a victim of his time and place in history—without the benefit of our increased knowledge of physiology, neurology, and psychology, with which we can now confidently attribute emotional states to nonverbal species. Armed with years of data on both wild and captive chimpanzees, we not only realize the breadth of their mental abilities but might now also begin to assess the depth of their mental needs. So what do we do with that knowledge? Surely, if contemporary research gleaned that chimpanzees required a particular nutritional element to maintain their health, we would feel obligated to provide it in their diets. How, then, does an expanded understanding of their psychological requirements shape the ways in which we care for them?

Today, research on the cognitive abilities of chimpanzees and other animals is flourishing. In the first seven years of the new millennium, more than 200 papers have been published on topics ranging from associative learning to an understanding of the number zero. Even considering that many of these various studies have been conducted on the same animals, it is not unreasonable to estimate that hundreds of individual chimpanzees have participated in cognitive research in a range of settings including research centers, zoos, and sanctuaries. Yet very few of these studies take an applied perspective, examining any proximate or long-term effects on their subjects. Though the fields of cognition and captive care have been intertwined for many decades, this chapter will be among the first to formally characterize this relationship and hopefully open new avenues of investigation.

From here, I will use several historical and contemporary examples to support the statement that advances in the study of primate cognition have helped to advance the care and management of chimpanzees. Many of the modern-day principles of chimpanzee husbandry—from determining appropriate group composition to providing supplementary enrichment—have grown out of research elucidating the complexities of their cognitive abilities. Next I will discuss evidence that addresses the proximate effects of these studies on the chimpanzee subjects involved. Though clearly this is an under-studied area of investigation, there are several hints that provide support for the enrichment value of cognitive studies. Finally I will identify areas in which further cognitive research may be of greatest value in continuing to influence captive care practices. In sum, the goal of this chapter is to provide the reader with the tools with which to evaluate past, present, and future cognitive studies from the perspective of applied animal well-being, and to advocate for continued voluntary cognitive research that provides proximate and long-term benefits for the subjects themselves.

An Overlapping History

Descartes' mechanistic doctrine held that animals were simply machines while men were "machines with minds"—a view clearly influenced by his voluntaristic, Christian views of humankind. To attribute mental capabilities to animals would have threatened the traditional dualistic religious belief system, as the concept of mind was tightly bound to the theological concept of soul. Among other reasons, Descartes felt justified in denying souls to nonhuman animals because of their failure "to indicate either by voice or signs that which could be accounted for by thought and not by natural impulse" (letter to Henry More, February 1649). This clear distinction, based on the cognitive ability for language, is one that has survived several centuries later, despite the work of the Gardners, Savage-Rumbaugh, Fouts, and others in teaching various type of language (e.g., symbols, sign language) to apes. Likewise, this is perhaps one of the first attempts to link concepts of cognition (the ability to communicate) with welfare considerations (the ability to have emotions).

As public and professional opinion on animal emotions and mental states has become more informed, so too has the advocacy for progressive improvement in animal care done the same. Chimpanzees are one species that has benefited from advances in captive care, and the way in which we manage them has changed radically over the past several decades. Specifically, a greater understanding of the complex nature of wild chimpanzee groups has led to a push to promote more complex groups in captive settings.

In Africa, chimpanzees live in complex social communities that split and fuse on a regular basis—requiring individuals to remember dozens of other group members, and necessitating an ever-changing understanding of a multilayered social hierarchy (see chapters 15 and 17). The mental skills to negotiate this social landscape are not unique to chimpanzees, of course, but they are formidable, and recreating the complexity of a wild social environment has been a challenging prospect for those managing captive chimpanzees. Smaller, less species-typical groupings may be easier to manage in some ways, but they do not provide the same depth of mental stimulation that is beneficial to chimpanzees.

Providing a cognitively challenging environment for captive chimpanzees is an important goal of contemporary management programs, but the influence of cognitive science on captive care might be best illustrated in the evaluation of rearing methods. In the early and mid-twentieth century there was a high degree of interest in the effects of the early rearing environment and how it related to later development—first with monkeys (Harlow 1958) and later with chimpanzees (Menzel et al. 1963; Rogers and Davenport 1969; Davenport and Rogers 1970). Chimpanzees and other primates were raised in physically and socially restricted environments with the goal of understanding the influences of these environments on behavior and cognition. While those studies might not be conducted today, given our increased concern for individual well-being, these data have proven to be tremendously important in understanding the effects of suboptimal early environments. It is important to note that early exposure to human environments unto itself does not necessarily diminish the cognitive abilities of apes. In fact there is evidence that when apes are raised by humans, or have broad access to human objects and artifacts, some cognitive abilities may actually be improved (for review, see Call and Tomasello 1996, Tomasello and Call 2004, Furlong et al. 2008). What is clear, however, is that *impoverished* environments, such as those without appropriate access to conspecifics, can have negative impacts on cognitive development, behavior, personality, and subsequent welfare. Restrictively reared chimpanzees did not apply skills to new situations as well as chimpanzees reared in more enriched conditions (Menzel et al. 1970), and nursery-reared chimpanzees did not improve their tool-use skills with practice, as did other chimpanzees (Brent et al. 1995). Likewise, socially deprived chimpanzees were uniformly unable to use a mirror to explore body parts that would not otherwise be visually accessible to them (Gallup et al. 1971). This "mark test" is a standard cognitive test of self-awareness that normally reared chimpanzees can pass easily (Gallup 1970). These and other investigations confirm that substandard early rearing, whether by grossly restrictive means or by the once-common practice of humans raising young chimpanzees in a nursery-like setting, often has a discrete and measurable negative impact on the cognitive development of chimpanzees. These collective findings have shaped the way in which captive facilities manage their chimpanzees today. Mother-rearing is now almost universally accepted as the "gold standard" toward which managers work, and advanced husbandry practices are formed with that goal in mind. This example is perhaps the most applicable in terms of how primate cognitive research has had a distinct influence on advancing captive care practices.

Let us now consider the concept of environmental enrichment. Though the origin of the term "enrichment" is unclear, some have attributed it to Robert Yerkes, a pioneer in the study of ape cognition. A commonly used definition is "an animal husbandry principle that seeks to enhance the quality of captive animal care by identifying and providing the environmental stimuli necessary for optimal psychological and physiological well-being" (Shepherdson 1998). The important consideration here is that these additions to the environment have some effect on the mental (internal) state of the animal, which results in changes in the behavioral (external) state. Over the years, various forms of enrichment have been used by those who care for chimpanzees in captive environments. Traditionally, they can roughly be categorized into four areas: social enrichment, such as increasing the complexity of the group; feeding enrichment, such as novel foods or food presentations; manipulable enrichment, such as destructible materials and puzzles; and sensory enrichment, such as novel olfactory or auditory additions. In general, the most successful nonsocial enrichment strategies have been in the manipulable enrichment category, in which chimpanzees are provided opportunities for manual interaction with materials or equipment (Pruetz and Bloomsmith 1992).

An extension of this type of enrichment is the provision of activities that promote problem-solving and/or tool use. Morimura (2006) termed these and other mentally challenging additions "cognitive enrichment." De-

spite the early seminal work of Kohler, Yerkes, and others into chimpanzee tool-using abilities, it was not until Jane Goodall reported the first instances of wild chimpanzees making and using tools to access termite mounds at Gombe National Park in Tanzania (Goodall 1964) that a wider community began to appreciate these skills. These accounts are likely to have led to the development of the first artificial termite mounds included in captive environments as means of enrichment (Poulson 1974; Gilloux et al. 1992; Nash 1982). Japanese zoos were the first to use rudimentary but functional puzzles made simply of perforated PVC pipes filled with peanut butter in the 1960s. They were quickly followed in this practice by American and European zoos. Today, elaborate apparatuses—both functional and natural-looking—are evident in progressive institutions that seek to provide apes with challenging and rewarding tasks. Maki et al. (1989) found that such additions enhanced the psychological well-being of laboratory-housed chimpanzees by increasing their activity, decreasing abnormal behaviors, and stimulating species-typical tool use. Linick et al. (2007) found similar effects in zoo-housed chimpanzees using a naturalistic termite mound for simulation (figure 24.1).

Despite the fact that these artificial termite-mounds elicit natural, species-typical tool-use, most applied scientists would agree that stimulation rather than simulation

Figure 24.1 Chimpanzees at Lincoln Park Zoo use an artificial termite mound.

is the key concept at play here. Enrichment that mimics an activity in which wild chimpanzees engage is easy to justify simply on those grounds, but "non-natural" enrichment and activities can be equally or more stimulating. Technology-based interventions may not look on the surface like anything wild chimpanzees encounter in equatorial Africa, but may in fact provide the types of complex, multidimensional choices and opportunities to solve problems that functionally reflect species-typical cognitive processes. Likewise, the carefully controlled nature of research tasks, computer-controlled or otherwise, provides an excellent setting in which to determine the potential benefit to chimpanzees.

The Proximate Effects of Cognitive Tasks

With this long and varied relationship between cognitive research and captive care, one might assume a robust literature describing with some precision the proximate and long-term effects of various cognitive tasks. In reality, there is only a smattering of dedicated empirical accounts of the behavioral and/or psychological effects of cognitive research in the extant literature. Whether the void exists because these questions are considered so divergent from the academic focus on mental capabilities or because of a general disinterest by participating scientists is unclear—but there are good reasons to care.

First, there is increasing public concern about the care of captive animals. Public interest in issues of animal well-being is typically expressed in broad concepts of animal health, emotions, cognition, and general welfare (Coleman 2004) and recent published surveys suggest that while myriad factors may influence attitude formation on animal issues (gender, class, race, and age—for review, see Kendall et al 2006) there is a trend towards increasing concern for the treatment of captive animals (Herzog et al. 2001). These changing attitudes have resulted in the enactment of legislation such as the Animal Welfare Act, decreased reliance on animal testing of consumer products, a decline in acceptance of the fur trade, and a dramatic increase in the number of Americans who are members of animal protection organizations. Among Europeans this negative attitude toward the use of animals in research is even stronger (Pifer et al. 1994). Likewise, the general public demonstrates higher levels of concern when considering high-profile animals such as dogs and primates (Herzog

et al. 2001). Developing and promoting research that might not only measure but *improve* captive animal welfare has the potential to affect the public's perception and support of cognitive research.

Secondly, there is evidence to suggest that animals living under enriched conditions are better research models for the types of cognitive investigations discussed throughout this volume. As was discussed in the previous section, providing unnatural or inappropriate physical or social environments can have short- and long-term cognitive effects that may influence performance in cognitive tasks. Even outside the mother-rearing condition, infant chimpanzees raised in a more emotionally responsive environment exhibit differences in cognition, behavior, and personality (Bard and Gardner 1996), including longer attention spans and greater persistence in attaining goals—both of which clearly could benefit performance in cognitive tasks. In short, an enriched subject performs better, and likely gives a more accurate representation of their mental capabilities. Although some have argued that environmental enrichment causes standardization problems for animal experiments (Watanabe 2007), the benefits of working with enriched and behaviorally appropriate subjects surely outweighs these concerns.

If consensus can be reached that there are good reasons to assess the value of cognitive research, we can move on to figuring out exactly what that value might be. The study of ape cognition extends from passive observational studies of learning and culture to experimental manipulations in which apes are given tools and devices with which to make choices and solve problems. The most technological cognitive testing paradigms involve the use of computers and various input devices such as joysticks and touch screens (figure 24.2). Though exceedingly rare, there are a few studies that do address welfare issues in the context of computer-based cognitive testing. The one that most directly addresses the question does so not with with apes but with a more commonly used research subject: the rhesus macaque.

In a series of studies, Washburn and Rumbaugh (1992) attempted to assess the "enrichment value" of their Language Research Center's Computerized Test System (LRC-CTS). Though their system was designed for administering an automated battery of psychological tests to human and nonhuman primate subjects, the authors asserted that the tasks themselves were enriching to the subjects. They

Figure 24.2 Vicky, a chimpanzee housed at Lincoln Park Zoo, engages in a computer task using an infrared touchframe.

tested 10 singly caged rhesus monkeys in their home cages with 24-hour access to the tasks and apparatus. The tasks involved manipulation of electronic stimuli with the use of a joystick, with correct responses being rewarded with a fruit-flavored chow pellet. They wisely took a multifaceted approach to measuring the subjects' welfare, including health checks (weights, coat quality) and activity patterns. They found that animals chose to interact with the system approximately 40% of the time that they were available—which indicated that the subjects preferred to spend their time engaged in these activities. These activity patterns were reported to be very stable across time, with two of the original subjects performing at similar rates even five years into the study. Secondly, subjects removed from access to the system "replaced" that activity with overgrooming, shaking/biting the mesh, and stereotypic behaviors such as pacing or rocking during the time formerly spent with the system. Supplementary enrichment provided to the monkeys during non-test times did not significantly affect levels of stereotypy and self-directed behavior.

One important criticism is that perhaps the motivation to engage in these tasks is completely tied to delivery of food as reinforcement. To investigate this, the same authors tested a baseline condition, in which the monkeys had constant access to the system but were fed supplementary chow at day's end, to a similar condition in which the supplementary chow was provided continuously throughout the day. They found that the subjects' motivation to engage in the task decreased considerably (from 1,000 trials per day down to 278) and their activity patterns returned to one dominated by eating and sleeping. In many ways this is hardly surprising; even if Drew

the chimpanzee left his precious apples behind him in the hope of engaging in joystick tasks with me that morning, we needn't expect that the enrichment value of computer tasks would trump his basic need for food in the long term. So Washburn and Rumbaugh allowed their subjects to demonstrate the degree to which they valued the tasks themselves. The monkeys could select between two forms of cognitive tasks (each about 60 seconds in duration, with about five reinforcement pellets) or a "free food" condition in which there were no contingencies upon receiving the same quantity of pellets. In this condition they chose the free food about two-thirds of the time, indicating that the food itself was a critical motivation. But when the "free food" selection resulted in an extended period (30 minutes) of non-task time, that preference dropped considerably (to 14%) even though the selection provided 150 "free" pellets. Given the choice between that of 150 free pellets and 30 minutes without access to the tasks, and that working on active tasks for some or all of that same 30-minute period, the free-food option was significantly *avoided*. Evidently there is some cost-benefit ratio at work here, but the important aspect is that there is "benefit" to the access to the tasks themselves above and beyond the value of the food alone.

The Case of Choice and Control

There are several characteristics of cognitive tasks that distinguish them from typical enrichment activities. These include the provision of problem-solving and learning opportunities, and the development of associated mental skills. Ultimately, though, what differentiates most cognitive tasks from standard enrichment interventions is the interactive nature that provides elements of choice and control. There is substantial evidence that control, or more precisely the perception of control, has effects on cognitive performance, attitude, and behavior for human subjects (Burger 1975, 1987; Winocur et al. 1987). Evidence in nonhuman animals is more sparse, but the general perception of the benefit of providing choice and control pervades the animal care and husbandry industry (Markowtiz 1982; Coe et al. 2001) and some advocate that control is a key aspect of effective environmental enrichment (Snowdon and Savage 1989). The International Guidelines for the Acquisition, Care, and Breeding of Nonhuman Primates distributed by the International Primatological Society state,

"There is strong evidence that control, or the perception of control, has powerful effects on cognitive, social and emotional functioning" (IPS 2007), and the Association of Zoo and Aquariums' *Chimpanzee Care Manual* advocates opportunities for choice and control in the context of environmental niches, yard access, operant conditioning training, and other human-ape interactions—such as the choice to interface with or have privacy from zoo visitors (Ross 2010). A greater number of contemporary chimpanzee facilities are finding ways to integrate these elements into their environments. For instance, the design of the Regenstein Center for African Apes at Lincoln Park Zoo in Chicago provides multiple opportunities for choice and control. A variety of animal-activated devices allow gorillas and chimpanzees to turn on water spritzers, air fans, and heaters. Apes can also freely choose between being indoors or outdoors (at least for the nine months of temperate weather in Chicago). Evaluating the effect of providing these choices is a difficult proposition given the large number of potential confounding factors, but preliminary data suggests there are some behavioral benefits to providing these sorts of choices to the apes (videotape viewing: Bloomsmith and Lambeth 2000; outdoor access: Vreeman and Ross 2007; termite mound: Linick et al. 2007).

What about the value of choice when engaging with cognitively challenging tasks? There is some benefit to engaging in these tasks, but is providing control over access to these activities an important feature of their benefit? To address this question, Washburn et al. (1991) provided a variety of computer tasks to rhesus monkeys in two conditions: one in which the experimenters chose the specific task, and one in which the monkey subject selected the task. They found that performance was substantially better when the monkeys could make their own choices, thus suggesting considerable support for the value of control. Likewise, in another study of the effects of choice and control, eight rhesus macaques were given the choice of freely accessing their food ration from a standard food box or from a custom-made food puzzle (Reinhardt 1994). The author describes how each monkey eagerly satisfied their initial appetite by gathering some "easy" food from the standard box, but then turned their attention to the food puzzle. Overall, they spent significantly more time at the food puzzle with considerably reduced efficiency in comparison to the standard feeder (60.2 sec per biscuit

vs. 1.1 sec per biscuit). From this, the author inferred that the puzzle's challenge was of high value to the monkeys, and that the expression of natural foraging behavior might be considered its own reward. The value of choice, control, and challenge has strong evidence in its favor, at least with monkeys.

But what of apes? Is there any reason to suspect that the value of choice and control is not generalizable to these primates? Bloomsmith et al. (2000) addressed this question by investigating the enrichment value of joystick-controlled computer tasks used by captive chimpanzees. As in similar studies, these subjects were required to complete a series of simple video tasks to receive a small food reward. There were three primary applied questions: (a) Would the chimpanzee use the system at all? (b) Was there any behavioral effect or benefit of simply having access to the system? (c) Was there any behavioral effect or benefit of having control of the system?

The answer to the first question was unequivocal and unsurprising. The chimpanzees used the system at relatively stable rates both within and across sessions. In fact, usage data compared favorably to those reported in previously published studies regarding use frequency of other types of enrichment, including rolls of paper-and-rubber "Kong" toys (Pruetz and Bloomsmith 1992), and the mere presence of the system also produced some behavioral benefit simply as an environmental addition. To test the effect of control, two groups of animals were given identical access to computer displays and food reward equipment. The experimental distinction between the two groups was that only one had control of the computer task—that is, both sets of animals could see the display, and both received identical rewards when a task was completed, but only one set was in control of the task itself. The other "yoked" group was considered to have only passive involvement in the task. The results, however, were surprising: having control over the tasks did not result in any change in levels of activity, self-directed behaviors, scratching, or stereotypes. The authors concluded that although the computer system may provide value as enrichment, giving chimpanzees control over the tasks did not—in this case at least—contribute to behavioral measures of well-being. It is worth noting that in a related study of providing control over videos as enrichment, chimpanzees who had control of the video player showed lower frequencies of scratching, a behavioral indicator of anxiety, than those who lacked such control (Bloomsmith et al. 2000). Likewise, subjects who not only were provided opportunities for control but took advantage of them (by operating the video controller often) showed fewer stress-related effects during mild challenge tests (Lambeth et al. 2001). In view of these mixed results, the jury remains out on the value of choice when it comes to cognitive-based enrichment.

Potential for Negative Effects

Another perspective toward the effects of these computer-based tasks is to consider the possibility that they may be in fact *stressful* for the apes. Certainly, the training stages in which new tasks are introduced are often novel and difficult experiences for the apes. Most researchers consider these challenges to be beneficial for psychological well-being, but could they be too much? Again, the data on assessing the effect of computer tasks is relatively scarce, but Leavens et al. (2001) examined the effect of cognitive challenge on self-directed behavior by chimpanzees. Self-directed behavior is associated with frustration, uncertainty, and anxiety in social contexts in a number of species, including chimpanzees (Maestripieri et al. 1992; Baker and Aureli 1997) and humans (Fairbanks et al. 1982; Troisi et al. 1998; Waxer 1977), and it may indicate some degree of compromised well-being. In this study, chimpanzees exhibited more self-directed behavior during difficult joystick-controlled tasks, but this effect was evident only with subjects who started at a low difficulty level. Those who started on a difficult task exhibited no differential rates of self-directed behavior. Several years later, Heintz and Parr (2007) completed a preliminary study with a group of computer-experienced chimpanzees. They compared the performance, behavior, and stress hormone levels of six chimpanzees working on easy and difficult tasks using a joystick interface. The chimpanzees performed significantly better on the easy tasks, and exhibited fewer behavioral signs of anxiety (scratching and displacement behaviors) in that condition. However, the hormonal data did not match the researcher's predictions: there was no difference in salivary cortisol levels between the two conditions—a disassociation of behavioral and hormonal responses that matches similar work with orangutans (Elder and Menzel 2001). The early indication of these studies is that despite some of the aforementioned behavioral data, these computer tasks are not likely to be stressful and may provide a range of positive enrichment effects. Ultimately, this is a potentially fruitful area for future research.

The Future Interface between Cognition and Care

In the preceding sections I have described several ways in which the study of primate cognition has influenced how chimpanzees are cared for in captive environments, and I have argued that provision of cognitively stimulating situations has the potential to improve welfare measures. While there has been a smattering of reports that use cognitive and behavioral measures to assess the effects of cognitive tasks, there remain infinite opportunities for future investigation. Like cognitive science, chimpanzee husbandry is a rapidly evolving field. Just a few decades ago, the standard of keeping single or paired animals in sterile indoor environments was pervasive—but as our knowledge of chimpanzee behavior and cognition has grown, so has the movement to create more appropriate captive environments. For instance, over the past 20 years the proportion of AZA-accredited zoos with multi-male groups has tripled and the average group size has grown by over 50% (Ross 2007). Likewise, the provision of choice-based enrichment is growing, and zoos are becoming increasingly involved in quantitative cognitive research.

This recent involvement in basic research allows zoological parks to influence the public's knowledge about chimpanzees in ways that traditional research centers may not be able to do. As venues of informal education for the general public, zoos are well positioned to convey scientific and academic advances to a wider audience in ways that are more accessible than peer-reviewed journals. These opportunities, especially in the field of ape cognitive science, have the potential for widespread impact on the way in which apes are perceived and valued by the general public. For instance, exhibits such as "Think Tank" at the National Zoo in Washington, D.C., were designed to educate visitors about animal thinking, and they included live demonstrations of orangutans engaged in cognitive research using touch-screen computers. The exhibit was assessed as being able to improve visitor knowledge about animal cognition and scientific methods while also positively affecting public attitudes towards animals (Bielick and Karns 1998). Understanding the impressive scope of chimpanzees' mental abilities can lead to broad-scale impact including a more pervasive expectation of high standards of care for chimpanzees in zoos and research centers. As more zoos become engaged in basic research including ape cognition, further opportunities to influence public opinion about the interface of care and cognition will become evident.

Beyond influencing public opinion on chimpanzees, there remain myriad opportunities for cognitive research to help inform future advances in chimpanzee care and management more directly. For example, there is the case of providing appropriate environments to elderly chimpanzees. Since the inclusion of chimpanzees in the CITES act of 1975, no chimpanzees have been legally imported from range countries. Since that time, captive populations have grown or shrunk on the basis of their own breeding potential. Now, more than 30 years later, the wild-born chimpanzees are moving quickly into their twilight years—and with advances in care and veterinary sciences, captive chimpanzees may be living longer. For instance, the proportion of elderly chimpanzees in the AZA population has gone from just 9% in the early 1990s to more than 26% in 2006 (Ross 2007). Not unlike the human "baby boomer" crisis, in which the large proportion of post–World War II births will soon threaten the capacity of Medicare insurance and retirement-based institutions in the United States, this increasing proportion of aged chimpanzees is a problem that requires attention. Furthermore, chimpanzee "retirement homes"—such as the federally funded Chimp Haven outside Shreveport, Louisiana—care for more than one hundred chimpanzees. Many are quite old—the average age at Chimp Haven is 31.2 years (Brent, personal communication). Many are retired from biomedical research, and may benefit greatly from progress in the field of elderly ape cognition. A greater understanding of whether and how the aging process affects the cognitive abilities of chimpanzees could provide valuable insight into their unique needs in a captive environment.

The possibilities for cognitive research to influence progress in captive care are endless. It is not difficult to imagine how advances in the study of apes' sensory perception might affect the ways in which progressive facilities for them are built. The vast potential of effectively interfacing humans—scientists, educators, or even the general public—with chimpanzees in ways that would prove mutually enriching for all is clear; imagine the potential impact of a setup in which chimpanzees could interact with zoo visitors via touch screens. Reading through the chapters of this volume, one can easily see other possibilities for the application of scientific findings. Understanding how chimpanzees communicate (see chapter 16), vocally or otherwise, could have important implications for assessing and predicting the course of social introductions and other crucial behavioral management activities. A greater understanding of the scope and abilities of chimpanzees to use tools (see chapters 10 and 11) will likely shape the form and function of future enrichment devices. Studying how chimpanzees socially learn tasks can facilitate progressive training techniques (see chapter 25) that ultimately have positive welfare consequences.

Ultimately, the care of chimpanzees is fundamentally connected with advances in cognitive science. The way in which they and other species conceptualize their social and physical environment is perhaps the most important step toward developing appropriate policies of captive care for animals that require mental stimulation and opportunities for choice and control. But while the potential for cognitive research to produce proximate and ultimate effects on chimpanzee care is limitless, our timeframe for action should not be. With this expansive body of knowledge about the ways in which chimpanzees think, feel and, learn comes an obligation to use the information effectively. Though the captive chimpanzees described throughout this volume have had opportunities to use their impressive mental abilities to demonstrate the scope of their abilities to us, many chimpanzees do not have such a life. We must continue to use these important studies to advance the state of chimpanzee husbandry and management, and ultimately to advocate for optimum care at research centers, zoos, and sanctuaries.

As a responsible research community, we are obliged to make decisions that positively influence chimpanzee care and well-being. For example, using chimpanzees from unaccredited or questionably managed sources—such as circuses or roadside zoos—in research projects may temporarily benefit those individual chimpanzees during the testing process, but the support given to gain access to those individuals may also facilitate an industry that may not have the best interests of chimpanzees in mind. Likewise, as explained earlier in this chapter, we should be aware of and interested in the potential negative consequences of cognitive testing, and avoid assuming that these tasks are enriching or benign. Scientists should report more about the early social history and physical environments in which their subjects are raised, in order to understand whether those factors may potentially confound published research. Given our shared interests in the internal states of animals, cognitive scientists, animal managers, and animal welfarists have common ground on which to work—and through these collaborations, both the care of animals and the study of their cognition can benefit.

The connection between cognitive research and captive care has not always been overt, but it has been productive. A greater emphasis on assessment of the potential applied value of cognitive studies such as those described throughout this volume could ultimately improve the lives of thousands of chimpanzees with whom we partner. As

Jane Goodall says in her foreword to *Great Apes and Humans* (Beck et al. 2001), "We should try to get inside the mind of the individual chimpanzees, and move slowly, a step at a time, toward a solution that is *best for them*."

Acknowledgments

Many thanks to my family, friends, and colleagues who contributed to this chapter through reading, editing, and support: Megan Ross, Elizabeth Lonsdorf, Rob Shumaker, Jaine Perlman, Mollie Bloomsmith, and all the research interns of the Lester E. Fisher Center for the Study and Conservation of Apes at Lincoln Park Zoo.

Literature Cited

Baker, K. C., and F. Aureli. 1997. Behavioural indicators of anxiety: An empirical test in chimpanzees. *Behaviour* 134:1031–50.

Bard, K. A., and K. H. Gardner. 1996. Influences on development in infant chimpanzees: Enculturation, temperament and cognition. In Russon, A. E., K. A. Bard, and S. T. Parker, eds., *Reaching into Thought: The Minds of the Great Apes*, 235–56. Cambridge: Cambridge University Press.

Beck, B. B., T. S. Stoinski, M. Hutchins, T. L. Maple, B. Norton, A. Rowan, E. F. Stevens, and A. Arluke, eds. 2001. *Great Apes and Humans: The Ethics of Coexistence*, 388. Washington, DC: Smithsonian Institution Press.

Bielick, S., and D. A. Karns. 1998. *Still Thinking about Thinking: A 1997 Telephone Follow-up Study of Visitors to the Think Tank Exhibition at the National Zoological Park*. Washington, DC: Smithsonian Institution Press.

Bloomsmith, M. A., K. C. Baker, S. P. Lambeth, S. K. Ross, and S. J. Schapiro. 2000. Is giving chimpanzees control over environmental enrichment a good idea? In *The Apes: Challenges for the 21st Century Conference Proceedings*, 88–89. Chicago: Brookfield Zoo.

Bloomsmith, M. A., and S. L. Lambeth. 2000. Videotapes as enrichment for captive chimpanzees (*Pan troglodytes*). *Zoo Biology* 19:541–51.

Bloomsmith, M. A., S. K. Ross, and K. C. Baker. 2000. Control over computer-assisted enrichment for socially housed chimpanzees. *American Journal of Primatology* 51:45.

Brent, L., M. A. Bloomsmith, and S. D. Fisher. 1995. Factors determining tool-using ability in two captive chimpanzee (*Pan troglodytes*) colonies. *Primates* 36:265–74.

Burger, J. M. 1975. Causality and anticipation. *Science* 189:194–98.

———. 1987. Increased performance with increased personal control: A self-presentation interpretation. *Journal of Experimental Social Psychology* 23:350–60.

Call, J., and M. Tomasello. 1996. The effect of humans on the cognitive development of apes. In A. E. Russon, K. A. Bard, and S. T. Parker, eds., *Reaching into Thought: The Minds of the Great Apes*, 371–402. Cambridge: Cambridge University Press.

Coe, J. C. C., R. Fulk, and L. Brent. 2001. Chimpanzee facility design. In L. Brent, ed., *The Care and Management of Captive Chimpanzees*, 39–82. San Antonio: American Society of Primatologists.

Coleman, G. 2004. Public attitudes to animal research. In Cragg, P., K. Stafford, D. Love, and G. Sutherland, eds., *Lifting the Veil: Finding Common Ground: Proceedings of the Australian and New Zealand Council for the Care of Animals in Research and Teaching Conference*, 78–88. New Zealand: Australian and New Zealand Council for the Care of Animals.

Davenport, R. K., and C. M. Rogers. 1970. Differential rearing of the chimpanzee: A project survey. In G. H. Bourne, ed., *The Chimpanzee: A Series of Volumes on the Chimpanzee*, 3:337–60. Baltimore: University Park Press.

Elder, C. M., and C. R. Menzel. 2001. Dissociation of cortisol and behavioral indicators of stress in an orangutan (*Pongo pygmaeus*) during a computerized task. *Primates* 42:345–57.

Fairbanks, L. A., M. T. McGuire, and C. J. Harris. 1982. Nonverbal interaction of patients and therapists during psychiatric interviews. *Journal of Abnormal Psychology* 91:109–19.

Furlong, E. E., K. J. Boose, and S. T. Boysen. 2008. Raking it in: The impact of enculturation on chimpanzee tool use. *Animal Cognition* 11: 83–97.

Gallup, G. G. 1970. Chimpanzee: Self-recognition. *Science* 167: 86–87.

Gallup, G. G., M. K. McClure, S. D. Hill, and R. A. Bundy. 1971. Capacity for self-recognition in differentially-reared chimpanzees. *Psychological Record* 21:69–74.

Gilloux, I., J. Gumell, and D. J. Shepherdson. 1992. An enrichment device for great apes. *Animal Welfare* 1:279–89.

Goodall, J. 1964. Tool-using and aimed throwing in a community of free-living chimpanzees. *Nature* 201:1264–66.

Harlow, H. F. 1958. The nature of love. *American Psychologist* 13: 673–85.

Heintz, M. R., and L. A. Parr. 2007. Are difficult computer tasks stressful for chimpanzees? Poster presented at the Mind of the Chimpanzee Conference, Chicago.

Herzog, H. A., A. N. Rowan, and D. Kossow, 2001. Social attitudes and animals. In Salem, D. J., and A. N. Rowan, eds., *State of the Animals*, 55–69. Washington, DC: HSUS Press.

Kendall, H. A., L. M. Lobao, and J. S. Sharp. 2006. Public concern with animal well-being: Place, social structural location, and individual experience. *Rural Sociology* 71:399–428.

Lambeth, S., M. Bloomsith, K. Baker, J. Perlman, M. Hook, and

S. Schapiro. 2001. Control over videotape enrichment for socially housed chimpanzees: Subsequent challenge tests. *American Journal of Primatology* 54 (supplement 1): 62–63.

Leavens, D. A., F. Aureli, W. D. Hopkins, and C. W. Hyatt. 2001. Effects of cognitive challenge on self-directed behaviors by chimpanzees (*Pan troglodytes*). *American Journal of Primatology* 55:1–14.

Linick, S. A., S. R. Ross, and E. V. Lonsdorf. 2007. How an artificial tool-use device affects behavior in captive chimpanzees (*Pan troglodytes*) and captive gorillas (*Gorilla gorilla gorilla*). Poster presented at the Annual Meeting of the Midwest Primate Interest Group, Carbondale, Illinois.

Maestripieri, D., G. Schino, F. Aureli, and A. Troisi. 1992. A modest proposal: Displacement activities as an indicator of emotions in primates. *Animal Behaviour* 44:967–79.

Maki, S., P. L. Alford, M. A. Bloomsmith, and J. Franklin. 1989. Food puzzle device simulating termite fishing for captive chimpanzees (*Pan troglodytes*). *American Journal of Primatology* 1:71–78.

Markowitz, H. 1982. *Behavioral Enrichment in the Zoo*. New York: Van Nostrand Reinhold Company.

Menzel, E. W., R. K. Davenport, and C. M. Rogers. 1963. The effects of environmental restriction upon the chimpanzee's responsiveness to objects. *Journal of Comparative and Physiological Psychology* 56:78–85.

———. 1970. The development of tool using in wild-born and restriction-reared chimpanzees. *Folia Primatologica* 12:273–83.

Morimura, N. 2006. Cognitive enrichment in chimpanzees: An approach of welfare entailing an animal's entire resources. In T. Matsuzawa, M. Tomonaga, and M. Tanaka, eds., *Cognitive Development in Chimpanzees*, 368–91. Tokyo: Springer-Verlag.

Nash, V. J. 1982. Tool use by captive chimpanzees on an artificial termite mound. *Zoo Biology* 1:211–21.

Pifer, L., K. Shimuzu, and R. Pifer. 1994. Public attitudes toward animal research: Some international comparisons. *Society and Animals* 2:95–113.

Poulson, H. 1974. Keeping chimpanzees occupied in captivity. *International Zoo News* 21:19–20.

Pruetz, J. D., and M. A. Bloomsmith. 1992. Comparing two manipulable objects as enrichment for captive chimpanzees. *Animal Welfare* 1:127–37.

Reinhardt, V. 1994. Caged rhesus macaques voluntarily work for ordinary food. *Primates* 35:95–98.

Rollin, B. E. 2004. Annual meeting keynote address: Animal agriculture and emerging social ethics for animals. *Journal of Animal Science* 82:955–64.

Rogers, C. M., and R. K. Davenport. 1969. Effects of restricted rearing on sexual behavior of chimpanzees. *Developmental Psychology* 1:200–204.

Ross, S. R. 2007. Balancing demographic, genetic and social factors in managing the North American zoo chimpanzee population. Paper presented at the annual meeting of Support for African/Asian Great Apes (SAGA10), Tokyo.

Ross, S. R. 2010. *Chimpanzee Care Manual*. Silver Springs: Zoo and Aquarium Association.

Shepherdson, D. 1998. Tracing the path of environmental enrichment in zoos. In Shepherdson, D., J. Mellen, and M. Hutchins, eds., *Second Nature: Environmental Enrichment for Captive Animals*, 1–12. Washington, DC: Smithsonian Institution Press.

Snowdon, C. T., and A. Savage. 1989. Psychological well-being of captive primates: General considerations and examples from callitrichids. In Segal, E. F., ed., *Housing, Care and Psychological Well-being of Captive and Laboratory Primates*, 75–88. Park Ridge, NJ: Noyes Publications.

Tomasello, M., and J. Call. 2004. The role of humans in the cognitive development of apes revisited. *Animal Cognition* 7: 213–15.

Troisi, A., G. Spalletta, and A. Pasini. 1998. Non-verbal behavior deficits in schizophrenia: An ethological study of drug-free patients. *Acta Psychologia Scandinavia* 97:109–15.

Vreeman, V. M., and S. R. Ross. 2007. Do seasonal changes affect activity in captive chimpanzees (*Pan troglodytes*) and gorillas (*Gorilla gorilla gorilla*) living in a northern zoo? *American Journal of Primatology* 69:89.

Washburn, D. A., W. D. Hopkins, and D. M. Rumbaugh. 1991. Perceived control in rhesus monkeys (*Macaca mulatta*): Enhanced video-task performance. *Journal of Experimental Psychology* 17:123–29.

Washburn, D. A., and D. M. Rumbaugh. 1992. Investigations of rhesus monkey video-task performance: Evidence for enrichment. *Contemporary Topics in Laboratory Animal Science* 31:6–10.

Watanabe, S. 2007. How animal psychology contributes to animal welfare. *Applied Animal Behaviour Science* 106:193–202.

Waxer, P. H. 1977. Nonverbal cues for anxiety: An examination of emotional leakage. *Journal of Abnormal Psychology* 86:306–14.

Winocur, G., M. Moscovitch, and J. Freedman. 1987. An investigation of cognitive function in relation to psychosocial variables in institutionalized old people. *Canadian Journal of Psychology* 41:257–69.

25

Positive Reinforcement Training, Social Learning, and Chimpanzee Welfare

Jaine E. Perlman, Victoria Horner, Mollie A. Bloomsmith, Susan P. Lambeth, and Steven J. Schapiro

For the past fifteen years, Betty has observed her human care-givers cleaning the indoor section of her chimpanzee enclosure. On countless occasions she has seen how they sweep and mop the floors and hose down the walls and surfaces. Many people would therefore be unsurprised to hear that on one occasion when a caregiver accidentally left a broom in her enclosure, she immediately picked it up and started to sweep the floors, despite never having been trained to do so.

Reports of spontaneous copying by captive chimpanzees are abundant, to the extent that the word "aping" has become synonymous with "copying" in everyday language. As previous chapters in this volume attest, chimpanzees appear to have a natural ability to learn from observing others, both in captivity and in the wild. Section 2 of this volume is dedicated to culture and social learning, and it illustrates that the ability of chimpanzees to learn from watching others is believed to underlie the transmission and maintenance of cultural variation among wild populations. In fact, as other chapters in this book indicate, the more we learn about chimpanzees, the more it becomes apparent that they have sophisticated minds which are capable of tool use (see chapters 10–12), complex social organization (see chapter 28), communication (see chapters 16 and 17), cooperation (see chapters 20 and 21), learning (see chapters 8, 9, and 19), and emotion (see chapters 23 and 26). This chapter will explore whether our growing

understanding of chimpanzee cognitive abilities can be used to improve how we care for them in captivity. There is a natural link between cognitive research and the quality of care that chimpanzee participants receive, such that advances made in one field can and should inform the other (see chapter 24). We will discuss how our increased knowledge of chimpanzee learning abilities can be incorporated into training procedures to improve chimpanzee welfare and improve cognitive research. We will begin with a background on chimpanzee training.

Positive Reinforcement Training

Over the many years in which chimpanzees have been housed in captivity, humans have recognized their innate ability to learn new behaviors and often have attempted to train them to perform a broad range of them. These range from simple husbandry tasks, such as moving from one location to another to facilitate the cleaning of a zoo enclosure, to more unnatural behaviors for human amusement, such as roller-skating or riding a bicycle. The ability of chimpanzees to learn these behaviors successfully is testimony to their impressive cognitive abilities, but the techniques used to train them have not always been positive. Historically some training has taken the form of punitive techniques to gain compliance, and aversive stimuli, such as spraying individuals with a hose, were often used to get chimpanzees to perform desired behaviors. However, advances in our understanding of chimpanzee intellect and emotion have rendered these practices inappropriate and unacceptable by modern ethical standards of care, and there has been a dramatic shift towards abolishing negative training techniques in favor of those that foster a relationship of trust between the human trainer and chimpanzee trainee. In place of aversive stimuli and threats, trainers at progressive animal facilities now encourage voluntary participation in training activities, whereby desired behaviors are rewarded and undesirable behaviors are simply ignored, not punished. The trainer therefore presents chimpanzees with opportunities to cooperate and earn extra rewards, treats, and praise rather than simply try to avoid a negative consequence. This form of training is known as positive reinforcement training (henceforth PRT), and the manner in which it interfaces with what we know about chimpanzee cognition will be the focus of this chapter.

Implementing Positive Reinforcement Training

At its most basic level, PRT is a form of operant conditioning whereby the participant receives a reward, such as food (called a primary reinforcer), for performing a particular behavior. Receiving this positive reward subsequently increases the likelihood that the chimpanzee will reproduce the reinforced behavior in the future. The reward is typically paired with a clicking sound or a whistle (called a conditioned reinforcer or "bridge") which serves as a marker to pinpoint the exact behavior for which the trainer aims to increase frequency. After multiple presentations of the bridge with the reward, the sound of the clicker develops a meaning of "good job" or "exactly right" (Pryor 1984; Bloomsmith 1998). Since the general principles of PRT are species-independent, these methodologies can be used with success across many taxa. Trainers who use positive reinforcement attest that behaviors trained using this technique can be recalled by the trainees for long periods of time, often measured in years. Negative reinforcement, another form of operant conditioning in which a successful behavior is marked by the removal of a stimulus that is aversive to the trainee (such as a loud noise) can also be highly effective, but is believed to cause stress and anxiety. Trainers may need to continually remind trainees of the aversive stimulus, and often must "raise the stakes" by intensifying the aversive stimuli, since trainees become habituated to the stimulus over time. Additionally, the voluntary nature of training is removed. Presenting the trainee with an aversive stimulus to motivate them to respond to a cue removes opportunities for that individual to choose or control some aspects of their environment. But when positive rewards are presented for responding appropriately to a cue, it gives the trainee more freedom to make choices. Choice and control are factors that are thought important in promoting welfare (see chapter 24).

Negative reinforcement techniques are distinguished from the use of punishment. Punishment is applied *after* an undesirable behavior occurs, and its purpose is to decrease or eliminate the behavior. It can be applied by either removing something positive or presenting something aversive to the individual (Ramirez 1999). An acceptable form of punishment is the "time-out," whereby something positive is removed. If a trainee exhibits a behavior that is unacceptable, such as grabbing or perhaps spitting, the trainer immediately withdraws anything that is reinforcing to the subject (e.g., food or attention) for a brief time. Then the trainer returns to the session as if this had not occurred, and starts fresh. Unacceptable forms of punishment include shouting, spraying of water on the trainee, or physical force; these are *never* appropriate responses, and almost certainly act to cause stress and anxiety.

PRT has been used with animals for many years to promote husbandry and veterinary behaviors, and its roots for these types of applications are in the captive marine mammal community dating back to the early 1980s (Gail Laule, personal communication; Ramirez 1999). In an aquatic environment, captive marine mammals often have the ability to avoid negative interactions with their trainers simply by swimming away. As such, trainers found that marine mammals were far more willing to participate in training if a strong and positive bond was first established through rewarding interactions. Prior to the 1980s, positive reinforcement training with whales and dolphins had been used primarily to train "show behaviors" such as jumping out of the water though hoops. At that time, routine husbandry and veterinary procedures were complicated, cumbersome, and stressful.

Take, for example, this account of the procedure used to acquire a blood sample from a dolphin at Marinepark in Palos Verdes, California in 1980 (Gail Laule, personal communication): "First the tank was drained to about 4 feet of water. Next, five to six staff entered the tank and corralled the individual and attempted to restrain the dolphin in their arms. Not an easy feat. Once the individual was restrained, the sample was taken, the dolphin was released, and the tank was refilled." Enormous resources and effort were often expended, and it was likely a stressful experience for the dolphin, as well as for other animals who shared the tank. Of course this example is not specific to Marinepark, but is likely indicative of that period in marine mammal management history (Gail Laule, personal communication). When PRT was incorporated into veterinary procedures at Marinepark in 1982, this same dolphin was trained to participate in the sample collection voluntarily. This was accomplished by gradually reinforcing her desired behaviors and ignoring her undesired ones. Following a simple hand cue, the dolphin learned to swim up next to a platform, turn over to present the ventral side of her tail fluke, and allow the trainer to hold the tail in position while a blood sample was taken. The dolphin could choose to swim away at any time, but instead she remained to allow her blood to be drawn and was rewarded afterwards with fish and positive human attention.

Despite these successes with marine mammals, it took several years for the use of PRT for veterinary and husbandry management to reach facilities housing terrestrial species. The literature reflects this trend starting in the late 1980s, with PRT being used to train primates, elephants, big cats, and other exotic animals in order to facilitate routine management such as feeding, cleaning, enrichment applications, reproduction, and veterinary care (Mellen and Ellis 1996). In the early 1990s, increasing numbers of animals were trained to show body parts for examination, to voluntarily move where they were needed to move, to allow the collection of biological samples (e.g., blood, urine, semen, saliva), to use enrichment devices, to behave better in group social interactions, and to do a host of other things (see Ramirez 1999). Moving away from a more aversive system of managing animals, PRT is now increasingly common throughout the zoological, animal research, and sanctuary communities and is generally considered necessary to the care and management of chimpanzees. Its methodologies are being used in a variety of captive environments to gain chimpanzees' participation in routine behaviors (Laule and Whittaker 2001) that are a vital part of their care and likely have positive consequences for their individual welfare. For example, PRT has been used in daily colony management, such as in encouraging chimpanzees to move within enclosures (Bloomsmith

et al. 1998). It has been used to improve the social dynamics within chimpanzee groups by reducing their competition and aggression during feeding times (Bloomsmith et al. 1994). It has also been employed to facilitate veterinary procedures, such as intramuscular injection for annual physical exams or the administration of antibiotics (Videan et al. 2005; Schapiro et al. 2005; Lambeth et al. 2006; and Russell et al. 2006), subcutaneous injection for the care of diabetes or research (Perlman et al. 2004; Schapiro et al. 2005), semen collection for reproductive evaluations (Perlman et al. 2003; Schapiro et al. 2005), urine collection for the monitoring of hormone levels (Stone et al. 1994; Laule et al. 1996), and presentation of an arm for voluntary and conscious blood collection (Laule et al. 1996; Lambeth et al. 2005; Coleman et al. 2008). The application of PRT makes it possible for trainees to remain in their home enclosure while providing samples, and it has reduced the need to anesthetize animals for simple veterinary procedures.

How Positive Reinforcement Training Influences Welfare

There is growing interest in determining whether PRT techniques improve the welfare of captive animals. Although defining and measuring animal welfare is difficult, most scientists agree that multiple measures should be applied to best capture the broad range of elements involved. These include physical health, behavioral and physiological evidence of stress, presence of species-typical behaviors, absence of abnormal behaviors, competence in responding to challenges, and access to choice and control (Meehan and Mench 2007; Novak and Suomi 1988; Swaisgood 2007, Markowitz 1982; Snowdon and Savage 1989; see also chapter 24). Several of these measures have been used in evaluations of the influence of PRT on chimpanzee welfare. Positive impacts on chimpanzee physiology have recently been demonstrated (Schapiro et al. 2007), and Lambeth et al. (2006) found that training chimpanzees for voluntary presentation for intramuscular injections of anesthesia significantly reduced some measures correlated with stress as compared to chimpanzees who did not voluntarily accept the anesthesia. Similarly, a reduction in stress indicators was reported when PRT was used to train chimpanzees to present an arm for blood collection (Lambeth et al. 2005). These studies demonstrate that PRT can elicit quantifiable reductions in physiological indicators of stress, and can thus ultimately improve individual welfare measures.

PRT has also been evaluated as a form of enrichment for chimpanzees. Individuals who participated in PRT were found to spend less time inactive or engaged in solitary activities during training sessions. They also showed improvements in their social interaction outside of training sessions, such as increased play behavior (Bloomsmith et al. 1993, 1997, 1999; Laule and Whittaker 2007). Similar findings have been reported with other nonhuman primate species as well. Desmond et al. (1987) successfully trained a group of drill monkeys to socialize with one another, and this trained behavior generalized to an increase in other affiliative behaviors, including reproductive behaviors, during their everyday interactions (Cox 1987; Desmond and Laule 1994). Schapiro and colleagues (2001) applied PRT to increase the affiliative behavior of group-housed rhesus macaques; they reported that individuals with very low rates of social behavior, who were trained to exhibit more of it, also showed increased sociality outside of their training sessions. Like more traditional enrichment interventions such as toys, feeders, and puzzles, these studies indicate that PRT can have a quantifiable positive impact on behavioral measures of well-being.

Another important element in increasing the welfare of captive animals is giving them opportunities for choice and the ability to control features of their environment (Markowitz 1982; Snowdon and Savage 1989; see also chapter 24). PRT is one way to do this, because it enables trainees to learn about the consequences of their actions and it relies on their voluntary participation (Bassett and Buchanan-Smith 2007). For example, if a chimpanzee has been trained to move to the outside portion of his enclosure upon hearing a verbal cue, then he chooses to respond to the request and thus controls the frequency of his reinforcement. Note that there is no negative consequence for choosing not to comply with the trainer's request. Some have even proposed that the ability to exert control over one's environment may in itself be enough of a reward to motivate participation in a PRT session without the need for further reinforcement (Laule and Desmond 1998).

Finally, encouraging behavioral diversity and reinforcing persistent problem-solving behavior are other ways in which positive reinforcement training may enhance chimpanzees' welfare. A trainer who strictly follows the tenets of PRT gives no negative consequences in response to any behavior. Some trainers report that this leads to increased behavioral experimentation and novel behavior in trainees

(Laule and Desmond 1998). Experienced trainers may reinforce animals not only for overt responses to commands, but also for less tangible actions such as problem-solving, offering creative solutions, or just trying hard (Laule and Desmond 1998).

Given all these benefits of PRT, it should come as no surprise that captive facilities are developing sustainable training programs that can be incorporated into their routine management practices. Bloomsmith and Else (2005) found that five of the six major research centers housing chimpanzees in the United States had implemented formal PRT programs to gain the cooperation of chimpanzees for routine management, veterinary, and research procedures. By 2006, all six institutions were expending monetary and personnel effort on chimpanzee training programs that are now maturing and being evaluated (Bloomsmith et al. 2006). Zoos are also rapidly developing PRT animal management systems, and there has been interest in developing training programs across and within institutions (Bloomsmith and Else 2005; Whittaker et al. 2007; McMillan et al. 2007) with more focus on determining the benefits and drawbacks of various approaches to their organizational structure (McMillan et al. 2007; Whittaker et al. 2007). Furthermore, training is being accomplished in more challenging circumstances, such as with chimpanzees while they remain in their social groups (Perlman et al. 2001; Laule and Whittaker 2007). Removing the need to physically separate individuals for training enables them to stay in their normal social environment and renders them likely to be more relaxed and motivated to participate. In sum, several lines of evidence suggest that PRT practices can increase the welfare of captive chimpanzees by reducing their stress, increasing their choice and control, and increasing their freedom to behave spontaneously, creatively, and—most important—representatively. With this knowledge that PRT can have positive consequences for the welfare of chimpanzees, we move now to a discussion of how these animals' advanced cognitive abilities, specifically their ability to learn through observation, facilitates such training techniques.

Positive Reinforcement Training and Social Learning

Studies of wild chimpanzees indicate that social learning plays an important role in their acquisition of new behavior. Indeed, previous chapters in this volume are dedicated to documenting and investigating the extent of their social learning abilities, and how these abilities may lead to cultural variation among populations (see section 2). Cognitive studies have shown that captive chimpanzees can employ a suite of different social learning mechanisms, including both simple and cognitively complex processes. To date, more than 30 such studies have been conducted (reviewed in Whiten et al. 2004), and while researchers may disagree about the exact mechanism responsible for the acquisition of new skills, these studies demonstrate the robust nature of chimpanzee social learning abilities. Previous researchers have hypothesized that these skills could be used to aid in reintroduction projects (Box 1991; Beck 1997; Custance et al. 2002; see also chapter 26). Primates may be able to learn adaptive behaviors such as predator avoidance by observing the avoidance behavior of knowledgeable conspecifics (Custance et al. 2002). Here we propose that a similar argument can be made for the application of social learning to aid in PRT and improve both captive care and cognitive research methodologies.

It is not uncommon that during training sessions, chimpanzees often attend to the human trainer even when they are not directly being trained themselves. What would they gain from observing but not directly interacting or receiving reinforcement? Trainers report that some chimpanzees who choose to stay and watch other animals being trained later respond more quickly in their own training, thereby maximizing the rewards they can receive during each session. This suggests that chimpanzees can benefit from watching other chimpanzees being trained (figure 25.1). This observation shares some similarities with the methodology used in group diffusion studies of cultural transmission, in which chimpanzees in a social group acquire a new skill by watching the actions of a trained conspecific (see chapters 8 and 9). It is also similar to a study by Matsuzawa (2002) in which a young chimpanzee, Ayumu, learned to interact with a touch-screen computer by observing his mother's training sessions.

Making use of chimpanzees' social learning skills may improve their training efficiency while providing them with behavioral enrichment and social stimulation. Trainees are likely to be more motivated to participate in training if they see other chimpanzees interacting positively with the trainer. Via simple forms of learning such as enhancement, in which the behavior of an individual draws

(A)

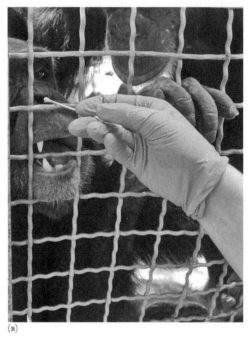

(B)

Figure 25.1 (a) Several chimpanzees from the FS2 social group at Yerkes National Primate Research Center observe a PRT session in which Amos (second from left) is being trained to exchange plastic "tokens" for food rewards in preparation for a cognitive study at the Living Links Center. Training takes place in their outdoor play area (520 m²) and normal social environment. (b) Several chimpanzees at the Michale E. Keeling Center for Comparative Medicine and Research at the University of Texas look on as Jake participates in target training to present his nose for visual inspection in their outdoor play area and normal social environment.

the attention of another to a stimulus or location (Spence 1937; Thorpe 1956), they also may be more likely to attend to the relevant location or apparatus being used in training, and thus experience more opportunities for learning than they would if they were trained alone.

Enhancement may also reduce chimpanzees' fear of novel stimuli or apparatus; trainees might respond more positively to such stimuli after seeing a conspecific respond to it positively. More cognitively complex forms of social learning might also be employed to produce specific results, such as teaching the animal to urinate in a collection container (goal emulation, cf. Whiten and Ham 1992) or present body parts for health assessments (imitation, cf. Custance et al. 1995). In some instances in the training of a complex behavior, its final form may not be obvious to the chimpanzee, and so training must be achieved by reinforcing successive approximations to the final form. However, giving trainees a chance to see the final behavior may enable them to use their social learning skills to reproduce the behavior, or achieve something close to it, like participants in a standard social-learning study (e.g., Whiten et al. 2005; Horner et al. 2006). The behavior could then be refined with PRT techniques ("shaping"). Thus the chimpanzees could speed through some steps in the training process and acquire new skills more efficiently.

Three Methodologies that Combine Social Learning and Positive Reinforcement Training

How might trainers apply what we know about the social learning abilities of chimpanzees? The first step is to provide trainees with opportunities to see the target behavior in some form. While chimpanzees are able to learn accurately from humans in some cases (e.g., Hayes and Hayes 1952; Tomasello et al. 1993; Custance et al. 1995; Horner and Whiten 2005), they are believed to learn more readily and spontaneously from conspecifics (Sumita et al. 1985; de Waal 2001; Whiten et al. 2005; Boesch 2007; de Waal et al. 2008). This may be particularly important if the desired behavior involves novel equipment or apparatus, because seeing a fellow chimpanzee interact positively with the apparatus is likely to reduce neophobia and the time spent training to overcome it, thus expediting acquisition of the new skill. In this section we will describe three methodologies that incorporate social learning into PRT techniques, although we acknowledge that there are likely to be many more. Across each of the three methodologies there is a reduction in social interaction between demonstrator and observer(s), thus decreasing opportunity for the close observation and physical social interaction that are thought to be important in creating a relaxed social learning environment (van Schaik 2003; see also chapter 9). Nevertheless, reduced social interaction decreases the po-

tential for negative social contact, such as aggression and competition for trainer attention. The most appropriate training methodology will likely be dictated by the quality of the relationship between the chimpanzees being trained, the nature of the target behavior (i.e., is close observation required?), and the physical layout of enclosures (i.e., can chimpanzees from different groups observe one another?). Therefore, it should be determined on a case-by-case basis.

Direct Interaction with a Trained Conspecific

In this case, a naïve chimpanzee observes and is given the opportunity to learn from a familiar conspecific with experience in the target behavior. For instance, a chimpanzee might learn to open a puzzle box by observing and interacting with a group member experienced in opening it. This provides social interaction with the demonstrator and gives the trainee a chance to witness the behavior at close range, examine it closely, and actually manipulate the apparatus, potentially learning the most about how to execute the behavior successfully. The most critical aspect is likely to be the quality of the social relationship between demonstrator and trainee. Chimpanzees are known to be selective about whom they learn from (Menzel 1973; Sumita et al. 1985, Horner et al. 2006), with factors such as age, sex, rank, and interpersonal relationships influencing whether an individual's behavior will be copied by others (Coussi-Korbel and Fragaszy 1985; Boesch and Tomasello 1998). For example, Horner et al. (2006) report that in a study involving dyadic learning between pairs of chimpanzees, certain individuals refused to participate if paired with a chimpanzee with whom they did not have a strong affiliative relationship, but later learned accurately when paired with a different social model (see chapter 9 for more details). Social tolerance between the demonstrator and the trainee is believed to influence the probability of learning because it allows close observation of a behavior and at the same time creates a relaxed learning environment without fear of aggression (van Schaik 2003). In this case, trainees could either observe a fellow group mate being trained (cf. Bloomsmith et al. 1999), or watch an already competent conspecific demonstrator (cf. Whiten et al. 2005). The time invested in training the original demonstrator using standard PRT could pay off in time saved from having to train the observing group members (who might thus be able to move more quickly through some training steps; see above).

Distanced Observation of a Conspecific

In some cases chimpanzees may not be able to be housed together and direct interaction with an experienced conspecific may be impossible. Whiten et al. (2007) have recently shown, however, that captive chimpanzees can accurately learn a novel behavior by observing the actions of chimpanzees in a neighboring group without physical interaction (see chapter 8). Since this methodology does not require contact between individuals, the problem of social incompatibility is reduced. This technique is likely to be more effective for behaviors that do not require close observation or detailed object manipulation, such as learning to reliably move to large targets. Depending on the design of enclosures, it may be possible to train one chimpanzee who is then visible to several other groups.

Video Presentation of Target Behavior

Previous studies have demonstrated that chimpanzees attend to videotapes played on a television monitor (Bloomsmith and Lambeth 2000), and that they and other primates can gain useful information via representational stimuli (Nagell et al. 1993; Kuhlmeier et al. 1999; Cook and Mineka 1989; Morimura and Matsuzawa 2001; Price and Caldwell 2007). It therefore seems that trainees might gain useful information about a target behavior from video just as they might by watching a live demonstration (methodologies 1 and 2), but with several key advantages. Video enables multiple novices to observe the same exemplar of the target behavior, making it possible for individual differences in learning ability to be assessed (Price and Caldwell 2007). It allows video stimuli to be shared among facilities and reduces the potential for negative social interactions (videos can be edited to remove dominance displays, threats, or gestures that might intimidate the observers). Finally, it allows rare but desirable behaviors to be presented repeatedly to novices without having to train a competent demonstrator for every social group. Below we describe one case study that was conducted to

investigate whether video presentation of a specific behavior could reduce the time needed to train chimpanzees to perform it.

CASE STUDY: CAN VIDEO PRESENTATION OF A TARGET BEHAVIOR EXPEDITE POSITIVE REINFORCEMENT TRAINING? Previous studies have shown that PRT can be used successfully to train chimpanzees to urinate into a container so that their hormone levels can be assessed and monitored (Stone et al. 1994; Laule et al. 1996). This study was conducted to determine whether training time for this task could be reduced by showing chimpanzees a video presentation of the target behavior before their training session. Participants were fifteen adult female chimpanzees living at the Michael E. Keeling Center for Comparative Medicine and Research (KCCMR), Department of Veterinary Sciences at the University of Texas M. D. Anderson Cancer Center. Eight of the chimpanzees participated as experimental subjects, and seven as control subjects. The experimental subjects were shown 10 minutes of video that showed female chimpanzees successfully urinating into a collection cup and receiving a food reward. This was followed immediately by a 15-minute PRT session in which the subjects were trained to perform the same behavior. The control subjects, on the other hand, received only PRT and were not shown the video presentation (figure 25.2). All subjects received up to 14 training sessions during the study period for the target behavior. Two dependent measures were calculated: time until first successful performance of the target behavior, and time taken to perform the target behavior reliably (urinating into the receptacle four out of five times).

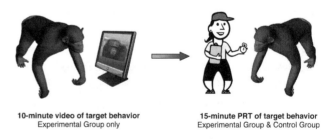

10-minute video of target behavior
Experimental Group only

15-minute PRT of target behavior
Experimental Group & Control Group

Figure 25.2 Schematic representation of the procedure for experiment 1. Chimpanzees in the experimental group observed a 10-minute video presentation of the target behavior, followed immediately by a 15-minute PRT session to train the same behavior. Chimpanzees from the control group received only the PRT session. Chimpanzee images by Devyn Carter.

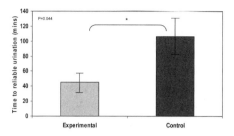

Figure 25.3 Experimental subjects achieved reliability (successful completion of the behavior four out of five times) to urinate into the receptacle significantly faster than subjects in the control group.

During the study period, only 64% of the subjects urinated into the receptacle. The experimental subjects did so significantly faster than the control subjects (mean experimental = 10 minutes, mean control = 62.5 minutes; df = 1, 8; F = 18.657; p = 0.003). Fifty-three percent of the participants (five from the experimental group and three from the control group) learned to perform the target behavior reliably within the study period. Those from the experimental group reached reliable performance (success four out of five times) significantly faster than those from the control group (mean experimental = 46 minutes, mean control = 106 minutes; df = 1, 6; F = 6.434; p = 0.044; see figure 25.3). The range of training time to urinate reliably into the receptacle was 17 to 70 minutes for the experimental group, compared to 66 to 157 minutes for the control group.

These data suggest that the pace of learning through PRT can be accelerated by providing an opportunity for social learning through a video presentation that presents a target behavior. While subjects from the experimental group were exposed to relevant video footage of the training task, the control group was not exposed to any video footage. To control this variable, a follow-up study is now in progress in which both experimental and control groups have access to video footage. The experimental group is exposed to relevant video footage while the control group is exposed to footage that is irrelevant to the training task, thus resulting in a more accurate assessment of how video presentation in general might affect the efficiency of training. Nevertheless, the results of the study are encouraging, and they indicate that social learning opportunities, in conjunction with PRT, can be beneficial to chimpanzee training.

Possible Contributions of Positive Reinforcement Training to Cognitive Studies

As evidenced by accounts throughout this book, the breadth of today's cognitive research with chimpanzees is astounding. Studies with captive chimpanzees in particular take a variety of forms including computer-mediated interactions, puzzle solving, cultural transmission studies, and tool-use tasks. Given the aforementioned advantages of PRT in stimulating subjects and streamlining their learning process, it is easy to imagine how scientists would be interested in their potential. The previous section has explored how chimpanzees' ability to learn socially can facilitate the training process. Here, we describe how the application of PRT can improve studies of chimpanzee cognition.

First, since chimpanzees demonstrate their most representative and hence informative learning abilities when learning from fellow chimpanzees rather than human experimenters, the role of trained demonstrator chimpanzees is beginning to be appreciated (de Waal 2001; Whiten et al. 2005; Horner et al. 2006; Boesch 2007; de Waal et al. 2008). Improvements and increased acceptance of PRT techniques in cognitive research studies can facilitate the training of competent chimpanzee demonstrators for such research. For example, chimpanzees can be trained to use devices or solve problems in particular ways (see, for example, chapters 8, 9, and 19). Second, many projects in cognitive research require subjects to be moved from one location to another for testing, or to be temporarily separated from other group members for a cognitive task. These behaviors can be accomplished more reliably and with less stress by the application of PRT techniques. Third, some cognitive research studies require the use of apparatus such as touch screens, joysticks, or foraging puzzles. The application of PRT to train chimpanzees to use novel testing equipment may be more precise and expedient than relying on simple trial-and-error learning without human intervention. In addition, when subjects respond fearfully to novel test environments or apparatus, desensitization training can be systematically applied to reduce this fear so that studies can progress more efficiently. Fourth, the ability of chimpanzees to learn from video presentation (as described above) might allow investigation of how chimpanzees of different ages and sexes, or in different social environments, vary in how they learn from the same video stimulus. Thus, the ability to standardize the model stimulus by showing the same training video to multiple subjects is a key benefit.

We also suspect that PRT programs could be used to assist in the selection and performance of subjects in cognitive research studies. Since all of the major chimpanzee research colonies in the United States have PRT programs in place (Bloomsmith and Else 2005) many of the chimpanzees at these facilities have been trained for behaviors that facilitate husbandry, veterinary, or research procedures, and also participate as subjects in a variety of behavioral and cognitive studies. Chimpanzees with PRT experience may thus be more motivated and representative participants for studies of their cognitive abilities, and therefore more suitable for such studies. We would expect that an individual who is trained to perform a large variety of behaviors using PRT techniques would be a better subject for cognitive research than would a subject with no training experience at all. This is not dissimilar to early findings by Harlow (1949) that found rhesus monkeys with experimental problem-solving experience learned to solve similar problems more quickly than non-experienced monkeys. Harlow termed this phenomenon "learning to learn."

Given the growing number of chimpanzees with training and task experience, one might imagine a comprehensive record that catalogues it for each individual. A "trainability index" could be derived that records the number of behaviors a chimpanzee has been trained to perform, his/her ability to reliably perform that behavior when cued, and perhaps other factors, such as the complexity of the learned behaviors and the training techniques applied. Individuals who are trained in a large number of behaviors and who perform them reliably would have a high trainability index, while individuals who have learned fewer behaviors would have a low trainability index. Such indices would allow for comparisons among chimpanzees both within and across colonies, and they could be used to help select subjects for cognitive studies.

Implications and Future Directions

There are many more opportunities to study the relationships between social learning and PRT comprehensively.

In an attempt to explore more thoroughly whether social learning outlets expedite the PRT process, a follow-up study to the urine collection study described earlier is in progress at the Yerkes National Primate Research Center. We plan to discover which specific information in a videotaped demonstration is most beneficial for accelerating the learning of chimpanzees by comparing their responses to videotapes depicting different elements of training. For example, do chimpanzees benefit from watching a video of the same target behavior they are being trained to perform? Or is observation of the general training process (i.e., training for a different target behavior) sufficient to produce such a benefit? What is the role of rewards in video presentation? Do chimpanzees need to see the target behavior being reinforced before they are motivated to learn it? (See Bonnie and de Waal 2007 for a related study performed with capuchin monkeys.)

At present, no one has calculated "trainability indices" for captive chimpanzees, or attempted to correlate such indices with the abilities of those chimpanzees in cognitive tasks. However, anecdotal evidence suggests that individuals who are the best training subjects (easiest to train for in most behaviors, and most likely to achieve and maintain reliable performance) are also among the most willing to participate in cognitive testing. These individuals are not necessarily the best at performing the cognitive tasks, but they are among the best at attending to the procedures. Perhaps the experience they have gained from attending to training procedures has enhanced their willingness and ability to attend to other tasks, such as solving puzzles and manipulating buttons, levers, or joysticks. Related arguments exist in the literature regarding the enculturation hypothesis, whereby some authors have suggested that human-reared apes have cognitive abilities more advanced than those of their mother-raised peers (Tomasello et al. 1993). Others have pointed out, however, that apes living in a human home have more exposure to human objects and novel situations than a typical captive ape, and so are better able to adapt to novel testing situations—particularly those that involve novel object manipulation—than mother-reared apes (de Waal 2001; Bering 2004). Similar experiential differences could affect how chimpanzees respond to cognitive studies, with individuals who have considerable training experience performing better than individuals who do not. Increased reporting of participants' background, in addition to a trainability index, could allow better interpretation of contradictory or divergent findings from different research populations. As a final point, we suggest that publications from cognitive studies should include descriptions of the training techniques applied (which now typically are not included), as this may improve other researchers' ability to replicate studies, and may also foster a greater awareness of the importance and possible influence of PRT on research results.

Humans who work around chimpanzees frequently tell stories of how they learn from watching other chimpanzees or familiar humans—as in the story at the beginning of this chapter, of Betty learning to clean her own enclosure. Even though chimpanzees' social learning skills have long been known, the information has not often been appropriately applied to improving their care or welfare. This chapter suggests one way this could be accomplished by exploiting the social learning abilities of chimpanzees in PRT programs. We have described possibilities for how this can best be studied, and how PRT techniques and programs can be applied to improve the quality of such studies of chimpanzee cognition. Given the wide range of these benefits, ranging from improving the efficiency of cognitive studies to having positive effects on individual welfare, PRT training should remain a viable tool in the care, management, and study of chimpanzees.

Literature Cited

Basset, L., and H. Buchanan-Smith. 2007. Effects of predictability on the welfare of captive animals. *Applied Animal Behaviour Science* 102:223–45.

Beck, B. 1997. Reintroduction, zoos, conservation and animal welfare. In *Ethics on the Ark*. Washington, DC: Smithsonian Institution Press.

Bering, J. M. 2004. A critical review of the "enculturation hypothesis": The effects of human rearing on great ape social cognition. *Animal Cognition* 7(4): 201–12.

Bloomsmith, M. A., K. C. Baker, S. K. Ross, and S. P. Lambeth. 1999. Comparing animal training to non-training human interaction as environmental enrichment for chimpanzees. *American Journal of Primatology* 49(1): 35–36.

Bloomsmith, M. A., and J. G. Else. 2005. Behavioral management of chimpanzees in biomedical research facilities: the state of the science. *ILAR Journal*, 46(2): 192–201.

Bloomsmith, M. A., and S. P. Lambeth. 2000. Videotapes as enrichment for captive chimpanzees. *Zoo Biology* 19:541–51.

Bloomsmith, M. A., S. P. Lambeth, G. E. Laule, and R. H. Thurston. 1993. Training as environmental enrichment for chimpanzees. *American Journal of Primatology* 33:299.

Bloomsmith, M. A., S. P. Lambeth, A. M. Stone, and G. E. Laule. 1997. Comparing two types of human interaction as enrichment in chimpanzees. *American Journal of Primatology* 42(2): 96.

Bloomsmith, M. A., G. E. Laule, P. L. Alford, and R. H. Thurston. 1994. Using training to moderate chimpanzee aggression during feeding. *Zoo Biology* 13:557–66.

Bloomsmith, M. A., S. J. Schapiro, and E. A. Strobert. 2006. Preparing chimpanzees for laboratory research. *ILAR Journal* 47(4): 316–25.

Bloomsmith, M. A., A. M. Stone, and G. E. Laule. 1998. Positive reinforcement training to enhance the voluntary movement of group-housed chimpanzees. *Zoo Biology* 17:333–41.

Boesch, C. 2007. What makes us human (*Homo sapiens*)? The challenge of cognitive cross-species comparison. *Journal of Comparative Psychology* 121(3): 227–240

Boesch, C., and M. Tomasello. 1998. Chimpanzee and human cultures. *Current Anthropology* 39(5): 591–614.

Bonnie, B. E., and F. B. M. de Waal. 2007. Copying without rewards: Socially influenced foraging decisions among brown capuchin monkeys. *Animal Cognition* 10:283–92.

Box, H. O. 1991. Training for life after release: Simian primates as examples. *Symposia of the Zoological Society of London* 62:111–23.

Coleman, K., L. Pranger, A. Maier, S. P. Lambeth, J. E. Perlman, E. Thiele, and S. J. Schapiro. 2008. Training rhesus macaques for venipuncture using positive reinforcement training techniques: A comparison with chimpanzees. *Journal of the American Association for Laboratory Animal Science* 47(1): 37–41.

Cook, M., and S. Mineka. 1990. Selective associations in the observational conditioning of fear in rhesus monkeys. *J of Experi Psycholo: Animal Beh Processes* 16(4): 372–89.

Coussi-Korbel, S., and D. M. Fragaszy. 1995. On the relation between social dynamics and social learning. *Animal Behaviour* 50:1441–53.

Cox, C. 1987. Increase in the frequency of social interactions and the likelihood of reproduction among drills. *Proceedings, American Association of Zoological Parks and Aquariums Western Regional Conference, Fresno*, 321–28.

Custance, D. M., A. Whiten, and K. A. Bard. 1995. Can young chimpanzees *(Pan troglodytes)* imitate arbitrary actions? Heyes and Heyes (1952) revisited. *Behaviour* 132: 837–38.

Custance, D. M., A. Whiten, and T. Fredman. 2002. Social learning and primate reintroduction. *International Journal of Primatology* 23(3): 479–99.

Desmond, T., and G. E. Laule. 1994. Use of positive reinforcement training in the management of species for reproduction. *Zoo Biology* 13:471–77.

Desmond, T., G. Laule, and J. McNary. 1987. Training for socialization and reproduction with drills. *Proceedings, American Association of Zoological Parks and Aquariums Annual Conference, Portland*, 435–41.

De Waal, F. B. M. 2001. *The Ape and the Sushi Master: Cultural Reflections by a Primatologist*. New York: Basic Books.

De Waal, F. B. M., C. Boesch, V. Horner, and A. Whiten. 2008. Comparing apes and children not so simple. *Science*, in press.

Harlow, H. 1949. The formation of learning sets. *Psychological Review* 56:51–65.

Hayes, K., and C. Hayes. 1952. Imitation in a home-raised chimpanzee. *Journal of Comparative and Physiological Psychology* 45:450–59.

Horner, V., and A. Whiten. 2005. Causal knowledge and imitation / emulation switching in chimpanzees *(Pan troglodytes)* and children *(Homo sapiens)*. *Animal Cognition* 8:164–81.

Horner, V., A. Whiten, E. Flynn, and F. B. M. de Waal. 2006. Faithful replication of foraging techniques along cultural transmission chains by chimpanzees and children. *Proceedings of the National Academy of Science USA* 103:13878–83.

Kuhlmeier, V. A., S. T. Boysen, and K. L. Mukobi. 1999. Scale-model comprehension by chimpanzees *(Pan troglodytes)*. *Journal of Comparative Psychology* 113(4): 396–402.

Lambeth, S. P., J. Hau, J. E. Perlman, M. Martino, and S. J. Schapiro. 2006. Positive reinforcement training affects hematologic and serum chemistry values in captive chimpanzees *(Pan troglodytes)*. *American Journal of Primatology* 68:245–56.

Lambeth, S. P., J. E. Perlman, E. Thiele, and S. J. Schapiro. 2005. Changes in hematology and blood chemistry parameters in captive chimpanzees *(Pan troglodytes)* as a function of blood sampling technique: Trained vs. anesthetized samples. *American Journal of Primatology* 65(1): 182.

Laule, G. and T. Desmond. 1998. Positive reinforcement training as an enrichment strategy. In *Second Nature: Environmental Enrichment for Captive Animals*, edited by David J. Shepherdson, Jill D. Mellen, and Michael Hutchins, 302–13. Washington DC: Smithsonian Institution Press.

Laule, G. E., R. H. Thurston, P. L. Alford, and M. A. Bloomsmith. 1996. Training to reliably obtain blood and urine samples from a young diabetic chimpanzee *(Pan troglodytes)*. *Zoo Biology* 15:587–91.

Laule, G. E., and M. A. Whittaker. 2001. The use of positive reinforcement techniques with chimpanzees for enhanced care and welfare. In *Care and Management of Captive Chimpanzees*, edited by Linda Brent, 243–65. San Antonio: American Journal of Primatology.

Laule, G. E., and M. A. Whittaker. 2007. Enhancing nonhuman primate care and welfare through the use of positive reinforcement training. *Journal of Applied Animal Welfare Science* 10(1): 31–38.

Markowitz, H. 1982. Behavioral Enrichment in the Zoo. New York: Van Nostrand Reinhold Company.

Matsuzawa, T. 2002. Chimpanzee Ai and her son Ayumu: An episode of education by master-apprenticeship. In *The Cognitive Animal: Empirical and Theoretical Perspectives on Animal Cognition*, edited by M. Bekoff, C. Allen, and G. M. Burghardt. Cambridge, MA: MIT Press, 189–95.

McMillan, J. L., J. E. Perlman, and M. A. Bloomsmith. 2007. Components of an animal training program at a large institution. *American Journal of Primatology* 69(1): 117.

Meehan, C. L., and J. A. Mench. 2007. The challenge of challenge: Can problem solving opportunities enhance animal welfare? *Applied Animal Behaviour Science* 102:246–61.

Mellen, J. D., and S. Ellis. 1996. Animal learning and husbandry training. *In Wild Mammals in Captivity: Principles and techniques*, edited by D. G. Kleiman, M. E. Allen, K. V. Thompson, and S. Lumpkin. Chicago and London: University of Chicago Press.

Menzel, E. W. Jr. 1973. Leadership and communication in young chimpanzees. In *Precultural Primate Behavior*: Basel: Karger.

Morimura, N. and T. Matsuzawa. 2001. Memory of movies by chimpanzees. *Journal of Comparative Psychology* 115(2): 152–58.

Nagell, K., R. Olguin, and M. Tomasello. 1993. Processes of social learning in the tool use of chimpanzees and human children. *Journal of Comparative Psychology* 107(2): 174–86.

Novak, M and S. J. Suomi. 1988. Psychological well-being of primates in captivity. *American Psychologist* 43:765–73.

Perlman, J. E., T. R. Bowsher, S. N. Braccini, T. J. Kuehl, and S. J. Schapiro. 2003. Using positive reinforcement training techniques to facilitate the collection of semen in chimpanzees *(Pan troglodytes)*. *American Journal of Primatology* 60(1): 77–78.

Perlman, J. E., S. P. Lambeth, M. A. Bloomsmith, G. E. Laule, S. J. Schapiro, and M. E. Keeling. 2001. Training captive chimpanzees: A focused look at the potential benefits of whole group training. In *Proceedings the Apes: Challenges for the 21st Century*. Brookfield, IL: Brookfield Zoo, 373.

Perlman, J. E., E. Thiele, M. A. Whittaker, S. P. Lambeth, and S. J. Schapiro. 2004. Training chimpanzees to accept subcutaneous injections using positive reinforcement training techniques. *American Journal of Primatology* 62(1): 96.

Price, E., and C. A. Caldwell. 2007. Artificially generated cultural variation between two groups of captive monkeys, *Colobus guereza kikuyuensis*. *Behavioural Processes* 74:13–20.

Pryor, Karen. 1984. *Don't shoot the dog! The new art of teaching and training*. New York: Bantam Books.

Ramirez, K. 1999. *Animal Training: Successful Animal Management through Positive Reinforcement*. Chicago, IL: Shedd Aquarium.

Russell, J. L., J. P. Taglialatela, and W. D. Hopkins. 2006. The use of positive reinforcement training in chimpanzees *(Pan troglodytes)* for voluntary presentation for IM injections. *American Journal of Primatology* 68(1): 122.

Schapiro, S. J., S. P. Lambeth, E. Thiele, and O. Rousset. 2007. The effect of behavioral management programs on dependant measures in biomedical research. *American Journal of Primatology* 69(1): 115.

Schapiro, S. J., J. E. Perlman, E. Thiele, and S. P. Lambeth. 2005. Training nonhuman primates to perform behaviors useful in biomedical research. *Lab Animal* 34:37–42.

Schapiro, S. J., J. E. Perlman, and B. Boudreau. 2001. Manipulating the affiliative interactions of group-housed rhesus macaques using positive reinforcement training techniques. *American Journal of Primatology* 55:137–49.

Snowdon, C. T., and A. Savage. 1989. General considerations and examples from Callitrichids in *Housing, Care and Psychological Well-Being of Captive and Laboratory Primates*, edited by E. F. Segal. Park Ridge, NJ: Noyes Publications.

Spence, K.W. 1937. Experimental studies of learning and higher mental processes in infra-human primates. *Psychological Bulletin* 34:806–50.

Stone, A. M., M. A. Bloomsmith, G. E. Laule, and P. L. Alford. 1994. Documenting positive reinforcement training for training chimpanzee urine collection. *American Journal of Primatology* 33(3): 242.

Sumita, K., J. Kitahara-Frisch, and K. Norikoshi. 1985. The acquisition of stone-tool use in captive chimpanzees. *Primates* 26:168–81.

Swaisgood, R.R. 2007. Current status and future directions of applied behavioral research for animal welfare and conservation. *Applied Animal Behaviour Science* 102:139–62.

Thorpe, W.H. 1956. *Learning and Instinct in Animals*. London: Methuen.

Tomasello, M., E. S. Savage-Rumbaugh, and A. Kruger. 1993. Imitative learning of actions on objects by children, chimpanzees and enculturated chimpanzees. *Child Development* 64:1688–1705.

Van Schaik, C. P. 2003. Local traditions in orangutans and chimpanzees: Social learning and social tolerance. In *The Biology of Traditions: Models and Evidence*, edited by D. M. Fragaszy and S. Perry. Cambridge: Cambridge University Press.

Videan, E. N., J. Fritz, J. Murphy, R. Broman, H. F. Smith, and S. Howell. 2005. Training captive chimpanzees to cooperate for an anesthetic injection. *Lab Animal* 34(5): 43–48.

Visalberghi, E., and D. M. Fragazy. 1990. Food washing behaviour in tufted capuchin monkeys *Cebus apella*, and crabeating macaques *Macaca fascularis*. *Animal Behaviour* 40:829–36.

Whiten, A., and R. Ham. 1992. On the nature and evolution of imitation in the animal kingdom: Reappraisal of a century of research. In *Advances in the Study of Behaviour*, edited by P. J. B. Slater, J. S. Rosenblatt, C. Beer, and M. Milinski. New York: Academic Press.

Whiten, A., V. Horner, and F. B. M. de Waal. 2005. Conformity to cultural norms of tool use in chimpanzees. *Nature* 437:737–40.

Whiten, A., V. Horner, C. A. Litchfield, and S. Marshall-Pescini. 2004. How do apes ape? *Learning and Behavior* 32(1): 36–52.

Whiten, A., A. Spiteri, V. Horner, K. E. Bonnie, S. P. Lambeth, S. J. Schapiro, F. B. M de Waal. 2007. Transmission of multiple traditions within and between chimpanzee groups. *Current Biology* 17(2): 1038–43.

Whittaker, M. A., J. E. Perlman, and G. E. Laule. 2007. Facing real-world challenges: Keeping behavioral management programs alive and well. *Proceedings from the 8th International Conference on Environmental Enrichment, Vienna, Austria.*

26

Chimpanzee Orphans: Sanctuaries, Reintroduction, and Cognition

Benjamin B. Beck

On the day of my retirement from the Smithsonian National Zoological Park in 2003, I walked around to say farewell and express thanks to the many people and animals with whom I had worked for 20 years. I saved two of my favorite animals, orangutans Azy and Indah, for last. I had known them since they were youngsters and worked with them collaboratively in cognitive research projects. We had close and trusting relationships, although Azy sometimes treated me as a competitor and performed impressive displays of his dominance. Indah was always sweet and curious, and was probably the brightest ape I have ever met. When I said goodbye to her, my emotions caught up and tears streamed down my cheeks. Indah had probably seen people cry before, but she had never seen me cry. She watched my tears closely, and then fixed my eyes. She reached out, and gently patted me on the shoulder and neck. She had never before done this, usually preferring to examine my pockets or nudge me toward hidden treats whose location she, but not I, knew. I have no doubt that Indah was able to recognize my state of sadness, and that she responded appropriately, as a friend, with consolation. Later I realized that I had never seen Indah, or any other ape, cry.

Introduction

I can find only one report of a great ape crying. Dian Fossey saw an orphaned infant mountain gorilla "sob and shed actual tears" as she, the gorilla, looked out at her former home on the Visoke volcano (Fossey 1983, p. 110). Fossey added this was "something I have never seen a gorilla do before or since." This case notwithstanding, are humans the only apes who cry? Are we *Homo lachrymosis*? Why would this conspicuous and effective communicative behavior be selected for in humans but not apes? How could Indah recognize a communicative behavior that she herself could not perform?

Our jobs would be easier if apes did cry, because then we would have an observable measure of pain and sadness. Instead of tears we must rely, in Jane Goodall's words (2001), on "eyes filled with pain and hopelessness" in an ill-treated, newly motherless chimpanzee orphan. But even without ape tears, most of us would know when an ape needs help. People of most cultures would want to help or rescue distressed chimpanzees, like the one described by Goodall, and like other apes subject to such suffering. We know that chimpanzees have remarkable, human-like cognitive capabilities, and that they are genetically the beings most like us. But it is important to remember that these are not the primary reasons for our helping them. We also rescue and help dogs, manatees, pandas, rabbits, hawks, turtles and myriad other less cognitively sophisticated and more distantly related animals. Nonetheless, cognition is important in this discussion and for me, Indah embodies a set of cognitive attributes that make apes special. In this chapter I discuss how cognition influences and intersects with conservation in the area of ape sanctuaries and reintroductions. I assert that we must critically examine the cognitive underpinnings of the way in which we manage apes in orphanages and, if we are to reintroduce ape orphans to the wild, the cognitive underpinnings of reintroduction management.

Four related sets of ethical questions are explored in this chapter. The first: Do ape sanctuaries serve conservation? The second: What are the options for the growing number of sanctuary apes, and is reintroduction a realistic, effective and humane option? The third: Can international reintroduction guidelines be adapted to facilitate great ape reintroduction? The fourth: Is the chimpanzee mind so similar to our own that orphan chimpanzees and other apes warrant disproportionate support for ape sanctuaries and ape reintroduction? I will focus on chimpanzees (*Pan troglodytes)* and bonobos (*Pan paniscus*) in this chapter, but most of the findings pertain also to gorillas (*Gorilla* spp.) and orangutans (*Pongo* spp.). I will use "apes" for "great apes" for brevity.

Ape Sanctuaries and Conservation

Ape Orphans: Definitions, Causes, and Numbers

I define an ape orphan as an infant or juvenile without a mother or conspecific caregiver, or an older ape with no familiar and safe place to live. Most chimpanzee and bonobo orphans are very young, having been taken recently from the wild after the death of their mothers. Since apes and monkeys are unusual vertebrates in that infants actually cling to their mothers, an ape or monkey infant is often found after its mother has been killed as a source of meat, and thus represents a form of "bycatch" for the people who killed the mother. The choice hunters typically make is to sell the living infant. There may be some cases in which a hunter specifically targets infants for sale to an animal dealer or prospective pet owner (Farmer 2002; Tutin et al. 2001) but since chimpanzees are often shot (Amman 2001), or caught in snares intended for antelope (Wilkie 2001), meat rather than an infant is usually the primary goal of such hunting. Some chimpanzee orphans are older and weaned, having spent longer periods in captivity. Some of these have been physically or psychologically abused, while others have been well cared for, and even pampered.

Many of these chimpanzee orphans are confiscated by authorities from their "owners" because personal ownership of an endangered chimpanzee is illegal, and/or because the chimpanzee was acquired illegally. Some are confiscated in countries other than those in which they were born, under international cooperative agreements and treaties, mainly the Convention on International Trade in Endangered Species (CITES). Other orphans are voluntarily surrendered by their "owners," either because of new awareness of the illegality or immorality of private "ownership," or because the "owner" finds that it is becoming too dangerous, expensive, or inconvenient to keep the ape. I have placed "owner" and "ownership" in quotation marks here to reflect the opinion that apes should not be regarded as property or as things that can be ethically or legally owned (Francione 1993; Wise 2000). Whether confiscated or surrendered, many of the orphaned chimpanzees are placed in range country orphanages or sanctuaries ("range country" refers to a country in which chimpanzees occur naturally, e.g. Uganda). An ape sanctuary is a facility whose primary purpose is to provide security and humane care for orphaned apes for as long as necessary (Beck et al. 2007; Rosen et al. 2002; Teleki 2001). Public education, local employment, tourism, and research are secondary goals for some sanctuaries (Farmer 2002).

Some ape orphans never spend time in private ownership but are rescued, usually just before or after dramatic, mechanized forest destruction that leaves them homeless. (Orangutans are now being more commonly victimized in this fashion than the African apes, by large-scale clearing of primary forest for conversion to palm oil plantations.) The rescuers' goal is to translocate these apes quickly to a suitable intact forest, but there are ever fewer such places. Thus they are often taken to a sanctuary to await identification of a translocation destination and, if necessary, medical treatment. Apes rescued as full adults are considered orphans because they do not have a safe place to live, and they add to the overall numbers in sanctuary populations.

At the time of the Mind of the Chimpanzee conference (March 2007), there were 729 chimpanzees, 84 gorillas, and 56 bonobos in member sanctuaries of the Pan African Sanctuary Alliance (PASA) in Africa, and approximately 130 more in other African sanctuaries and orphanages (Doug Cress, personal communication). As there is no single umbrella organization for Indonesian and Malaysian orangutan sanctuaries, the figures for orangutans are less specific, but personal communications, sanctuary websites, and newsletters suggest that there are conservatively 1,200 of them in sanctuaries. Thus there are currently at least 2,200 great apes in range country sanctuaries. For comparison, there are about 1,106 chimpanzees, 752 gorillas, 167 bonobos and 585 orangutans (a total of 2,610 great apes) in all of the world's zoos combined (ISIS 2007).

An important issue for defining sanctuary populations now and into the future is how quickly they are growing. This is somewhat more difficult to estimate because historical sanctuary population sizes are not well documented. From 2000 to 2006, the number of chimpanzees in the 13 PASA chimpanzee sanctuaries grew 63%—a rate of about 10% per year (Doug Cress, personal communications). According to annual reports of the Lola ya Bonobo sanctuary in the Democratic Republic of the Congo (DRC), the population grew from 45 in 2005 to 56 in 2007. Two of the 56 were born at the sanctuary, but the nine newly arrived orphans represent a growth of about 10% per year. We are currently modeling the growth of ape sanctuaries in range countries over the next 50 years (Faust et al. 2007) and hope to be able to better define the problem. Until then, the evidence we do have suggests that sanctuary populations for all four types of great apes are growing at an alarming rate.

The Importance of Ape Sanctuaries for Conservation

Some in the wildlife conservation community have questioned the role of sanctuaries. This is the first of the questions we will explore in this chapter. In my opinion, and there seems to be little disagreement on this, the major purpose of sanctuaries is to offer a safe and caring place for these wonderful beings. A more contentious issue is whether and how ape sanctuaries serve ape conservation. First and foremost, they provide authorities with a reputable place to take apes that they confiscate, thus giving them the option of actively enforcing conservation laws. It is not clear, however, that the authorities always exercise this option or follow up with prosecution of the "owner" if they do. Furthermore, there is no evidence that confiscation reduces the number of apes being taken alive from the forest and kept illegally in captivity. Nevertheless, it seems safe to conclude that without the option of confiscation and prosecution, many more apes would be taken from the wild to be bought and sold. Farmer and Courage (2009)

summarize other concerns about sanctuaries. For example, they could be misperceived as markets for apes rather than rescue centers, and thus could stimulate the illegal taking of wild apes in the hopes of selling them to sanctuaries. Further, they could create resentment among local people who might perceive that the apes are receiving better care and more attention than they themselves receive. In contrast, others argue that sanctuaries can serve conservation through environmental educational programs for local audiences (e.g., Cartwright and Bettinger 2006; Farmer and Courage 2009; Weber 1995). However, there is little measurable evidence that they either do or don't promote public awareness of and support for animal welfare and conservation of biodiversity. Finally, sanctuaries can provide local employment and may serve as venues for noninvasive research (Farmer 2002; Miller 2007), both of which may be beneficial.

Despite the potential critiques, sanctuaries do have real opportunities to serve conservation efforts. The primary way is by preparing apes for reintroduction and by actually conducting reintroduction programs. I can point to only three successful sanctuary-based reintroduction programs for great apes. Project HELP (Habitat Ecologique et Liberté des Primates) has reintroduced 37 chimpanzees into the Conkouati-Douli National Park in the Republic of Congo (Goosens et al. 2005; Tutin et al. 2001). Sanctuaries managed by the John Aspinall Foundation have reintroduced 40 wild-born gorilla orphans (and 10 captive-borns) in Gabon and the Republic of Congo (King et al. 2006). The Sumatran Orangutan Conservation Programme has reintroduced about 65 orangutan orphans (plus one zoo-born adult) in the Bukit Tigapulah National Park in Sumatra (Pratje and Singleton 2006). Each of these reintroduction projects can claim success in terms of high post-release survivorship and some post-release reproduction. There have been some other reintroductions of sanctuary chimpanzees and orangutans, with less successful outcomes (see Carter 2003; Yeager 1997). Several more ape sanctuaries are also currently planning reintroductions, and are committed to implementing best practices and following international guidelines (Carlsen et al. 2006).

Options for Sanctuary Apes

Given that there are sanctuaries for great apes, the second question to be considered here is what to do with the more than 2,200 apes they contain. Reintroducing them all is an attractive option, but the very conditions that are producing ape orphans make it extremely difficult to locate suitable reintroduction sites. Some sanctuary apes suffer injuries and illnesses that would make it unlikely that they would survive in the wild, and communicable illnesses might pose a threat to existing wild populations. These and other obstacles are discussed at length in the World Conservation Union (IUCN) guidelines on great ape reintroduction (Beck et al. 2007), but it is clear that options other than reintroduction must be considered. It is also clear, and universally reaffirmed, that reintroduction cannot responsibly be undertaken just to dispose of "surplus apes," as much as sanctuaries would welcome relief from the overcrowding.

In 1992, Rwandan officials seized an eastern lowland gorilla, later named "Amahoro", as she was being illegally exported. Amahoro was taken to a sanctuary and the International Gorilla Conservation Programme commissioned Sandy Harcourt and Jose Kalpers to summarize the options for her future. This resulted in two widely circulated but unpublished papers which, with Harcourt's earlier paper on options for confiscated primates (Harcourt 1987), serve as a basis for this section.

One option is euthanasia. Note that this would be killing a healthy ape, and must be distinguished from euthanasia of an ape with a painful, debilitating, untreatable injury or illness. Also, this option is seen as not providing the body, body parts, or bodily products for any use whatsoever; the ape's remains would instead be incinerated or buried. The chief advantage of euthanasia is that it is inexpensive and technically simple. Nobody profits financially, and there could not even be a perception of profit. It signals strict (some would say harsh) enforcement of conservation laws, although the impact would be reduced if the poacher and smuggler were not also punished. The chief disadvantage is that euthanasia of a healthy ape is difficult to reconcile with high standards of respect for animals, animal welfare, and the genetic and cognitive similarities of apes to humans. It squanders the opportunity for the ape to be used in public education and research programs, as well as the opportunity for it to reproduce in captivity or in the wild. Euthanasia of a healthy ape might even decrease public support for animal welfare and conservation by seeming to devalue its life.

A second option would be to sell the ape orphan to raise funds for conservation or to promote biological knowledge. For example, the ape could be sold as a pet, or to a biomedical research laboratory. It could be euthanized, and its body could be sold as a museum specimen. Body parts could be sold for use in traditional medicine or as artifacts, as long as the appropriate permit processes were followed (e.g., CITES). The only advantage of this option is generation of funds that could support protection and conservation of wild conspecifics. All of the disadvantages of euthanasia would pertain, but sale of the ape additionally would encourage illegal commercial trade when others, noting that a living or dead ape was sold, might try to kill or capture another wild ape and sell it.

A third option would be to send the ape to a zoo or research institution abroad. If it were a research institution, the understanding would be that ape would not be killed prematurely. The destination institution might buy the ape or make a contribution to conservation, or the transaction could be a donation. Again, the appropriate permits would be required. The advantages are that the ape would remain alive and could potentially have the opportunity to reproduce. Funds could be generated for in situ conservation, and there also could be benefits to research or public education. The main disadvantage is possible facilitation of the illegal commercial trade. Some might add that life in a zoo or research institution would be inhumane; others might counter that life in a good zoo would surely be more humane than life in an overcrowded sanctuary. Citizens of the ape's home country, however, might resent exportation of their natural heritage and lose confidence in their local and national conservation efforts.

A fourth option would be long-term, probably lifetime captivity in the ape's country of origin. This is by far the most expensive and technically complicated of the first four options. However, it would be less expensive to build and operate a sanctuary or other captive facility in most ape range countries than to build one in North America, Europe, Asia, or Australia. Such a facility could also be a venue for research and education. Since the primary educational audience would be local, there is the potential for positive effects for conservation. The facility could also be a venue for tourism which, if well managed, could produce a significant income stream without adverse effects on the apes. The facility could offer local employment opportunities, which in turn would support conservation. Finally, the apes could readily be reintroduced to the wild if a suitable site became available (see below). The Ngamba Island Chimpanzee Sanctuary in Uganda is an excellent example of a well-managed sanctuary in which orphaned and confiscated chimpanzees could spend their entire lives, and which serves all of these functions.

The fifth option is reintroduction to the wild. This is the preferred option (IUCN 2002) if it can be done properly, following the guidelines for great ape reintroduction (Beck et al. 2007). However, this is an expensive and technically demanding option. The Conkouati chimpanzee reintroduction (see below) has cost about US$34,000 per individual, and costs mount with ongoing monitoring. This compares with about US$20,000 per reintroduced individual of the smaller golden lion tamarin (Kierulff et al. 2002). But the cost is front-loaded; it decreases once the ape is reintroduced and monitored until it dies, disappears, or begins to live independently in the wild. Some would say that the money would be better spent on protecting wild apes and their habitats, but this implies implementation of the first, second, or third option, which I and many others find unacceptable. The main benefit of ape reintroduction is that an ape can be assimilated into a natural evolving ecosystem. Thus, the size and genetic variability of the wild population can be augmented. The local and international conservation message is strongly positive, and there can be no perception of private economic gain. However, there is disagreement as to whether reintroduction enhances individual ape welfare more than the other options do; presumably this is testable. As some individual reintroduced apes will without doubt die prematurely, probably in pain or distress, reintroduction is not necessarily a humane option (see below). Finally, if a wild population of conspecifics lives in the release area, it will be subjected to intensified resource competition, disease exposure, and disruption from human activity.

Of these five options, I find only the fourth (lifetime care in well-managed sanctuaries) and the fifth (properly conducted reintroduction) acceptable. Great apes are the animals most like ourselves; they have advanced cognitive capacities that may constitute a mind. Thus I find the first option (euthanasia) and the second (commercial exploitation) unacceptable on ethical grounds. I also do not favor the third option (shipment abroad) because it does not encourage, and may actually discourage, support for wild apes and their habitats.

An Abbreviated History of Chimpanzee Reintroductions

Since reintroduction is such an appealing option for great apes living in sanctuaries, it is useful to examine some case studies of chimpanzee reintroduction. This is not an exhaustive survey (see Carter 2003 for a more exhaustive treatment), but it illustrates approaches, techniques, problems, and outcomes, and thus provides a basis to evaluate feasibility.

RUBONDO ISLAND, TANZANIA. Representatives of the Frankfurt Zoological Society released 17 chimpanzees on Rubondo Island off the southwest shore of Lake Victoria in Tanzania between 1966 and 1969 (Borner 1985). All of the Rubondo chimpanzees had lived in European zoos for three months to nine years, and were between four and 12 years of age when released. Few of the apes knew each other and as such, there were no established groups. As their geographic origins are unspecified, we do not know which of the four currently recognized subspecies of *Pan troglodytes* may have been represented. Rubondo, which became a national park in 1977, is 240 square kilometers (93 square miles) and has ample evergreen deciduous forest. No other apes lived on the island. Potential predators included crocodiles, pythons, and martial and crowned eagles, but Borner did not consider these to be capable of hunting and killing a chimpanzee. Rubondo has subsequently been documented to have reliable, abundant and high quality food for chimpanzees (Moscovice et al. 2007).

The chimpanzees were released in four separate cohorts. The first—four males and seven females—was provisioned with food for two months, but quickly began to eat wild foods. They began building nests within a year of release. Two males were subsequently released individually, and a cohort of two males and two females was released last. There was little post-release monitoring and the chimpanzees were not individually identified, so we do not know which survived and reproduced, the causes of most losses, the composition and dynamics of newly formed groups, or the details of how they began to eat natural foods and build nests. This is unfortunate because the chimpanzees' ages, time in captivity, and individual histories varied and might have suggested correlates of reintroduction success. Some had lived in "good zoo conditions" and others in inadequate cages. Some had known one or more of the other chimpanzees before reintroduction, while others had lived alone in captivity. Their health had ranged from poor to good, and their behavior from abnormal to normal.

At least two of the reintroduced chimpanzees, both females, were known to be alive in 1985. One or perhaps two males were shot by unknown parties because of aggressive behavior toward humans. The chimpanzees continued to harass people and invade homes with decreasing frequency at least through 1985. Island-born infants were first observed in 1968, and by 1985 the population numbered about 20. Most of these had been born on the island but some of the original captive chimpanzees probably survived longer, as suggested by two particular events: in 1998 a female was observed stealing a blanket and wrapping it around her body (Pusey in Matsumodo-Oda 2000), and in the same year a male snatched a bottle of whiskey from a tourist camp and drank it (Matsumoto-Oda 2000). Moscovice et al. (2007) estimate that there are currently between 27 and 35 chimpanzees on Rubondo. This demonstrates that some chimpanzees taken from the wild as infants and juveniles can survive reintroduction into the wild with little prerelease preparation or postrelease support. But significantly, chimpanzees on Rubundo have had no competition from a resident chimpanzee population. This has reduced concern about communicable diseases (although monkeys and humans may have been exposed), but it is also likely that the current population includes subspecific hybrids. The chimpanzees are unhabituated and difficult to observe, but habituation for ecotourism is underway.

NIOKOLA KOBA, SENEGAL. Eddie Brewer, then director of wildlife conservation in the Gambia, and his daughter Stella Brewer began a rehabilitation-and-release program for confiscated chimpanzees in 1971. There were six wild-born and two captive-born rehabilitants, all less than six years old (Brewer 1978; Carter 1988, 2003; Hannah and McGrew 1991). Stella Brewer began their prerelease training in the Abuko Nature Reserve in the Gambia, leading them into the forest and teaching them how to select appropriate foods, how to use stones as hammers to open hard fruits, and how to use sticks to extract termites. She encouraged them to sleep in trees and showed them how to build nests. The eight chimpanzees—in a later article (Marsden et al. 2006) there are said to have been nine—

were reintroduced at Mount Asserik in Niokola Koba National Park in Senegal in 1973. Stella Brewer monitored them, provided supplemental food, and continued to train them in survival-critical skills after their release. They were slow to leave camp, and they imitated many human behaviors. The oldest male was destructive and sometimes aggressive. There were also wild resident chimpanzees at Mount Asserik, who attacked the rehabilitants when they were separated from Brewer or her staff, and this resulted in serious wounding. One chimpanzee had to be removed and placed back in captivity, and two others disappeared completely. Brewer then moved the survivors to an island in the River Gambia National Park in the Gambia, where they were provisioned with food. By 1985, when Brewer's group was merged with another group of rehabilitants (see below), there were one adult male, four adult females, five subadults, and four infants.

BABOON ISLAND, THE GAMBIA. The other group in the River Gambia National Park was established by Janice Carter in 1979. Carter worked on Baboon Island—an island neighboring the one on which Brewer worked, with an area of about 500 hectares. Baboon Island was uninhabited by people or apes, but it had hippopotamuses, baboons, vervet and colobus monkeys, hyenas, antelopes, leopards and other unspecified feline species, reptiles (including crocodiles, pythons and venomous snakes), and a rich bird fauna (Carter 1981, 1988; Marsden et al. 2006). Carter's original group comprised seven chimpanzees (three males and four females) who had been wild-born, captured illegally, and confiscated. All were younger than six years old. The group additionally included Lucy, an 11-year-old captive-born female who had been a participant in a celebrated language research project in the United States. Lucy had been born in captivity and intensively hand-reared in a household environment. Her companion, a captive-born four-year-old female chimpanzee, was also included. The language project had ended, and release of the chimpanzees to "freedom in the wild" was seen as preferable, in terms of their well-being, to a lifetime in captivity. Carter had first acclimatized these nine chimpanzees in a large naturalistic enclosure within the Abuko Nature Reserve for 18 months, tutoring them in appropriate foods, nest building, tool use, and predator avoidance. The chimpanzees were tutored further after their release on the island, and were given supplemental food and medical treatment. There was no prerelease medical or genetic screening.

Normal feeding, nesting, antipredator behavior, and social behavior emerged slowly, but the older wild-born chimpanzees needed less instruction than the captive-born individuals. Each of the captive-born females adopted one of the three-year-old males, and showed good maternal-like behavior. Lucy's adopted son died of endoparasitism, and Lucy herself died in 1987—perhaps killed by a human. The other captive-born female gave birth on the island but needed assistance from Carter in rearing her baby. This same female also captured, killed, and ate a colobus monkey (now known to be a common chimpanzee behavior), and learned to detect venomous snakes and give the appropriate alarm call.

Carter's and Brewer's projects are distinctive in successfully rehabilitating four captive-born chimpanzees, two of whom had been hand-reared. Their adaptation took a long time and a lot of tutoring, and they might not have survived without provisioning, but they did finally show many normal survival-critical behaviors.

Carter's and Brewer's groups were merged in 1985, and they then separated on their own into two new groups. Other orphans were added periodically; a total of 50 have been released (Marsden et al. 2006). There have been 50 first-generation and 8 second-generation births (Marsden et al. 2006). Carter's and Brewer's projects have continued for almost three decades, and as of 2006 the total reintroduced chimpanzee population was 69, living on three islands in four social groups. They are provisioned every other day, and are provided some veterinary care.

CONKOUATI, REPUBLIC OF CONGO. Aliette Jamart, a French expatriate, founded a sanctuary for confiscated chimpanzees in Pointe Noire, Republic of Congo (Congo-Brazzaville) in 1989. For many years she provided much of the funding and animal care herself. In 1992, having founded the private not-for-profit organization Habitat Ecologique et Liberte des Primates (HELP), Jamart began to move the chimpanzees from her sanctuary and the Pointe Noire Zoo to a new sanctuary on the shore of the Conkouati Lagoon, 180 km north of Pointe Noire. The new sanctuary also included three forested offshore islands, the largest of which is 50 hectares in size. Forty-eight chimpanzees have lived on the islands, and another 12 have remained in the onshore facility since being deemed

physically or psychologically unprepared for life on the islands. Between 1996 and 2001 a total of 37 chimpanzees were released from these islands into an area adjacent to the Conkouati Reserve (Goosens et al. 2005). These individuals probably originated from different areas, and thus different communities, within the Republic of Congo (Goosens et al. 2001). The release site is known as the Triangle—an area of 21 square km bounded by the lagoon and two rivers. The Triangle has a small natural chimpanzee population and is connected to the rest of the reserve by natural bridges. (Aczel 1993; Farmer 2000; Goosens et al. 2005; Karlowski 1996; Tutin et al. 2001).

While on the islands, the chimpanzees had the opportunity to acquire some skills in locomotion, navigation, nest building, and foraging. They could also spatially separate themselves from one another, and thus had different experiences and possibly communicated about objects and experiences removed in space and time. Although they found some natural food on the islands, they had to be fed by caretakers each day. They also experienced natural weather conditions on the islands. They were kept in large social groups where they could develop normal intra-group social behaviors and relationships and could reproduce. The chimpanzees were given routine preventive veterinary care at the sanctuary, including tuberculosis tests, examination of fecal samples for endoparasites, and vaccination against polio and tetanus. In addition, hematological and blood chemistry analyses; serological screening for retroviruses, filoviruses, hepatitis A and B viruses, and blood parasites; vaccination boosters; and antibiotic and vitamin injections were administered before reintroduction. DNA analyses from hair samples were done to verify the chimpanzees' geographic origin and to be sure that there were no close relatives among them. During the prerelease physical examinations, four chimpanzees were deemed physically unfit for reintroduction (Tutin et al. 2001).

The 37 chimpanzees have been released in groups of three or more (n = 4 groups, 19 individuals), seven pairs, and singly (n = 3) (Goosens et al. 2005). (I don't know why the numbers cited in different parts of this paper do not always correspond.) There were 27 females and 10 males. This mirrors the gender ratio of the original Conkouati population (12 males, 36 females), and it seems to favor reintroduction success because adult male chimpanzees are intensely xenophobic while cycling females are generally accepted by males of many neighboring communities.

Additionally, Tutin (personal communication) notes that "male orphans [chimpanzees] tend to form excessively strong bonds with human caretakers which can hinder the independent spirit essential for successful adaptation to return to the wild."

All released individuals except an infant and two large males carried collar-mounted radio transmitters and have been monitored closely on a daily basis. A preestablished, GPS-mapped trail system has facilitated monitoring. Systematic observations have been made of social, ranging, nesting, and feeding behavior. Supplemental feeding has been minimal, though observers have attempted to lead the chimpanzees to fruit trees (Tutin et al. 2001). Tutin et al. (2001) and Goosens et al. (2005) provide details on postrelease outcomes for the reintroduced chimpanzees. As of early 2004 there had been five confirmed deaths, nine disappearances, and 23 survivors. Some of the individuals who disappeared may well be alive. The chimpanzees ate a wide variety of wild plant and animal foods, had an activity budget resembling that of wild chimpanzees, made night nests, and appeared healthy. They ranged widely, and most made excursions into neighboring forests. They have also had many interactions with the resident wild chimpanzees. Nine of the ten released males were attacked by wild chimpanzees, some repeatedly. Four died or disappeared following attacks, and the others required veterinary intervention, leading Goosens et al. (2005) to conclude that male chimpanzees should not be reintroduced at sites where there is a resident wild population. Nine females were also attacked by wild chimpanzees; two died and the dependent infant of another was killed. Many of the serious and fatal wounds to both males and females have been to the anogenital area. Some of the reintroduced females formed sexual consorts with wild males, disappearing for days at a time. Five infants were born, at least one of whom was sired by a reintroduced male. Several of these females were integrated into wild groups for extended periods, and then returned to the release zone pregnant and gave birth in their released group.

The Conkouati reintroduction effort has from the outset adhered closely to the guidelines for reintroduction published by the IUCN Reintroduction Specialist Group (IUCN 1998). The rationale for decision making is clearly explicated, and is rooted in a scientific understanding of chimpanzees' natural history and a careful review of other reintroduction efforts. Individual chimpanzees underwent veterinary and genetic screening and were behaviorally assessed before reintroduction. They experienced a long period of prerelease conditioning in a nearby naturalistic environment, and their release site was carefully chosen and exhaustively surveyed. There has also been intensive postrelease monitoring of the reintroduced chimpanzees, and peer-reviewed scientific documentation of their outcomes. Furthermore, there has been an associated community education program with participatory management, and local people have been employed by the project; these factors are correlated with reintroduction success (Beck et al. 1994). Not only has the survival rate of reintroduced individuals been relatively high, but reductions in poaching and deforestation in the release area have been proposed as an additional benefit (Tutin et al. 2001).

Some Cognitive Lessons from Reintroduction

These examples demonstrate that chimpanzees from a variety of backgrounds can be reintroduced to the wild. Sanctuary or zoo chimpanzees who were born in the wild appear to adjust to life in the wild and demonstrate survival-critical behaviors. I hypothesize that chimpanzees (and probably orangutans) who were born in the wild, and who lived with their mother for at least a year before being orphaned, appear to enjoy a "behavioral inoculation" that facilitates their later reintroduction into the wild, even after years in captivity. This would not be predicted from the long period (six years or more) of maternal dependency during which young apes are thought to slowly acquire knowledge of appropriate foods and foraging techniques, recognition and avoidance of predators and dangerous animals, and other survival-critical skills such as nest building, locomotion, and appropriate social behavior. In contrast to wild-born individuals, captive-born chimpanzees appear to require more tutoring from humans or conspecifics, as well as more postrelease support, especially if they have been hand-reared by humans.

Chimpanzees and orangutans show "traditions" comprising intergenerational group-specific foraging and communicative behaviors that are not dependent on specific environmental affordances, and they may differ from other groups of conspecifics living in similar environments (van Schaik et al. 2003; see also chapters 8, 9, and 14). These traditions are based on social learning, mainly by infants from mothers and peers, and most of this learning occurs from birth through adolescence. The chimpanzee reintroductions described above mixed chimpanzees from different social and geographic origins. Whatever behavioral traditions they had acquired were homogenized and diluted by their lives in captivity, and generations may be required for new traditions to be established. The Conkouati reintroductees may have introduced new behavioral variants to the resident wild population. Cultural considerations may be less important than conservation, welfare, or medical considerations in ape reintroduction—but if possible, we should avoid either introducing novel cultural variants or eradicating ones that might survive from reintroductees' lives before captivity.

There may be another, longer-term effect of orphanage life on reintroduced female chimpanzees that pertains to their rearing techniques. Human and chimpanzee mothers rear their offspring quite differently (Matsuzawa 2007). For example, chimpanzee infants cling to their mothers for at least the first four months of life and rarely have direct face-to-face contact with them, while humans are commonly separated from their mothers in this time frame and are often placed in a supine position from which they have opportunities for face-to-face interactions with their mothers. Furthermore, unlike human infants, chimpanzees learn little if anything through direct teaching by the mother, but instead rely almost completely on observational learning. Infant chimpanzee orphans are usually assigned surrogate human mothers who tenderly attend to all of the infants' needs in, from what I have been able to learn, *the style of human mothers*. The surrogate mother—and, later, special human "tutors"—actively mold or teach survival-critical behaviors as well as passively demonstrating them. What are the effects of these differences on reintroductions? Could these "human" rearing techniques be learned by chimpanzee mothers-to-be? If so, is it possible that ape orphanages might keep a population of willing surrogate ape mothers to rear the incoming infants? We know that chimpanzees readily adopt infants and juveniles who have lost their mothers, especially if they are kin (Goodall 1986). A surrogate mother can easily be trained to present the infant for bottle feeding or medical examination and, with proper environmental affordances, might be able to raise and tutor infants in ape, rather than human, fashion.

The main threat to reintroduced chimpanzees, especially males, appears to be life-threatening aggression from wild conspecifics. Severe intraspecific aggression may be less prominent among gorillas and orangutans. The descriptions of intraspecific aggression experienced by reintroduced chimpanzees (Goosens et al. 2005) should dispel the romantic notion that reintroduction increases their well-being. Still to be discovered is the cognitive basis of chimpanzee territoriality in general (see chapter 28), as well as more specific aspects of aggression such as male chimpanzees' practice of targeting the anogenital region, particularly the testicles, of males they attack. Even if we understood the development and causation of male chimpanzees' aggressive behavior, however, it would seem inadvisable to try to control or lessen it in potential reintroduction candidates.

Guidelines Facilitating Ape Reintroduction

Reintroduction is feasible, and is the preferred option for orphaned chimpanzees. But it is still uncommon because of its high cost and its varied challenges that require multidisciplinary expertise. The most serious obstacle is the lack of suitable reintroduction sites. The IUCN has published a set of general guidelines for the reintroduction of all taxa of animals and plants (IUCN 1998), and another set of guidelines specifically for nonhuman primates (Baker 2002). Both sets stipulate that reintroduction should occur only in *suitable habitat* within the *historic range* of a species. In my opinion, the welfare issues that surround the great apes, and the critically imperiled status of at least some subspecies of gorillas and orangutans, compel us to consider the possibility of releasing apes into habit that is less than suitable and/or outside of the historic range of a given taxon's historic range. This is the third set of questions we explore in this chapter.

A new set of IUCN reintroduction guidelines written specifically for great apes (Beck et al. 2007) provides greater flexibility, mainly in terms of the selection of release sites. Before considering this flexibility, however, we should reiterate the guidelines that cannot responsibly be stretched. Every set of reintroduction guidelines stresses the Precautionary Principle, which is a phrase often attributed to the Hippocratic Oath: "Above all, do no harm." A reintroduction should "not cause adverse side effects of greater impact" than the conservation benefit. Such side effects would include crowding, social stress, genetic swamping, disease, exaggerated resource competition, preventable suffering or death, and unmanageable human conflict. The main concern here is the welfare of a remnant wild population, if one exists, and the environment in the reintroduction area—but there could also be harm to the wild or captive donor population.

Many of the other reintroduction guidelines also should not be stretched. The threats that have originally caused the extirpation or reduction of the wild population must be eliminated or substantially reduced. The long period of dependency and cognitive development of great apes necessitates extensive and expert prerelease preparation, especially of orphans who have spent little time in the wild. Skills in locomotion, food finding, spatial orientation, predator avoidance, nest building and appropriate social behavior must all be learned (e.g., Russon 2000). There must also be extensive programs of veterinary health to protect not only the reintroduced apes but also the wild apes and other animals they may contact, as well as their caretakers and the other people they might contact after reintroduction (see Beck et al. 2007 for details on these and other requirements and best practices).

The reintroduced apes should also be monitored for at least a year after their reintroduction, which is especially challenging. A group's home range may be as large as 300 km^2, suggesting that an ape could easily travel 10 km per day. Apes, particularly those habituated to humans, might attack humans or enter settlements and cultivated fields, creating ape-human conflict (see chapter 27). There is also resistance to using radiotelemetry with great apes. Some primatologists, rightfully, oppose elective immobilization to attach radiotelemetry devices, but this objection would not pertain to sanctuary apes which are immobilized anyway for prerelease veterinary examinations. Others object that such devices could alter the apes' behavior, or that wearing them is aesthetically objectionable. However, telemetry was used successfully in the Conkouati chimpanzee reintroduction.

The new guidelines provide for release of sanctuary or wild great apes *outside* the historic range of the species or subspecies in question. These would technically be called *introductions* rather than *reintroductions*. The guidelines describe and set conditions for *conservation introductions* and *welfare introductions* (Beck et al. 2007). Much of the following wording is borrowed from the new guidelines, to which the reader is directed for further information.

Conservation introductions of great apes would be conducted only as a last resort to save a genus, species, or subspecies—perhaps under emergency circumstances such as a natural disaster, war, or epidemic. When great apes are introduced outside of their historic range for conservation purposes, there should be clear agreement among all parties that the introduced population will be repatriated to an area within its historic range as soon as habitat is available and threats have been ameliorated.

Welfare introductions would be considered only when it is no longer possible to provide the apes with humane care in a sanctuary, or when there is strong reason to believe that moving them from a sanctuary or zoo to a free-ranging habitat would increase their well-being substantially. Many sanctuaries already have such habitats (e.g., Ngamba Island Sanctuary), and welfare introductions might be said to have already occurred in some cases (there currently is no accepted standard as to whether a large forested enclosure that is part of a sanctuary is actually an introduction site). Note that welfare introductions are not to be conducted solely to dispose of surplus animals or to relieve overcrowding. Welfare introductions should take place only when there is no realistic prospect of reintroducing the apes to suitable habitat within their historic range.

Another example of release outside of historic range involves islands. The new ape guidelines allow for and describe introduction on islands that may never have had living apes. Indefinite provisioning with food and water, along with active population management, is likely to be necessary on islands smaller than 500 ha and/or with densities of more than 0.1 individual per hectare (based on experiences with chimpanzees in Africa; density might vary with ape type and age/sex distribution). This would not be a self-sustaining population; the introduction might better be characterized as a seminaturalistic sanctuary. The island would probably have to be at least 50,000 ha in order to support a self-sustaining population. Islands smaller than 5 ha are not likely to support an ape population of any size, even with intense provisioning.

There is also a temporal dimension to "historic range." Apes are known to have occurred at locations where they no longer occur now and may not have occurred for centuries. For example, wild orangutans have been absent from mainland Asia for at least 500 years (Rijksen and Meijaard 1999). There are no guidelines about the maximum time a taxon has to have been absent from an area for scientists to conclude that it can no longer be considered the taxon's historic range for purposes of reintroduction. Therefore, one might responsibly consider reintroducing orangutans in Thailand.

Reintroduction guidelines stipulate that animals should be reintroduced into suitable habitat of sufficient size to support a self-sustaining population. But chimpanzees and Bornean orangutans are known to be able to survive in partially cleared forests, at least for short periods. Apes also have been found at higher altitudes than those permanently occupied by wild conspecifics. Some individuals have been able to survive in marginal habitats with intensive support. Great apes might be reintroduced in habitats that are suboptimal in quality, location, or size if there could be extensive postrelease support, or if there were a real prospect of more or better habitat becoming available—from reforestation, for example. Another guideline stipulates that at any reintroduction site, wild conspecifics must be absent or well below the site's carrying capacity. Estimating a site's carrying capacity, estimating the size and distribution of an existing wild population, or even confirming the local extinction of a species are daunting scientific exercises, and beyond the expertise and resources of most reintroduction planners. It might be sufficient to rely on qualitative judgments by ecologists who have no personal position on whether or not to conduct the reintroductions. Conservation and welfare introductions, introductions into "stretched" definitions of historic range and suitable habitats, and reliance on informed qualitative judgments of a site's carrying capacity and population size are all intended to help planners overcome some major obstacles to the reintroduction of sanctuary apes, but they are not invitations to conduct these reintroductions irresponsibly, or convenient solutions for reducing sanctuary populations.

We may see more great ape reintroductions in the next decade, but it is unlikely that they will reduce sanctuary populations significantly. The growth of those populations appears inevitable, and the options for sanctuary apes have not changed in the past 20 years. Although results are still preliminary, there is some basis to predict that the number of sanctuary chimpanzees will continue to increase even if each of our pilot sanctuaries could reintroduce or introduce 8 to 10 apes every other year (Faust et al. 2007). Therefore we can no longer avoid having to choose from among the options for sanctuary apes described above, and

we should no longer avoid providing the resources to support implementation of one or more options. Refusing to choose from among the options is in itself a choice to leave the apes in already overcrowded range country sanctuaries without assured long-term support. One obvious source of support for range country sanctuaries would be zoos in North America, Europe, Asia, and Australia. Zoos admittedly are not sanctuaries, but they share the sanctuaries' mission of furthering animal welfare, and they also might have constituencies that would contribute to the maintenance of sanctuaries. Zoos have expertise in animal care, nutrition, veterinary medicine, public education, behavioral research, and animal training, all of which could be contributed in the form of staff exchange. Some zoos already do this, but many others contribute nothing to range-country ape conservation. One disincentive to financial support may be that many ape-range countries have recent histories of political instability, and at least one sanctuary has had to be abandoned in the midst of armed conflict. But there are many other range-country sanctuaries where such disruption is unlikely today. Support of these sanctuaries appears to be an appropriate and appealing way for zoos in the developed world to fulfill their commitments to conservation.

The Chimpanzee Mind and Ape Conservation

Ape orphans appear to receive disproportionate consideration from humans. As I have noted above, their clinging response may be partly responsible for this attention, but our fourth ethical question is whether the cognitive and genetic similarities of apes to humans also play a role.

Boesch (2007) stresses that we must be cautious in our comparisons of apes and humans, specifically with regard to their cognitive capacities. While humans may be the fifth ape, they are decidedly different from the other apes. Ape behavior is not always inspiring (but then, human behavior is not always inspiring), and in comparisons with humans the apes always seem somehow lacking (e.g., Premack 2004), or at least different (Herrmann et al. 2007). Of course the comparisons are rarely really fair since the apes are often tested on cognitive tasks designed by humans, and are tested by humans in a human-configured setting.

Herrmann et al. (2007) exemplify these confounds. They presented young human children, orangutans and chimpanzees with several tasks in both social and physical cognitive domains. Their major finding was that the children's performance was basically equivalent to that of chimpanzees in tasks of physical cognition, but "far outstripped" those of both both chimpanzees and orangutans in tasks of social cognition (p. 1365). The chimpanzee and orangutan subjects lived in very good sanctuaries, but their very presence in sanctuaries meant that they had been orphaned, probably at a young age and probably after witnessing violent events involving their mothers and other group members. They must have been emotionally traumatized, since we know that chimpanzees grieve deeply for lost kin. After their initial capture, the ape subjects had lived in a variety of captive conditions where they were forced to cope socially with a variety of humans and unfamiliar chimpanzees. But Herrmann et al. optimistically dismiss this background and state that "All of these apes live in the richest social and physical environments available to captive apes . . ." (p. 1361). Compare this to Matsuzawa's (2006) straightforward assessment: "An ape that was isolated from its conspecifics and raised by humans cannot be a real ape" (p. 5). Even without maternal deprivation, as Sue Savage-Rumbaugh (personal communication) asserts: "captivity builds captive brains"—and captive brains build captive minds. In contrast to the apes in the Herrmann et al. study, the human subjects lived with their parents in a German city, in a natural physical and social environment, and presumably had been spared childhood trauma. De Waal et al. (2008) identify additional confounds in this study: the apes were tested by another species (humans) and received no verbal instructions while the humans were tested by conspecifics and did receive verbal instructions. Each of the apes was tested alone, while each of the human children was tested in a room in the presence of a parent. It is not surprising that the human children "far outstripped" these apes on social tasks. Like Boesch (2007) and de Waal et al. (2008), I would assert that the findings do not represent the social capabilities of wild great apes raised in a natural environment. Herrmann et al. did administer a test battery designed to measure "the comfort level of the participants in the test situation," but the dependent variable was test performance, not comfort level.

Of course, despite the confounds of subject history and experimental context, the major conclusion of Herrmann et al. may prove to be correct. But it is not totally consistent with other findings on chimpanzee social cognition. Chimpanzees recognize their own mirror image, and thus are thought to be self-aware (Gallup 1970). They can understand the perceptions, the knowledge, and perhaps the intentions of other chimpanzees and humans, and they respond appropriately (Whiten and Byrne 2007; see also chapters 19 and 21). However, they may not be able to understand the full range of emotional states of others (Tomasello et al. 2003). Despite their lack of tears, they can understand the distress of another and, like Indah, they can console each other to alleviate distress (see chapter 17). They are more likely to do so with familiar individuals (Preston and de Waal 2002). They have a conception of the relative social ranks of other chimpanzees and of themselves, behaving differently in social interactions depending on whether a more dominant individual is present (see chapters 15 and 16). They know the relative number of adult males, the sex ratio of adults, and the home ranges of neighboring groups (see chapter 15). As we have discussed above, they learn socially and appear to have intergenerational group traditions or cultures.

But when it comes to food, individual chimpanzees do not commonly behave in a way that will benefit other chimpanzees more than themselves, especially when the others are neither kin, nor familiar, nor of equal rank. They can cooperate to solve problems that one individual cannot solve alone, sometimes playing different roles in the cooperative relationship if the rewards are distributed equally (see chapter 20; also Jensen et al. 2007). Otherwise, though, they tend to behave in a way that benefits themselves more than others (see chapter 22; also Silk et al. 2005). Wild chimpanzees do share meat, but they seem to be motivated by the social and sexual benefits of the exchange (Stanford 1998; but see also chapter 18). Chimpanzees can and do actively deceive others (de Waal 1982; Hare et al. 2006). Bonobos are not as well studied as robust chimpanzees, but there are indications that they may have more regard for the interests of others than chimpanzees do (Miller 2007). All of these studies are commonly thought to demonstrate intentionality in chimpanzees.

In summary, the social profile of chimpanzees suggests that they have some recognition of the interests and perceptions of others, and that they grieve and console each other. They recognize their kin, and they have a keen understanding of their own social group and their place in it.

But for the most part they are selfish, favoring their own interests as well as those of their kin and others who have helped them or are likely to do so. They are also capable of brutality against like-sexed strangers, potential rivals, and the infants of rivals. This is not an especially flattering profile, but it is equally applicable to many humans.

Outside of the social realm, chimpanzees show sophisticated knowledge of the physical world (Herrmann et al. 2007; Tomasello and Call 1997). They use and make a variety of tools (Beck 1980; Goodall 1986). They understand that objects that have been removed from their surroundings still exist (object permanence), and in some situations they also show understanding of causality. They possess cognitive maps of their own group's home range, and probably of important features within the range (see chapter 15). Some chimpanzees (as well as bonobos) receptively comprehend hundreds of non-iconic symbols and accurately associate them with corresponding objects, individuals, and actions. They can also produce these symbols, combine them according to grammatical rules, and combine them to label novel objects. These operations are language-like (Savage-Rumbaugh and Lewin 1994). They can make comparative judgments of the quantities of objects, recognize cardinal numerals up to 9, associate the appropriate cardinal numeral with the number of objects in a display, and they may be able to sum simple numerals and the numbers of objects (Beran and Rumbaugh 2001; Biro and Matsuzawa 2001; Boysen and Berntson 1989).

While chimpanzees appear to have elemental skills with regard to tool use, language, and quantitation, they are clearly not engineers, novelists, or mathematicians. Nor are human children and many human adults. Premack (2007) asserts that the *similarities* with human cognitive function do not demonstrate *equivalence* with human cognitive function, and that the dissimilarities between chimpanzee and human cognition outnumber and outweigh the similarities. Even the magnitude and meaning of the genetic similarities and differences between humans and chimpanzees are being questioned, and our close genetic affinity is being called mythical (Cohen 2007).

Further, many monkeys, and even other mammals and some birds, exhibit sophisticated social, foraging, communicative, and tool behaviors that are comparable to those of chimpanzees. A complete review is beyond the scope of this chapter, but see Cheney and Seyfarth (2007) and Beck (1973) for information on the behaviors of baboons.

The last common ancestor of humans, apes, and modern monkeys lived millions of years ago. We often uncritically accept the similarities of Old World monkeys with apes and humans as being homologous (shared by common descent), while those of the more distantly related New World monkeys and other mammals and birds as being convergent (selected by similar environmental conditions). This is plausible, given the length of these species' evolutionary separation, but it is entirely unproven. The similarities between baboons and chimpanzees, and even between chimpanzees and humans, may also be convergent. But the sheer volume of similarities between chimpanzees and humans, combined with their close genetic similarity, points to homology. This, in turn, consciously or unconsciously causes many people to grant special conservation status to orphan apes, and to support ape reintroduction and conservation.

Implications and Future Directions

I acknowledge being a "speciesist" with regard to great apes. I see them and their conservation as being more compelling than that of other animals and plants, although I recognize that this makes no sense ecologically. They are special, sentient beings, as is reflected in the movement to make great apes World Heritage Species. I argue that we are ethically obligated to bring resources, compassion, and resolve to the issues of ape orphans and sanctuaries. If we can't stop the destruction of wild ape populations and their habitats, we are ethically obligated to find ways to reintroduce or introduce these orphans into the wild and/or to expand sanctuary capacity, continue improving sanctuary management, and provide lifetime care in range-country sanctuaries. Note that this is explicitly *not* an effort to acquire more apes for zoos in the developed world. Sanctuaries serve great ape conservation. The remarkable cognitive abilities of apes suggest that the only palatable options for sanctuary apes are to reintroduce them to the wild or retain them in good psychological and physical health in sanctuaries. Both of these options are expensive. Reintroduction is feasible in some cases, especially with the flexibility provided in newly published guidelines, but reintroduction will not be an effective, safe, or humane option in most situations. While every sanctuary manager wants nothing more than to go out of business, that may not be possible for the next century, at best.

Literature Cited

Aczel, P. 1993. Pointe Noire Zoo chimpanzees rescued. *IPPL Newsletter*, August 1993: 15–16.

Amman, Karl. 2001. Bushmeat hunting and the great apes. In *Great Apes & Humans: The Ethics of Coexistence*, ed. Benjamin Beck, Tara Stoinski, Michael Hutchins, Terry Maple, Bryan Norton, Andrew Rowan, Elizabeth Stevens, and Arnold Arluke, 71–85. Washington DC: Smithsonian Institution Press.

Baker, L. R. 2002. IUCN/SSC Re-introduction specialist group: Guidelines for nonhuman primate re-introductions. *Re-Introduction NEWS* 21:29–57.

Beck, B. 1980. *Animal Tool Behavior*. New York, Garland.

———. 1973. Cooperative tool use by captive hamadryas baboons. *Science* 182:594–97.

Beck, B., L. G. Rapaport, M. R. Stanley Price, and A. C. Wilson. 1994. Reintroduction of captive-born animals. In *Creative Conservation: Interactive Management of Wild and Captive Animals*, ed. G. Mace, P. Olney, and A. Feistner, 265–86. London: Chapman and Hall.

Beck, B., T. Stoinski, M. Hutchins, T. Maple, B. Norton, A. Rowan, E. Stevens, and A. Arluke. 2001. *Great Apes & Humans: The Ethics of Coexistence*. Washington, DC: Smithsonian Institution Press.

Beck, B., K. Walkup, M. Rodrigues, S. Unwin, D. Travis, T. Stoinski, and L. Williamson. 2007. *Guidelines for Great Ape Re-Introduction*. Gland, Switzerland: International Union for the Conservation of Nature and Natural Resources.

Beran, M. J., and D. M. Rumbaugh. 2001. "Constructive" enumeration by chimpanzees (*Pan troglodytes*) on a computerized task. *Animal Cognition* 4:81–89.

Biro, A. and T. Matsuzawa. 2001. Chimpanzee numerical competence: Cardinal and ordinal skills. In *Primate Origins of Human Cognition and Behavior*, ed. T. Matsuzawa, 199–225. Tokyo: Springer.

Boesch, C. 2007. What makes us human (*Homo sapiens*)? The challenge of cognitive cross-species comparison. *Journal of Comparative Psychology* 121(3): 227–40.

Borner, M. 1985. The rehabilitated chimpanzees of Rubondo Island. *Oryx* 19:151–54.

Boysen, S. T., and G. G. Berntson. 1989. Numerical competence in a chimpanzee (*Pan troglodytes*). *Journal of Comparative Psychology* 103:23–31.

Brewer, S. 1978. *The Chimps of Mt. Asserik*. New York: Alfred A. Knopf.

Carlsen, F., D. Cress, N. Rosen, and O. Byers. 2006. *African Primate Reintroduction Workshop Final Report*. Apple Valley, MN: IUCN/SSC Conservation Breeding Specialist Group.

Carter, J. 1981. A journey to freedom. *Smithsonian* 12(1): 90–101.

———. 1988. Freed from keepers and cages. *Smithsonian* 19(3): 36–49.

———. 2003. Orphan chimpanzees in West Africa: Experiences and prospects for viability in chimpanzee rehabilitation. In *West African Chimpanzees*, ed. R. Kormos, C. Boesch, M. Bakarr, and T. Butynski, 157–67. Gland, Switzerland: International Union for Conservation of Nature and Natural Resources.

Cartwright, B., and T. Bettinger. 2006. Overview of conservation education at PASA and the program evaluation process. *International Journal of Primatology* 27, Supplement 1, Abstract #194.

Cheney, D. L., and R. M. Seyfarth. 2007. *Baboon Metaphysics*. Chicago: University of Chicago Press.

Cohen, J. 2007. Relative differences: The myth of 1%. *Science* 316:1836.

De Waal, F. 1982. *Chimpanzee Politics: Power and Sex among Apes*. New York: Harper & Row.

De Waal, F., C. Boesch, V. Horner, and A. Whiten. 2008. Comparing social skills of children and apes. *Science* 319:569.

Farmer, K. H. 2000. The final step to freedom: Conkouati chimpanzees returned to freedom. *IPPL News* 27(2): 17–20.

———. 2002. Pan-African Sanctuary Alliance: Status and range of activities for great ape conservation. *American Journal of Primatology* 58:117–32.

Farmer, K. H. and A. Courage. 2009. Sanctuaries and reintroduction: A role in gorilla conservation? In *Conservation in the 21st Century: Gorillas As a Case Study*, ed. T. S. Stoinski, H. D. Steklis, and P. Mehlman. Springer.

Faust, L., B. Beck, and D. Cress. 2007. Estimating future capacity needs for PASA sanctuary ape populations. Paper presented as the annual managers' meeting of the Pan African Sanctuary Alliance, June 21–24, in Kigali, Rwanda.

Fossey, D. 1983. *Gorillas in the Mist*. Boston: Houghton Mifflin.

Francione, G. L. 1993. Personhood, property and legal competence. In *The Great Ape Project*, ed. P. Cavalieri and P. Singer, 248–57. New York: St. Martin's Press.

Gallup, G. G. 1970. Chimpanzees: Self-recognition. *Science* 167:86–87.

Goodall, J. 1986. *The Chimpanzees of Gombe: Patterns of Behavior*. Cambridge, MA: Belknap Press of Harvard University Press.

———. 2001. Problems faced by wild and captive chimpanzees: Finding solutions. In *Great Apes & Humans: The Ethics of Coexistence*, ed. B. Beck, T. Stoinski, M. Hutchins, T. Maple, B. Norton, A. Rowan, E. Stevens, and A. Arluke, xiii–xxiv. Washington, DC: Smithsonian Institution Press.

Goosens, B., S. M. Funk, C. Vidal, S. Latour, A. Jamart, M. Ancrenaz, E. J. Wickings, C. E. G. Tutin, and M. W. Bruford. 2001. Measuring genetic diversity in translocation programmes: Principles and application to a chimpanzee release project. *Animal Conservation* 5:225–36.

Goosens, B., J. M. Setchell, E. Tchidongo, E. Dilambaka, C. Vidal, M. Ancrenaz, and A. Jamart. 2005. Survival, interactions with conspecifics and reproduction in 37 chimpanzees released into the wild. *Biological Conservation* 123:461–75.

Hannah, A. C. and W. C. McGrew. 1991. Rehabilitation of captive chimpanzees. In *Primate Responses to Environmental Change*, ed. H.O. Box, 167–86. London: Chapman and Hall.

Harcourt, A. H. 1987. Options for unwanted or confiscated primates. *Primate Conservation* 8:111–13.

Hare, B., J. Call, J, and M. Tomasello. 2006. Chimpanzees deceive a human competitor by hiding. *Cognition* 101:495–514.

Herrmann, E., J. Call, M. V. Hernandez-Lloreda, B. Hare, and M. Tomasello. 2007. Humans have evolved specialized skills of social cognition: The cultural intelligence hypothesis. *Science* 317:1360–366.

Husband, S., R. Meyerson, and W. Shellabarger. 2007. Successful orangutan infant rearing at the Toledo Zoo using multiple training and rearing techniques. Paper presented at the Orangutan SSP Husbandry Workshop, Brookfield, Illinois, 16–18 October.

ISIS. 2007. Taxon reports for orangutans, chimpanzees, bonobos, and gorillas. http://www.isis.org (membership required).

IUCN. 1998. *IUCN guidelines for re-introductions*. Gland, Switzerland, International Union for the Conservation of Nature and Natural Resources.

———. 2002. *IUCN guidelines for the placement of confiscated animals*. Gland, Switzerland, International Union for the Conservation of Nature and Natural Resources.

Jensen, K., J. Call, and M. Tomasello. 2007. Chimpanzees are rational maximizers in an ultimatum game. *Science* 318:107–9.

Karlowski, U. 1996. The Conkouati Chimpanzee Refuge: A new chance for orphans. *Gorilla Journal* 12:20.

Kierulff, M. C. M., P. P. de Oliveira, B. B. Beck, A. and Martins. 2002. Reintroduction
and translocation as conservation tools for golden lion tamarins. In *Lion Tamarins: Biology and Conservation*, ed. Devra Kleiman and Anthony Rylands, 271–82. Washington, DC: Smithsonian Institution Press.

King, T., C. Chamberlan, E. Pearson, and A. Courage. 2006. Western gorilla (*Gorilla g. gorilla*) reintroduction to the Bateke Plateaux and the challenge of tourism. *International Journal of Primatology* 27, Supplement 1, Abstract #486.

Lyttle, J. 1997. *Gorillas in Our Midst*. Columbus: Ohio University Press.

Marsden, S. B., D. Marsden, and M. E. Thompson. 2006. Demographic and female life history parameters of free-ranging chimpanzees at the Chimpanzee Rehabilitation Project, River Gambia National Park. *International Journal of Primatology* 27(2): 391–410.

Matsumoto-Oda, A. 2000. Chimpanzees in the Rubondo Island National Park, Tanzania. *Pan Africa News*: 7(2): 16–17.

Matsuzawa, T. 2006. Sociocognitive development in chimpanzees: A synthesis of laboratory work and fieldwork. In *Cognitive Development in Chimpanzees*, ed. T. Matsuzawa, M. Tomonaga, and M. Tanaka, 3–33. Tokyo: Springer-Verlag.

———. 2007. Comparative cognitive development. *Developmental Science* 10(1): 97–103.

Melis, A. P., B. Hare, and M. Tomasello. 2006. Chimpanzees recruit the best collaborators. *Science* 311: 1297–1300.

Miller, G. 2007. All together now—pull! *Science* 317:1338–40.

Moscovice, L. R., M. H. Issa, K. J. Petrzelkova, N. S. Keuler, C. T. Snowdon, and M. A. Huffman. 2007. Fruit availability, chimpanzee diet, and grouping patterns on Rubondo Island, Tanzania. *American Journal of Primatology* 69:487–502.

Pratje, P., and I. Singleton. 2006. Reintroduction of orangutans in Sumatra. *International Journal of Primatology* 27, Supplement 1, Abstract #490.

Preston, S. D., and F. B. M. de Waal. 2002. Empathy: Its ultimate and proximate basis. *Behavioral and Brain Sciences* 25:1–72.

Premack, D. 2004. Is language the key to human intelligence? *Science* 303:318.

———. 2007. Human and animal cognition: Continuity and control. *Proceedings of the National Academy of Sciences* 104:13861–67.

Rijksen, H. D., and E. Meijaard. 1999. *Our Vanishing Relative: The Status of Wild Orang-utans at the Close of the Twentieth Century*. Dordrecht, the Netherlands: Kluwer Academic Publishers.

Rosen, N, D. Cox, C. Montgomery, and D. Cress. 2002. *Pan-African Sanctuaries Workshop Report*. Apple Valley, MN: IUCN/SSC Conservation Breeding Specialist Group.

Russon, Anne. 2000. *Orangutans: Wizards of the Rainforest*. Buffalo, NY: Firefly Books.

Savage-Rumbaugh, S., and R. Lewin. 1994. *Kanzi: The Ape at the Brink of the Human Mind*. New York: Wiley.

Silk, J. B., S. F. Brosnan, J. Vonk, J. Henrich, D. J. Povinelli, A. S. Richardson, S. B. Lambeth, J. Mascaro, and S. J. Schapiro. 2005. Chimpanzees are indifferent to the welfare of unrelated group members. *Nature* 437:1357–59.

Shumaker, R. W., and B. B. Beck. 2003. *Primates in Question*. Washington, DC: Smithsonian Institution Press.

Stanford, C. B. 1998. *Chimpanzee and Red Colobus*. Cambridge, MA: Harvard University Press.

Teleki, G. 2001. Sanctuaries for ape refugees. In *Great Apes & Humans: The Ethics of Coexistence*, ed. B. Beck, T. Stoinski, M. Hutchins, T. Maple, B. Norton, A. Rowan, E. Stevens, and A. Arluke, 133–49. Washington, DC: Smithsonian Institution Press.

Tomasello, M., and J. Call. 1997. *Primate Cognition*. New York: Oxford University Press.

Tomasello, M., J. Call, and B. Hare. 2003. Chimpanzees understand emotional states—the question is which ones and to what extent? *Trends in Cognitive Sciences* 7(4):153–56.

Tutin, C., M. Ancrenaz, J. Paredes, M. Vacher-Vallas, C. Vidal, B. Goosens, M. Bruford, and A. Jamart. 2001. Conservation biology framework for the release of wild-born orphaned chimpanzees into the Conkouati Reserve, Congo. *Conservation Biology* 15(5): 1247–57.

Van Schaik, C. P., M. Ancrenaz, G. Borgen, B. Galdikas, C. D. Knott, I. Singleton., A. Suzuki, S. S. Utami, and M.Y. Merrill. 2003. Orangutan cultures and the evolution of material culture. *Science* 299:102–5.

Weber, W. 1995. Monitoring awareness and attitude in conservation education: the mountain gorilla project in Rwanda. In *Conserving Wildlife: International Education and Communication Approaches*, ed. S. K. Jacobson, 28–48. New York: Columbia University Press.

Whiten, A., and Byrne, R. W. 2007. *Machiavellian Intelligence II: Extensions and Evaluations*. Cambridge, UK: Cambridge University Press.

Wilkie, David. 2001. Bushmeat trade in the Congo Basin. In *Great Apes & Humans: The Ethics of Coexistence*, ed. B. Beck, T. Stoinski, M. Hutchins, T. Maple, B. Norton, A. Rowan, E. Stevens, and A. Arluke, 86–109. Washington, DC: Smithsonian Institution Press.

Yeager, C. 1997. Orangutan rehabilitation in Tanjung Putting National Park, Indonesia. *Conservation Biology* 11:1997.

27

Human-Chimpanzee Competition and Conflict in Africa: A Case Study of Coexistence in Bossou, Republic of Guinea

Kimberley J. Hockings

"The Bossou chimpanzees will never leave the forests surrounding our village; they are our ancestors and are different from other chimpanzees. They come to the forest edge and scream to us when an elder in the village is about to die. They are very intelligent and like us; the males protect the females and young from danger. They often visit the village and fields where we cultivate, forming an orderly queue when they enter. We let them feed from our fields so they will never go hungry."

Elderly Bossou villager, 2005

Increasing forest degradation and expanding land cultivation bring human and nonhuman species into more frequent direct contact and competition, demanding that various aspects of human-wildlife relationships be reexamined (Lee et al. 1986; Cowlishaw and Dunbar 2000; Fuentes and Wolfe 2002; Patterson and Wallis 2005; Lonsdorf 2007). For the most part throughout Africa, chimpanzees *(Pan troglodytes)* inhabiting large protected areas such as the Niokolo-Koba National Park in Senegal or Taï National Park in Côte D'Ivoire rarely come into contact with people (McGrew et al. 1981; Boesch and Boesch-Acherman 2000). However, between 45 and 81% of the surviving chimpanzee populations exist outside designated protected areas (Kormos et al. 2003), and those adjacent to agricultural land and human settlements are frequently reported to raid crops (Hill 1997; Naughton-

Treves et al. 1998; Hockings et al. 2009) and compete with humans for natural resources such as wild fruits and water (Pruetz 2002).

In this chapter I provide a general overview of past research pertaining to situations of human-primate competition and conflict over resources, with particular focus on chimpanzees. I then present a case study of a chimpanzee community living in close proximity to humans, and argue that in-depth ecological, behavioral, and cognitive knowledge of a species is essential for understanding how they perceive and show adaptations to changes in their environment. Using this case study, the importance of accurately identifying a situation of conflict is highlighted, along with the need for collecting data from a range of different perspectives.

Defining Conflict

There are inherent difficulties with defining a "situation of conflict." Human cultural attitudes and practices vary greatly between areas, and therefore the perception of conflicts between humans and wildlife will differ accordingly. The broad definition of human-wildlife conflict (modified from the IUCN/SSC African Elephant Specialist Group) employed in this chapter is "any human-wildlife interaction which results in negative effects on human social, economic or cultural life, wildlife social, ecological or cultural life or the conservation of wildlife and their environment." There are many different situations in which wildlife and humans come into conflict (see chapters 26 and 28). They can be directly created by humans, and they include disease transmission, the deliberate killing and capture of wildlife for the bushmeat and pet trades (Teleki 1989), and the accidental killing or wounding of wildlife by traps and snares (Reynolds 2005). Furthermore, conflicts might be a consequence of wildlife activity: namely crop raiding (Naughton-Treves et al. 1998; Hockings et al. 2009), natural resource competition (e.g., wild foods and water; Pruetz 2006), domestic animal predation (Wallis, personal communication 2007), and aggressive interactions with humans (Wrangham 2001; McLennan 2008).

When a wild species crosses into agricultural land to raid crops, or is in close contact with humans, it is important not to assume automatically that conflict always results (Siex and Struhsaker 1999; Hill et al. 2002). The term conflict, which is regularly used to describe these associations, may in fact provoke the situation into one of hostility (Lee 2004). It is necessary to accurately determine whether human-wildlife relationships actually lead to a conflict situation before making negative assumptions (Priston 2005). It has also been argued that referring to wildlife feeding on human-cultivated foods as "crop raiding" will elicit or exacerbate negative perceptions, and therefore the term resource competition may be more appropriate. However, the term resource competition is also ambiguous, as humans and wildlife share and compete for different resources in various ways. For consistency with past research the expression crop raiding will be employed in this chapter to describe guarded crop-feeding forays.

The Human-Primate Frontier

As our evolutionary cousins, nonhuman primates form an integral part of human mythologies, diets, and scientific concepts (Fuentes and Wolfe 2002). However, primates in particular pose severe problems as crop raiders, owing to their ecological flexibility and behavioral plasticity, along with the large quantity of crops that they may damage and consume (Strum 1984; Newmark et al. 1994; Patterson and Wallis 2005). Nearly all families of primates have been noted to include cultivated foods in their feeding repertoire. Around African and Asian reserves, primates are considered responsible for more than 70% of the damage

events and 50% of the area damaged (Naughton-Treves 1998; Hill 2000, 2005).

Certain primate species prosper in, and have adapted to, areas of human-wildlife interface, as illustrated by widespread complaints about members of the *Macaca* genus in Asia, and the *Papio* and *Chlorocebus* genera in Africa. Crop raiding is a foraging strategy, and therefore it has dynamic costs and benefits. These genera share traits that enable them to successfully exploit the agricultural landscape surrounding them: they live in complex social organizations, are highly omnivorous, and are partly or primarily terrestrial with the ability to use arboreal habitats. In addition to common crop-raiding species, there are growing numbers of complaints about ecologically more specialized primates such as the endangered Zanzibar red colobus *(Procolobus kirkii)* and the Angolan black and white colobus *(Colobus angolensis)* feeding in agricultural areas (Siex and Struhsaker 1999; Anderson et al. 2007). Long-term studies have shown that subpopulations of the critically endangered western purple-faced langur *(Semnopithecus vetulus nestor)* have adapted to a diet high in ripe fruit, a feature which is not recorded for any other colobine; these langurs have become dependent on fruits cultivated by humans (Mittermeier et al. 2006).

Although it is less common than crop raiding by monkeys, crop raiding by the great apes is being increasingly reported (Hockings and Humle 2009). The gorillas of Bwindi *(Gorilla gorilla berengei)* have begun visiting farms and plantations as a result of increased habituation levels, and gorillas in Gabon *(Gorilla gorilla gorilla)* have been recorded raiding crops such as cassava; farmers now perceive this highly endangered great ape to be a "problem species" (Lahm 1996). Orangutans in Borneo *(Pongo pygmaeus)* are persecuted as crop pests, as some take to raiding oil-palm plantations and feeding on the young palms during shortages of natural food (Yuwono et al. 2007). The Sumatran orangutan *(Pongo abelli)* is also coming increasingly into conflict with people. Due to habitat loss through encroachment and the conversion of land for agriculture, a decline in forest fruits has forced this critically endangered great ape to feed from fruit crops and oil-palm plantations surrounding the forest, where it is often killed (SOCP 2002). Bonobos *(Pan paniscus)* at Kokolopori, Demographic Republic of Congo, live in close proximity to villages and crop-raid pineapple, sugarcane, palm pith, and banana (Georgiev, personal communication 2007).

Therefore, as increasingly important amounts of climax forest are lost and fragmented, crop raiding by bonobos is likely to increase.

Despite little detailed information on the ecology of most crop-raiding primates, even fewer data exist on behavioral adaptations of primates living in human-influenced environments, especially in the context of cognition. Maples et al. (1976) reported that several aspects of the movements and social organization of crop-raiding yellow baboon *(Papio cynocephalus)* populations in Kenya were of particular advantage for crop raiding, and were significantly different from those of non-raiding populations. Baboons located near the edges of farms showed enhanced vigilance behaviors compared to the baboons that were protected by forests, thereby reducing the potential costs associated with crop foraging. Furthermore, one baboon troop divided into subunits to conduct simultaneous raids, and possibly used vocalizations to distract the farmers' attention while other individuals crop-raided, though it remains unclear whether cognitively complex diversionary tactics are used by baboons when crop-raiding. Priston (2005) reported that Buton macaques *(Macaca ochreata)* in Sulawesi showed vigilance behaviors when crop raiding, and also decreased their risks by transporting food obtained in cultivated areas back to the relative safety of the forest. However, Priston (2005) made no inferences regarding the use of diversionary tactics or cooperative behaviors by the monkeys during raids.

Chimpanzees on the Interface

Wild chimpanzees' diet is dominated by ripe fruit, which accounts for 50 to 75% of their feeding effort (Wrangham 1977; Goodall 1986; Morgan and Sanz 2006; Pruetz 2006). Availability of, and fluctuations in, wild forest foods and fruits in particular may affect crop-raiding levels: certain crops might be crucial for chimpanzees' subsistence, while others are exploited because they are more palatable or nutritious than wild foods. For example, in Uganda abundant forest fruit did not diminish the Kibale chimpanzees' appetite for protein-rich maize, whereas increased consumption of banana pith was associated with forest fruit shortages (Naughton-Treves et al. 1998). Greengrass (2000) described how the Gombe chimpanzees in Tanzania fed on banana, mango, and other cultivated fruit situated on the forest edge, due to a loss of the habitat that had previously provided all their dietary requirements. Hill (1997)

reports that chimpanzees in the Budongo Forest, Uganda, do not crop-raid as much as other species such as baboons, although different chimpanzee communities within the reserve do vary in their crop-raiding propensities (Reynolds 2005).

The Sonso chimpanzees that inhabit the main Budongo Forest block occasionally raid crops, namely mango from surrounding orchards and sugar cane from commercial fields on the forest edge. The adjacent Nyakafunjo community, however, which lives in closer proximity to human settlements, exhibits higher crop-raiding levels than their Sonso neighbors, possibly due to recent reductions in the size of their home range. The newly studied Kasokwa forest chimpanzees inhabit a riverine strip of forest to the south of Budongo. This small community subsists mainly on forest foods, but on occasion, due to forest food shortages, the apes engage in crop-raiding activities, feeding upon papaya, mango, and sugar cane (Reynolds 2005). Approximately 30 km south of Budongo, chimpanzees at Bulindi in the Hoima District of Uganda also live in very small forest patches and regularly raid human crops including sugar cane (McLennan 2008). These studies touch upon the importance of access to wild fruits for determining levels of crop raiding, underlining the likelihood that the two are closely linked.

Relatively few data exist on the pockets of chimpanzees that live in proximity to humans and are offered protection by local human beliefs. The villagers of Bossou in the Republic of Guinea regard the chimpanzee as a sacred totem and as a reincarnation of their ancestors (Kortlandt 1986; Yamakoshi 2005). As a result of such beliefs, humans and chimpanzees have been close neighbors for many generations, and the habitual usage patterns of wild and cultivated foods by the Bossou chimpanzees are inextricably connected (Hockings et al. 2009). The Fongoli community of savannah-dwelling chimpanzees in south-eastern Senegal overlaps substantially with humans in its use of forest resources (Pruetz 2002). The chimpanzees have peacefully coexisted with humans due to a cultural taboo against the hunting of chimpanzees, but concerns over potential rising levels of crop raiding and elevated levels of natural resource competition now threatens such relations. In Guinea-Bissau, there is a similarly strong local taboo against the hunting of chimpanzees, which has enabled them to survive in proximity to humans (Gippoliti and Sousa 2004) despite the regular occurrence of crop raiding.

Chimpanzee Behavior and Cognition

As more chimpanzee populations are forced into human-influenced environments, they will be increasingly at the mercy of their human neighbors. The resulting interplay between risk and vulnerability can produce a complex set of adaptive behaviors (Hockings et al. 2006; Miller and Treves 2006). Cooperation, defined as "joint action for mutual benefit" (Gilby et al. 2006), by adult male chimpanzees has been documented in certain situations in the wild; it includes hunting and consequent meat sharing (Boesch and Boesch-Achermann 2000) as well as border patrols (Wilson et al. 2001). Cooperative behaviors by chimpanzees may prove beneficial in a human-influenced environment, for example during crop raids and road crossings, by increasing their protection and access to resources. Chimpanzees have flexible fission-fusion social systems, and the social organization they employ in such human-influenced risky situations may be comparable to some aspects of their strategies for predator avoidance (Sakura 1994). Chimpanzees in the Taï forest suffer from heavy predation by leopards and show specific social responses to such risks. Boesch (1991, p.236) reports that "Taï chimpanzees are mostly found in parties with the best defense capacities (mixed and all-male) that allow both sexes to profit from others' support." Given that adult males are usually the most physically powerful group members, they might be expected to be the most willing to enter unexplored areas. As a consequence, they may be expected to exhibit higher crop-raiding levels than other age and sex classes, and take up more high-risk positions during group movements in human-dominated environments (De-Vore and Washburn 1963; Hamilton 1971; Matsuzawa and Sakura 1988). Few studies have quantitatively analyzed human-chimpanzee competition and coexistence from the apes' ecological perspective, and much less from the cognitive and behavioral perspective.

Understanding the behavior of chimpanzees whose home ranges border agricultural land and human settlements is central to answering questions about how they cognitively perceive and adjust to such environments. It is foreseeable that crop raiding by the great apes will become more widespread and prevalent throughout Africa as levels of deforestation and human population growth rates continue to soar. Effective management schemes require an understanding of the complete behavioral ecology of the raiding species involved, especially as the adaptive responses of large-brained primates mean that solutions are often not straightforward.

The Chimpanzees of Bossou: A Case Study

This section provides a case study of both the feeding and social behavioral adaptations demonstrated by a small community of chimpanzees at Bossou, emphasizing the cognitive implications of particular behaviors pertaining to human-chimpanzee coexistence. Two broad research topics are addressed in this chapter. First I describe what is known about the ecological determinants of crop raiding in the Bossou chimpanzees. I discuss the temporal variations and the degree to which they relate to wild fruit availability. Secondly I document some behavioral adaptations by the chimpanzees to a heavily human-influenced environment.

The Chimpanzee Community and Environment

The Bossou chimpanzees have been studied intensively since 1976 (for a historical perspective, see Matsuzawa 2006) and are well habituated to being observed by researchers (see chapters 10 and 12). During this study the Bossou chimpanzee community size ranged from 12 to 14 individuals, always with the same 3 adult males (Hockings 2007). The social rank of the males has varied over the years, but during this study the relative status of the alpha male (Yolo), the second-ranking male (Foaf), and the third-ranking male (Tua) was stable (Biro et al. 2003; Sugiyama 2004).

The village of Bossou is situated in the forest region in the southeastern part of the Republic of Guinea, west Africa (latitude 7° 38′ 71.7″ N; and longitude 8° 29′ 38.9″ W), approximately 6 km from the Nimba Mountain range (figure 27.1). The small hills (70 to 150 m high) that constitute the chimpanzees' home range are covered in primary, secondary, and scrub forest, cultivated and abandoned fields, and orchards (figure 27.2). Primary forest accounts for just 1 km² of the 15 km² home range, and is predominantly located at the summit of the largest hill (Gban). There is a noticeable occurrence of certain tree species within this secondary forest, including the umbrella tree (*Musanga cecropioides*) and the oil palm tree (*Elaeis guineensis*).

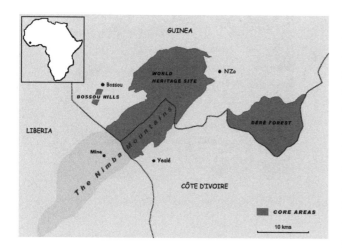

Figure 27.1   A map of the field study site Bossou in the Republic of Guinea, and the neighboring Nimba Mountain range (map drawn by N. Granier)

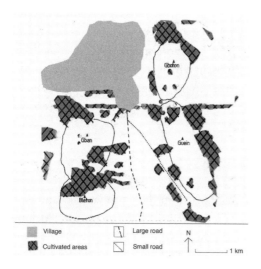

Figure 27.2   A map of the field study site Bossou in the Republic of Guinea, showing the village and the three main hills (forest) of Gban, Guein, and Gboton. The presence of cultivated areas within and surrounding the Bossou chimpanzees' core area and the large and small roads are highlighted.

Human Impact and Local Beliefs

In general, human utilization in forested ecosystems in the Republic of Guinea is extensive; the area around the village of Bossou is no exception. There are approximately 2000 people living in Bossou, although numbers have fluctuated due to the establishment of temporary refugee camps during civil wars in Liberia and the Ivory Coast (Humle 2003a; Yamakoshi 2005). Although Bossou has been declared a reserve area, the majority of people living there are subsistence farmers practicing swidden ("slash

and burn") agriculture. Consequently the chimpanzees' home range is fragmented and surrounded by cultivated and abandoned orchards, fruit trees, fields, and farms (see figure 27.2). The locals rely heavily on rice and cassava for their carbohydrate intake, but they also produce a wide variety of fruits including pineapple, papaya, orange, mandarin, and mango for their own consumption and for retail. The village of Bossou and the chimpanzees' home range are also bisected by one large road (approximately 12 m wide) that stretches from the Guinea-Liberia border into the forested region of Guinea. A narrower dirt road (approximately 3 m wide) branches off from the large road and is used by pedestrians. The chimpanzees must cross both of these roads to move from one forested area to the next.

The village of Bossou is home to the Manon people, who hold the neighboring chimpanzees sacred as the reincarnation of their ancestors, and believe that the souls of those ancestors rest on the sacred hill of Gban (Kortlandt 1986). As the chimpanzee is a totem of the most influential family of Bossou, it is strictly forbidden to hunt or eat the chimpanzee (Yamakoshi 2005). Yamakoshi (2005) proposes that Gban was an important location for village protection, serving as a refuge for women and children during periods of tribal conflict. The current peaceful coexistence between man and chimpanzee may have historical war roots, and it is "firmly embedded in the political and environmental history of the village" (Yamakoshi 2005, p.96). In more recent times, people and chimpanzees have come into contact regularly, either by the roadside or around cultivated areas, and many villagers are afraid of the chimpanzees.

Approach

Phenological Surveys

A total of six transect lines (total distance 4739 m) were set up to monitor all trees in the three principal hill forests of Bossou. Each transect line was 10 m wide and included every tree and liana greater than 5 cm in diameter at breast height (DBH). Every second and fourth week of each month the transects were monitored (Humle 2003b), and each tree was scored for fruit (ripe and unripe) and flower availability (for further details see Hockings et al. 2009).

Behavioral Sampling

I conducted focal sampling to collect data on an individual's feeding patterns and associated behaviors. A focal individual was randomly selected from a predetermined list prior to the observation session, and followed from 0630h to nest. Only adults (n = 8 − 9) were selected as focal individuals. Throughout this study, infants or juveniles less than eight years old were classified as immature. Instantaneous point sampling every five minutes was used to record the individual's activity and behavior. On the same five-minute mark, scan sampling was used to record the presence of party members. As with previous studies at Bossou, a party was defined as the chimpanzees within a 30-m radius from the focal individual; beyond this, visibility was restricted (Sugiyama and Koman 1979; Sakura 1994). Data were collected over 12 months (observations were recorded during each month of the year in three periods, from May 2004 to December 2005), and 187 focal samples were recorded, totaling 1,673 hours of focal observation. All-occurrence sampling was also employed to record all incidents of crop raiding, road crossing, and rough self-scratching, a self-directed behavioral pattern exhibited by chimpanzees possibly in response to anxiety (van Lawick-Goodall 1972; Aureli and de Waal 1997).

Crop Raiding

Certain cultivated species (e.g., mango fruit) were only consumed by chimpanzees from abandoned orchards or fields (see Hockings et al. 2009; figure 27.3). As these areas were not guarded, acquiring this food was not classified as crop raiding. Therefore, cultivated foods were divided into two groups depending on where they were obtained: abandoned (crops that were not guarded) and raided (crops that were still guarded by humans). A crop-raid "event" was defined as a foray by a single individual to obtain guarded cultivated food, and classified by exit from and return to natural vegetation (Naughton-Treves et al. 1998).

For each group of cultivated food, the presence or absence (either auditory or visual) of local people and the location of the field or orchard were recorded, as was the first chimpanzee to enter. I defined the counting line as the edge of the forest to the field or orchard, and it therefore changed depending on the crops' location. However, I did not record instances of chimpanzees visually scanning

Figure 27.3   Adult male chimpanzee at Bossou raiding papaya fruit on the edge of the village

the field or orchard due to observation difficulties from the back of the party.

All instances of food sharing (defined as an individual holding a food item but allowing another individual to consume part of that item) were recorded (see Hockings et al. 2007). Females were classified as "of reproductive age," "cycling," or, more specifically, "maximally swollen." Consortships, in which an adult female and an adult male move together to the periphery of their community range so that the male gains exclusive mating access, were also recorded.

Road Crossing

Study 1 was carried out between January and April 2005 following the widening of the road (see Hockings et al. 2006), and a follow-up study (study 2) was carried out from November to December 2005. Both roads have forest cover to the roadside and are separated by a middle zone of secondary forest and coffee plantations that normally takes two to three minutes to cross. Except for researchers and field assistants, people were never observed in this area during the study period.

I recorded the exact order in which chimpanzees left the forest and moved onto the road at regular crossing points. It was possible for them to scan the road visually without crossing the counting line. The first chimpanzee in each party to scan the road was recorded; this individual was not necessarily the first to cross the counting line. The latency between arrival of the first individual to scan the road and the last party member to cross the forest-road edge was termed *waiting time*. Presence or absence (either auditory or visual) of humans and vehicles at each road-crossing event was recorded. The expected frequency of being first to scan, first to cross, or last to cross was calculated from the mean number of adult males per progression divided by the mean number of group members present (excluding the dependent infants).

Results and Discussion

When and What Do Chimpanzees Crop-Raid?

The Bossou chimpanzees consume a wide range of cultivated foods which are found extensively within their small, fragmented home range. Throughout the 12-month study period, a total of 786 crop-raiding events were observed and the chimpanzees of Bossou consumed 17 different varieties of cultivated foods at varying intensities. These included sugar fruits (total of 320 crop-raiding events) such as papaya, orange, pineapple, and banana, and carbohydrates and proteins (total of 466 crop-raiding events) such as cassava, maize, and papaya leaf. The most frequently raided crops and the rates at which they were raided are shown in figure 27.4a.

There is significant variation in the importance of various cultivated foods to the chimpanzees' diet. In general, sugar fruits were taken in response to wild fruit scarcity (May to November, see figure 27.4b; for further details see Hockings et al. 2009). However, when abandoned mango fruits that are found abundantly within the forest were available to the chimpanzees, crop-raiding rates of simple-sugar fruits decreased, with mango fruits being consumed preferentially (total of 149 events). The high rate of mango consumption in May therefore effectively resembles a month of high wild-fruit availability. The chimpanzees also fed on other abandoned foods when available, namely banana fruits and pith (total of 25 and 26 events respectively) and cassava tuber (total of 26 events).

Conservation of primary forest throughout Africa is obviously of paramount importance for ape conservation. However, this finding suggests that conserving and encouraging certain wild fallback species such as the umbrella tree (*Musanga cecropioides)* and the oil palm tree *(Elaeis guineensis)*, which are fast-growing and thrive in secondary forest (Humle 2003b), may provide a potentially "rapid" means of alleviating crop raiding during periods of low wild-fruit availability (Naughton-Treves et al. 1998). In addition, Yamakoshi (1998) highlighted the relationship between elevated levels of tool use by the Bossou

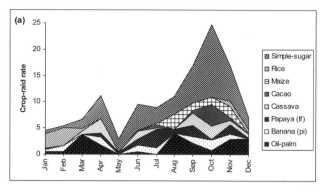

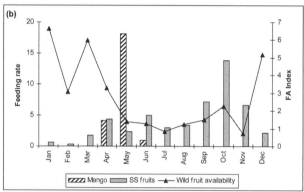

Figure 27.4 Monthly (a) crop-raid event rates of frequently raided cultivars and (b) crop-raid event rates of sugar fruits and feeding rates of abandoned mango fruits, in relation to wild fruit availability (FA index; Hockings et al. 2009).

chimpanzees and wild-fruit scarcity. The data presented throughout this chapter permits speculation regarding the future relationship between tool-use and crop-raiding levels in this community. Does the community need to maintain complex feeding behaviors, such as tool use (see chapter 10) when energy-rich cultivated foods are more readily available?

Do the Sexes Show Differences in Subsistence Behaviors?

From focal follows of five hours or more, the Bossou chimpanzees engaged in feeding behaviors for 22.5% of the day, an average of 129 minutes. Of that daily feeding time, males spent significantly longer than females feeding on crops (one-way ANOVA; $F_{1,167} = 24.0$, $p < 0.001$): Males spent on average 25.6 minutes per day feeding on crops (range: 0 – 105 mins, SE ± 3.7 mins), whereas females spent only 9.5 minutes per day (range: 0 – 55 mins, SE ± 1.4 mins).

To further investigate sex differences, party compositions were divided into the categories of male-only, mixed (at least one male, female, and immature present), male

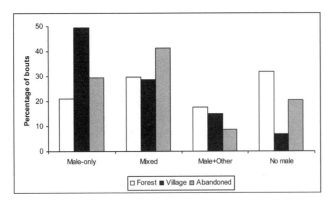

Figure 27.5 Party compositions during sugar fruit crop-raiding bouts in the forest and the village, and abandoned mango fruit feeding bouts

and other (at least one adult male and at least one adult female or immature), or no male (females and immatures only). Party composition during crop-raiding bouts was significantly affected by the location of the crop (forest (n = 148) vs. village (n = 119), $X^2(3) = 36.40$, p < 0.001); male-only parties were much more common during village crop raids than during raiding bouts in the forest (plate 1; figure 27.5). When feeding on abandoned mango fruits, which presented little associated risk, no-male and mixed-party compositions were much more common than they were in village raids on other sugar fruits, suggesting that they chimpanzees perceived the need for greater security (figure 27.5).

The presence of local people also had a significant effect on party composition during sugar-fruit raiding bouts $(X^2(3) = 13.10$, p < 0.01): parties with no males raided in the presence of people significantly less than expected. In addition, as the number of adult males in a raiding party increased, so did their time spent feeding on crops in the presence of local people (in parties including only adult males, $X^2(2) = 13.01$, p < 0.001). When they were sugar-fruit raiding in the presence rather than absence of local people, adult males also exhibited elevated levels of rough self-scratching, a potential indicator of anxiety $(X^2(1) = 4.10$, p < 0.05). Chimpanzees were observed traveling a maximum of 120 meters from the forest edge to crop-raid, and males traveled significantly further from the forest edge to crop-raid than did other age and sex classes (Mann-Whitney U-test: Z = −3.71, p < 0.001).

These results indicate that in determining whether to crop-raid, the Bossou chimpanzees appear to evaluate the degree of risk inherent in obtaining different cultivated

foods. Understanding the range of subsistence behaviors of adult males is therefore critical for determining the extent to which a community raids, and will continue to raid, cultivated foods. Stanford (2001, p.117) notes that "males and females in both human and non-human primate societies have a strong vested interest in obtaining key foods, but both the means and the goal of obtaining the food differ between the sexes. Studying these differences leads us to many of the most interesting features of our humanity and of the origins of human cognition."

Does Crop Raiding Provide Any Social Benefits?

In addition to nutritional reasons, crop raiding provides adult male chimpanzees at Bossou with highly desirable food commodities that can be traded for other currencies. Adult chimpanzees at Bossou transferred wild plant foods very rarely (1 out of 59 food-sharing events, excluding transfers of wild foods from mother to infant—and they were never observed sharing food from abandoned sources. In contrast, they shared cultivated plant foods raided from orchards and fields much more frequently (58 out of 59 food-sharing events). Raided foods were always transported to the forest before being shared. Papaya, the largest and most easily divisible cultivated fruit available, was the crop type shared most frequently (36 out of 58 instances of sharing). Other cultivated plant foods—including pineapples, oranges, and cassava—were also shared, but less often than papaya (for detailed descriptions see Hockings et al. 2007).

Mother-offspring sharing was observed in 20 instances and male-immature sharing was observed in 9 instances; both sharing patterns can be explained on the basis of kin selection. Adult females never shared crops with unrelated adults, but they did share with unrelated immatures on 3 occasions. Males rarely shared crops with one another (only in 1 out of 58 sharing events), despite failing to obtain a fruit in more than 33% of all papaya raids. Thus, it is unlikely that crop sharing at Bossou enhances cooperative raiding, as has been proposed for meat sharing at other sites (Boesch and Boesch-Acherman 2000).

The exploitation of human-cultivated foods affected the sociosexual behavior of the Bossou chimpanzees. Sharing consisted primarily of adult males allowing reproductively cycling females to take food—mostly papaya fruit—that they possessed. Papaya raids occurred independently of these females, and adult males shared particularly with one cycling female (14 out of 23 events) who took part in 83% of all consortships with males. The second-ranking adult male, who shared most with this cycling female, was also her most frequent consort and grooming recipient. In contrast, the alpha male shared less frequently with this female and, in spite of his dominance, was less likely than the second-ranking male to consort and receive grooming from her. Males shared crops with a maximally swollen female in 16% of sharing events, but were never observed mating with that female immediately after sharing. No aggressive interactions were observed during crop-sharing episodes. In addition, clusters of individuals begging the possessor for a share of raided crops were rare; males shared mostly with a single female. Although further research is required on the influence of begging behaviors, it appears that chimpanzees shared the fruits obtained during crop raids to enhance affiliative relationships with reproductively valuable females (Hockings et al. 2007).

The cognitive implications of sophisticated behaviors such as food sharing invite us to consider how and when such changes may have occurred in our own evolutionary history. The present data suggest that in addition to meat sharing, the acquisition of difficult-to-obtain plant food items may have also played an important role (Goodall 1986; Boesch and Boesch-Acherman 2000; Mitani and Watts 2001) in the advancement of complex food-sharing behaviors in hominoids.

Do Chimpanzees Exhibit Socio-Spatial Organization When Dealing with Anthropogenic Aspects of Their Environment?

Regularities in spatial patterns are a well-known occurrence in the animal kingdom. For example, through adaptive spatial patterning during group movements, monkeys reduce their risk of being attacked by predators (Rhine and Tilson 1987). Here we extend data presented on group compositions during crop raids to assess whether chimpanzees exhibit any specific changes in their spatial organization when dealing with anthropogenic aspects of their environment, such as when raiding crops and crossing roads. To correctly analyze the socio-spatial organization of a party, it is necessary to have data from a combination of age and sex classes. Therefore, only data from mixed parties were included when assessing both crop-raiding and road-crossing progressions.

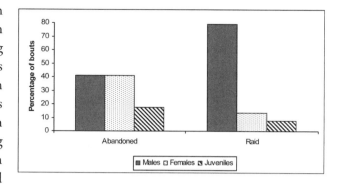

Figure 27.6   Percentage of bouts in which adult males, adult females, and juveniles took the frontward position in abandoned and raid conditions

WHO CROP RAIDS FIRST? The mean party size and number of adult males present during feeding on abandoned crops were very similar to the corresponding size and number during feeding on guarded crops (mean party size: 5.3 versus 5.4 individuals; mean number of males: 1.9 versus 2.0 individuals). As there was no significant difference between the two feeding conditions in the number of males present, these conditions were directly compared. During mixed-party raids, in which one to three adult males were present, the adult males were significantly more likely than females and juveniles to take a forward position when crop raiding than when feeding on abandoned crops (see figure 27.6: $X^2(1) = 9.30$, p < 0.01). Adult males took up a forward position during 55 out of 69 bouts of crop raiding. However, the males took up a forward position during only 8 out of 17 bouts of feeding on abandoned crops. In addition, rough self-scratching (RSS), a likely behavioral indicator of anxiety especially in adult males (Hockings et al. 2007), was substantially higher when chimpanzees were acquiring and feeding on cultivated food as opposed to wild or abandoned foods (crop raid: 1.16 bouts/hr; wild food: 0.25 bouts/hr; abandoned crops: 0.20 bouts/hr). Vigilance behaviors, such as visual scanning when crop raiding, were difficult to distinguish quantitatively from vigilance behaviors within and around the forest itself, and were therefore not recorded.

ARE ROAD CROSSINGS RISKY? In addition to the data presented on crop raiding, I assessed the degree of risk posed by road crossing. Road crossings in which party members were the same between the large and small roads were used to analyze RSS and party waiting time. RSS was significantly more likely to occur when party members crossed the large road than when they crossed the small road (X^2 (1) = 9.06, p<0.01). Correspondingly, during both studies 1 and 2 they waited a significantly longer time before crossing the large road than before crossing the small one (study 1: medians: 80 vs. 4 s; study 2: medians: 40 vs. 10 s). During study 1 the presence of people or vehicles on the large road significantly increased the party's waiting times; such analyses were not conducted during study 2 due to the small number of samples. However, despite people being present during 8 out of the 9 large road crossings during study 2, the median waiting time on the large road was much less than during study 1 (40 vs. 80 s). It is possible that this difference is due to the absence of vehicles during large road crossings in study 2.

To investigate how chimpanzees respond to the risk of crossing roads, I analyzed the positions of chimpanzees during road crossing. The analysis of road-crossing progressions during study 1 focused on 28 crossings by mixed parties (17 small and 11 large road crossings) in which all three adult males were present. The second analysis focused on the data from 25 crossings by mixed parties (12 small and 13 large road crossings), in which all three adult males were present.

In this study, the positions of chimpanzees during road crossings were differentially assumed by individuals in ways that maximized group protection, with the dominant and bolder individuals tending either to be first to enter anthropogenic environments or protective of others when they perceived possible hazards. Overall, adult males mostly took up forward positions during both study periods (study 1: 18 out of 28 progressions, binomial test; p < 0.001; study 2: 23 out of 25 progressions, binomial test: p < 0.001). However, variations existed in the positioning of individuals during the two study periods. During progressions shortly after the widening of the large road (study 1), the second- and third-ranking males maintained their high forward presence when crossing both roads, and the alpha male was mostly rearward, increasing his rearward presence on large road crossings (23.5% of small road crossings and 45.5% of large road crossings). In study 2, however, the alpha male increased his forward presence particularly on the large road (study 1: 9.1% vs. study 2: 69.2%), and only infrequently was last to cross either road. The second- and third-ranking males correspondingly increased their rearward positioning—meaning that across both studies, adult males were forward and rearward in

a similar percentage of progressions. During both study periods, the alpha female was last in approximately 40% of small-road progressions, but in only 8% of large-road progressions. Each non-alpha female and each juvenile was infrequently first or last during road crossings during both periods. The differences between study 1 and study 2 in both waiting time and positioning of dominant individuals show that chimpanzees exhibited caution towards novel features of their environment, and were flexible in their responses over time.

DO THE CHIMPANZEES SHOW ANY OTHER PROTECTIVE BEHAVIORS DURING ROAD-CROSSINGS? The occurrence of guarding behaviors (defined as standing in a quadrupedal posture on the road for more than five seconds without moving) whilst road-crossing was not recorded systematically during study 1, but during study 2 guarding occurred on 9 out of 25 road crossings (3 on the large road and 6 on the small road). As is shown in figure 27.7, guarding behavior—predominantly by the alpha male—appeared to be a response to the presence of local people on both roads (8 out of 9 occurrences), although it did not occur every time people were present. The individual to exhibit guarding behavior was usually also the first individual to cross the road (7 out of 9 instances).

Many aspects of the socio-spatial organization of the Bossou chimpanzees are congruent with findings on progression orders in other primates; namely terrestrial baboons (Rhine and Tilson 1987), Sulawesi macaques (Priston 2005), and arboreal gray-cheeked mangabeys (Waser 1985). However, the considerable flexibility of responses by chimpanzees when dealing with anthropogenic aspects of their environment at Bossou highlights the possibility that dominant and bolder individuals may be cooperat-

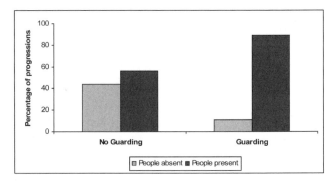

Figure 27.7 Occurrence of guarding behavior and the presence or absence of people

Figure 27.8   Alpha male exhibiting guarding behavior on the large road

ing to maximize party protection (see Hockings et al. 2006)—something that has not been concluded from studies of wild monkeys. This finding has obvious implications regarding the importance of altruistic behaviors, which are often difficult to observe in wild chimpanzee communities (figure 27.8).

In contrast to previous findings reported in captive chimpanzees (e.g., Silk et al. 2005), these data indicate that chimpanzees are not indifferent to the welfare of other group members. The chimpanzees of Bossou appear to have an understanding of other group members' vulnerability, and they take action in response to this knowledge. The present research shows that greater insights into the cognitive capacities of chimpanzees may be obtained through studying the ecological and social adaptations of primates in the wild. A changing human-dominated landscape provides chimpanzees with fresh challenges, and provides researchers with opportunities to study innovation and flexibility in a free-ranging context.

Implications and Future Directions

As our closest living phylogenetic relatives, chimpanzees occupy a special position in terms of their complex relationship with humans, spanning the border of the human-animal divide (Yamakoshi 2005). Despite this, of the four recognized subspecies of chimpanzees, the western chimpanzee *(Pan troglodytes verus)* is the second-most threatened, having been extirpated from at least two countries and sent to the verge of extinction in five others (Kormos et al. 2003). Elevated levels of crop raiding by chimpanzees and other nonhuman primates are a result of natural resources becoming less available while the nutritional

riches of agricultural production are increasingly known to them. It is in this framework that complex human social and ethical issues become increasingly important when discussing primate conservation. Whether rural people prize, fear, or consume chimpanzees is a function of culture and context, and is also a key consideration for any conservation program. In many places in west Africa, chimpanzees have complex relationships with humans and are afforded protection as a function of local customs (Kortlandt 1986; Pruetz 2002; Gippoliti and Sousa 2004). Some authors (e.g., Lee and Priston 2005; Kiss 2002) suggest that despite the long-term vulnerability of such relations, it is within these attitudes that conservation efforts have the greatest chance of being successful. However, relatively few data exist on the communities of chimpanzees that enjoy such protection.

Although the Bossou chimpanzees do raid crops, and although this discussion proposes some management strategies, it would be wrong to assume that the phenomenon of crop raiding is a problem that needs to be solved. As is highlighted at the beginning of this chapter, Priston (2005, p.335) emphasizes that "by using the term conflict, and initiating management strategies and plans, we could be elevating people's perceptions to a state of increased conflict." We need to be very careful when considering such issues. Although relatively little conflict exists now at Bossou, the potential for an increase in human-ape conflict throughout Africa is huge. Basic types of information should be assessed prior to the implementation of any conflict mitigation strategy (Hockings and Humle 2009). For example, does one actually know what the conflict problem is? What is the source and reliability of the conflict information? Who is affected, and what are their perceptions and attitudes? What constraints might one face when addressing a conflict situation?

The issues concerning conflict between humans and wildlife are often complex and site-specific, and must be viewed from a range of different perspectives. Conflict issues should be placed in the context of local community and individual needs as well as conservation objectives. Cost-benefit analyses of economics, interactions, risks, and rewards to each party can help tease out the significant ecological and cultural dimensions to human-chimpanzee relations. For example, if farmers perceive direct benefits from chimpanzee conservation, they may be more likely to accept crop damage. In addition to traditional beliefs that protect the chimpanzees, resources from foreign-based partners (such as research centers and conservation groups) may be invested in improving local health and education facilities, which in turn could contribute to social pressure to accept a certain degree of crop raiding by the chimpanzees. If they are to succeed, management strategies need to be designed intelligently.

As deforestation and human population growth rates continue to rise, such situations will become more widespread and prevalent throughout Africa. Although the competition between humans and chimpanzees may not be an even one, both species will exhibit adaptations and behavioral strategies that can only be properly understood when both are included in the analyses. More quantitative data that detail both human and chimpanzee perspectives are required, focusing on how rural people throughout Africa perceive and value chimpanzees, and how this in turn affects chimpanzees' behavior and capacity to survive in increasingly human-dominated landscapes. These human-ape interactions, and the range of human percep-tions of our nearest living relatives, are important because the behavior and choices of rural people who live in daily contact with wildlife ultimately will play a crucial role in its survival (Kiss 1990; Sicotte and Uwengeli 2002). These are human problems, the solutions to which will benefit both people and chimpanzees.

Acknowledgments

I am grateful to the Direction Nationale de la Recherche Scientifique et Technique and to Soumah Ali Gaspard, director of the Institut de Recherche Environementale de Bossou, the Republic of Guinea. I would also like to thank all the local assistants and Bossou villagers who helped during this research period. Thanks to V. Fishlock and reviewers for helpful comments. This work was supported by a Stirling University studentship and post-doctoral research grant from Fundação para a Ciência e a Tecnologia, Portugal.

Literature Cited

Anderson, J., J. M. Rowcliffe, and G. Cowlishaw. 2007. Does the Matrix matter? A forest primate in a complex agricultural landscape. *Biological Conservation* 135:212–22.

Aureli, F., and F. B. M. de Waal. 1997. Inhibition of social behaviour in chimpanzees under high-density conditions. *American Journal of Primatology* 41:213–28.

Biro, D., N. Inoue-Nakamura, R. Tonooka, G. Yamakoshi, C. Sousa, and T. Matsuzawa. 2003. Cultural innovation and transmission of tool use in wild chimpanzees: Evidence from field experiments. *Animal Cognition* 6:213–23.

Boesch, C. 1991. The effect of leopard predation on grouping patterns in forest chimpanzees. *Behaviour* 117:220–42.

Boesch, C., and H. Boesch-Acherman. 2000. *The Chimpanzees of the Tai Forest*. Oxford: Oxford University Press.

Cowlishaw, G., and R. Dunbar. 2000. *Primate Conservation Biology*. Chicago: University of Chicago Press.

DeVore, I., and S. L. Washburn. 1963. Baboon ecology and human evolution. In *African Ecology and Human Evolution*, edited by F. C. Howelland and F. Bourliere, 335–67. Chicago: Aldine.

Fuentes, A., and Wolfe, L. D. 2002. *Primates Face to Face: The Conservation Implications of Human-Nonhuman Primate Interconnections*. Cambridge: Cambridge University Press.

Gilby, I. C., L. E. Eberly, L. Pintea, and A. E. Pusey. 2006. Ecological and social influences on the hunting behaviour of wild chimpanzees, *Pan troglodytes schweinfurthii*. *Animal Behaviour* 72:169–80.

Gippoliti, S., and C. Sousa. 2004. The chimpanzee, *Pan troglodytes*, as an "umbrella" species for conservation in Guinea-Bissau, West Africa: Opportunities and constraints (abstract). *Folia Primatologica* 75:385–414.

Goodall, J. 1986. *The Chimpanzees of Gombe: Patterns of Behavior*. Cambridge, MA: Harvard University Press.

Greengrass, E. 2000. The sudden decline of a community of chimpanzees at Gombe National Park: A supplement. *Pan Africa News* 7:25–26.

Hamilton, W. D. 1971. Geometry for the selfish herd. *Journal of Theoretical Biology* 7:295–311.

Hill, C. M. 1997. Crop-raiding by wild vertebrates: The farmer's perspective in an agricultural community in western Uganda. *International Journal of Pest Management* 43:77–84.

Hill, C. M., F. Osborn, and A. J. Plumptre. 2002. Human-wildlife conflict: Identifying the problem and possible solutions. *Albertine Rift Technical Report Series Vol. 1*. New York: Wildlife Conservation Society.

Hill, C. M. 2005. People, crops and primates: A conflict of interests. In *Commensalism and Conflict: The Human-Primate Interface*, edited by J. D. Paterson and J. Wallis, 40–59. Oklahoma City: American Society of Primatologists.

Hockings, K. J. 2007. Human-Chimpanzee Coexistence at Bossou, the Republic of Guinea: A Chimpanzee Perspective. PhD dissertation, University of Stirling.

Hockings, K. J., J. R. Anderson, and T. Matsuzawa. 2006. Road-crossing in chimpanzees: A risky business. *Current Biology* 16:668–70.

———. 2009. Use of wild and cultivated foods by chimpanzees at Bossou, Republic of Guinea: Feeding dynamics in a human-influenced environment. *American Journal of Primatology* 71:636–46.

Hockings, K. J., and T. Humle. 2009. *Best Practice Guidelines for the Mitigation of Conflict between Great Apes and Humans*. Gland, Switzerland: SSC Primate Specialist Group of the World Conservation Union.

Hockings, K. J., T. Humle, J. R. Anderson, D. Biro, C. Sousa, G. Ohashi, and T. Matsuzawa. 2007. Chimpanzees share forbidden fruit. *PLoS ONE* 2:e886.

Humle, T. 2003a. Chimpanzees and crop raiding in West Africa. In *West African Chimpanzees: Status Survey and Conservation Action Plan*, edited by R. Kormos, C. Boesch, M. I. Bakarr, and T. M. Butynski, 147–55. IUCN/SSC Primate Specialist Group. IUCN, Gland, Switzerland, and Cambridge, UK.

———. 2003b. Culture and variation in wild chimpanzee behaviour: A study of three communities in West Africa. PhD dissertation, University of Stirling.

Inoue, S., and T. Matsuzawa. 2007. Working memory of numerals in chimpanzees. *Current Biology* 17:1004–5.

Kiss, A. 1990. *Living with Wildlife: Wildlife Resource Management with Local Participation in Africa*. World Bank. World Bank Technical Paper: 130.

Kormos, R., C. Boesch, M. I. Bakarr, and T. Butynski. 2003. *West African Chimpanzees: Status Survey and Conservation Action Plan*. IUCN/SSC Primate Specialist Group. IUCN, Gland, Switzerland, and Cambridge.

Kortlandt, A. 1986. The use of stone tools by wild-living chimpanzees and earliest hominids. *Journal of Human Evolution* 15:77–132.

Lahm, S. A. 1996. A nationwide survey of crop-raiding by elephants and other species in Gabon. *Pachyderm* 21:69–77.

Lee, P. C. 2004. Who wins? Human-primate conflict in the context of conservation, development and gender. *Primate Eye* 84:15–16.

Lee, P. C., E. J. Brennan, J. G. Else, and J. Altmann. 1986. Ecology and behaviour of vervet monkeys in a tourist lodge habitat. In *Primate Ecology and Conservation*, edited by J. G. Else and P. C. Lee, 229–35. Cambridge: Cambridge University Press.

Lee, P. C., and N. E. C. Priston. 2005. Human attitudes to primates: Perception of pests, conflict and consequences for primate conservation. In *Commensalism and Conflict: The Human-Primate Interface*, edited by J. D. Paterson and J. Wallis, 1–23. Oklahoma City: American Society of Primatologists.

Lonsdorf, E. V. 2007. The role of behavioural research in the conservation of chimpanzees and gorillas. *Journal of Applied Animal Welfare Sciences* 10:71–78.

Maples, W. R., M. K. Maples, W. F. Greenhood, and M. L. Walek. 1976. Adaptations of crop-raiding baboons in Kenya. *American Journal of Physical Anthropology* 45:309–16.

Matsuzawa, T. 2006. Sociocognitive development in chimpanzees: A synthesis of laboratory work and fieldwork. In *Cognitive Development in Chimpanzees*, edited by T. Matsuzawa, M. Tomonaga, and M. Tanaka, 3–33. Tokyo: Springer-Verlag.

Matsuzawa, T., and O. Sakura. 1988. Choice of foraging sites in wild chimpanzees: Analysis by observing progressions and foot print identification. *Reichoru Kenkyu / Primate Research* 4:155.

McGrew, W. C., P. J. Baldwin, and C. E. G. Tutin. 1981. Chimpanzees in a hot, dry and open habitat: Mt. Assirik, Senegal, West Africa. *Journal of Human Evolution* 10:227–44.

McLennan, M. R. 2008. Beleaguered chimpanzees in the agricultural district of Hoima, western Uganda. *Primate Conservation* 23:45–54.

Miller, L. E., and A. Treves. 2006. Predation on primates. In *Primates in Perspective*, edited by C. Campbell, A. Fuentus, K. MacKinnon, K. Panger, and S. K. Bearder, 525–36. New York: Oxford University Press.

Mitani J. C., and D. P. Watts. 2001. Why do chimpanzees hunt and share meat? *Animal Behaviour* 61:915–24.

Mittermeier, R. A., C. Valladares-Padua, A. B. Rylands, A. A. Eudey, T. M. Butynski, J. U. Ganzhorn, R. Kormos, J. M. Aguiar, and S. Walker. 2006. Primates in peril: The world's 25 most endangered primates, 2004–2006. *Primate Conservation* 20:1–28.

Morgan, D., and C. Sanz. 2006. Chimpanzee feeding ecology and comparisons with sympatric gorillas in the Goualougo Triangle, Republic of Congo. In *Feeding Ecology in Apes and Other Primates*, edited by G. Hohmann, M. Robbins, and C. Boesch, 97–122. Cambridge: Cambridge University Press.

Naughton-Treves, L. 1998. Predicting patterns of crop damage by wildlife around Kibale National Park, Uganda. *Conservation Biology* 12:156–68.

Naughton-Treves, L., A. Treves, C. Chapman, and R. Wrangham. 1998. Temporal patterns of crop-raiding by primates: Linking food availability in croplands and adjacent forest. *Journal of Applied Ecology* 35:596–606.

Patterson, J. D., and J. Wallis. 2005. *Commensalism and Conflict: The Human-Primate Interface*. Oklahoma City: American Society of Primatologists.

Priston, N. 2005. Crop-raiding by *Macaca ochreata brunnescens* in Sulawesi: Reality, Perceptions and Outcomes for Conservation. PhD dissertation, University of Cambridge.

Pruetz, J. D. 2002. Competition between savanna chimpanzees and humans in southeastern Senegal (abstract). *American Journal of Physical Anthropology* 34:128.

———. 2006. Feeding ecology of savanna chimpanzees *(Pan troglodytes verus)* at Fongoli, Senegal. In *Feeding Ecology in Apes and Other Primates*, edited by G. Hohmann, M. Robbins, and C. Boesch, 326–64. Cambridge: Cambridge University Press.

Reynolds, V. 2005. *The Chimpanzees of the Budongo Forest: Ecology, Behaviour, and Conservation*. Oxford: Oxford University Press.

Rhine, R. J. and R. Tilson. 1987. Reactions to fear as a proximate factor in the sociospatial organization of baboon progressions. *American Journal of Primatology* 13:119–28.

Rose, A. L. 2002. Conservation must pursue human-nature biosynergy in the era of social chaos and bushmeat commerce. In *Primates Face to Face: The Conservation Implications of Human-Nonhuman Primate Interconnections*, edited by A. Fuentes and L. D. Wolfe, 208–39. Cambridge: Cambridge University Press.

Sakura, O. 1994. Factors affecting party size and composition of chimpanzees *(Pan troglodytes verus)* at Bossou, Guinea. *International Journal of Primatology* 15:167–81.

Sicotte, P., and P. Uwengeli. 2005. Reflections on the concept of nature and gorillas in Rwanda: Implications for conservation. In *Primates Face to Face: The Conservation Implications of Human-Nonhuman Primate Interconnections*, edited by A. Fuentes and L. D. Wolfe, 163–81. Cambridge: Cambridge University Press.

Siex, K. S., and T. T. Struhsaker. 1999. Colobus monkeys and coconuts: A study of perceived human-wildlife conflicts. *Journal of Applied Ecology* 36:1009–20.

Silk, J. B., S. F. Brosnan, J. Vonk, J. Henrich, D. J. Povinelli, A. S. Richardson, S. P. Lambeth, J. Mascaro, and S. J. Schapiro. 2005. Chimpanzees are indifferent to the welfare of unrelated group members. *Nature* 437:1357–59.

SOCP .2002. News from the field. *Sumatran Orangutan Conservation Programme News* 2.

Stanford, S. 2001. The ape's Gift: Meat-eating, meat-sharing, and human evolution. In *Tree of Origin: What Primate Behavior Can Tell Us about Human Social Evolution*, edited by F. de Waal, 95–118. Cambridge, MA: Harvard University Press.

Strum, S. C. 1984. The pumphouse gang and the great crop raids. *Animal Kingdom* 87:36–43.

Sugiyama, Y., and J. Koman. 1979. Social structure and dynamics of wild chimpanzees at Bossou, Guinea. *Primates* 20:323–39.

Sugiyama, Y. 2004. Demographic parameters and life history of chimpanzees at Bossou, Guinea. *American Journal of Physical Anthropology* 124:154–65.

Takemoto, H. 2004. Seasonal change in terrestriality of chimpanzees in relation to microclimate in the tropical forest. *American Journal of Physical Anthropology* 124:81–92.

Teleki, G. 1989. Population status of wild chimpanzees *(Pan troglodytes)* and threats to survival. In *Understanding Chimpanzees*, edited by P. G. Heltne and L. A. Marquardt, 312–53. Cambridge, MA: Harvard University Press.

Van Lawick-Goodall, J. 1972. A preliminary report on expressive movements and communication in the Gombe Stream chimpanzees. In *Primate Patterns*, edited by P. Dolhinow, 25–84. New York: Holt, Rinehart and Winston.

Waser, P. M. 1985. Spatial Structure in mangabey groups. *International Journal of Primatology* 6:569–80.

Wilson, M. L., M. D. Hauser, and R. W. Wrangham. 2001. Does participation in intergroup conflict depend on numerical assessment, range location, or rank for wild chimpanzees? *Animal Behaviour* 61:1203–16.

Wrangham, R. W. 1977. Feeding behaviour of chimpanzees in Gombe National Park, Tanzania. In *Primate Ecology: Studies of Feeding and Ranging Behaviour in Lemurs, Monkeys and Apes*, edited by T. H. Clutton-Brock, 504–38. London: London Academic Press.

Wrangham, R. 2001. Moral decisions about wild chimpanzees. In *Great Apes and Humans: The Ethics of Coexistence*, edited by B. B. Beck, T. S. Stoinski, M. Hutchins, T. L. Maple, B. Norton, A. Rowan, E. F. Stevens, and A. Arluke, 230–44. Washington and London: Smithsonian Institutional Press.

Yamakoshi, G. 1998. Dietary responses to fruit scarcity of wild chimpanzees at Bossou, Guinea: Possible implications for ecological importance of tool use. *American Journal of Physical Anthropology* 106:283–95.

———. 2005. What is happening on the border between humans and chimpanzees? Wildlife conservation in West African rural landscapes. In *Coexistence with Nature in a "Globalising" World: Field-Science Perspectives*, edited by K. Hiramatsu, 91–97. Kyoto: Proceedings of the 7th Kyoto University International Symposium, 2005.

Yuwono, E. H., P. Susanto, C. Saleh, N. Andayani, D. Prasetyo, and S. S. U. Atmoko. 2007. Guidelines for the better management practices on avoidance, mitigation and management of human-orangutan conflict in and around oil palm plantations. WWF–Indonesia.

28

Chimpanzee Mind, Behavior, and Conservation

Elizabeth V. Lonsdorf

A female chimpanzee travels slowly through her forest, her youngest offspring riding on her back as she looks for the ripe fruit of the season. Suddenly, she hears a sharp, metallic crack that echoes through the valley. The startled female responds by scurrying into a dense fig tree and huddling there quietly. Without a sound, she scans the dense vegetation around her. Her infant seems to take her cue and peers, silent and wide-eyed, over her shoulder. For a long time the pair sits quietly vigilant high above the forest floor as they appear to cogitate the source and consequences of the unnatural and foreboding sounds—sounds that recently have grown both in frequency and proximity to their home range.

Chimpanzees are our closest living relatives, sharing most of our genetic code and many similarities in anatomy, physiology, and behavior. These apes have the capacity to make and use tools as humans do, have strong family bonds like humans, and have population-specific behaviors similar to human cultures. But populations of chimpanzees are in dramatic decline due to hunting for bushmeat, disease outbreaks (such as Ebola), loss of habitat, and the varied risks of small, isolated populations (Butynski 2001). Recognizing and understanding the complexities of these threats is the first step in conserving the world's wild chimpanzee populations, but mitigating the risks takes a deeper understanding of the chimpanzee mind. In this chapter I will review current chimpanzee conservation threats and

provide examples of how the time is ripe for a synergistic link between studies focusing on cognition and on conservation.

Background

Perhaps the most often cited commonality is the chimpanzee's "intellect," which has been of interest since Kohler began his first investigations into chimpanzee intelligence in the early twentieth century (Kohler 1927). Building on the work of Kohler and others, long-term behavioral studies began in the 1960s at Gombe National Park and Mahale Mountains National Park in Tanzania (Goodall 1986; Nishida 1990). Both studies sought an in-depth understanding of the ecology, behavior, and natural history of our closest living relatives. They focused from the start on individuals and their personal histories and personalities, and included observations of events that presupposed emotions and feelings. At the time, these ideas were criticized by other scientists taking a more "Skinnerian" approach to animal behavior (Goodall 1986). However, the field of animal cognition has changed dramatically since the early studies of Kohler, Goodall, and Nishida. Research into the chimpanzee mind has exploded in the past decade, with the result that chimpanzee cognitive capacities are now being well documented. These capacities, detailed by contributors to this volume, include complex problem-solving capabilities (see chapters 1, 10, 11, and 12), varied forms of social learning (see chapters 8 and 9), advanced perceptual abilities (see chapters 5, 7, 16, and 22), and navigation of an extremely complex social system and its politics (see chapters 15, 17, and 18).

Over the years we have also learned more about the natural history characteristics of chimpanzees that make their conservation particularly challenging. Conservation is ultimately a numbers game—that is, one needs viable numbers of animals to maintain a population or inevitably the population will go extinct. However, chimpanzees' basic natural history characteristics make them particularly vulnerable to population declines and less likely to be able to recover from them (Tutin 2001). Chimpanzees live in "communities" (van Lawick-Goodall 1968) or "unit-groups" (Nishida 1968) that range in size from 20 to 150 individuals. These communities are multi-male, promiscuous, and have a male-dominance hierarchy in which males form the stable core of the community and defend a group territory (Goodall 1986). Chimpanzee society is termed "fission-fusion" as members of a community can join or leave traveling parties at any time (Goodall 1986; Wrangham 1979). An individual's presence in a party is determined by a combination of factors which may include food availability, sexual state of females, and social relationships with other individuals. Neighboring chimpanzee communities are highly territorial and intercommunity interactions can have fatal results (Goodall 1986; Wilson and Wrangham 2003). Chimpanzees are also slow to reproduce in the wild, having their first offspring at 13 years of age, on average, with a three- to five-year interval between births (Goodall 1986; Nishida et al. 2003). Single offspring are the most common result of pregnancy, although some sets of twins have been reported (Goodall 1986; Nishida et al. 2003). Offspring are nutritionally dependent on their mother until age three to five, but remain behaviorally dependent (i.e., continually traveling and socializing with the mother) until at least the age of eight (Pusey 1990). As I will describe below, these natural history and behavioral characteristics have important consequences for ape conservation, in terms of the apes' ability to recover from catastrophes and/or the implementation of conservation strategies (Lonsdorf 2007).

Populations of chimpanzees across Africa are in decline due to several primary threats, such as habitat destruction, poaching, the bushmeat trade, and epidemic disease (Tu-tin 2001; Butynski 2001). Despite the efforts of many conservation professionals, assessments of precisely how many chimpanzees remain in the wild are extremely hard to generate. In a recently published action plan for ape conservation in western equatorial Africa, the authors specifically declined to estimate the number of apes remaining in that area due to recognized problems with current survey and extrapolation methodologies (Tutin et al. 2005). Many reports estimate a reduction in chimpanzee numbers over the last 100 years from one to two million at the turn of the century to a few hundred thousand today, but such estimates are largely acknowledged to be little more than educated guesses (Oates 2006).

In this chapter I will detail specific sources of threat and how they relate to understanding the mind of the chimpanzee. Further, I will discuss how the chimpanzee's cognitive complexity may inform and/or complicate proposed strategies for mitigating conservation threats. Many arguments presented here are likely not unique to chimpanzees; however, the cognitive abilities of other taxa have have been studied and documented much less, so I will focus largely on chimpanzees. I contend that the fields of cognition and conservation have much to offer each other, and that by furthering studies of the chimpanzee mind, we will ultimately be better able to provide for their conservation.

The Chimpanzee Mind in Relation to Threats and Their Mitigation

Habitat Destruction

The African forests in which apes live are becoming smaller and more fragmented due to the intense pressure of human land needs for farming, logging, or other uses (Wilkie, et al. 2000; Tutin 2001; Tutin et al. 2005). This fragmentation changes the forest's carrying capacity, and resident animal populations are inevitably affected. Basic population biology principles tell us that smaller and more isolated populations are at higher risk of decline from random catastrophes such as forest fires, civil unrest, and disease outbreaks (Shaffer 1987). Chimpanzees' slow reproduction rates, detailed above, mean that they are also less able to rebound from these events which can drive small populations to extinction.

Mechanized logging is perhaps the single largest cause of chimpanzee habitat destruction, and is especially sig-

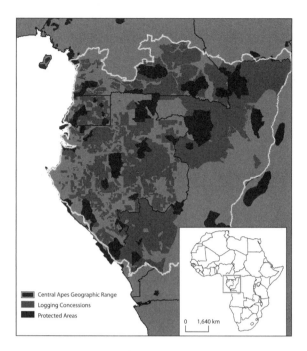

Figure 28.1   Logging concessions and priority great ape populations in western equatorial Africa. Priority populations as identified during the Regional Action Plan Workshop for Great Ape Conservation in Western Equatorial Africa, Brazzaville, May 2005 (Tutin et al. 2005). Map reprinted with permission from Morgan and Sanz (2007).

nificant in western equatorial Africa, where the largest remaining populations of wild chimpanzees reside. A recent report details how 36% of priority ape conservation areas in that region are within logging concessions (see figure 28.1—reprinted with permission, Morgan and Sanz 2007). Studies of the effect of logging on chimpanzee densities have produced conflicting results, largely due to scientists (1) trying to compare results that have been gathered from different sites and/or by different methodologies, and (2) attempting to infer effects after logging without proper baseline data from the period before logging. Studies like those being conducted by Morgan and Sanz in the Goualougo Triangle of the Republic of Congo—in which careful standardized transect data is collected before, during, and after logging—are sorely needed. Such investigations will hopefully provide solid evidence for the assumption that logging can substantively alter ape habitats and thereby affect food resources, disrupt social groups, fragment populations, and increase exposure to disease (Morgan and Sanz 2007).

The negative effects of logging are likely to be compounded by aspects of chimpanzees' advanced cognitive complexity. We know that chimpanzees are territorial, and that male chimpanzees will defend their territory from in-

cursions by neighboring groups—to the death, in some cases (Goodall 1986; Wilson and Wrangham 2003). This suggests that despite their very fluid social dynamics, chimpanzees have strong perception of who "belongs" to their group versus "strangers." Chimpanzees are not unique in this high level of territoriality, as fatal intergroup aggression is also seen in other taxa (e.g., wolves: Mech 1994) but the interaction of abrupt, large-scale habitat destruction and chimpanzee territoriality increases the likelihood of these lethal interactions. In some cases, entire communities of chimpanzees may be displaced into the territory of other communities, which might increase mortality rates due to aggression. These indirect effects of logging were reported by White and Tutin (2001), who proposed that such a scenario reduced chimpanzee densities after logging at Lopé, Gabon.

Little quantitative data exists with regard to how chimpanzees perceive spatial intrusion and/or changes in habitat. However, from combining observations from the field and experimental studies, it is clear that complex processing, weighing of costs and benefits, and decision-making are at least cognitively possible and likely play a large role in cases of habitat encroachment. From studies in Uganda we know that chimpanzees avoid human presence (Plumptre and Johns 2001) and change ranging patterns in response to logging in their range (Reynolds 2005). As detailed in Hockings (see chapter 27), chimpanzees living in a highly human-dominated landscape in Guinea showed flexible responses to changes in their perceived risk during road crossing and crop raiding, and positioned themselves to minimize risk to more vulnerable individuals in the group. Further, new studies in captivity on the chimpanzee vocal repertoire and its use for communication about specific objects (e.g., food) or events (e.g., agonistic encounters) (see chapter 16) suggest the possibility that chimpanzees may be able to communicate about risky situations before deciding on a response to a change in their surroundings (such as road crossing). Indeed, Wilson et al. (2007) suggest that chimpanzees do alter their vocal behavior in response to location-specific risk. Future studies in this area could help greatly to clarify how chimpanzees perceive and respond to the inevitable change of their habitat by human encroachment.

Until we understand how chimpanzees respond to these habitat alterations, it is difficult to predict the success of mitigation strategies aimed at protecting ape landscapes.

The complex intellect of chimpanzees suggests that they may be able to adapt to habitat change better than other animals. Recent reports from the Fongoli study site in Senegal, detailing previously undescribed behaviors such as tool use for foraging on other primates, have opened our eyes to the potential for such behavioral flexibility and adaptability in harsher environments (Pruetz and Bertolani 2007). When an element of risk is added, however, the picture becomes less clear. Take the case of selective logging, which aims to take only high-value trees while leaving low-value trees and related vegetation intact. If enough chimpanzee foods and nesting sites remain after such logging, chimpanzees theoretically should not be affected. However, the disturbance caused by humans and machines in the forest may cause chimpanzees to abandon an area altogether, and put them in conflict with a neighboring community. The ideal solution is to end logging in ape range countries and protect as much forest as possible. Unfortunately that option isn't realistic or practical in most ape range states today, and a clearer understanding of chimpanzees' response to these changes is needed to help chart the most effective conservation management strategies.

The Bushmeat Trade

One of the primary threats to chimpanzee populations is the hunting by humans for bushmeat. Indigenous forest people historically hunted apes for meat, but did so at sustainable levels that did not threaten the survival of ape populations. Now, the combined effects of expanding human populations and the perception of bushmeat as a delicacy by city-dwellers, both in Africa and abroad, have resulted in catastrophic losses of chimpanzee populations (Ammann 2001). The logging industry compounds the problem by opening roads into pristine forests and setting up camps for thousands of employees (Wilkie et al. 2000). Forestry teams working in a logging concession are often encouraged to provision themselves, which results in hunting and/or gathering within the concession. Logging companies may further facilitate the bushmeat trade by providing firearms, ammunition, and transport of bushmeat to local markets. In addition, the roads built by logging companies give poachers access to formerly inaccessible forests (Wilkie et al. 2000). Walsh et al. (2003) suggest that gorilla and chimpanzee numbers in Gabon declined by more than half between the years of 1983 and

2000, with commercial hunting attributed as the primary cause. These results have been debated due to concern over survey methodology; however, it is clear that the commercialization of bushmeat and the practices of the logging industry result in the opening of pristine forests and an increase in the harvesting of forest animals, including chimpanzees.

An understanding of chimpanzee behavior helps us to understand more fully the effects of the bushmeat trade. As described above, chimpanzees are known to have very long periods of offspring dependence. In addition, Goodall has provided evidence that chimpanzee offspring appear to be emotionally dependent on their mothers well after they become nutritionally independent, as is evidenced by the fact that two of six young chimpanzees who lost their mothers when they were between four and six years old died within 18 months (Goodall 1986). Three of the four other survivors showed various detrimental physiological and behavioral effects, such as potbellies (perhaps parasite infestations or malnutrition), lethargy, and marked delays in the appearance of the first sexual swelling (in females) and scrotum development (in males). The implication of this lengthy maternal dependence in relation to the impact of the bushmeat trade is clear: if an adult female is shot for bushmeat, her offspring are likely to be casualties as well. In practice, mothers may be shot for bushmeat and their infants sold live into the pet trade where their chances for survival are small. As in the above discussion of logging and habitat fragmentation, little is known about chimpanzees' ability to perceive the humans they may encounter, and potentially avoid the risks of hunting. In areas where chimpanzees are totally naïve about human presence, they may have learned no anti-predator behavior and may thus be at a higher risk (Morgan and Sanz 2003). However, one would expect that anti-predator tactics can be learned quickly in areas that are heavily hunted, when the option of simply leaving the area has already disappeared. While direct experiments to test such learning behavior in captivity would be unethical in practice, detailed behavioral observations of heavily hunted populations could add much to our knowledge in this area.

Even when chimpanzees are not the focus of bushmeat hunting they may become bycatch from hunting for other animals; such is often the case with snare hunting. Snares set for forest antelope, pigs, and other species often entangle chimpanzees and can cause severe injury, deformity, and infection (Quiatt et al. 2002). However, chimpanzees injured by snares appear to be able to adapt to their disabilities and deformities, perhaps in part because of their advanced problem-solving abilities. In the Sonso community of chimpanzees in the Budongo Forest, Uganda, 20% (10 individuals) of the study population were disabled by snare injuries, each resulting in the loss of use of a hand, a foot, or multiple limbs. By analyzing a complex manual food-processing task in detail, Stokes and Byrne (2001) found that the injured individuals were able to adapt their feeding techniques so as to "work around" their injuries, and only the two most severely injured subjects showed any decrease in feeding efficiency. Like logging, bushmeat hunting is a threat that chimpanzees would appear able to manage better than other animals because of their intelligence and behavioral flexibility. In the case of adults that may well be true, but we cannot ignore the fact that a long period of offspring dependence is the norm for this big-brained, cognitively complex species. A catch-22 then exists in that cognitively complex animals such as chimpanzees may be able to adapt and somewhat mitigate the risks of bushmeat hunting, but the fact that these big-brained primates need their mothers for upwards of five to ten years in order to survive puts youngsters at high risk of being secondary casualties of the trade.

Disease Risk

Many chimpanzee and gorilla study sites have reportedly been affected by epidemic disease, and most of these outbreaks are suspected to be the result of close contact with humans (Goodall 1983; Guerrera et al. 2003; Homsy 1999; Leendertz et al. 2006; Köndgen et al. 2008). In fact, at all three chimpanzee study sites for which comprehensive analyses on causes of death have been performed, illness was the primary cause of mortality (Taï: Boesch and Boesch-Achermann 2000; Mahale: Nishida et al. 2003; Gombe: Williams et al. 2008). In addition, a 2003 outbreak of the Ebola virus is thought to have killed a significant number of gorillas and chimpanzees across a large area in western equatorial Africa (Walsh et al. 2003). Because of these emerging disease risks, more and more ape behaviorists are becoming involved in trying to understand the disease threats to their populations and potential ways to reduce such threats.

We actually know very little about disease transmission and its impact on wild chimpanzees. Often a specific pathogen is difficult to identify due to the lack of noninvasive techniques developed for that purpose. However, these methodologies are being improved rapidly (see Chi et al. 2007; Goldberg et al. 2007; Köndgen et al. 2008). We often have only anecdotal information or assumptions about how diseases are introduced and transmitted, and in less habituated populations it is often impossible to estimate the population impact of disease (Lonsdorf et al. 2006). Finally, without knowing much about pathogen type and pathogen flow, it is very hard to understand which behaviors may put chimpanzees at risk.

These issues are conservation and epidemiological concerns, but more germane to the topic of the chimpanzee mind is the following question: How do chimpanzees perceive illness in themselves and in others? Many anecdotes exist in the popular literature about individuals being attacked and losing dominance when they are in a weakened state, and removing themselves from the larger group when they are ill (see Goodall 1986). But do group members consistently choose to stay away from a sick individual or not? Does it depend on what type of illness the individual has, and does this relate to the virulence of the disease? These are all questions for future study, but one well-known example that provides some insight into chimpanzees' self-perception of illness is that of the leaf-swallowing behavior at Gombe National Park and Mahale Mountains National Park in western Tanzania (Wrangham and Nishida 1983). The behavior is described as follows (source: http://jinrui.zool.kyoto u.ac.jp/ChimpHome/ Mahale/MedPlant.html): "Chimpanzees use their lips to carefully remove one leaf from the plant at a time and pull it into the mouth with the tongue. This causes the rough, hairy leaves to fold up accordion-style. Each folded leaf is then swallowed whole without being chewed. Leaves are evacuated whole and undigested in their feces." Follow-up over many years by Huffman and colleagues (see Huffman 1997 for an overview) has found medicinal qualities in the plants chimpanzees eat when they appear ill with such symptoms as diarrhea and general lethargy. In terms of cognitive processes, medicinal plant knowledge exists at the juncture between food categorization and illness experience, and in offspring it likely requires complex types of learning that these foods are eaten not just for satiation but for treatment.

Chimpanzees' use of medicinal plants suggests the abil-

Figure 28.2 Knuckles, top, often behaves unusually due to his cerebral palsy, but the other chimpanzees afford him a level of tolerance that suggests they understand that he is different. Photo courtesy Devyn Carter.

ity to perceive their own illness state. However, the manner in which they perceive illness in others is still an open field of inquiry. One striking example is that of Knuckles, a chimpanzee at the Center for Great Apes sanctuary in Florida who has been diagnosed with cerebral palsy (see figure 28.2). J. D. Carter (personal communication) provides this report of Knuckles and his interactions with group mates:

Knuckles displays a range of physical and mental impairments which result in highly atypical interactions with others. Although his group mates showed great interest in his behavior, their actions towards him were characterized by a high degree of tolerance; they were seen to regularly groom him, but never to act aggressively towards him. In addition, they were highly tolerant of his acts of minor aggressions, such as occasional biting. This level of tolerance is usually reserved for infant chimpanzees, suggesting that although Knuckles's physical size is representative of his age, his group mates were aware of his impairments and were able to modify their behavior towards him.

As detailed above, this description suggests that Knuckles's group mates understood that something was "different" about him and that he should be afforded a certain level of tolerance. This stands in stark contrast to written descriptions by Goodall (1971) of chimpanzee reactions to other group members afflicted with polio. In those cases, the typical reaction to an individual with some level of paralysis (and therefore, abnormal movement) was outright fear—which would eventually give way to either accep-

tance, in the case of minor paralysis, or total ostracism for the most severe cases. Knuckles's youth may have afforded him more tolerance as opposed to the adult victims of polio described by Goodall, but the degree to which these results can be generalized remains in question. The ramifications of chimpanzees' perception of illness in others are germane to various discussions current in the conservation community about potentially treating and/or vaccinating wild chimpanzees. The question of whether chimpanzees avoid seemingly ill individuals, sequester themselves, or appear to behave differently depending on their illness symptoms is an important factor when considering which individuals would be the subjects of intervention.

Another way in which cognition researchers and conservation professionals could join forces to mitigate the risk of disease is in the development of strategies for outbreak control. In the case of the Ebola virus, researchers are currently trying to develop ways to deliver vaccines to wild chimpanzees and gorillas in the Congo Basin (Sanz, personal communication). When a vaccine is developed, the question then arises as to how to deliver it in a forest full of non-target animals (elephants, pigs, etc.). The effectiveness of oral baiting programs targeting chimpanzees and gorillas could be greatly enhanced by knowledge of species-specific responses to novel stimuli and food characteristics that would be preferred by both of these sympatric apes. In addition, the apes' cognitive abilities could be harnessed in such an endeavor through the use of puzzle-box or bi-manual task paradigms to provide baits that would be inaccessible to non-ape species (Ross, personal communication).

Chimpanzee Mind in Relation to Proposed Conservation Strategies

In the above section I have described the major conservation threats facing apes and how their cognitive abilities interact with these threats. In addition, I have provided examples of how the chimpanzee mind could help or hinder strategies to mitigate those threats. But how might the advanced cognitive abilities of chimpanzees come into play in relation to conservation strategies already in place for at-risk ape populations? In this section I will describe how the fields of cognition and conservation interact in the case of two strategies that have been proposed to help support or augment existing ape populations.

Ecotourism

Perhaps because of the close evolutionary relationship and behavioral similarities between humans and apes, people the world over are fascinated by gorillas and chimpanzees and wish to see them in the wild. Ecotourism has been proposed as a strategy to make apes more valuable to local communities alive than they would be in the bushmeat or pet trade (Butynski and Kalina 1998; Wilkie and Carpenter 1998). However, the process of habituating apes for ecotourism has come under criticism for the risks it imposes on the apes. These may include changes in behavior (Butynski and Kalina 1998; Goldsmith 2000) and increased risk of disease transmission (Homsy 1999; Woodford et al. 2002). In a recent study, Blom et al. (2004) found that gorillas showed negative behavioral effects from the habituation process, including increase in daily path length and reactions of aggression and fear during contacts. Similar studies of wild chimpanzees are needed to assess behavioral changes under different levels of tourist observation. In addition, captive studies could build on the work of Parr and Slocombe (see chapters 5 and 16) to fully investigate categorization and generalization abilities in chimpanzees, especially in terms of delineating different types of individuals (familiar/unfamiliar, male/female, etc.). In a recent study, Murai et al. (2005) found that both human and chimpanzee infants could form categorical representations of three categories of objects (mammals, furniture, and vehicles) without any training. Extending such a study to investigate chimpanzees' abilities to delineate or prefer different types of individuals would ultimately help us to understand how they perceive human presence and whether they categorize humans into different risk levels in the wild. Such studies would contribute greatly to the various cost/benefit discussions surrounding the ecotourism issue. Whether or not ecotourism is a profitable, ethical, or feasible conservation strategy is outside the scope of this book. The important point here is that measures of the impact of ecotourism on chimpanzees could benefit from a better understanding of chimpanzee cognition.

Reintroduction

Reintroduction of captive-born or orphaned animals into their native habitat is often proposed as an option for the restocking of endangered species. For primates it is an exceedingly difficult task because of their behavioral and social complexity. As such, primate-specific guidelines for reintroduction have been published by the International Union for the Conservation of Nature (IUCN) (Baker 2002). Chimpanzees and the other members of the ape taxa are considered especially difficult cases because of, among other reasons, their highly advanced cognitive abilities. For example, the territorial nature of chimpanzees makes it a risky prospect to reintroduce them into areas where wild chimpanzees reside. Their behavioral competence is also a concern, given that much of their species-typical behavior may be socially learned from adult conspecifics (Custance et al. 2002). Finally, if previously captive chimpanzees are released in an area close to human settlements, their familiarity with former caretakers may result in attraction to nearby humans and an increase in human-wildlife conflict. New guidelines specific to apes have recently been published under the auspices of the IUCN (Beck et al. 2007; see also chapter 26). That document states that "there must be careful assessment of individual histories and behavioural competence before re-introduction. This will require advice from a group of experts with different specializations (for example, cognition, social interaction and temperament)" (p. 13).

There are several opportunities for cognition and conservation to overlap in this case. For example, studies on cognitive development can set "baselines" or "standards" for age-typical cognitive milestones, which can be used to assess the appropriateness of particular individuals for release (see chapter 3). Similarly, studies on theory of mind (see chapter 19) help us to understand the complex cognitive situations that a reintroduced individual may face in social interactions. A small number of ape reintroductions to mainland forest have been attempted with both chimpanzees and gorillas (Courage et al. 2001; Goossens et al. 2005; King et al. 2005). The most successful has been the reintroduction of 36 orphaned chimpanzees into the forests of the Republic of Congo, of whom 26 survived (Goossens et al. 2005). The group of scientists involved with this reintroduction outlined a strict decision-making process and methodology for this project that relied heavily on knowledge about the behavior of chimpanzees gathered from long-term field studies (Tutin et al. 2001). Perhaps it would be appropriate in the future to integrate cognitive testing into the assessments of candidates for reintroduction. Indeed, what larger problem-solving task could one give an animal than to transport him or her to a wholly new environment and expect them to survive and reproduce? These assessments would not take place in a vacuum, but would add to a broad set of data informing the decision as to which candidates for reintroduction were most appropriate and likely to succeed. One could envision a scenario in which the decision were made to reintroduce four chimpanzees out of eight possible candidates. If all other measures of suitability were equal, those chimpanzees who were more successful at cognitive tasks might be selected over those who were not. There recently has been an increase in active cognition research programs taking place in African sanctuaries (see Hare et al., 2007, and chapter 21), so the time seems especially ripe for a synergistic use of cognition research to inform reintroduction practices for apes.

Implications and Future Directions

Behavioral studies that began in the early and mid-twentieth century laid the foundation for understanding the complexity of behaviors shown by chimpanzees and the mental processes underlying those behaviors. Many wild chimpanzee populations are now relatively small and require proactive conservation management to prevent them from going extinct. I have proposed here that a deeper understanding of the functioning and complexities of the chimpanzee mind not only allows us better to understand our own evolutionary history and cognitive architecture, but actually provides important and profound insights into better facilitating chimpanzee conservation. On one hand, chimpanzees' highly advanced cognitive capacities can compound conservation problems, as may be the case with their territoriality exacerbating the impact of habitat fragmentation. On the other hand, chimpanzees may be able to use their minds to learn to avoid hunters or compensate for damage done by poachers' snares. Finally, the synergy of cognition and conservation could provide new opportunities, such as integrating cognitive testing into reintroduction assessments. The boundaries between disciplines are becoming ever more blurred, and we need to use all the tools in our scientific toolbox to protect the last remaining populations of chimpanzees. I firmly believe that as the conservation of wild chimpanzees becomes more scientific and evidence-based, cognitive studies can and should play an important role.

Literature Cited

Ammann, K. 2001. Bushmeat hunting and the great apes. In B.B. Beck, T.S. Stoinski, M. Hutchins, T. L. Maple, B. Norton, A. Rowan, E. F. Stevens, and A. Arluke, eds., *Great Apes and Humans: The Ethics of Coexistence*, 71–85. Washington, DC: Smithsonian Institution Press.

Baker, L. R. 2002. Guidelines for non-human primate reintroductions. *Reintroduction News: Newsletter of the Reintroduction Specialist Group*, Volume 21.

Beck, B., K. Walkup, M. Rodrigues, S. Unwin, D. Travis, and T. Stoinski. 2007. Best practice guidelines for the re-introduction of great apes. IUCN/SSC Primate Specialist Group, Gland, Switzerland.

Blom, A., C. Cipolletta, A. M. H. Brunsting, and H. H. T. Prins. 2004. Behavioral responses of gorillas to habituation in the Dzanga-Ndoki National Park, Central African Republic. *International Journal of Primatology* 25:179–96.

Boesch, C., and H. Boesch-Achermann. 2000. *The Chimpanzees of the Tai Forest*. New York: Oxford University Press.

Butynski, T. M. 2001. Africa's great apes. In B. B. Beck, T. S. Stoinski, M. Hutchins, T. L. Maple, B. Norton, A. Rowan, E. F. Stevens, and A. Arluke, eds., *Great Apes and Humans: The Ethics of Coexistence*, 71–85. Washington, DC: Smithsonian Institution Press.

Butynski, T. M., and J. Kalina. 1998. Gorilla tourism: A critical look. In E. J. Milner-Gulland and R. Mace, eds., *Conservation of Biological Resources*, 280–300. Oxford: Blackwell Science.

Chi, F., M. Leider, F. Leendertz, C. Bergmann, C. Boesch, S. Schenk, G. Pauli, H. Ellerbrok, and R. Hakenbeck. 2007. New *Streptococcus pneumoniae* clones in deceased wild chimpanzees. *Journal of Bacteriology* 189:6085–88.

Courage, A., I. Henderson, and J. Watkin. 2001. Orphan gorilla reintroduction: Lesio-Louna and Mpassa. *Gorilla Journal* 22:33–35.

Custance, D. M., A. Whiten, and T. Fredman. 2002. Social learning and primate reintroduction. *International Journal of Primatology* 23:479–99.

Goldberg, T. L., T. R. Gillespie, I. B. Rwego, E. Wheeler, E. L. Estoff, and C. S. Chapman. 2007. Patterns of gastrointestinal bacterial exchange between chimpanzees and humans involved in research and tourism in western Uganda. *Biological Conservation* 135:511–17.

Goldsmith, M. L. 2000. Effects of ecotourism on the behavioral ecology of Bwindi gorillas, Uganda: Preliminary results. *American Journal of Physical Anthropology*, Supplement 30, 161.

Goodall, J. 1971. *In the Shadow of Man*. Boston: Houghton Mifflin.

———. 1983. Population dynamics during a 15-year period in one community of free-living chimpanzees in the Gombe National Park, Tanzania. *Zeitschrift fur Tierpsychologie* 61:1–60.

———. 1986. *The Chimpanzees of Gombe: Patterns of Behavior*. Cambridge, MA: Harvard University Press.

Goossens, B., J. M. Setchell, E. Tchidongo, E. Dilambaka, C. Vidal, M. Ancrenaz, and A. Jamart. 2005. Survival, interactions with conspecifics and reproduction in 37 chimpanzees released into the wild. *Biological Conservation* 123:461–75.

Guerrera, W., J. M. Sleeman, S. B. Jasper, L.B. Pace, T. Y. Ichinose, and J. S. Reif. 2003. Medical survey of the local human population to determine possible health risks to the mountain gorillas of Bwindi Impenetrable Forest National Park, Uganda. *International Journal of Primatology* 24:197–207.

Hare, B., A. Melis, V. Woods, S. Hastings, and R. Wrangham. 2007. Tolerance allows bonobos to outperform chimpanzees on a cooperative task. *Current Biology* 17:619–23.

Homsy, J. 1999. Ape tourism and human diseases: How close should we get? A critical review of the rules and regulations governing park management and tourism for the wild mountain gorilla, *Gorilla gorilla beringei*. In *Report to the International Gorilla Conservation Programme*.

Huffman, M. A. 1997. Current evidence for self-medication in primates: A multidisciplinary perspective. *American Journal of Physical Anthropology* 104(s25): 171–200.

King, T., C. Chamberlan, and A. Courage. 2005. Reintroduced gorillas: Reproduction, ranging, and unresolved issues. *Gorilla Journal* 30:30–32.

Köndgen, S., H. Kuhl, P. K. N'Goran, P. S. Walsh, S. Schenk, N. Ernst, R. Biek, P. Formenty, K. Matz-Rensing, B. Schweiger, S. Junglen, H. Ellerbrok, A. Nitsche, T. Briese, W. I. Lipkin, G. Pauli, C. Boesch, and F. H. Leendertz. 2008. Pandemic human viruses cause decline of endangered great apes. *Current Biology* 18:1–5.

Kohler, W. 1927. *The Mentality of Apes*. London: Routledge & Kegan Paul.

Leendertz, F. H., G. Pauli, K. Maetz-Rensing, W. Boardman, C. Nunn, S. Jensen, S. Junglen, and C. Boesch. 2006. Pathogens as drivers of population declines: The importance of systematic monitoring in great apes and other threatened mammals. *Biological Conservation* 131:325–37.

Lonsdorf, E. V. 2007. The role of behavioral research in the conservation of chimpanzees and gorillas. *Journal of Applied Animal Welfare Science* 10:71–78.

Lonsdorf, E. V., D. Travis, A. E. Pusey, and J. Goodall. 2006. Using retrospective health data from the Gombe chimpanzee study to inform future monitoring efforts. *American Journal of Primatology* 68:897–908.

Mech, D. 1994. Buffer zones of territories of gray wolves as regions of intraspecific strife. *Journal of Mammalogy* 175:199–202.

Morgan, D., and C. Sanz. 2003. Naïve encounters with chimpanzees in the Goualougo Triangle, Republic of Congo. *International Journal of Primatology* 24:369–81.

———. 2007. Best Practice Guidelines for Reducing the Impact of Commercial Logging on Great Apes in Western Equatorial Africa. IUCN/SSC Primate Specialist Group, Gland, Switzerland.

Murai, C., D. Kosugi, M. Tomonaga, M. Tanaka, T. Matsuzawa, and S. Itakura. 2005. Can chimpanzee infants (*Pan troglodytes*) form categorical representations in the same manner as human infants (*Homo sapiens*)? *Developmental Science* 8:240–54.

Nishida, T. 1968. The social group of wild chimpanzees in the Mahale Mountains. *Primates* 9:167–224.

———, ed. 1990. *The Chimpanzees of the Mahale Mountains: Sexual and Life History Strategies*. Tokyo: University of Tokyo Press.

Nishida, T., N. Corp, M. Hamai, T. Hasegawa, M. Hiraiwa-Hasegawa, K. Hosaka, K. D. Hunt, N. Itoh, K. Kawanaka, A. Matsumoto-Oda, J. C. Mitani, M. Nakamura, K. Norikoshi, T. Sakamaki, L. Turner, S. Uehara, and K. Zamma. 2003. Demography, female life history, and reproductive profiles among the chimpanzees of Mahale. *American Journal of Primatology* 59:99–121.

Plumptre, A. J., and A. G. Johns. 2001. Changes in primate communities following logging disturbance. In R. Fimbel, A. Grajal, and J. G. Robinson, eds., *The Cutting Edge: Conserving Wildlife in Logged Tropical Forests*, 71–92. New York: Columbia University Press.

Plumptre, A. J., and V. Reynolds. 1994. The effect of selective logging on the primate populations in the Budongo Forest, Uganda. *International Journal of Applied Ecology* 31:631–41.

Pruetz, J., and P. Bertolani. 2007. Savanna chimpanzees, *Pan troglodytes verus*, hunt with tools. *Current Biology* 17:412–17.

Pusey, A. E. 1990. Behavioral changes at adolescence in chimpanzees. *Behavior* 115:203–46.

Quiatt, D., V. Reynolds, and E. J. Stokes. 2002. Snare injuries to chimpanzees (*Pan troglodytes*) at 10 study sites in East and West Africa. *African Journal of Ecology* 40:303–5.

Reynolds, V. 2005. *The Chimpanzees of the Budongo Forest: Ecology, Behaviour, and Conservation*. Oxford: Oxford University Press.

Shaffer, M. 1987. Minimum viable populations: Coping with uncertainty. In M. Soulé, ed., *Viable Populations for Conservation*, 69–86. Cambridge: Cambridge University Press.

Stokes, E. J., and R. W. Byrne. 2001. Cognitive capacities for behavioural flexibility in wild chimpanzees (*Pan troglodytes*): The effect of snare injury on complex manual food processing. *Animal Cognition* 4:11–28.

Tutin, C. E. G. 2001. Saving the gorillas (*Gorilla gorilla gorilla*) and chimpanzees (*Pan troglodytes troglodytes*) of the Congo Basin. *Reproduction, Fertility, and Development* 13: 469–76.

Tutin, C. E. G., M. Ancrenaz, J. Paredes, M. Vacher-Vallas, C. Vidal, B. Goossens, M. W. Bruford, and A. Jamart. 2001. Conservation biology framework for the release of wild-born orphaned chimpanzees into the Conkouati Reserve, Congo. *Conservation Biology* 15:1247–57.

Tutin, C. E. G., E. Stokes, C. Boesch, D. Morgan, C. Sanz, T. Reed, A. Blom, P. Walsh, S. Blake, and R. Kormos. 2005. *Regional Action Plan for the Conservation of Chimpanzees and Gorillas in Western Equatorial Africa*. Washington, DC: Conservation International.

Van Lawick-Goodall, J. 1968. Behavior of free-living chimpanzees of the Gombe Stream area. *Animal Behavior Monographs* 1:163–311.

Walsh, P. D., K. A. Abernethy, M. Bermejo, R. Beyers, P. De Wachter, M. E. Akou, B. Huijbregts, D. I. Mambounga, A. K. Toham, A. M. Kilbourn, S. A. Lahm, S. Latour, F. Maisels, C. Mbina, Y. Mihindou, S. N. Obiang, E. N. Effa, M. P. Starkey, P. Telfer, M. Thibault, C. E. G. Tutin, L. J. T. White, and D. S. Wilkie. 2003. Catastrophic ape decline in western equatorial Africa. *Nature* 422:611–14.

White, L. J. T., and C. E. G. Tutin. 2001. Why chimpanzees and gorillas respond differently to logging: A cautionary tale from Gabon. In W. Weber, L. J. T. White, A. Vedder and L. Naughton-Treves, eds., *African Rain Forest Ecology and Conservation: An Interdisciplinary Perspective*, 449–62. New Haven: Yale University Press.

Wilkie, D., E. Shaw, F. Rotberg, G. Morelli, and P. Auzel. 2000. Roads, development, and conservation in the Congo Basin. *Conservation Biology* 14:1614–22.

Wilkie, D. S., and J. F.Carpenter. 1999. Can nature tourism help finance protected areas in the Congo Basin? *Oryx* 33:332–38.

Williams, J. M., E. V. Lonsdorf, M. L. Wilson, J. Schumacher-Stankey, J. Goodall, and A. E. Pusey. 2008. Causes of death in the Kasekela chimpanzees of Gombe National Park, Tanzania. *American Journal of Primatology* 70:766–77.

Wilson, M. L., M. D. Hauser, and R. W. Wrangham. 2007. Chimpanzees (*Pan troglodytes*) modify grouping and vocal behaviour in response to location-specific risk. *Behaviour* 144:1621–53.

Wilson, M. W., and R. W. Wrangham. 2003. Intergroup relations in chimpanzees. *Annual Review of Anthropology* 32:363–92.

Woodford, M. H., T. M. Butynski, and W. B. Karesh. 2002. Habituating the great apes: The disease risks. *Oryx* 36(2): 153–60.

Wrangham, R. W. 1979. Sex differences in chimpanzee dispersion. In D.A. Hamburgh and E.R. McCown, eds., *The Great Apes*, 480–89. Menlo Park, CA: Benjamin/Cummings.

Wrangham, R. W., and T. Nishida. 1983. *Aspilia spp.* leaves: A puzzle in the feeding behavior of wild chimpanzees. *Primates* 24:276–82.

Meanings of Chimpanzee Mind

Richard Wrangham

In the 1960s, when Jane Goodall and Toshisada Nishida began their field studies, the fact that chimpanzees (*Pan troglodytes*) were one of humans' closest living relatives lent their discoveries instant meaning. The implications of chimpanzees modifying tools, having cultural traditions, deceiving each other, hunting collectively, and sharing meat were obvious. Given that such abilities were found in apes, these human-like traits had probably arisen in the prehuman lineage long before our ancestors had language, at a time when brains were still relatively small and ape-sized. The realization promised new insights into both the course of evolution and the adaptive significance of mental processes.

The continuing fulfillment of that promise is shown throughout the present volume with a remarkable breadth of studies on chimpanzees. Thanks to a rapidly growing set of explanations for similarities and differences in cognition among chimpanzees, humans, and other animals, the prospect of a sophisticated theory of cognitive evolution lies strikingly closer than it did in the 1960s. Such a theory does not yet exist, however—so in this afterword, after briefly surveying the special significance of chimpanzees for understanding the evolution of cognition, I suggest some of the challenges on the road to an integrated theory of how cognition has evolved in complex species like chimpanzees and humans. I do so from the perspective of a behavioral ecologist who studies chimpanzees almost

entirely in the wild, where cognitive research has still been very limited.

Why Chimpanzees Are Worth Special Attention

Although chimpanzees are sometimes accused of receiving more than their fair share of attention (Sayers and Lovejoy 2008), there are two reasons why their cognition merits special study. First, they show a greater variety of human-like behaviors and abilities than any other species known to date. This is not to say that chimpanzees necessarily perform better than other animals at any given task. For example, they are clearly inferior to bottle-nosed dolphins (*Tursiops truncatus*) at imitation (Connor 2007) and they appear to have less impressive generalized problem-solving ability than orangutans (*Pongo pygmaeus*) (Deaner et al. 2006). Furthermore, in various domains chimpanzee abilities are matched so closely by other species that it is difficult to decide which performs better (e.g., parrots, *Psittacus erithacus*, linguistic abilities, Pepperberg 2007; New Caledonian crows, *Corvus moneduloides*, tool invention, Taylor et al. 2007; scrub jays, *Aphelocoma californica*, theory of mind, Emery and Clayton 2001; see also Emery and Clayton 2008). Nevertheless, according to present data chimpanzees outperform all other nonhumans in some areas requiring advanced cognitive abilities, such as cultural diversity or self-medication—and more importantly, the range of human-like behaviors that they show in the wild exceeds that of any other species. This means that chimpanzees give us many opportunities to explore theories for the adaptive significance of aspects of cognition that we humans find especially interesting.

More detailed study of species with advanced cognitive abilities—such as bonobos (*Pan paniscus*), orangutans, bottle-nosed dolphins, or corvids—might one day erode the idea that chimpanzees are exceptional in showing so many human-like behaviors. But even if chimpanzees lose their uniqueness in this respect, the second reason for intense research on their cognition would still apply, which is that chimpanzees provide the best model among living species for the human ancestor living around five to seven million years ago.

This conclusion comes from a combination of genetic and morphological evidence (Pilbeam 1996). The celebrated similarity of human and chimpanzee genes, around 98.4%, shows that chimpanzees and bonobos are related more closely to humans than to gorillas. Yet both chimpanzees and bonobos resemble gorillas (*Gorilla gorilla*) in most aspects of their biology more closely than they do humans. There are two explanations for this mismatch between genes and morphology. The most likely is that the common ancestor of today's African apes looked like the living apes—that is, it was a black-haired, knuckle-walking species that lived in rainforests, living off ripe fruits when they were available and using its thin-enameled teeth to eat leaves and stems when fruits were scarce. The less probable hypothesis is that the numerous resemblances between gorillas and the chimpanzee-bonobo lineage, such as in teeth, limbs, locomotion, and ecology, arose by convergence. The convergence idea is challenged by the details of the similarities and their uniqueness to this lineage, so we can have considerable confidence that the common ancestor of African apes was indeed in the gorilla-chimpanzee-bonobo mold, rather than a fundamentally dissimilar type.

These three species are in many ways versions of the same phenotype expressed at different body sizes, so if the ancestor was relatively large it would have been more like a gorilla, whereas if it was relatively small it would have been more like a chimpanzee or bonobo. Since all the other great apes in this lineage, including both orangutans and the early australopithecines that evolved from the rain forest apes, had females that were in the size range of chimpanzees more than gorillas, the ape ancestor was

almost certainly more like a chimpanzee or bonobo than a gorilla.

A final point establishes why chimpanzees give us a better model of our five- to seven-million-year ancestor than bonobos do. Bonobos are as closely related to humans as chimpanzees are, but their morphological and behavioral differences from chimpanzees are idiosyncrasies compared to other great apes—unique patterns that argue for them being changed relatively strongly from the common ancestor they share with chimpanzees. Most notably, their crania grow slowly and reach maturity in a pedomorphic form compared to those of chimpanzees and every other ape. The chimpanzee morphology therefore provides a better candidate to resemble the ancestor, with differences in behavior and cognition between bonobos and chimpanzees resulting from an unusual pattern of selection on bonobos (Wrangham and Pilbeam 2001).

This logic means that the chimpanzee is a good model for the species that lived around six million years ago and gave rise to the australopithecines before the later emergence of *Homo* (around two million years ago). More than research with any other ape, the study of chimpanzees allows us to tentatively assign a minimum time for the evolutionary appearance of some of our own cognitive abilities, and to speculate about the kind of habitat and social system in which they evolved.

The Problem of Characterizing Cognitive Abilities

To explain how and when cognition evolves, we first need a confident description of species traits. Several important cognitive characteristics of chimpanzees are still subject to lively debate, including how well they understand their physical world, how well they imitate, and how altruistic or cooperative they are. Recent findings suggest that the difficulty in reconciling contrary results comes partly from differences between studies in the experience of the chimpanzees. Furlong et al. (2008) provide an example. They gave captive chimpanzees an opportunity to choose one of two kinds of tool to bring food towards them: a functional rake with a wooden head, or a useless rake with a flimsy head. Individual chimpanzees differed in their success at solving the problem, and the variation was easily explained. The successful individuals were "enculturated," meaning that they were reared with much human contact in a rich environment. The unsuccessful chimpanzees were "semi-enculturated," having had less human contact. The contrast shows that unless chimpanzees reared under a wide variety of conditions all prove to show the same abilities, it is meaningless to talk of "the chimpanzee mind" because the results from any particular context, such as a captive study of orphans, cannot be assumed to apply to chimpanzees in general. Until we know more about the variation in chimpanzee cognitive abilities among zoos, sanctuaries, and the wild, or in enculturated, semi-enculturated, and natural social groups, we must remain open-minded about whether a particular trait is typical of the species. A crucial question for evolutionary analysis is that of whether an ability found in captivity is expressed in the wild, where unfortunately experimentation is most difficult.

Some might object to this cautious approach on the basis that natural habitats are the ideal social and physical environments for chimpanzees, so that any ability found in captivity would likely be found in the wild. Against this, however, captivity can prompt new kinds of behavior. Consider the case of Binti Jua. This young adult female gorilla was filmed rescuing an unconscious boy who had fallen into her enclosure at Chicago's Brookfield Zoo. She cradled him and carried him to a door where keepers took him from her. Her behavior suggested that "cognitive empathy" occurs in gorillas (Preston and de Waal 2002). But since Binti Jua was hand-raised by humans after being rejected by her own mother, and was trained in maternal skills (including retrieving a doll-like object and bringing it to people), the question of whether her response was characteristic of gorillas in general is undecided—let alone that of whether it was based on empathy, or whether gorillas show empathy in the wild (Silk 2008). Although the fact that humans deliberately trained Binti Jua makes this case extreme, it serves to remind us that conditions in captivity are different from nature. However good the evidence that chimpanzees can collaborate in sanctuaries, for example, we cannot conclude that they do so in the wild without direct evidence—a challenging problem, as Gilby and Connor (see chapter 18) remind us. Ultimately our characterization of chimpanzee cognition needs to be in terms of which abilities are displayed under given conditions, rather than a claim that "chimpanzees" in general have a particular type of mind.

Which Mental Processes Are Evolutionarily Adapted, and Which Are Incidental Consequences?

The principal evolutionary framework for understanding species distributions of cognitive ability sees minds as being adapted to a set of ecological and behavioral challenges, such as the nature of the food supply, the pattern of grouping, or the complexity of the social system. This approach looks for shared causes of advanced cognition regardless of phylogenetic relationship, and it has had considerable success. For example, cognitive performance generally tends to be positively correlated with relative brain size, and brains tend to be larger where groups are larger, including in primates, carnivores, ungulates, insectivores, bats, and cetaceans (Byrne and Bates 2007). Such findings suggest that natural selection has had particularly strong effects on mental abilities in species living in larger groups, and that in general, the minds of chimpanzees and other animals will be explicable in terms of their adaptive functions.

While the adaptive approach is undoubtedly powerful, data on the great apes also indicate that some of the variation in cognitive ability is an incidental consequence rather than an adaptive trait. For example, great apes share the ability to learn to communicate in captivity through symbols. Since no symbolic communication has been reported in the wild, this probably results from their general intelligence or from the use of other cognitive abilities, rather than having been subject to direct selection. The same can obviously be said for self-recognition in mirrors since it has no application in the wild, but for many other traits the question of adaptation is harder to be sure about. Chimpanzees use tools skillfully, for example, and in Taï they can obtain as many as 3,000 calories per day by tool use (Boesch and Boesch-Achermann 2000), but this does not mean that tool use is a product of selection. In some chimpanzee populations, tools are used only trivially, and there is no indication that tool use is necessary for survival (unlike in humans, where tool use appears critical). Rather, it appears to be an option that individuals can use to their competitive advantage. In sum, given the variation in tool use among populations, we cannot assume that tool-using ability has been a target of natural selection. The same can be said for cooperation. As Hirata et al. (see chapter 20) show, chimpanzees can become effective cooperators under the right conditions, but they are not necessarily quick to learn how to do so. Like tool use, cooperation may or may not have been favored by natural selection.

The extent to which cognitive abilities have arisen as a consequence of the incidental evolution of other traits will not be easily resolved. Until better data are available from the wild and on the neural mechanisms underlying different abilities, the verdict should be open. For example, relative brain size is strongly correlated with longevity, suggesting that larger brains allow longer lives (Barrickman et al. 2008). If a principal mechanism is that larger brains protect against dementia by providing a cognitive reserve, as Allen et al. (2005) suggest, many of the mental faculties of chimpanzees, humans, and other animals could result from selection for a limited number of traits—such as the computing power that results from having a relatively large brain selected to protect against premature dementia.

How Important Is Temperament for Cognitive Performance?

One of the clearest illustrations of a cognitive ability arising as an incidental consequence comes from the finding that signal-reading ability is more proficient among domesticated silver foxes (*Vulpes vulpes*) than among undomesticated individuals of the same species (Hare et al. 2005). In this comparison the undomesticated population were captives that had not been subject to any deliberate selection, whereas the domesticated population were chosen for breeding if they showed reduced emotional reactivity—that is, they tolerated close approach by humans. Improved signal-reading by the domesticated foxes was explained by their having calmer temperaments and improved ability to pay attention to other individuals, traits that result from selection against aggressiveness. An evolutionary change in temperament thus appeared to have led to increased communicative performance.

A similar effect applies to the comparison of chimpanzee and bonobo cooperation. In the wild, chimpanzees appear to cooperate more than bonobos do, whether in coalitions against conspecifics or in hunting. An adaptive hypothesis would therefore suggest that they should cooperate better than bonobos in captivity. On the other hand

bonobos tend to be more tolerant of each other than chimpanzees, which suggests that they might find cooperation easier than chimpanzees. Experiments show that bonobos indeed cooperate in obtaining food more easily than chimpanzees, in support of the incidental-consequence hypothesis (Hare et al. 2007).

Temperamental differences among species have been studied relatively little with respect to cognition, but these examples suggest that they may have pervasive effects. In particular the evident differences in emotional reactivity between chimpanzees and humans seem likely to contribute substantially to human communicative abilities. This suggests that theories of cognitive evolution will benefit from efforts not only to understand the cognitive abilities that underlie reason and attention, but also to explain the emotional propensities that regulate an individual's willingness to interact. The evolution of cognition may thus be dependent on both adaptive and nonadaptive mechanisms, including aspects of temperament that evolve independently of the cognitive abilities that they make possible.

Synthesis

"Mind," according to *Merriam-Webster's Collegiate Dictionary*, is "the element or complex of elements in an individual that feels, perceives, thinks, wills, and especially reasons." The philosopher Rene Descartes thought animals mindless because they could not speak, and the same attitude can occasionally be found today in people who are blind to the emotional needs of other species. But Charles Darwin challenged Descartes by discussing mental evolution in *The Expression of the Emotions in Man and Animals* (1872); Wolfgang Koehler (1925) found evidence of insightful problem solving in captive chimpanzees; and now we have hundreds of studies showing how chimpanzees feel, perceive, and reason. The revolution in thinking was legitimized by Griffin (1976), who argued that the study of animal mental processes was a valid area of inquiry, but it has owed most to the dramatic and surprising revelations about chimpanzees that began in Gombe and Mahale in the 1960s. Similar discoveries will continue for some time, as this book clearly indicates, and they eventually should answer many of the provocative questions still remaining, such as how much of the chimpanzee's cognitive ability is used in the wild, why the great apes are so intelligent, and what accounts for humans being so temperamentally and cognitively different from other apes. Darwin (1871) wrote, "When we confine our attention to any one form, we are deprived of the weighty arguments derived from the nature of the affinities which connect together whole groups of organisms." The research on chimpanzees represented in this book surely indicates how right he was.

Literature Cited

Allen, J. S., J. Bruss, and H. Damasio. 2005. The aging brain: The cognitive reserve hypothesis and hominid evolution. *American Journal of Human Biology* 17:673–89.

Barrickman, N. L., M. L. Bastian, K. Isler, and C. P. Van Schaik. 2008. Life history costs and benefits of encephalization: A comparative test using data from long-term studies of primates in the wild. *Journal of Human Evolution* 54:568–90.

Boesch, C., and H. Boesch-Achermann. 2000. *The Chimpanzees of the Taï Forest: Behavioral Ecology and Evolution*. Oxford: Oxford University Press.

Byrne, R. W., and L.A. Bates. 2007. Sociality, evolution and cognition. *Current Biology* 17:R714–23.

Connor, R. C. 2007. Dolphin social intelligence: Complex alliance relationships in bottlenose dolphins and a consideration of selective environments for extreme brain size evolution in mammals. *Philosophical Transactions of the Royal Society B* 362:587–602.

Darwin, C. 1871 (1990). *The Descent of Man and Selection in Relation to Sex*. Chicago: Encyclopaedia Britannica Inc.

———. 1872. *The Expression of the Emotions in Man and Animals*. London: John Murray.

Deaner, R. O., C. P. Van Schaik, and V. Johnson. 2006. Do some taxa have better domain-general cognition than others? A meta-analysis of nonhuman primate studies. *Evolutionary Psychology* 4:149–96.

Emery, N. J., and N. S. Clayton. 2001. Effects of experience and social context on prospective caching strategies in scrub jays. *Nature* 414:443–46.

———. 2008. Comment on Sayers and Lovejoy (2008). *Current Anthropology* 49:100.

Furlong, E. E., K. J. Boose, and S. T. Boysen. 2008. Raking it in: The impact of enculturation on chimpanzee tool use. *Animal Cognition* 11:83–97.

Griffin, D. R. 1976. *The Question of Animal Awareness: Evolutionary Continuity of Mental Experience*. New York: Rockefeller University Press.

Hare, B., A. P. Melis, V. Woods, S. Hastings, and R. Wrangham. 2007. Tolerance allows bonobos to outperform chimpanzees on a cooperative task. *Current Biology* 17:619–23.

Hare, B., I. Plyusnina, N. Ignacio, O. Schepina, A. Stepika, R. Wrangham, and L. Trut. 2005. Social cognitive evolution in captive foxes is a correlated by-product of experimental domestication. *Current Biology* 15:1–20.

Koehler, W. 1925. *The Mentality of Apes*. London: Routledge and Kegan Paul.

Pepperberg, I. M. 2007. Grey parrots do not always "parrot": The roles of imitation and phonological awareness in the creation of new labels from existing vocalizations. *Language Sciences* 29:1–13.

Pilbeam, D. 1996. Genetic and morphological records of the Hominoidea and hominid origins: A synthesis. *Molecular Phylogenetics and Evolution* 5:155–68.

Preston, S. D., and F. B. M. de Waal. 2002. Empathy: Its ultimate and proximate bases. *Behavioral and Brain Sciences* 25:1–72.

Sayers, K., and C. O. Lovejoy. 2008. The chimpanzee has no clothes: A critical examination of *Pan troglodytes* in models of human evolution. *Current Anthropology* 49:87–99.

Silk, J. B. 2008. Social preferences in primates. In *Neuroeconomics: Decision Making and the Brain*, edited by Glimcher, P., Camerer, C., Fehr, E. & Poldrack, R., 269–84. New York: Academic Press.

Taylor, A. H., G. R. Hunt, J. C. Holzhaider, and R. D. Gray. 2007. Spontaneous metatool use by New Caledonian crows. *Current Biology* 17:1504–7.

Wrangham, R. W., and D. Pilbeam. 2001. African apes as time machines. In *All Apes Great and Small. Volume 1: Chimpanzees, Bonobos, and Gorillas*, edited by B. M. F. Galdikas, N. Briggs, L. K. Sheeran, G. L. Shapiro, and J. Goodall, 5–18. New York: Kluwer Academic / Plenum.

Major Chimpanzee Research Sites

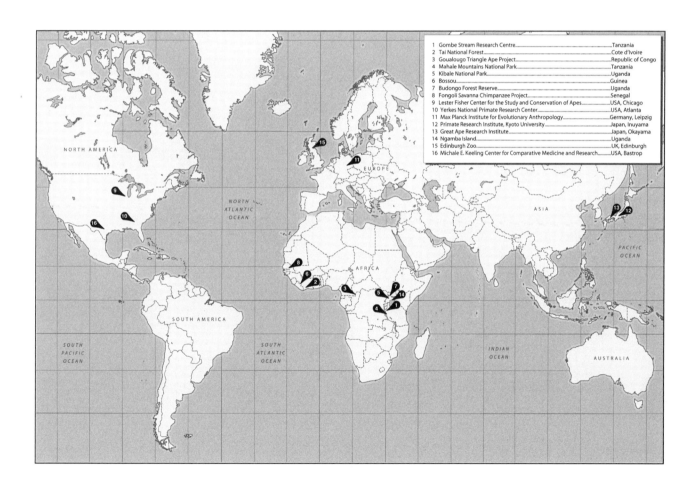

1 Gombe Stream Research Centre...Tanzania
2 Tai National Forest..Cote d'Ivoire
3 Goualougo Triangle Ape Project..Republic of Congo
4 Mahale Mountains National Park...Tanzania
5 Kibale National Park...Uganda
6 Bossou...Guinea
7 Budongo Forest Reserve..Uganda
8 Fongoli Savanna Chimpanzee Project...Senegal
9 Lester Fisher Center for the Study and Conservation of Apes..................................USA, Chicago
10 Yerkes National Primate Research Center...USA, Atlanta
11 Max Planck Institute for Evolutionary Anthropology...Germany, Leipzig
12 Primate Research Institute, Kyoto University..Japan, Inuyama
13 Great Ape Research Institute..Japan, Okayama
14 Ngamba Island..Uganda
15 Edinburgh Zoo...UK, Edinburgh
16 Michale E. Keeling Center for Comparative Medicine and Research.........................USA, Bastrop

Index